Chemical Kinetics and Dynamics

Chemical Kinetics and Dynamics

Jeffrey I. Steinfeld
Massachusetts Institute of Technology

Joseph S. Francisco
Wayne State University

William L. Hase
Wayne State University

Prentice Hall, Upper Saddle River, New Jersey 07458

Library of Congress Cataloging-in-Publication Data
Steinfeld, Jeffrey I.
    Chemical kinetics and dynamics / Jeffrey Steinfeld, Joseph
Francisco, William Hase.
        p.   cm.
    Bibliography.
    Includes index.
    ISBN 0-13-129479-2
    1. Chemical reaction, Rate of.   2. Molecular dynamics.
I. Francisco, Joseph   II. Hase, William.   III. Title.
QD502.S74 1989   88-21842
541.3'94--dc19   CIP

Editorial/production supervision: Karen Winget/Wordcrafters
Cover design: Joel Mitnick Design, Inc.
Manufacturing buyer: Paula Massenaro

 © 1989 by Prentice Hall, Inc.
Simon & Schuster / A Viacom Company
Upper Saddle River, New Jersey 07458

Printed in the United States of America
10 9 8 7

ISBN 0-13-129479-2

PRENTICE-HALL INTERNATIONAL (UK) LIMITED, *London*
PRENTICE-HALL OF AUSTRALIA PTY. LIMITED, *Sydney*
PRENTICE-HALL CANADA INC., *Toronto*
PRENTICE-HALL HISPANOAMERICANA, S.A., *Mexico*
PRENTICE-HALL OF INDIA PRIVATE LIMITED, *New Delhi*
PRENTICE-HALL OF JAPAN, INC., *Tokyo*
SIMON & SCHUSTER ASIA PTE. LTD., *Singapore*
EDITORA PRENTICE-HALL DO BRASIL, LTDA., *Rio de Janeiro*

Contents

Chapter 3

KINETIC MEASUREMENTS   109

Chapter 4

REACTIONS IN SOLUTION   156

Chapter 5

CATALYSIS   178

Contents

Chapter 10

STATISTICAL APPROACH TO REACTION DYNAMICS: TRANSITION STATE THEORY   308

Chapter 11

UNIMOLECULAR REACTION DYNAMICS   342

Contents   ix

Contents

Preface

In recent years, the principal focus of the study of chemical reactions has shifted from the macroscopic treatment of empirical kinetics to the microscopic, molecular viewpoint of chemical dynamics. This microscopic approach has stimulated new experimental and theoretical developments, making chemical dynamics one of the most active fields of physical chemistry. Unfortunately, most of the textbooks treating chemical kinetics emphasize either the macroscopic kinetics or the microscopic dynamics aspect, often at the expense of the other. In this book, we have tried to cover both older and newer aspects of chemical reaction processes as comprehensively as possible, with the aim of illustrating the interconnections between phenomenological chemical kinetics and molecular reaction dynamics.

The book, which is based on a series of lectures given over a number of years to students at the Massachusetts Institute of Technology and at Wayne State University, is aimed at advanced undergraduate and graduate students who possess a background in thermodynamics, a little quantum and statistical mechanics, and some familiarity with basic chemical kinetics at the level of an undergraduate physical chemistry course. The mathematics required is essentially the ability to set up and solve linear differential equations; when other techniques, such as Laplace transform and matrix methods, are utilized, a brief introduction is provided in the text. Following an exposition of the basic principles of chemical kinetics, current experimental and analytic techniques in kinetics are described. The treatment includes reactions in the gas phase, in liquids, and at catalytic surfaces. The transition to the microscopic level of description is made by introducing molecular scattering and potential energy surfaces. Following a description of statistical reaction rate theories (transition-state and RRKM treatments for bimolecular and uni-molecular reactions, respectively), the analysis of multilevel and multicomponent kinetic systems is carried out using master equation and information-theoretical methods. The text concludes with a treatment of some important real-world kinetic systems, viz., atmospheric and combustion chemistry.

The book should provide the student with the necessary background to delve into current research topics using the journal and review literature. To this end, extensive chapter references and bibliographies are provided. There is also a large number of problems and exercises, some of which involve numerical procedures which can be carried out using the computer software available as an adjunct to the text.

Our collective thanks go to those students at M.I.T., Wayne State, and elsewhere who endured the various stages of development of this material. Many of our colleagues provided material and problem sets for incorporation into the text, as well as insightful criticism of the resulting product. Among these individuals are S. Chapman, B. J. Garrison, R. G. Gilbert, D. T. Gillespie, D. M. Hirst, J. T. Hynes, S. R. Leone, R. D. Levine, W. H. Miller, B. S. Rabinovitch, H. Rabitz, G. C. Schatz, D. G. Truhlar, J. C. Tully, and several reviewers chosen by the publishers. Dan Joraanstad of Prentice Hall is to be commended for his enthusiasm and encouragement, which contributed significantly to the completion of the project. Thanks are also due to P. Ellison and V. Spanos for their skillful typing of the manuscript, and to Xiche Hu and Duan Tian for assistance with the preparation of figures. Preparation of the accompanying software was assisted by Project Athena at M.I.T. Finally, we wish to express our thanks to Messrs. Ed Bethel and Sidney Stubbs of Nassau, Bahamas; to Rolf and Eve Villmann of Lochaber Mines, Nova Scotia; and to our colleagues, friends, and families for their forbearance while we were writing the book.

J. I. STEINFELD
J. S. FRANCISCO
W. L. HASE

CHAPTER 1
Basic Concepts of Kinetics

Among .... n is its capacity for *chemical cha*.. two clear liquids and obtains a .... grow, reproduce, and die. Ever.. spheres consists of a set of chemi.................................... be anywhere from a few femto.................................... 10^{+16} sec).

The .... stems at equilibrium, which by .... with time. Most real systems are .... changes as they seek to approach the e.. *...netics* deals with changes in chemical properties in tim........................ ...rmodynamics, chemical kinetics can be understood in terms of a co...inuum model, without reference to the atomic nature of matter. The interpretation of chemical reactions in terms of the interactions of atoms and molecules is frequently called *reaction dynamics*. A knowledge of the dynamic basis for chemical reactions has, in fact, permitted us to design and engineer reactions for the production of an enormous number of compounds which we now regard as essential in our technological society.

We begin our study of chemical kinetics with definitions of the basic observable quantities, which are the chemical changes taking place in a system, and how these changes depend on time.

1.1 DEFINITION OF THE RATE OF A CHEMICAL REACTION

Broadly speaking, chemical kinetics may be described as the study of chemical systems whose composition changes with time. These changes may take place in the gas, liquid, or solid phase of a substance. A reaction occurring in a single phase is usually referred to as a *homogeneous* reaction, while a reaction which takes place at an interface between two phases is known as a *heterogeneous* reaction. An example of the latter is the reaction of a gas adsorbed on the surface of a solid.

The chemical change that takes place in any reaction may be represented by a *stoichiometric equation* such as

$$a\mathrm{A} + b\mathrm{B} \longrightarrow c\mathrm{C} + d\mathrm{D} \tag{1-1}$$

where a and b denote the number of moles of reactants A and B that react to yield c and d moles of products C and D. Various symbols are used in the expression which relates the reactants and products. For example, the formation of water from hydrogen and oxygen may be written as the balanced, irreversible chemical reaction

$$2\mathrm{H}_2 + \mathrm{O}_2 \longrightarrow 2\mathrm{H}_2\mathrm{O} \tag{1-2}$$

In this simple example, the single arrow is used to indicate that the reaction proceeds from the left (reactant) side to the right (product) side as written: water does not spontaneously decompose to form hydrogen and oxygen. A double arrow in the stoichiometric equation is often used to denote a reversible reaction, that is, one which can proceed in either the forward or the reverse direction; an example is

$$\mathrm{H}_2 + \mathrm{I}_2 \rightleftharpoons 2\mathrm{HI} \tag{1-3}$$

While each of equations 1–2 and 1–3 describes an apparently simple chemical reaction, it so happens that neither of these reactions proceeds as written. Instead, the reactions involve the formation of one or more *intermediate* species, and include several steps. These steps are known as *elementary* reactions. An elementary reaction is one in which the indicated products are formed directly from the reactants, for example, in a direct collision between an A and a B molecule; intuitively, they correspond to processes occurring at the molecular level. In the hydrogen-oxygen reaction, a key elementary reaction is the attack of oxygen atoms on hydrogen molecules given by

$$\mathrm{O} + \mathrm{H}_2 \longrightarrow \mathrm{OH} + \mathrm{H}$$

while in the hydrogen-iodine reaction it is

$$2\mathrm{I} + \mathrm{H}_2 \longrightarrow \mathrm{H}_2\mathrm{I} + \mathrm{I}$$

The details of these reactions are discussed in sections 15.2 and 2.3.2, respectively. In the meantime, note here that they involve atoms (O, I), free radicals (OH), and/or unstable intermediates ($\mathrm{H}_2\mathrm{I}$); this is often the case with elementary reactions.

The change in composition of the reaction mixture with time is the *rate of reaction, R*. For reaction 1–1, the rate of consumption of reactants is

$$R = -\frac{1}{a}\frac{d[\mathrm{A}]}{dt} = -\frac{1}{b}\frac{d[\mathrm{B}]}{dt} \tag{1-4}$$

A standard convention in chemical kinetics is to use the chemical symbol enclosed in brackets for species concentration; thus, [X] denotes the concentration of X. The negative signs in equation 1–4 indicate that during the course of the reaction the concentration of reactants decreases as the reactants are consumed; conversely, a positive sign indicates that the concentration of products increases as those species are formed. Consequently, the rate of formation of products C and D can be written as

$$R = +\frac{1}{c}\frac{d[C]}{dt} = +\frac{1}{d}\frac{d[D]}{dt} \qquad (1-5)$$

The factors a, b, c, and d in equations 1–4 and 1–5 are referred to as the *stoichiometric coefficients* for the chemical entities taking part in the reaction. Since the concentrations of reactants and products are related by equation 1–1, measurement of the rate of change of any one of the reactants or products would suffice to determine the rate of reaction R. In the reaction 1–2, the rate of reaction would be

$$R = -\frac{1}{2}\frac{d[H_2]}{dt} = -\frac{d[O_2]}{dt} = +\frac{1}{2}\frac{d[H_2O]}{dt} \qquad (1-6)$$

A number of different units have been used for the reaction rate. The dimensionality of R is

$$[\text{amount of material}][\text{volume}]^{-1}[\text{time}]^{-1}$$

or

$$[\text{concentration}][\text{time}]^{-1}$$

The standard SI unit of concentration is moles per cubic decimeter, abbreviated mol dm^{-3}. In the older literature on kinetics, one frequently finds the units mol $liter^{-1}$ for reactions in solution and mol cm^{-3} for gas phase reactions. The SI unit is preferred, and should be used consistently. Multiplying moles cm^{-3} by Avogadro's Number (6.022×10^{23}) gives the units molecules cm^{-3}, which is still extensively used and, indeed, is convenient for gas phase reactions.

A subcommittee of the International Union of Pure and Applied Chemistry chaired by Laidler has attempted to standardize units, terminology, and notation in chemical kinetics.[1] We have attempted to follow the subcommittee's recommendations in this text.

1.2 ORDER AND MOLECULARITY OF A REACTION

In virtually all chemical reactions that have been studied experimentally, the reaction rate depends on the concentration of one or more of the reactants. In general, the rate may be expressed as a function f of these concentrations,

$$R = f([A],[B]) \qquad (1-7)$$

In some cases the reaction rate also depends on the concentration of one or more intermediate species, e.g., in enzymatic reactions (see chapter 5). In other cases the rate expression may involve the concentration of some species which do not appear in the stoichiometric equation 1–1; such species are known as *catalysts*, and will be discussed in chapter 5. In still other cases, the concentration of product molecules may appear in the rate expression.

The most frequently encountered functional dependence given by equation 1–7 is the rate's being proportional to a product of algebraic powers of the individual concentrations, i.e.,

$$R \propto [A]^m[B]^n \qquad (1-8)$$

The exponents m and n may be integer, fractional, or negative. This proportionality can be converted to an equation by inserting a proportionality constant k, thus:

$$R = k \, [A]^m [B]^n \tag{1-9}$$

This equation is called a *rate equation* or *rate expression*. The exponent m is the *order* of the reaction with respect to reactant A, and n is the order with respect to reactant B. The proportionality constant k is called the *rate constant*. The overall order of the reaction is simply $p = m + n$. A generalized expression for the rate of a reaction involving K components is

$$R = k \prod_{i=1}^{K} c_i^{n_i} \tag{1-10}$$

The product is taken over the concentrations of each of the K components of the reaction. The reaction order with respect to the ith component is n_i, $p = \Sigma_{i=1}^{K} n_i$ is the overall order of the reaction, and k is the rate constant.

In equation 1–10, k must have the units

$$[\text{concentration}]^{-(p-1)}[\text{time}]^{-1}$$

so for a second-order reaction, i.e., $m = n = 1$ in equation 1–8, the units would be $[\text{concentration}]^{-1}[\text{time}]^{-1}$, or $dm^3 \, mol^{-1} \, sec^{-1}$ in SI units. Note that the units of liter $mol^{-1} \, sec^{-1}$ are frequently encountered in the older solution-kinetics literature, and $cm^3 \, mol^{-1} \, sec^{-1}$ or $cm^3 \, molecule^{-1} \, sec^{-1}$ are still encountered in the gas-kinetics literature.

Elementary reactions may be described by their *molecularity*, which specifies the number of reactants that are involved in the reaction step. If a reactant spontaneously decomposes to yield products in a single reaction step, given by the equation

$$A \longrightarrow \text{products} \tag{1-11}$$

the reaction is termed *unimolecular*. An example of a unimolecular reaction is the dissociation of N_2O_4, represented by

$$N_2O_4 \longrightarrow 2NO_2$$

If two reactants A and B react with each other to give products, i.e.,

$$A + B \longrightarrow \text{products} \tag{1-12}$$

the reaction is termed *bimolecular*. An example of a bimolecular reaction would be a metathetical atom-transfer reaction such as

$$O + H_2 \longrightarrow OH + H$$

or

$$F + H_2 \longrightarrow HF + H$$

Both of these reactions are discussed in subsequent chapters.

Three reactants that come together to form products constitute a *termolecular* reaction. In principle, one could go on to specify the molecularity of four, five, etc., reactants involved in an elementary reaction, but such reactions have not been encountered in nature. The situation reflects the molecular basis of elementary reactions. A single, suitably energized molecule can decompose according to equation 1–11; such unimolecular processes are discussed in chapter 11. A collision between two molecules can lead to a bimolecular reaction according to equation 1–12; this is further discussed in chapters 8 and 10. At moderate to high gas pressures, termolecular processes can occur, such as three-body recombination, i.e.,

$$A + B + M \longrightarrow AB + M \tag{1–13}$$

However, physical processes involving simultaneous interaction of four or more independent particles are so rare in chemical kinetics as to be completely negligible.

An elementary reaction is one in which the molecularity and the overall order of the reaction are the same. Thus, a bimolecular elementary reaction is second order, a termolecular reaction third order, and so on. The reverse is not always true, however. For example, the hydrogen-iodine reaction 1–3 is second order in both directions, but bimolecular reactions between H_2 and I_2, and between two HI molecules, are thought not to occur. Instead, the reaction consists of several unimolecular, bimolecular, and possibly termolecular steps (see chapter 2).

A further distinction between molecularity and reaction order is that, while molecularity has only the integer values 0, 1, 2, and 3, order is an experimentally determined quantity which can take on noninteger values. In principle, these values could be any number between $-\infty$ and $+\infty$, but values between -2 and 3 are usually encountered in practice. Negative orders imply that the component associated with that order acts to slow down the reaction rate; such a component is termed an *inhibitor* for that reaction. Fractional values of the reaction order always imply a complex reaction mechanism (see section 1.6). An example of a fractional-order reaction is the thermal decomposition of acetaldehyde given by

$$CH_3 CHO \xrightarrow{300–800°C} CH_4 + CO \tag{1–14}$$

which has a $\frac{3}{2}$ reaction order, i.e.,

$$\frac{d[CH_4]}{dt} = (\text{constant})[CH_3 CHO]^{3/2} \tag{1–15}$$

Similarly, under certain conditions the reaction of hydrogen with bromine

$$H_2 + Br_2 \longrightarrow 2HBr \tag{1–16}$$

has a $\frac{3}{2}$ reaction order, first order in $[H_2]$ and $\frac{1}{2}$ order in $[Br_2]$:

$$\frac{d[HBr]}{dt} = (\text{constant})[H_2][Br_2]^{1/2} \tag{1–17}$$

Under other conditions, reaction 1–16 can display an even more complicated behavior, viz.,

$$\frac{d[HBr]}{dt} = \frac{(\text{constant})[H_2][Br_2]^{1/2}}{1 + (\text{constant}')[HBr]} \tag{1–18}$$

The constants in equations 1–15, 1–17, and 1–18 are clearly not identifiable with an elementary reaction, but instead are phenomenological coefficients obtained by fitting the rate expression to experimental data. Such coefficients are more properly termed *rate coefficients,* rather than rate constants. The latter term should be reserved for the coefficients in rate expressions for elementary reactions, which follow a rate expression having the form of equation 1–10.

1.3 ELEMENTARY REACTION RATE LAWS

Thus far, we have defined the rate of reaction in terms of concentrations, orders, and reaction rate constants. Next, we consider the time behavior of the concentration of reactants in elementary reactions with simple orders. The time behavior is determined by integrating the rate law for a particular rate expression.

1.3.1 Zero-Order Reaction

The rate law for a reaction that is zero order is

$$R = -\frac{d[A]}{dt} = k[A]^0 = k \tag{1–19}$$

Zero-order reactions are most often encountered in heterogeneous reactions on surfaces (see chapter 5). The rate of reaction for this case is independent of the concentration of the reacting substance. To find the time behavior of the reaction, equation 1–19 is put into the differential form

$$d[A] = -kdt \tag{1–20}$$

and then integrated over the boundary limits t_1 and t_2. Assuming that the concentration of A at $t_1 = 0$ is $[A]_0$, and at $t_2 = t$ is $[A]_t$, equation 1–20 becomes

$$\int_{[A]_0}^{[A]_t} d[A] = -k \int_{t_1=0}^{t_2=t} dt \tag{1–21}$$

Hence,

$$[A]_t - [A]_0 = -k(t - 0) \tag{1–22}$$

Consequently, the integrated form of the rate expression for the zero-order reaction is

$$[A]_t = [A]_0 - kt \tag{1–23}$$

A plot of [A] versus time should yield a straight line with intercept $[A]_0$ and slope k.

1.3.2 First-Order Reactions

A first-order reaction is one in which the rate of reaction depends only on one reactant. For example, the isomerization of methyl isocyanide, CH_3NC, is a first-order unimolecular reaction:

$$CH_3NC \longrightarrow CH_3CN \tag{1–24}$$

This type of equation can be represented symbolically as

$$A \longrightarrow B$$

and the rate of disappearance of A can be written as

$$R = -\frac{1}{a}\frac{d[A]}{dt} = k[A]^1 \tag{1-25}$$

Note that the reaction is of order one in the reactant A. Thus, since only one A molecule disappears to produce one product B molecule, $a = 1$ and equation 1–25 becomes

$$-\frac{d[A]}{dt} = k[A] \tag{1-26}$$

Integration of equation 1–26 leads to

$$-\int \frac{d[A]}{[A]} = k \int dt$$

$$-\ln[A] = kt + \text{constant} \tag{1-27}$$

If the boundary conditions are such that at $t = 0$ the initial value of A is $[A]_0$, the constant of integration in equation 1–27 can be eliminated if we integrate over the boundary limits as follows:

$$-\int_{[A]_0}^{[A]_t} \frac{d[A]}{[A]} = k \int_0^t dt \tag{1-28}$$

This gives

$$-(\ln[A]_t - \ln[A]_0) = kt \tag{1-29}$$

and hence

$$-\ln[A]_t = kt - \ln[A]_0 \tag{1-30}$$

Thus, the constant in equation 1–27 is just

$$\text{constant} = -\ln[A]_0 \tag{1-31}$$

Equation 1–30 can be written in various forms. Some that are commonly used are

$$\ln\left(\frac{[A]_t}{[A]_0}\right) = -kt \tag{1-32a}$$

$$[A]_t = [A]_0 e^{-kt} \tag{1-32b}$$

and

$$\frac{[A]_t}{[A]_0} = e^{-kt} \tag{1-32c}$$

These forms of the integrated rate expression for the first-order reaction are worth remembering. From the exponential form of equations 1–32b and 1–32c, one can determine a *time constant* τ which is called the *decay time* of the reaction. This quantity is defined as the time required for the concentration to decrease to $1/e$ of its initial value $[A]_0$, where $e \approx 2.7183$ is the base of the natural logarithm. The time τ is given by

$$\tau = \frac{1}{k} \qquad (1\text{–}33)$$

In experimental determinations of the rate constant k, the integrated form of the rate law is often written in decimal logarithms as

$$\log_{10}[A]_t = \log_{10}[A]_0 - \frac{kt}{2.303} \qquad (1\text{–}34)$$

and a semilog plot of $[A]_t$ versus t will yield a straight line with $k/2.303$ as slope and $[A]_0$ as intercept.

1.3.3 Second-Order Reactions

There are two cases of second-order kinetics. The first is a reaction between two identical species, viz.,

$$A + A \longrightarrow \text{products} \qquad (1\text{–}35)$$

The rate expression for this case is

$$R = -\frac{1}{2}\frac{d[A]}{dt} = k[A]^2 \qquad (1\text{–}36)$$

The second case is an overall second-order reaction between two unlike species, given by

$$A + B \longrightarrow \text{products} \qquad (1\text{–}37)$$

In this case, the reaction is first order in each of the reactants A and B and the rate expression is

$$R = -\frac{d[A]}{dt} = k[A][B] \qquad (1\text{–}38)$$

Note the appearance of the stoichiometric coefficient $\frac{1}{2}$ in equation 1–36, but not in equation 1–38.

Let us consider the first case, given by equations 1–35 and 1–36. Although not an elementary reaction, the disproportionation of HI (equation 1–3) is a reaction which is exactly second order in a single reactant. Another example is the recombination of two identical radicals, such as two methyl radicals:

$$2CH_3 \longrightarrow C_2H_6 \qquad (1\text{–}39)$$

We integrate the rate law, equation 1–36, to obtain

$$-\int_{[A]_0}^{[A]_t} \frac{d[A]}{[A]^2} = 2k \int_0^t dt \qquad (1-40)$$

which gives

$$\frac{1}{[A]_t} = \frac{1}{[A]_0} + 2kt \qquad (1-41)$$

A plot of the inverse concentration of A ($[A]^{-1}$) versus time should yield a straight line with slope equal to $2k$ and intercept $1/[A]_0$.

To integrate the rate law for the second case, equations 1–37 and 1–38, it is convenient to define a progress variable x which measures the progress of the reaction to products as

$$x = ([A]_0 - [A]_t) = ([B]_0 - [B]_t) \qquad (1-42)$$

where $[A]_0$ and $[B]_0$ are the initial concentrations. The rate expression given by equation 1–38 can then be rewritten in terms of x as

$$\frac{dx}{dt} = k([A]_0 - x)([B]_0 - x) \qquad (1-43)$$

To find the time behavior, we integrate equation 1–43 thus:

$$\int_{x(0)}^{x(t)} \frac{dx}{([A]_0 - x)([B]_0 - x)} = k \int_0^t dt \qquad (1-44)$$

To solve the integral on the left-hand side of equation 1–44, we separate the variables and use the method of partial fractions:

$$\int \frac{dx}{([A]_0 - x)([B]_0 - x)}$$
$$= \int \frac{dx}{([A]_0 - [B]_0)([B]_0 - x)} - \int \frac{dx}{([A]_0 - [B]_0)([A]_0 - x)} \qquad (1-45)$$

Solving the right-hand side of equation 1–45 and equating it to the left-hand side of equation 1–44, we obtain, as the solution to the rate expression for the second case,

$$\frac{1}{([A]_0 - [B]_0)} \ln \left(\frac{[B]_0[A]_t}{[A]_0[B]_t} \right) = kt \qquad (1-46)$$

In this case the experimental data may be plotted in the form of the left-hand side of the equation against t.

1.3.4 Third-Order Reactions

From the definition of overall reaction order in equation 1–10, we see that there are three possible types of third-order reactions: (1) $3A \rightarrow$ products; (2) $2A + B \rightarrow$ products; and (3) $A + B + C \rightarrow$ products. In the first case, in which the rate law depends on the third power of one reactant, the rate expression is

$$R \equiv -\frac{1}{3}\frac{d[A]}{dt} = k[A]^3 \tag{1-47}$$

This rate law can be integrated readily to obtain the solution

$$-\frac{1}{2}\left(\frac{1}{[A]^2} - \frac{1}{[A]_0^2}\right) = 3kt \tag{1-48}$$

Rearranging gives

$$\frac{1}{[A]^2} = \frac{1}{[A]_0^2} + 6kt \tag{1-49}$$

A plot of the inverse squared concentration of A ($[A]^{-2}$) with time should yield slope $6k$ and intercept $1/[A]_0^2$.

The second case,

$$2A + B \longrightarrow C \tag{1-50}$$

which is second order in reactant A and first order in reactant B, has the overall order 3. The rate law for this reaction is

$$R \equiv -\frac{1}{2}\frac{d[A]}{dt} = k[A]^2[B] \tag{1-51}$$

This rate expression can be integrated for two possible subcases. The first is when the concentration of B is so much greater than that of the reactant A ($[B] \gg [A]$), that the concentration of B does not change during the course of the reaction. Under this condition, the rate expression can be rewritten as

$$R \equiv -\frac{1}{2}\frac{d[A]}{dt} = k'[A]^2 \tag{1-52}$$

so that the third-order expression reduces to a "pseudo second-order" expression. The solution for this case is equation 1-41, i.e.,

$$\frac{1}{[A]} = \frac{1}{[A]_0} - 2k't \tag{1-53}$$

A plot of $[A]^{-1}$ vs. time, for a fixed $[B]$, should then yield slope $2k'$ and intercept $1/[A]_0$; but note that the resulting rate coefficient is a function of the concentration of B, that is,

$$k' = k[B] \tag{1-54}$$

Forgetting that this rate coefficient contains an added concentration term can lead to errors in interpretation of data. A simple example of this type of reaction is the three-body recombination process, such as $I + I + M \rightarrow I_2 + M$ and $O + O_2 + M \rightarrow O_3 + M$. In these cases the third body acts to remove the excess energy from the recombining reactants, thereby stabilizing the molecular products.

The other instance in which equation 1-51 can be easily integrated is when the initial concentrations of the dissimilar reactants A and B are equal. In integrating the rate law for this case, subject to the stated initial conditions, it is once again convenient to introduce a progress variable as we did in solving the second-order reaction case. Accordingly, we define a progress variable y by

$$[A]_t = [A]_0 - 2y$$

and

$$[B]_t = [B]_0 - y$$

and with this condition we can rewrite equation 1–51 in terms of y as

$$\frac{dy}{dt} = k([A]_0 - 2y)^2([B]_0 - y) \tag{1-55}$$

Upon rearranging, we obtain

$$\frac{dy}{([A]_0 - 2y)^2([B]_0 - y)} = k\,dt \tag{1-56}$$

This equation can be integrated by the method of partial fractions to yield

$$\frac{1}{([A]_0 - 2[B]_0)}\left(\frac{1}{[A_0]} - \frac{1}{[A]_t}\right) + \frac{1}{([A]_0 - 2[B]_0)^2}\ln\left(\frac{[A]_t[B]_0}{[A]_0[B]_t}\right) = kt \tag{1-57}$$

An example of such a reaction is the gas phase reaction between nitric oxide and oxygen

$$2NO + O_2 \longrightarrow 2NO_2 \tag{1-58}$$

The third type of third-order reaction is first order in three different components, i.e.,

$$A + B + C \longrightarrow \text{product} \tag{1-59}$$

The rate law for this reaction is

$$R = \frac{d[A]}{dt} = k[A][B][C] \tag{1-60}$$

To solve for the integrated rate law expression in this case, we use the method of partial fractions as before. The solution is left as an exercise at the end of this chapter.

1.3.5 Reactions of General Order

There are no known examples of fourth-, fifth-, or higher order reactions in the chemical literature. The highest order which has been empirically encountered for chemical reactions is third order. Nevertheless, in this section we develop the general solution for a reaction which is nth order in one reactant, for n equal to any integer or noninteger value. The rate expression for such a reaction is

$$R = -\frac{d[A]}{dt} = k[A]^n \tag{1-61}$$

A simple integration of this expression yields the result

$$\frac{1}{(n-1)}\left(\frac{1}{[A]_t^{n-1}} - \frac{1}{[A]_0^{n-1}}\right) = kt \qquad (1\text{--}62)$$

which can be rewritten as

$$\frac{1}{[A]_t^{n-1}} - \frac{1}{[A]_0^{n-1}} = (n-1)kt \qquad (1\text{--}63)$$

Equation 1–63 is valid for any value of n except $n = 1$, in which case it is undefined and equation 1–32 must be used instead. Figure 1–1 shows several plots of concentration vs. time for various values of n.

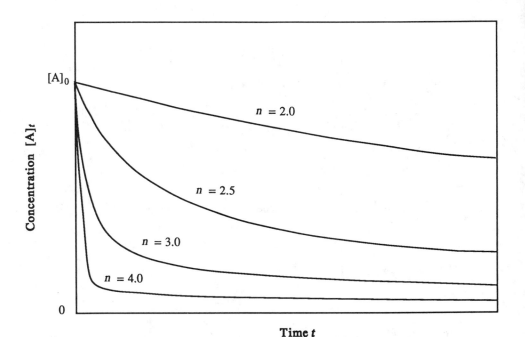

Figure 1–1. Plot of $[A]_t$ versus time for reaction of general order. The plot shows various functional behaviors for $n = 2$, 2.5, 3, 3.5, and 4. Concentration and time are in arbitary units.

In the general case there is no simple plot that can be constructed to test the order of the reaction, as can be done for the first- and second-order cases. When the order n is unknown, a van't Hoff plot can be constructed as an aid to deducing the order of the reaction. In a van't Hoff plot, the logarithm of the rate is plotted against the logarithm of the concentration of the reactant A. This is equivalent to making a plot of equation 1–63 on log-log graph paper. The slope of such a plot gives the order of the reaction n. Examples of van't Hoff plots for several reaction orders are shown in Figure 1–2.

Basic Concepts of Kinetics   Chap. 1

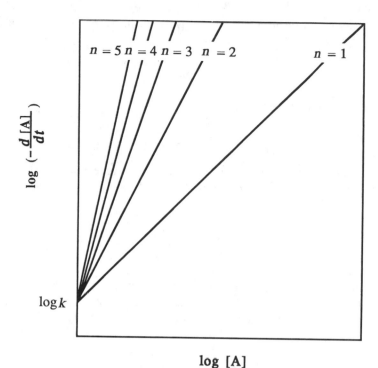

Figure 1–2. Van't Hoff plot of $\log\left(-\dfrac{d[A]}{dt}\right)$ versus $\log[A]$ for various reaction orders.

1.4 DETERMINATION OF REACTION ORDER: REACTION HALF-LIVES

Thus far, we have introduced the concept of a rate law and have shown that experimental rate laws can frequently be written as the product of concentrations of reacting species raised to some power; the exponents in such a rate law then define the order of the particular reaction. These reaction orders are empirically determined and may be nonintegral. We have also considered some simple rate laws and, by integrating them, have shown how experimental data can be plotted to enable the reaction order to be determined. The van't Hoff method has been introduced as a method for determining a general reaction order.

An alternative to the van't Hoff method, and one of the more popular methods for determining reaction order, is the *half-life method*. The reaction half-life $t_{1/2}$ is defined as the period of time necessary for the concentration of a specified reactant to reach one-half of its initial concentration. Measurement of $t_{1/2}$ as a function of initial reactant concentration can help establish the order with respect to that reactant. Consider the simple first-order reaction 1–11. The integrated rate equation is given by equation 1–32a,

$$\ln([A]_t/[A]_0) = -kt$$

By definition, at $t = t_{1/2}$, $[A]_t = [A]_0/2$; therefore, the rate equation can be rewritten as

$$-\ln\left(\frac{[A]_0/2}{[A]_0}\right) = kt_{1/2} \qquad (1\text{–}64)$$

The half-life is then

$$t_{1/2} = \frac{\ln 2}{k} = \frac{0.693}{k} \qquad (1\text{–}65)$$

Thus, for a first-order reaction, $t_{1/2}$ is independent of concentration. On the other hand, for a reaction of order $n > 1$ in a single reactant, the reaction half-life is

$$t_{1/2} = \frac{(2^{n-1} - 1)}{k(n - 1)[A]_0^{n-1}} \qquad (1\text{–}66)$$

where k is the rate constant and $[A]_0$ is the initial concentration of the reactant. Thus, the half-life for orders $n > 1$ is a function of the initial concentration of the reactant; consequently, a plot of the logarithm of $t_{1/2}$ against the logarithm of $[A]_0$ should enable one to determine the reaction order. To illustrate this, let us take the logarithm of both sides of equation 1–66. We obtain

$$\log t_{1/2} = \log \frac{(2^{n-1} - 1)}{k(n - 1)} - (n - 1)\log[A]_0 \qquad (1\text{–}67)$$

It is clear from this expression that the plot will be linear with a slope equal to $n - 1$, from which the order can be determined. With this information, and one or more absolute values of $t_{1/2}$, the rate constant can also be calculated. The reader is reminded that this procedure is valid only for reactions which are nth order in a single reactant.

Other methods of determining reaction orders are discussed in some earlier kinetics textbooks, including those by Hammes[3] and Benson[4].

1.5 TEMPERATURE DEPENDENCE OF RATE CONSTANTS: THE ARRHENIUS EQUATION

We have seen that rate expressions are often simple functions of reactant concentrations with a characteristic rate constant k. If the rate expression is correctly formulated, the rate constant should indeed be a constant—that is, it should not depend on the concentrations of species appearing in the rate law, or of any other species which may be present in the reaction mixture.* The rate constant should also be independent of time. It does, however, depend strongly on temperature. This behavior was described by Svante Arrhenius in 1889[5] on the basis of numerous experimental rate measurements. Arrhenius found that rate constants varied as the negative exponential of the reciprocal absolute temperature, that is,

*The rate constant for reactions in liquid solutions may depend on pH, that is, H^+ concentration; this will be discussed in chapter 4.

$$k(T) = A \exp(-E_{act}/RT) \qquad (1-68)$$

This relationship is now known as the *Arrhenius equation,* and a plot of $\ln k$ (or $\log_{10} k$) vs. $1/T$ is called an *Arrhenius plot*. In the Arrhenius equation, the temperature dependence comes primarily from the exponential term, although the quantity A, referred to as the *pre-exponential* or the *frequency factor,* may have a weak temperature dependence, no more than some fractional power of T. The units of A are the same as the units of the rate constant k, since the exponential term has no units. In the case of a first-order reaction, A has units of sec^{-1}; for reactions of higher order, A has units $(concentration)^{1-p} sec^{-1}$. The temperature T is, of course, in absolute or Kelvin units (degrees Celsius + 273.16).

The key quantity in the Arrhenius equation is the *activation energy* E_{act}. The activation energy can be thought of as the amount of energy which must be supplied to the reactants in order to get them to react with each other. Since this is a positive energy quantity, the majority of reactions have k increasing with temperature. For some reactions, however, the rate decreases with temperature, implying a negative activation energy. Such reactions are generally complex, involving the formation of a weakly bound intermediate species. An example is the recombination of iodine atoms in the presence of a molecular third body M, which proceeds via the following steps:[6]

$$I + M \longrightarrow IM$$

$$IM + I \longrightarrow I_2 + M$$

The IM species is a van der Waals complex whose stability decreases with increasing temperature.

The standard method for obtaining E_{act} is to graph experimental rate constant data on an Arrhenius plot, i.e., $\log_{10} k$ vs. $1/T$. The slope gives $E_{act}/2.303R$, where $R = 8.3145$ J $mol^{-1}K^{-1}$. The units of E_{act} are thus J mol^{-1}, but since the magnitudes of activation energies are typically in the range of a few to several hundred thousand J mol^{-1}, it is customary to report their values in kJ mol^{-1}. The older (non-SI) unit of calories or kcal mol^{-1} is still often encountered, but should be discouraged. The conversion factor[7] is 1 calorie = 4.184 J.

The origin of the activation energy is a barrier on the potential energy surface between the reactants and products; this is discussed in detail in chapters 7 and 10. F\ 'he time being, the *enthalpy diagram* shown in Figure 1–3 may be instructive. This is simply a sketch of the thermodynamic energies associated with the reactants, the products, and a (for the moment) hypothetical *transition state* connecting the two. The energy difference between the reactants and products is the difference in their heats of formation and is given by

$$\Delta H^0_{reaction} = \Delta H^0_f(products) - \Delta H^0_f(reactants) \qquad (1-69)$$

A reaction which is highly *endothermic,* that is, which has a large positive $\Delta H^0_{reaction}$, is not likely to proceed spontaneously except at very high temperatures. A highly *exothermic* reaction, however, may do so unless the activation energy required to reach the transition state is very high; in that case, the reaction will be slow at other than very high temperatures. The calculation of reaction exothermicities is often a

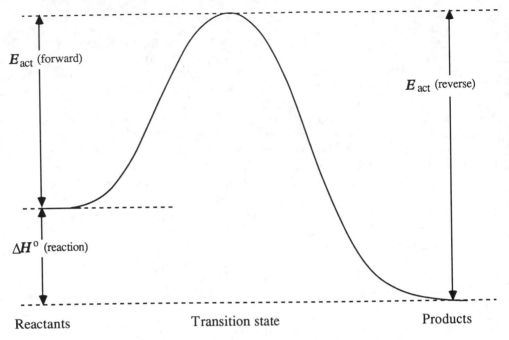

Figure 1–3. Enthalpy diagram for a chemical reaction.

helpful guide in assessing the relative importance of individual reactions in a complex mechanism.

An important relationship which may be established by inspection of Figure 1–3 is that between forward and reverse activation energies for a given reaction. Clearly,

$$E_{act}(\text{forward}) - E_{act}(\text{reverse}) = \Delta H^0_{\text{reaction}} \tag{1–70}$$

Thus, exothermic reactions have a larger activation energy for the reverse reaction, while the opposite is true for endothermic reactions.

1.6 REACTION MECHANISMS, MOLECULAR DYNAMICS, AND THE ROAD AHEAD

In this chapter we have concentrated mainly on elementary chemical reactions. We have examined ways in which the chemical kineticist can determine the reaction order from measurements of the time dependence of the reactant and product concentrations, and the reaction activation energy from the temperature dependence of the rate constant. Unfortunately, with just this amount of information it is not possible to achieve a complete understanding of most chemical reactions. This is because reactions that are studied experimentally rarely occur in a single step. More often, there is a complex series of elementary reactions which take place during the transformation of reactants into products. Reactions that consist of several elementary reaction steps are called *complex*, or *composite*, reactions. The *mechanism* of a com-

16   Basic Concepts of Kinetics   Chap. 1

plex reaction is the assembly of individual steps that make it up. One of the principal aims of experimental studies in chemical kinetics is to elucidate mechanisms and to describe the overall observed reaction process in terms of the elementary reaction steps which constitute the mechanism. In chapter 2, we consider various types of complex reactions together with methods for analyzing their kinetics. Chapter 3 deals with some of the experimental methods that can be used to carry out rate measurements on complex reaction systems, with emphasis on gas-phase reactions. Chapter 4 then goes on to treat reactions in liquid solutions, and in chapter 5 we consider catalyzed reactions, including those which occur at the gas-solid interface.

Chapters 2 through 5 deal primarily with chemical *kinetics,* that is, the phenomenological behavior of reactions. In chapters 6 through 13, we turn our attention to chemical *dynamics,* the description of chemical reactions at the molecular level. The approach there will be to attempt to isolate elementary reaction steps experimentally, to examine them in great detail, and to relate the findings to microscopic molecular properties. Finally, chapters 14 and 15 present some examples of kinetic systems, that is, large coupled reaction sets which describe complex real-world phenomena.

REFERENCES

1. K. J. Laidler, *Pure App. Chem. 53,* 753 (1981).
2. G. Kornfeld and E. Klinger, *Z. physik. Chem. B4,* 37 (1929).
3. G. G. Hammes, *Principles of Chemical Kinetics* (New York: Academic Press, 1978).
4. S. W. Benson, *Foundations of Chemical Kinetics* (New York: McGraw-Hill, 1960).
5. S. Arrhenius, *Z. physik. Chem. 4,* 226 (1889).
6. G. Porter and J. A. Smith, *Proc. Roy. Soc. A261,* 28 (1961).
7. I. Mills, ed., *Quantities, Units, and Symbols in Physical Chemistry* (Oxford: Blackwell, 1988).

BIBLIOGRAPHY

A᷾ ᴵR, I., and HAMMES, G. G. *Chemical Kinetics: Principles and Selected Topics.* New ᴸ ᴷ: McGraw-Hill, 1966.

CAPELLOS, C., and BIELSKI, B. *Kinetic Systems.* New York: Wiley, 1972.

ESPENSON, J. H. *Chemical Kinetics and Reaction Mechanisms.* New York: McGraw-Hill, 1981.

EYRING, H., and EYRING, E. M. *Modern Chemical Kinetics.* New York: Reinhold, 1963.

FLECK, G. M. *Chemical Reaction Mechanisms.* New York: Holt, Rinehart and Winston, 1971.

FROST, A. A., and PEARSON, R. G. *Kinetics and Mechanism.* 2d ed. New York: Wiley, 1961.

GARDINER, W. C., JR. *Rates and Mechanisms of Chemical Reactions.* Menlo Park, CA: W. A. Benjamin, 1969.

GLASSTONE, S., LAIDLER, K. J., and EYRING, H. *The Theory of Rate Processes.* New York: McGraw-Hill, 1941.

JOHNSTON, H. S. *Gas Phase Reaction Rate Theory*. New York: Ronald Press, 1961.

JORDAN, P. C. *Chemical Kinetics and Transport*. New York: Plenum, 1979.

KONDRAT'EV, V. N. *Chemical Kinetics of Gas Reactions*. Translated by J. M. Crabtree and S. N. Carnethes. Reading, MA: Addison-Wesley, 1964.

KONDRAT'EV, V. N., and NIKITIN, E. E. *Gas-Phase Reactions*. Berlin: Springer-Verlag, 1981.

LAIDLER, K. J. *Chemical Kinetics*. 3d ed. New York: Harper and Row, 1987.

MOORE, J. W., and PEARSON, R. G. *Kinetics and Mechanism*. 3d ed. New York: Wiley, 1981.

NICHOLAS, J. *Chemical Kinetics*. New York: Wiley, 1976.

PIMENTEL, G. C. (Chairman, Committee to Survey Opportunities in the Chemical Sciences). *Opportunities in Chemistry*. Washington, DC: National Academy Press, 1985.

WESTON, R. W., JR. and SCHWARZ, H. A. *Chemical Kinetics*. Englewood Cliffs, NJ: Prentice-Hall, 1972.

PROBLEMS

1.1 **(a)** What fractions of the molecules in H_2 and UF_6 have kinetic energies greater than 100 kJ/mole at 300 K and at 3,000 K (neglect dissociation)?

  (b) In what proportion of binary collisions does the kinetic energy along the line of centers exceed 50 kJ/mole at 100 K, at 300 K, and at 2,000 K?

1.2 The units in the preceding problem were in kJ/mole, which is an example of SI units. Give values for the following quantities both in kJ/mole and in the indicated alternative units in parentheses:

  (a) Heat capacity of liquid water at 15°C and 1 atm (calories/mole K).

  (b) Vibrational fundamental frequency of $H^{35}Cl$ (cm^{-1}).

  (c) Ionization potential of H atom (eV).

  (d) R.M.S. average translational kinetic energy of Br_2 at 300 K (ergs/molecule).

  (e) energy of CO_2 laser photons having a wavelength of 10.59 μm (Hz).

1.3 **(a)** Derive the integrated rate equation for a reaction of $\frac{3}{2}$ order. Derive the expression for the half-life of such a reaction. Can you think of an example of such a reaction?

  (b) Derive the integrated rate equation for a reaction of order n.

1.4 The first-order gas reaction $SO_2Cl_2 \rightarrow SO_2 + Cl_2$ has $k_1 = 2.20 \times 10^{-5}$ s^{-1} at 593 K. What percent of a sample of SO_2Cl_2 would be decomposed by heating at 593 K for 1 hour? For 3 hours? How long will it take for half the SO_2Cl_2 to decompose?

1.5 The kinetics of the formation of ethyl acetate from acetic acid and ethyl alcohol as homogeneously catalyzed by a constant amount of HCl has been studied by titrating 1-cc aliquots of the reaction mixture with 0.0612 N base at various times. The following data have been obtained at 25°C [O. Knoblauch, *Z. physik. Chem.* 22, 268 (1897)]:

Initial Concentrations

$$[CH_3COOH] = 1.000 \text{ M}$$
$$[C_2H_5OH] = 12.756 \text{ M}$$
$$[H_2O] = 12.756 \text{ M}$$
$$[CH_3COOC_2H_5] = 0$$

t, min	Base, cc	t, min	Base, cc
0	24.37	148	18.29
44	22.20	313	14.14
62	21.35	384	13.40
108	19.50	442	13.09
117	19.26	∞	12.68

The overall reaction can be written as

$$CH_3COOH + C_2H_5OH \underset{k_{-1}}{\overset{k_1}{\rightleftharpoons}} CH_3COOC_2H_5 + H_2O$$

The reaction has been found to be first order with respect to each of the four reactants. Calculate the specific rate constants k_1 and k_{-1}. What is the equilibrium constant K_1 at 25°C?

1.6 Nitrogen pentoxide decomposes according to the reaction $N_2O_5 \longrightarrow 2NO_2 + \frac{1}{2}O_2$ with rate constant k. The measured rates between 273 K and 338 K are as follows:

T (K)	273	298	308	318	328	338
k ($\times 10^5$ sec^{-1})	0.0787	3.46	13.5	49.8	150	487

Make an Arrhenius plot of these data, and determine E_{act} and A for the first-order decomposition of nitrogen pentoxide [F. Daniels and F. H. Johnston, *J. Am. Chem. Soc.* **43**, 53 (1921)].

1.7 The reaction

$$NO_3 + NO \longrightarrow 2NO_2$$

is known to be an elementary process.
(a) Write the rate expression for the rate of disappearance of NO_3 and NO.
(b) Write the rate expression for the appearance of NO_2.
(c) Show how the rate constants in (a) and (b) are related.

1.8 T-butyl bromide is converted into t-butyl alcohol in a solvent containing 90 percent acetone and 10 percent water. The reaction is given by

$$(CH_3)_3CBr + H_2O \longrightarrow (CH_3)_3COH + HBr$$

The following table gives the data for the concentration of t-butyl bromide versus time [L. C. Bateman, E. D. Hughes, and C. K. Ingold, *J. Chem. Soc.* 960, 1940]:

t (min)	$(CH_3)_3 CBr(mol\ \ell^{-1})$
0	0.1056
9	0.0961
18	0.0856
27	0.0767
40	0.0645
54	0.0536
72	0.0432
105	0.0270

What is the order of the reaction? What is the rate constant of the reaction? What is the half-life of the reaction?

1.9 The reaction

$$A + B + C \longrightarrow products$$

has the rate expression

$$R = -\frac{d[A]}{dt} = k[A][B][C]$$

Derive the integral form of the rate expression.

1.10 **(a)** What are the units of the rate constants of first-, second-, and third-order reactions if the concentrations are expressed in moles cm^{-3} and the time is given in seconds? What are the conversion functions that must be used to convert to concentration units of molecules/cm^3 in each case?

(b) If a reaction obeyed the rate law

$$R = k[A]^{1/2}[B]^{2/3}$$

what would be the units of k?

1.11 Write the rate expression for the reaction

$$2A + 3B \longrightarrow X$$

and derive the integrated rate expression for the reaction.

Basic Concepts of Kinetics   Chap. 1

CHAPTER 2

Complex Reactions

2.1 EXACT ANALYTIC SOLUTIONS FOR COMPLEX REACTIONS

2.1.1 Introduction

In chapter 1, we considered primarily reactions involving a single step; but most chemical processes are complex, i.e., they consist of a number of coupled elementary reactions. These complex reactions can be divided into several classes: (1) opposing or reversible reactions, (2) consecutive reactions, (3) parallel reactions, and (4) mixed reactions.

In this chapter we examine methods for determining exact analytic solutions for the time dependence of concentrations of species involved in complex reactions. Since it is difficult to obtain such exact solutions for many complex reactions, approximate treatments, such as the steady-state or pseudo-first-order methods, may need to be used to obtain analytic results. Accordingly, we discuss both of these methods in our treatment. Two other methods that are receiving increased use in solving systems of kinetic equations for complex reactions are the Laplace transform method and determinant method. Using these methods, exact analytic solutions may be obtained for many complex reactions. However, when analytic solutions are impossible to obtain either directly or by using approximation methods, a numerical solution is generally possible.

In our discussion, we introduce the reader to the basic theory behind various numerical methods, discuss some of the limitations of these methods, and explain what must be considered in choosing a numerical method for solving kinetic equations for complex reactions. Finally, we introduce a more advanced topic, stochastic methods. This topic explores the origin of kinetic equations from a statistical view and gives a justification for the use of differential equations modeling the time evolution of species involved in complex reactions.

2.1.2 Reversible Reactions

Reversible or opposing reactions are reactions in which the products of the initial reaction can proceed to re-form the original substance. A chemical example of this is the cis-trans isomerization of 1,2-dichloroethylene:[1]

$$\begin{array}{ccc} \text{Cl} \quad \text{Cl} & \rightleftharpoons & \text{Cl} \\ & & \text{Cl} \end{array} \tag{2-1}$$

From this example, we can see that the simplest reversible reaction is of the form

$$A_1 \underset{k_r}{\overset{k_f}{\rightleftharpoons}} A_2 \tag{2-2}$$

and is first order in each direction. The differential equation for this mechanism is

$$-\frac{d[A_1]}{dt} = k_f[A_1] - k_r[A_2] \tag{2-3}$$

$$-\frac{d[A_2]}{dt} = k_r[A_2] - k_f[A_1] \tag{2-4}$$

If it is assumed that both A_1 and A_2 are present in the system at time $t = 0$, that is, $[A_1] = [A_1]_0$ and $[A_2] = [A_2]_0$, then at any time afterwards the total amount of reactant remaining and new product formed must equal the initial amount of the reactants before reaction. (This is the law of conservation of mass.) Hence,

$$[A_1]_0 + [A_2]_0 = [A_1] + [A_2] \tag{2-5}$$

Solving for $[A_2]$, we obtain

$$[A_2] = [A_1]_0 + [A_2]_0 - [A_1] \tag{2-6}$$

and substituting this into equation (2-3) yields

$$-\frac{d[A_1]}{dt} = + k_f[A_1] - k_r([A_1]_0 + [A_2]_0 - [A_1])$$

$$-\frac{d[A_1]}{dt} = -k_r([A_1]_0 + [A_2]_0) + (k_f + k_r)[A_1]$$

$$-\frac{d[A_1]}{dt} = (k_f + k_r)\left\{ \frac{-k_r([A_1]_0 + [A_2]_0)}{(k_f + k_r)} + [A_1] \right\} \tag{2-7}$$

To find the solution of equation (2-7) we introduce a variable m, defined as

$$-m = \frac{-k_r([A_1]_0 + [A_2]_0)}{k_f + k_r} \tag{2-8}$$

This allows us to rewrite equation (2-7) as

$$-\frac{d[A_1]}{dt} = (k_f + k_r)\{-m + [A_1]\} \tag{2-9}$$

which we may then integrate

$$\int \frac{d[A_1]}{m - [A_1]} = (k_f + k_r)\int dt \tag{2-10}$$

The solution is

$$\ln\left(\frac{k_f[A_1] - k_r[A_2]}{k_f[A_1]_0 - k_r[A_2]_0}\right) = -(k_f + k_r)(t - t_0) \tag{2-11}$$

If only $[A_1]$ is present in the system initially, at $t = 0$, then the solution reduces to

$$\ln\left(\frac{k_f[A_1] - k_r[A_2]}{k_f[A_1]_0}\right) = -(k_f + k_r)t \tag{2-12}$$

which is just

$$[A_1] = \frac{[A_1]_0}{(k_f + k_r)}[k_r + k_f e^{-(k_f + k_r)t}] \tag{2-13}$$

Using the mass conservation constraint $[A_1]_0 = [A_1] + [A_2]$, we can obtain the solution for $[A_2]$:

$$[A_2] = \frac{k_f[A_1]_0}{(k_f + k_r)}[1 - e^{-(k_f + k_r)t}] \tag{2-14}$$

When equilibrium is reached, the individual reactions must be balanced; in other words, the reaction A → B must occur just as frequently as the reverse reaction. Consequently, detailed balancing is achieved by considering each reversible reaction to be independently able to bring about equilibrium between the reactant and the products. More formally, the principle of *detailed balance* (see Chapter 6), says that when a reaction system reaches equilibrium the forward and reverse reactions occur at the same rate. Consequently, for the reaction $A_1 \rightleftarrows A_2$ at equilibrium,

$$\frac{d[A_1]}{dt} = \frac{-d[A_2]}{dt} = 0 \tag{2-15}$$

and

$$-k_f[A_1]_e + k_r[A_2]_e = 0 \tag{2-16}$$

and we have the following definition of the equilibrium constant K_{eq} expressed in terms of rate constants:

$$\frac{k_f}{k_r} = \frac{[A_2]_e}{[A_1]_e} = K_{eq} \tag{2-17}$$

The same argument can be extended to a reversible reaction that occurs in multiple stages. In the next section we consider a simple two-stage (-step) reaction.

2.1.2.1 First-order reversible reactions involving two steps. Reversible reactions may be distinguished by the number of stages and the number of initial reactants involved in the reaction. Here, we consider the complete derivation for first order reversible reactions involving only two stages, that is,

$$A_1 \underset{k_{-1}}{\overset{k_1}{\rightleftharpoons}} A_2 \underset{k_{-2}}{\overset{k_2}{\rightleftharpoons}} A_3 \tag{2-18}$$

The kinetic equations for the system are

$$\frac{d[A_1]}{dt} = k_{-1}[A_2] - k_1[A_1] \tag{2-19}$$

$$\frac{d[A_2]}{dt} = k_1[A_1] - k_{-1}[A_2] - k_2[A_2] + k_{-2}[A_3] \tag{2-20}$$

and

$$\frac{d[A_3]}{dt} = k_2[A_2] - k_{-2}[A_3] \tag{2-21}$$

We assume that initially, at $t = 0$, $[A_1] = [A_1]_0$ and $[A_2]_0 = [A_3]_0 = 0$ and that the amounts of A_1, A_2 and A_3 which have reacted at a later time satisfy the equation:

$$[A_1]_0 = [A_1] + [A_2] + [A_3] \tag{2-22}$$

By the principle of detailed balance, we have

$$\frac{k_1}{k_{-1}} = \frac{[A_2]_e}{[A_1]_e} \tag{2-23}$$

and

$$\frac{k_2}{k_{-2}} = \frac{[A_3]_e}{[A_2]_e} \tag{2-24}$$

Using equations (2–22)–(2–24) gives

$$[A_1]_0 = \left(\frac{k_{-1}}{k_1} + \frac{k_2}{k_{-2}} + 1\right)[A_2]_e \tag{2-25}$$

so

$$[A_2]_e = \frac{k_1 k_{-2}}{k_{-1}k_{-2} + k_1 k_2 + k_1 k_{-2}} [A_1]_0 \tag{2-26}$$

Substituting equation (2–26) into equation (2–19) we obtain the new first-order differential equation

$$\frac{d[A_1]}{dt} = \frac{k_{-1}k_1 k_{-2}}{k_{-1}k_{-2} + k_1 k_2 + k_1 k_{-2}} [A_1]_0 - k_1[A_1] \tag{2-27}$$

Solving this equation using standard methods, we obtain

$$[A_1] = \frac{[A_1]_0}{k_{-1}k_{-2} + k_1 k_2 + k_1 k_{-2}} [k_{-1}k_{-2} + (k_1 k_2 + k_1 k_{-2})e^{-k_1 t}] \tag{2-28}$$

Since $[A_3] = [A_1]_0 - [A_1] - [A_2]$ by conservation of mass, if we substitute equations (2–26) and (2–28) into this expression, we have, upon simplification,

$$[A_3] = \frac{[A_1]_0}{k_{-1}k_{-2} + k_1k_2 + k_1k_{-2}} [k_1k_2 - (k_1k_2 + k_1k_{-2})e^{-k_1t}] \quad (2\text{–}29)$$

As $t \rightarrow \infty$,

$$[A_1]_\infty = [A_1]_0 \frac{k_{-1}k_{-2}}{k_{-1}k_{-2} + k_1k_2 + k_1k_{-2}} \quad (2\text{–}30)$$

$$[A_2]_\infty = [A_1]_0 \frac{k_1k_{-2}}{k_{-1}k_{-2} + k_1k_2 + k_1k_{-2}} \quad (2\text{–}31)$$

and

$$[A_3]_\infty = [A_1]_0 \frac{k_1k_2}{k_{-1}k_{-2} + k_1k_2 + k_1k_{-2}} \quad (2\text{–}32)$$

and the system in then in a state of equilibrium.

The detailed balance method thus allows us to solve an otherwise complicated system of kinetic equations easily; but the method holds only when the system is in a state of equilibrium. Furthermore, equations (2–26), (2–28), and (2–29) do not describe the time evolution of $[A_1]$, $[A_2]$, and $[A_3]$ at short times. To explore what happens early in the reaction or before equilibrium is reached, we must solve the system of kinetic equations exactly. We can do so using Laplace transforms as discussed in Sec. 2–4. Nevertheless, the detailed balance treatment can be used to develop equilibrium relations and to find equilibrium concentrations. A word of caution, however, is appropriate to avoid one misuse of the detailed balance principle, particularly in investigating complex reactions. We can use the example just discussed to illustrate this point.

The misuse of the principle arises when the intermediate A_2 is difficult to detect, so that an experimentalist might think that $A_1 \rightarrow A_3$ is an elementary reaction. Thus, if the loss of A_1 is assumed in one experiment to determine k_1 and the loss of A_3 in another to find k_{-2}, the equilibrium constant for $[A_3]_e/[A_1]_e$ might be taken to be k_1/k_{-2}. However, since $A_1 \rightarrow A_3$ is not an elementary reaction, this conclusion is incorrect.

To determine $K = [A_3]_e/[A_1]_e$ correctly, the equilibrium of each elementary reaction must be considered, that is, equation (2–23) and equation (2–24) must be used. From each of these measurements, the correct expression for K, can be found:

$$K = \frac{[A_3]_e}{[A_1]_e} = \frac{k_1k_2}{k_{-1}k_{-2}} \quad (2\text{–}33)$$

Another example of a first-order reversible reaction involving only two stages is the cyclic reaction

The detailed balance approach can be used to solve the kinetic equations for this system (See Problem 2–7).

2.1.2.2 First and second order reversible reactions.

A reversible reaction may be of mixed order, such as

$$A_1 \underset{k_{-1}}{\overset{k_1}{\rightleftharpoons}} A_2 + A_3$$

An example is $N_2O_4 \rightleftharpoons 2NO_2$. The rate expression for this type of reaction is

$$\frac{d[A_1]}{dt} = -k_1[A_1] + k_{-1}[A_2][A_3] \tag{2–34}$$

To find the solution, we introduce a progress variable,

$$x = [A_1]_0 - [A_1] \tag{2–35}$$

Introducing this variable into equation (2–34) transforms that equation into

$$\frac{dx}{dt} = k_1([A_1]_0 - x) - k_{-1}([A_2]_0 + x)([A_3]_0 + x)$$

or

$$\frac{dx}{dt} = k_1[A_1]_0 - k_{-1}[A_2]_0[A_3]_0 - (k_1 + k_{-1}[A_3]_0 + k_{-1}[A_2]_0)x - k_{-1}x^2 \tag{2–36}$$

We let

$$\alpha = k_1[A_1]_0 - k_{-1}[A_2]_0[A_3]_0$$

$$\beta = -(k_1 + k_{-1}[A_3]_0 + k_{-1}[A_2]_0) \tag{2–37}$$

and

$$\gamma = -k_{-1}$$

so that

$$\frac{dx}{dt} = \alpha + \beta x + \gamma x^2 \tag{2–38}$$

Then

$$\int \frac{dx}{\alpha + \beta x + \gamma x^2} = \int dt \tag{2–39}$$

The solution of equation (2-39) is

$$\ln\left[\frac{x + (\beta - q^{1/2})/2\gamma}{x + (\beta + q^{1/2})/2\gamma}\right] - \ln\left[\frac{x(t_0) + (\beta - q^{1/2})/2\gamma}{x(t_0) + (\beta + q^{1/2})/2\gamma}\right] = q^{1/2}(t - t_0) + \theta \tag{2–40}$$

where $q = \beta^2 - 4\alpha\gamma$ and $\theta = \ln((\beta - q^{1/2})/(\beta + q^{1/2}))$. At $t_0 = 0$, $x(0) = 0$, and the solution simplifies to

$$\ln\left[\frac{x + (\beta - q^{1/2})/2\gamma}{x + (\beta + q^{1/2})/2\gamma}\right] = q^{1/2}t + 2\theta \tag{2–41}$$

2.1.3 Consecutive Reactions

Irreversible reactions can be defined as those which start with an initial reactant and produce products or intermediates generally in only one direction. Consecutive reactions are sequential irreversible reactions. There are two classes of consecutive reactions: those which are first order and those which are mixed first order and second order. Let us consider first the simplest consecutive reaction of first order.

2.1.3.1 First-order consecutive reactions.
Consecutive reactions are further classified by the number of steps involved or the number of substances initially present, which define their order. We begin by considering the two-step (or -stage) first-order reaction.

2.1.3.1.1. First-Order with Two Steps. Consider the consecutive first-order reaction involving two stages:

$$A_1 \xrightarrow{k_1} A_2 \tag{2-42}$$
$$A_2 \xrightarrow{k_2} A_3$$

This mechanism can be described by the following set of rate expressions:

$$\frac{d[A_1]}{dt} = -k_1[A_1] \tag{2-43}$$

$$\frac{d[A_2]}{dt} = k_1[A_1] - k_2[A_2] \tag{2-44}$$

$$\frac{d[A_3]}{dt} = k_2[A_2] \tag{2-45}$$

Hence, the concentration of A_1 is obtained after integration as

$$[A_1] = [A_1]_0 e^{-k_1 t} \tag{2-46}$$

The concentration of A_2 is computed from equation (2–44), which is a linear differential equation of first order. Substituting equation (2–46) into equation (2–44) we obtain

$$\frac{d[A_2]}{dt} + k_2[A_2] = k_1[A_1]_0 e^{-k_1 t} \tag{2-47}$$

This general differential equation can be solved using standard methods. Doing so, we obtain the time dependence of $[A_2]$:

$$[A_2] = [A_2]_0 e^{-k_2 t} + \frac{k_1[A_1]_0}{k_2 - k_1} (e^{-k_1 t} - e^{-k_2 t}) \tag{2-48}$$

If $[A_2]_0 = 0$ at $t = 0$, then

$$[A_2] = \frac{k_1[A_1]_0}{k_2 - k_1} (e^{-k_1 t} - e^{-k_2 t}) \tag{2-49}$$

The concentration of $[A_3]$ can be determined from conservation of mass, i.e.,

$$[A_1]_0 = [A_1] + [A_2] + [A_3] \tag{2-50}$$

By substituting the expressions for $[A_1]$ and $[A_2]$ into equation (2–50) we obtain

$$[A_3] = [A_1]_0[1 - e^{-k_1 t} - \frac{k_1}{k_2 - k_1}(e^{-k_1 t} - e^{-k_2 t})] \tag{2-51}$$

Simplifying this expression gives

$$[A_3] = [A_1]_0\left[1 - \frac{k_2}{k_2 - k_1}e^{-k_1 t} + \frac{k_1}{k_2 - k_1}e^{-k_2 t})\right] \tag{2-52}$$

Figure 2–1 shows typical kinetic curves of the accumulation of A_2 and A_3 for this case. As $[A_2]$ approaches its maximum value in the initial period, the rate of production of A_3 increases, but the rate of accumulation of A_3 is slow enough to allow an initial accumulation of A_2 to occur before it is converted to A_3, as shown in Fig. 2–1a. In reactions involving unstable intermediates such as free radicals, the production rate of the intermediate is usually slow; however, once the intermediate is formed, it rapidly transforms to products. In this case the rate of production of A_3 is faster than the rate of accumulation of the intermediate A_2, i.e., $k_2 \gg k_1$. In Fig. 2–1b, the concentration of the intermediate A_2 does not have sufficient time to build up. Consequently, its concentration remains small throughout the reaction.

Some caution must be exercised in interpreting kinetic plots such as those shown in Fig. 2–1. While it is natural to interpret the rising portion of the curve as the formation rate of the short-lived intermediate and the falling portion as its decay rate, that is not necessarily the case. If $k_2 > k_1$, that is, the consumption rate of the unstable intermediate A_2 is faster than its formation rate (as is frequently the case), then for small times equation (2–49) approximates $\{k_1[A_1]/(k_2 - k_1)\} \times \{1 - \exp(-k_2 t)\}$, while the formation rate appears in the long-time behavior, which is proportional to $\exp(-k_1 t)$. If, on the other hand, $k_1 > k_2$, then we can rewrite equation (2–49) as

$$[A_2] = \frac{k_1[A_1]_0}{k_1 - k_2}(e^{-k_2 t} - e^{-k_1 t})$$

Now the rise time of $[A_2]$ goes as $(1 - e^{-k_1 t})$, while its decay time goes as $e^{-k_2 t}$; this is the case illustrated in Fig. 2-1a.

2.1.3.1.2 First Order with Three Steps.
The system of differential equations for a first-order consecutive reaction involving three steps, viz.:

$$A_1 \xrightarrow{k_1} A_2$$
$$A_2 \xrightarrow{k_2} A_3 \tag{2-53}$$
$$A_3 \xrightarrow{k_3} A_4$$

can be integrated in a way similar to the two-step case. The differential equations for

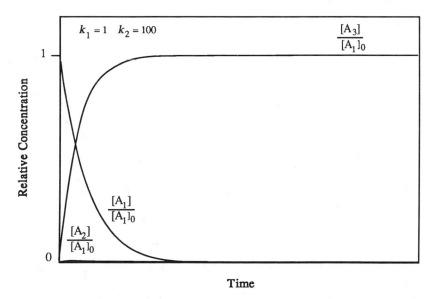

Figure 2–1 (a) Plot of the relative concentrations in the consecutive reaction $A_1 \rightarrow A_2 \rightarrow A_3$ for the case where $k_1/k_2 = 10$. (b) Plot of the relative concentrations for the case where $k_1/k_2 = 0.01$.

A_1 and A_2 and their solutions do not differ from those obtained in the two-step case, while the differential equation for A_3 is

$$\frac{d[A_3]}{dt} = k_2[A_2] - k_3[A_3] \tag{2-54}$$

Substituting equation (2–49) for $[A_2]$ into equation (2–54) gives

$$\frac{d[A_3]}{dt} + k_3[A_3] = \left[\frac{k_1 k_2 [A_1]_0}{(k_2 - k_1)}\right][e^{-k_1 t} - e^{-k_2 t}] \tag{2-55}$$

Integrating this linear equation subject to the initial condition that $[A_3] = 0$ at $t = 0$, we obtain

$$[A_3] = \left[\frac{k_1 k_2 [A_1]_0}{(k_2 - k_1)(k_3 - k_1)}\right]e^{-k_1 t} + \left[\frac{k_1 k_2 [A_1]_0}{(k_1 - k_2)(k_3 - k_2)}\right]e^{-k_2 t}$$
$$+ \left[\frac{k_1 k_2 [A_1]_0}{(k_1 - k_3)(k_2 - k_3)}\right]e^{-k_3 t} \tag{2-56}$$

By conservation of mass

$$[A_1]_0 = [A_1] + [A_2] + [A_3] + [A_4] \tag{2-57}$$

Therefore,

$$[A_4] = [A_1]_0 - [A_1] - [A_2] - [A_3] \tag{2-58}$$

Substituting the appropriate expressions for A_1, A_2, and A_3 into equation (2–58) we get

$$[A_4] = [A_1]_0\left[1 - \frac{k_2 k_3 e^{-k_1 t}}{(k_2 - k_1)(k_3 - k_1)} - \frac{k_1 k_3 e^{-k_2 t}}{(k_1 - k_2)(k_3 - k_2)} - \frac{k_1 k_2 e^{-k_3 t}}{(k_1 - k_3)(k_2 - k_3)}\right] \tag{2-59}$$

In a similar manner the system of differential equations corresponding to $n - 1$ consecutive reactions

$$A_1 \longrightarrow A_2$$
$$A_2 \longrightarrow A_3$$
$$A_3 \longrightarrow A_4 \tag{2-60}$$
$$\vdots$$
$$A_{n-1} \longrightarrow A_n$$

can be integrated. We leave this as an exercise in the problems at the end of this chapter.

2.1.3.2 Higher Order Consecutive Reactions

Some higher order consecutive reactions may be first order in one step and second order in the second, such as

$$A_1 \xrightarrow{k_1} A_2$$

$$A_1 + A_2 \xrightarrow{k_2} A_3 \qquad (2\text{--}61)$$

or both steps may be second order consecutive reactions, as in

$$A_1 + A_2 \longrightarrow A_3$$

$$A_1 + A_3 \longrightarrow A_4 \qquad (2\text{--}62)$$

The differential equations for most of these systems of reactions are nonlinear and generally they have no exact solutions. Nevertheless, analytic solutions for such systems can be obtained if time is eliminated as a variable[2]. To illustrate this method, consider the reaction sequence described by the system of equations (2–61):

$$\frac{d[A_1]}{dt} = -k_1[A_1] - k_2[A_1][A_2] \qquad (2\text{--}63)$$

$$\frac{d[A_2]}{dt} = +k_1[A_1] - k_2[A_1][A_2] \qquad (2\text{--}64)$$

$$\frac{d[A_3]}{dt} = k_2[A_1][A_2] \qquad (2\text{--}65)$$

To solve this system of equations we divide equation (2–64) by (2–63)

$$\frac{d[A_2]}{d[A_1]} = \frac{k_1[A_1] - k_2[A_1][A_2]}{-k_1[A_1] - k_2[A_1][A_2]} = -\left(\frac{k_1 - k_2[A_2]}{k_1 + k_2[A_2]}\right) \qquad (2\text{--}66)$$

If we let $K = k_1/k_2$, then

$$\frac{d[A_2]}{d[A_1]} = -\frac{K - [A_2]}{K + [A_2]} \qquad (2\text{--}67)$$

Equation (2–67) can be rearranged to

$$-\frac{(K - [A_2])d[A_2]}{K - [A_2]} + \frac{2Kd[A_2]}{(K - [A_2])} = -d[A_1] \qquad (2\text{--}68)$$

or

$$-d[A_2] + \frac{2K}{(K - [A_2])}d[A_2] = -d[A_1] \qquad (2\text{--}69)$$

Integrating equation (2–69) over the respective limits, i.e.,

$$-\int_{[A_2]=0}^{[A_2]} d[A_2] - 2K\int_{[A_2]=0}^{[A_2]} \frac{d(K - [A_2])}{(K - [A_2])} = -\int_{[A_1]_0}^{[A_1]} d[A_1] \qquad (2\text{--}70)$$

gives

$$\frac{[A_2]}{K} + 2\ln\left(1 - \frac{[A_2]}{K}\right) = \left(\frac{[A_1]_0}{K}\right)\left(\frac{[A_1]}{[A_1]_0} - 1\right) \qquad (2\text{--}71)$$

In the case of consecutive first- and second-order reactions, the only way that this system may be solved exactly is by eliminating time as an independent variable. In this manner, $[A_1]$ and $[A_2]$ are given at any time, as in equation (2–71); and by conservation of mass, the concentration relationship between $[A_3]$ and the other components is obtained. There are some special cases, however, of first- and second-order reactions that can be solved exactly without eliminating time as a variable[3]. One such reaction is

$$A_1 \xrightarrow{k_1} A_2$$
$$2A_2 \xrightarrow{k_2} A_3$$

(2–72)

The rate equations for this reaction system are

$$\frac{d[A_1]}{dt} = -k_1[A_1]$$

(2–73)

$$\frac{d[A_2]}{dt} = k_1[A_1] - 2k_2[A_2]^2$$

(2–74)

and

$$\frac{d[A_3]}{dt} = k_2[A_2]^2$$

(2–75)

Integration of equation (2–73) yields

$$[A_1] = [A_1]_0 e^{-k_1 t}$$

(2–76)

Substituting equation (2–76) into equation (2–74) yields

$$\frac{d[A_2]}{dt} = [A_1]_0 k_1 e^{-k_1 t} - 2k_2[A_2]^2$$

(2–77)

If we make the further substitution

$$[A_2] = \frac{1}{k_2 u(t)} \frac{du(t)}{dt}$$

(2–78)

we can transform the nonlinear first order equation (2–77) in $[A_2]$ into a linear second order equation in $u(t)$

$$\frac{d^2 u}{dt^2} - [A_1]_0 k_1 k_2 e^{-k_1 t} u = 0$$

(2–79)

A further change of the independent variable t to $\tau = e^{-k_1 t}$ yields a Bessel equation

$$\frac{d}{d\tau}\left(\tau \frac{du}{d\tau}\right) - K u = 0$$

(2–80)

the solution of which is

$$u = \alpha J_0(2i\sqrt{K\tau}) + \beta' i H_0^{(1)}(2i\sqrt{K\tau})$$

(2–81)

where $K = [A_1]_0 k_2 / k_1$, α and β' are constants of integration, $i = \sqrt{-1}$ and J_0 and

$H_0^{(1)}$ are Bessel functions of the first and third kind, respectively, of order zero. Then by the recurrence relations of Bessel functions, we have

$$\frac{dJ_0(2i\sqrt{K\tau})}{d\tau} = -\sqrt{K/\tau}\,iJ_1(2i\sqrt{K\tau}) \tag{2-82}$$

and

$$i\frac{dH_0^{(1)}(2i\sqrt{K\tau})}{d\tau} = \sqrt{K/\tau}\,H_1^{(1)}(2i\sqrt{K\tau}) \tag{2-83}$$

and the solution for $[A_2]$ is

$$[A_2] = [A_1]_0\sqrt{\tau/K}\,\frac{iJ_1(2i\sqrt{K\tau}) - \beta'H_1^{(1)}(2i\sqrt{K\tau})}{J_0(2i\sqrt{K\tau}) + \beta'H_0^{(1)}(2i\sqrt{K\tau})} \tag{2-84}$$

The constant β' is determined by the initial condition $[A_2]_0 = 0$, which gives

$$\beta' = \frac{iJ_1(2i\sqrt{K})}{H_1(2i\sqrt{K})} \tag{2-85}$$

The full solution to this problem is

$$[A_1] = [A_1]_0\tau$$

$$[A_2] = [A_1]_0\sqrt{\tau/K}\,\frac{iJ_1(2i\sqrt{K\tau}) - \beta'H_1^{(1)}(2i\sqrt{K\tau})}{J_0(2i\sqrt{K\tau}) + \beta'H_0^{(1)}(2i\sqrt{K\tau})} \tag{2-86}$$

Finally, using conservation of mass, $[A_3]$ can be determined as:

$$[A_3] = \frac{1}{2}([A_1]_0 - [A_1] - [A_2])$$

The definitions used in equation (2–86) are summarized below:

$$\tau = e^{-k_1 t}$$

$$K = [A_1]_0 k_2/k_1$$

$$\beta' = iJ_1(2i\sqrt{K})/H_1^{(1)}(2i\sqrt{K})$$

2.1.4 Parallel Reactions

Parallel reactions are defined as two or more processes in which the same species participate in each reaction step. The most common cases of parallel reactions are: (1) those in which the initial reactant decomposes into several different products; (2) those in which the initial reactants are different, but yield the same products; and (3) those in which a substance reacts with two or more initial reactants.

2.1.4.1 First order decay to different products. Consider the mechanism

$$
\begin{align*}
A_1 &\xrightarrow{k_2} A_2 \\
A_1 &\xrightarrow{k_3} A_3 \qquad (2-87)\\
A_1 &\xrightarrow{k_4} A_4
\end{align*}
$$

which describes a kinetic system in which a single reactant decomposes into several different products. At time $t = 0$ the initial concentrations of the four components are

$$[A_1] = [A_1]_0 \tag{2-88}$$

$$[A_2] = [A_3] = [A_4] = 0 \tag{2-89}$$

The differential equations for the system can be written, as exemplified for $[A_1]$,

$$\frac{d[A_1]}{dt} = -k_2[A_1] - k_3[A_1] - k_4[A_1]$$

$$= -(k_2 + k_3 + k_4)[A_1] \tag{2-90}$$

$$= -k_T[A_1]$$

where $k_T = (k_2 + k_3 + k_4)$ which is the sum of the rates for each of the independent parallel paths.

Integrating equation (2-90) yields

$$[A_1] = [A_1]_0 e^{-k_T t} = [A_1]_0 e^{-(k_2 + k_3 + k_4)t} \tag{2-91}$$

To solve for $[A_2]$ we substitute equation (2-91) into equation (2-92)

$$\frac{d[A_2]}{dt} = k_2[A_1] \tag{2-92}$$

and upon integrating we obtain

$$[A_2] = [A_2]_0 + \frac{k_2[A_1]_0}{k_T}[1 - e^{-k_T t}]$$

or

$$[A_2] = \frac{k_2[A_1]_0}{k_T}[1 - e^{-k_T t}] \tag{2-93}$$

Expressions for $[A_3]$ and $[A_4]$, when derived in a similar way are, respectively

$$[A_3] = \frac{k_3[A_1]_0}{k_T}[1 - e^{-k_T t}] \tag{2-94}$$

and

$$[A_4] = \frac{k_4[A_1]_0}{k_T}[1 - e^{-k_T t}] \tag{2-95}$$

It follows that the relative rate constants can be determined by measuring the relative product yields:

$$\frac{[A_3]}{[A_4]} = \frac{k_3}{k_4} \tag{2-96}$$

This ratio defines the *branching ratio* for the reaction; note that this branching ratio is independent of time.

The above example can easily be generalized to reactions of a single reactant into n different products:

$$A_1 \xrightarrow{k_2} A_2$$

$$A_1 \xrightarrow{k_3} A_3$$

$$A_1 \xrightarrow{k_4} A_4 \qquad (2\text{-}97)$$

$$\vdots$$

$$A_1 \xrightarrow{k_n} A_n$$

The rate expression for the disappearance of $[A_1]$ is

$$\frac{-d[A_1]}{dt} = k_T[A_1]_0 \qquad (2\text{-}98)$$

where $k_T = \sum_{i=2}^{n} k_i$; upon integration we find that

$$[A_1] = [A_1]_0 e^{-k_T t} \qquad (2\text{-}99)$$

To find the production of a given product A_n

$$\frac{d[A_n]}{dt} = k_n[A_1] \qquad (2\text{-}100)$$

we substitute equation (2–99) into (2–100) and integrate the resulting expression

$$[A_n] = \frac{k_n[A_1]_0}{k_T}(1 - e^{k_T t}) \qquad (2\text{-}101)$$

2.1.4.2 First order decay to the same products. Consider the reaction sequence

$$A_1 \xrightarrow{k_1} A_2$$

$$A_3 \xrightarrow{k_3} A_2 \qquad (2\text{-}102)$$

The rate expressions for the disappearance of $[A_1]$ and $[A_3]$ and the appearance of A_2 are

$$-\frac{d[A_1]}{dt} = k_1[A_1] \qquad (2\text{-}103)$$

$$+\frac{d[A_2]}{dt} = k_1[A_1] + k_3[A_3] \qquad (2\text{-}104)$$

$$-\frac{d[A_3]}{dt} = k_3[A_3] \qquad (2\text{-}105)$$

with the initial condition that $[A_1] = [A_1]_0$, $[A_3] = [A_3]_0$, and $[A_2] = 0$ at time $t = 0$. The equations describing the time dependence for each component are

$$[A_1] = [A_1]_0 e^{-k_1 t} \qquad (2\text{-}106)$$

and

$$[A_3] = [A_3]_0 e^{-k_3 t} \qquad (2\text{--}107)$$

Since

$$\frac{d[A_2]}{dt} = k_1[A_1]_0 e^{-k_1 t} + k_3[A_3]_0 e^{-k_3 t} \qquad (2\text{--}108)$$

upon integration we obtain

$$[A_2] = [A_1]_0 - [A_1]_0 e^{-k_1 t} + [A_3]_0 - [A_3]_0 e^{-k_3 t} \qquad (2\text{--}109)$$

Rearranging equation (2–109) yields

$$[A_2] = ([A_1]_0 + [A_3]_0) - ([A_1]_0 e^{-k_1 t} + [A_3]_0 e^{-k_3 t}) \qquad (2\text{--}110)$$

2.1.4.3 Parallel second-order reactions. In the case of parallel second-order reactions,

$$\begin{aligned} A_1 + A_2 & \xrightarrow{\ k_1\ } A_4 \\ A_1 + A_3 & \xrightarrow{\ k_2\ } A_5 \end{aligned} \qquad (2\text{--}111)$$

two different equations can be written:

$$\frac{d[A_2]}{dt} = -k_1[A_1][A_2] \qquad (2\text{--}112)$$

$$\frac{d[A_3]}{dt} = -k_2[A_1][A_3] \qquad (2\text{--}113)$$

If we consider conservation of mass, we obtain

$$[A_4] = [A_2]_0 - [A_2] \qquad (2\text{--}114)$$

$$[A_5] = [A_3]_0 - [A_3] \qquad (2\text{--}115)$$

and

$$[A_1]_0 - [A_1] = [A_4] + [A_5]$$
$$= [A_2]_0 - [A_2] + [A_3]_0 - [A_3] \qquad (2\text{--}116)$$

Eliminating time as a variable, as in Sec. 2.1.3.2, yields

$$\frac{d[A_3]}{d[A_2]} = \frac{k_2[A_3]}{k_1[A_2]} \qquad (2\text{--}117)$$

which can be integrated under the initial condition that $[A_3] = [A_3]_0$ and $[A_2] = [A_2]_0$ at time $t = 0$. This gives

$$\frac{[A_3]}{[A_3]_0} = \left(\frac{[A_2]}{[A_2]_0}\right)^{k_2/k_1} \qquad (2\text{--}118)$$

From the mass conservation relation we can determine $[A_5]$ by substituting equation (2–115) into equation (2–118). Rearranging then gives

$$[A_5] = [A_3]_0 \left\{ 1 - \left(\frac{[A_2]}{[A_2]_0} \right)^{k_2/k_1} \right\} \tag{2-119}$$

or

$$[A_5] = [A_3]_0 \left\{ 1 - \left(1 - \frac{[A_4]}{[A_2]_0} \right)^{k_2/k_1} \right\} \tag{2-120}$$

The expression for $[A_1]$ is obtained from equation (2-116), thus:

$$[A_1] = [A_1]_0 - [A_2]_0 - [A_3]_0 + [A_2] + [A_3] \tag{2-121}$$

Substituting equation (2-118) into equation (2-121), we obtain

$$[A_1] = [A_1]_0 - [A_2]_0 - [A_3]_0 + [A_2] + [A_3]_0 \left(\frac{[A_2]}{[A_2]_0} \right)^{k_2/k_1} \tag{2-122}$$

Using this expression for $[A_1]$, we obtain a differential equation for $[A_2]$:

$$\frac{d[A_2]}{dt} = k_1[A_2] \left\{ [A_1]_0 - [A_2]_0 - [A_3]_0 + [A_2] + [A_3]_0 \left(\frac{[A_2]}{[A_2]_0} \right)^{k_2/k_1} \right\} \tag{2-123}$$

If we now let $\alpha = [A_1]_0 - [A_2]_0 - [A_3]_0$ and $\beta = + [A_3]_0([A_2]_0)^{-k_2/k_1}$ we can rewrite equation (2-123) as

$$\frac{d[A_2]}{dt} = k_1[A_2]\{\alpha + [A_2] + \beta[A_2]^{k_2/k_1}\} \tag{2-124}$$

Integrating equation (2-124), we obtain

$$\int_{[A_2]_0}^{[A_2]_t} \frac{d[A_2]}{[A_2]\{\alpha + [A_2] + \beta[A_2]^{k_2/k_1}\}} = \int_0^t k_1 \, dt \tag{2-125}$$

There is no explicit solution to this integral, but we can consider some limiting cases.

Case I. $k_1 \gg k_2$

Under this condition the ratio k_2/k_1 is approximately zero, thus equation (2-125) reduces to the simple form

$$\int_{[A_2]_0}^{[A_2]_t} \frac{d[A_2]}{[A_2](\alpha + \beta + [A_2])} = \int_0^t k_1 \, dt \tag{2-126}$$

whose solution is

$$\ln \frac{(\alpha + \beta + [A_2]_t)[A_2]_0}{(\alpha + \beta + [A_2]_t)[A_2]_t} = -(\alpha + \beta)k_1 t \tag{2-127}$$

Rearranging and solving for $[A_2]_t$ yields

$$[A_2]_t = [A_2]_0 \frac{\alpha + \beta}{(\alpha + \beta + [A_2]_0)e^{-(\alpha+\beta)k_1 t} - [A_2]_0} \tag{2-128}$$

Case II. $k_1 = k_2$

When k_1 and k_2 are equal, the ratio of k_1 to k_2 is unity so the integral equation (2–125) simplifies once again to an expression which is integrable, viz.:

$$\int_{[A_2]_0}^{[A_2]_t} \frac{d[A_2]}{[A_2](\alpha + (1 + \beta)[A_2])} = \int_0^t k_1 \, dt \qquad (2\text{--}129)$$

The solution to equation (2–129) is

$$\ln \frac{(\alpha + (1 + \beta)[A_2]_t)[A_2]_0}{(\alpha + (1 + \beta)[A_2]_0)[A_2]_t} = e^{-\alpha k_1 t} \qquad (2\text{--}130)$$

Solving for $[A_2]_t$, we obtain

$$[A_2]_t = [A_2]_0 \frac{\alpha}{(\alpha + (1 + \beta)[A_2]_0)e^{\alpha k_1 t} - (1 + \beta)[A_2]_0} \qquad (2\text{--}131)$$

Other cases can also be considered; however, when $k_1 = 2k_2$ and $k_1 \ll k_2$, it is difficult to obtain a solution for equation (2–125), and moreover the solutions that are obtained do not have a simple form.

2.2. APPROXIMATION METHODS

2.2.1 Steady-State Approximation

In section 2.1, we have derived analytic solutions for a few complex kinetic systems. As we have seen, some reaction sequences can lead to very cumbersome solutions which cannot be simplified to a tractable form. In some cases no analytic solution can be obtained using any of the conventional mathematical methods. Nevertheless, a simplified solution can often be obtained by application of the *steady-state approximation*. This method was originally proposed by Bodenstein[4] and further developed by Semenov[5]. It is particularly useful when intermediates which are present in small concentrations are involved in the reaction scheme. In that case, the rate of change of their concentration is negligible provided that their concentration remains small compared to the major species in the kinetic scheme. Mathematically,

$$d[A_i]dt \cong 0 \qquad (2\text{--}132)$$

where $[A_i]$ is the concentration of the intermediate A_i. This condition must be met to validate the steady-state method. In order to demonstrate the use of the steady-state method, let us consider the simple reaction

$$A_1 \longrightarrow A_2 \longrightarrow A_3 \qquad (2\text{--}133)$$

which involves the conversion of species A_1 to an intermediate A_2 and finally to the product A_3. The kinetic equations for the species in this scheme are equations (2–43)–(2–45). Solving these equations exactly gives equations (2–46), (2–49), and (2–52), i.e.,

$$[A_1] = [A_1]_0 e^{-k_1 t}$$

$$[A_2] = \frac{k_1 [A_1]_0}{k_2 - k_1} (e^{-k_1 t} - e^{-k_2 t})$$

$$[A_3] = [A_1]_0 \left\{ 1 + \frac{1}{k_1 - k_2} (k_2 e^{-k_1 t} - k_1 e^{-k_2 t}) \right\}$$

under the initial condition that at $t = 0$, $[A_1] = [A_1]_0$ and $[A_2] = [A_3] = 0$.

Under steady-state conditions, the rate of change in the concentration of A_2 is zero, that is,

$$\frac{d[A_2]}{dt} = -k_2[A_2] + k_1[A_1] = 0 \qquad (2\text{--}134)$$

In this case the steady-state condition yields the concentration of the intermediate as

$$[A_2]_{ss} = \frac{k_1}{k_2}[A_1] \qquad (2\text{--}135)$$

which yields the following much simpler solutions for $[A_2]$ and $[A_3]$,

$$[A_2]_{ss} = \frac{k_1}{k_2}[A_1]_0 e^{-k_1 t} \qquad (2\text{--}136)$$

$$[A_3]_{ss} = [A_1]_0 (1 - e^{-k_1 t}) \qquad (2\text{--}137)$$

Comparison of equations (2–52) and (2–137) shows that $[A_2]$ and $[A_3]$ approximate closely the steady-state solution only if $k_2 \gg k_1$. In other words, the intermediate must be so reactive that it has little time to accumulate. This is illustrated quite clearly in Fig. 2-1(b). The difference between the two expressions provides information on the extent of deviation from the steady state approximation, and can be used to examine the validity of the method. Volk et al.[6] have developed another method for studying the validity of the steady-state approximation, which considers the effect of the deviation from the steady state for reactions of various orders.

One complex reaction mechanism which has wide application to many important kinetic processes is the first-order consecutive reaction with a reversible first step:

$$A_1 \underset{k_{-1}}{\overset{k_1}{\rightleftarrows}} A_2 \xrightarrow{k_2} A_3 \qquad (2\text{--}138)$$

This reaction sequence is used in the Lindemann mechanism to describe how a reactant molecule is activated and reacts by first-order kinetics to give products (see Chapter 11). The same sequence of reactions is used to describe enzyme-catalyzed reactions, which are discussed in detail in Chapter 5. We choose this reaction to illustrate how the steady state treatment can simplify the mathematical complexity and yield an explicit solution of the rate equations for the reaction represented by equation (2–138). The rate equations are

$$\frac{d[A_1]}{dt} = -k_1[A_1] + k_{-1}[A_2] \qquad (2\text{--}139)$$

$$\frac{d[A_2]}{dt} = k_1[A_1] - k_{-1}[A_2] - k_2[A_2] \tag{2-140}$$

$$\frac{d[A_3]}{dt} = k_2[A_2] \tag{2-141}$$

Unfortunately, it is not possible to obtain an explicit solution for these equations by straightforward integration. Using either the Laplace transform (Sec. 2–4) or matrix methods (Sec. 2–5), we can find an exact but very complex solution. Suppose that the steady-state assumption is valid for A_2 so that

$$\frac{d[A_2]}{dt} = k_1[A_1] - k_{-1}[A_2] - k_2[A_2] = 0 \tag{2-142}$$

Then the steady state concentration $[A_2]_{ss}$ is given by

$$[A_2]_{ss} = \frac{k_1}{(k_{-1} + k_2)}[A_1] \tag{2-143}$$

Substituting this expression into equations (2–139) and (2–141) gives a linear first-order differential equation for $[A_1]$ and $[A_3]$

$$\frac{d[A_1]}{dt} = \left(\frac{-k_1 k_2}{k_{-1} + k_2}\right)[A_1]$$

$$\frac{d[A_3]}{dt} = \left(\frac{k_1 k_2}{k_{-1} + k_2}\right)[A_1] \tag{2-144}$$

The solution to these equations is straightforward using standard methods. Note that the validity of the steady state treatment requires either that A_2 is always small or that $[A_2]$ changes very slowly in time.

Depending on the relative magnitude of the rate constants in the reaction mechanism, either the reversible first step or the consecutive second step in equation (2–138) describing the mechanism can be *rate-determining;* in other words, the rate of one of the elementary reaction steps in the mechanism can limit the overall rate of product transformation. If the steady state assumption is valid and $k_2 \gg k_{-1}$, the intermediate is transformed to product A_3 much more rapidly than it is able to go back to being a reactant. In that case, the overall rate of product formation is given as

$$\frac{d[A_3]}{dt} \simeq k_1[A_1] \tag{2-145}$$

Thus, the first step in the reaction mechanism is rate-determining.

Now suppose that the ratio k_{-1}/k_2 is very large, so that the reaction of the intermediate in the second step is too slow to perturb the equilibrium of $A_1 \rightleftarrows A_2$ in the first step. Then, the overall rate of product formation is

$$\frac{d[A_3]}{dt} = \frac{k_1 k_2}{k_{-1}}[A_1] \tag{2-146}$$

For this case, the rate-determining step is the reaction of the intermediate given by the second reaction step.

In applying the steady state method to more complex reaction mechanisms, care must be taken, since there are conditions in which departure from the steady-state can lead to serious deviation from the exact solution.[7]

2.2.2 Pseudo-First-Order Method

For complex reactions with consecutive steps involving more than one reactant, finding an exact solution can become prohibitively difficult. Consider, for example, the following reaction sequence:

$$A_1 + A_2 \xrightarrow{k_1} \text{products}$$
$$A_3 + A_2 \xrightarrow{k_2} \text{products} \tag{2-147}$$

The second-order rate equations are given by

$$\frac{d[A_1]}{dt} = -k_1[A_1][A_2] \tag{2-148}$$

and

$$\frac{d[A_3]}{dt} = -k_2[A_2][A_3] \tag{2-149}$$

If the conditions of the experimental design are such that $[A_2] \gg [A_1]$ and $[A_2] \gg [A_3]$, then during the course of reaction the concentration of A_2 remains nearly constant; this implies that

$$\frac{d[A_1]}{dt} = -\kappa_1[A_1] \tag{2-150}$$

and

$$\frac{d[A_3]}{dt} = -\kappa_2[A_3] \tag{2-151}$$

where $\kappa_1 = k_1[A_2]$ and $\kappa_2 = k_2[A_2]$ are constants.

Equations (2-148) and (2-149) reduce to first order in the reactant; thus, the reaction follows pseudo-first-order kinetics. It should be pointed out that the pseudo-first-order method is not a mathematical approximation method so much as it is an experimental condition that must be adjusted so that second-order rate constants can be examined. The experimental technique is often referred to as *flooding*. In principle, even third-order consecutive reactions can be reduced to pseudo-second or -first order using this technique by merely having all reactants in considerable excess except one.

2.3 EXAMPLE OF A COMPLEX REACTION MECHANISM: THE HYDROGEN + HALOGEN REACTION

The reactions between molecular hydrogen and the halogens, first studied by Bodenstein over a century ago, continue to engage the attention of chemical kineticists. Since these reactions illustrate several of the principles discussed in this chapter, we shall examine them in some detail. Moreover, since the $H_2 + F_2$ reaction, in particular, has been thoroughly studied from the point of view of microscopic, or state-to-state kinetics, we shall return to this system frequently throughout the text.

The overall reaction can be written as

$$H_2 + X_2 \rightleftharpoons 2 HX, \text{ where } X = F, Cl, Br, \text{ or } I$$

With fluorine, the reaction can occur explosively and go to completion in a matter of seconds; with iodine, reaction times are typically an hour or so, eventually reaching an equilibrium mixture of H_2, I_2, and HI. This variety of behavior reflects both the range of stabilities of the diatomic halogens, and the varied kinetics which can occur in these systems.

It is now generally believed that the reaction does not occur between the diatomic molecules themselves, although in the case of $H_2 + I_2$, Bodenstein's original work seemed to imply that the reaction did indeed proceed in this way. Rather than this simple bimolecular reaction, the mechanism appears to require initial dissociation of the halogen, either thermally or photochemically. The halogen atom then reacts with the hydrogen molecule, initiating a *chain reaction* which proceeds until the reactants are consumed or the reactive species are deactivated, e.g., by atomic recombination.

2.3.1 The Hydrogen-Bromine Reaction

The rate of appearance of the HBr product during the initial stages of the reaction, when $[\text{HBr}] \ll [\text{H}_2]$ and $[\text{Br}_2]$, depends on concentration as follows:

$$\text{Rate} = \frac{d[\text{HBr}]}{dt} = (\text{const.})[\text{H}_2][\text{Br}_2]^{1/2} \tag{2-152}$$

At later times, there is inhibition of the rate by the HBr product:

$$\text{Rate} = \frac{d[\text{HBr}]}{dt} = (\text{const.})\frac{[\text{H}_2][\text{Br}_2]^{1/2}}{1 + k'[\text{HBr}]} \tag{2-153}$$

The appearance of a fractional order in equation (2–152) suggests the existence of a *free radical chain reaction* mechanism. The inhibition expressed in Eq. (2–153) further suggests that the product can react with one of the free radical species which carry the chain. Initiation of the reaction requires dissociation of one of the molecular species involved; under thermal (i.e., nonphotochemical) conditions, the species with the lowest bond dissociation energy will be the principal source of atoms. In the hydrogen-bromine system, the bond dissociation energies are as follows:

$$H_2: D_0^0 = 4.476 \text{ eV} = 430 \text{ kJ/mole}$$

$$HBr: D_0^0 = 3.75 \text{ eV} = 360 \text{ kJ/mole}$$

$$Br_2: D_0^0 = 1.97 \text{ eV} = 190 \text{ kJ/mole}$$

For reference, RT at 500 K is about 4 kJ/mole. Clearly, Br_2 will undergo the greatest extent of dissociation, although this will be only a small fraction at moderate temperatures. The reaction mechanism consists of this initial dissociation, plus the following sequence of reactions:

Initiation:

$$Br_2 \xrightarrow{k_1} 2Br \qquad\qquad (2\text{-}154)$$

Propagation:

$$Br + H_2 \xrightarrow{k_2} HBr + H$$
$$H + Br_2 \xrightarrow{k_3} HBr + Br \qquad\qquad (2\text{-}155)$$

Inhibition:

$$H + HBr \xrightarrow{k_4} H_2 + Br \qquad\qquad (2\text{-}156)$$

Termination:

$$2Br + M \xrightarrow{k_5} Br_2 + M \qquad\qquad (2\text{-}157)$$

The reader may wonder why the inhibition step involves a reaction of product with H and not with Br atoms. The reason is that the reaction Br + HBr → Br_2 + H is approximately 170 kJ/mole *endothermic,* as may readily be found by taking differences of the preceding bond dissociation energies. This reaction therefore proceeds with a negligible rate, particularly as compared with the H + HBr reaction, which is 67 kJ/mole *exothermic.* The termination reaction is written as a three-body combination, so that its rate coefficient k_5 is pressure-dependent. The termination step may also include contributions from wall recombination processes.

An analytic rate law may be derived from this mechanism by applying the steady-state approximation (see section 2.2.1) to the reactive intermediates, which in this case are the atomic free radicals H and Br. The steady-state conditions are

$$\frac{d[Br]}{dt} = +2k_1[Br_2] - k_2[Br][H_2] + k_3[H][Br_2] + k_4[H][HBr]$$

$$-2k_5[Br_2]^2 = 0 \qquad\qquad (2\text{-}158)$$

and

$$\frac{d[H]}{dt} = +k_2[Br][H_2] - k_3[H][Br_2] - k_4[H][HBr] = 0 \qquad (2\text{-}159)$$

To solve this system of equations, note that equation (2–159) is just the negative of the middle three terms of equation (2–158) and is stipulated as equal to zero. Thus,

$$2k_1[Br_2] = 2k_5[Br]^2$$

and

$$[Br]_{ss} = \left(\frac{k_1}{k_5}[Br_2]\right)^{1/2} \qquad\qquad (2\text{-}160)$$

Although the relation between [Br] and $[Br_2]$ superficially resembles an equilibrium coefficient, these species are not necessarily in thermodynamic equilibrium, because k_1 and k_5 (which is pressure-dependent) are not related by detailed balancing. Similarly, the steady-state hydrogen atom concentration is found to be

$$[H]_{ss} = \frac{k_2(k_1/k_5)^{1/2}[H_2][Br_2]^{1/2}}{k_3[Br_2] + k_4[HBr]} \qquad (2\text{--}161)$$

The rate of HBr formation is given by

$$\frac{d[HBr]}{dt} = +k_2[Br][H_2] + k_3[H][Br_2] - k_4[H][HBr] \qquad (2\text{--}162)$$

At this point, we could just substitute the previous expressions for steady-state H and Br concentrations in equation (2–162) and grind away; but it is easier just to note that, from equation (2–159), $k_2[Br][H_2] = k_3[H][Br_2] + k_4[H][HBr]$, so a simple substitution yields

$$\frac{d[HBr]}{dt} = 2k_3[H][Br_2]$$

$$= \frac{2k_2(k_1/k_5)^{1/2}[H_2][Br_2]^{1/2}}{1 + (k_4/k_3)[HBr][Br_2]^{-1}} \qquad (2\text{--}163)$$

Note that equation (2–163) gives the correct concentration dependences, in accordance with equations (2–152) and (2–153), and provides explicit expressions for the empirical coefficients in terms of elementary rate processes.

The reaction between hydrogen and chlorine proceeds in a manner similar to that in the hydrogen-bromine system, except that the larger bond dissociation energy of chlorine (as compared to bromine) makes the initiation step more difficult, and the overall mechanism more susceptible to inhibition, especially by impurities such as oxygen. Dissociation of the chlorine molecule by photolysis, or at the surface of the reaction vessel, frequently dominates the kinetics. The higher exothermicity of the propagation steps makes the overall $H_2 + Cl_2$ reaction quite vigorous.

2.3.2 The Hydrogen-Iodine Reaction

A direct bimolecular reaction between H_2 and I_2,

$$H_2 + I_2 \; \underset{k_r}{\overset{k_f}{\rightleftharpoons}} \; 2\,HI \qquad (2\text{--}164)$$

would have the rate law

$$\frac{d[HI]}{dt} = 2k_f[H_2][I_2] - 2k_r[HI]^2 \qquad (2\text{--}165)$$

If we define the progress variables

$$[H_2]_t = H_0 - y(t)$$

and

$$[I_2]_t = X_0 - y(t)$$

then $[HI]_t = 2y(t)$, and the differential equation can be written as

$$2\frac{dy}{dt} = 2k_f[(H_0 - y)(X_0 - y)] - 2k_r(2y)^2$$

or

$$\frac{dy}{dt} = k_f[H_0X_0 - (H_0 + X_0)y + y^2] - 4k_ry^2 \qquad (2\text{-}166)$$

$$= k_f[H_0X_0 - (H_0 + X_0)y + (1 - 4/K_{eq})]y^2$$

where $K_{eq} = k_f/k_r = [HI]^2_{eq}/[H_2]_{eq}[I_2]_{eq}$, by detailed balancing.

Integrating equation (2-166) as in Sec. 1-2-2 gives

$$\ln\left[\frac{y + (H_0 + X_0 - q^{1/2})/2(1 - 4/K_{eq})}{y + (H_0 + X_0 + q^{1/2})/2(1 - 4/K_{eq})}\right] - \ln\left[\frac{H_0 + X_0 - q^{1/2}}{H_0 + X_0 + q^{1/2}}\right]$$

$$= k_fq^{1/2}(t - t_0) \qquad (2\text{-}167)$$

where $q = (H_0 + X_0)^2 - 4H_0X_0(1 - 4/K_{eq})$. Although mathematically a bit complicated, equation (2-167) predicts a straightforward dependence of the concentration of HI on both time and the initial concentrations of H_2 and I_2.

Bodenstein's measurements on this system[8-10] indicated that the concentrations of the various species did indeed appear to follow equation (2-167) accurately over a wide range of temperatures. A typical experimental result is shown in Fig. 2-2. The activation energy for the bimolecular reaction deduced from the data was approximately 170 kJ/mole. Thus, for the better part of a century, the $H_2 + I_2$ reaction was taken to be the paradigmatic example of a bimolecular reaction with a "*four-center*" transition state,

$$\overset{\displaystyle{}_{\scriptstyle\backslash\backslash}H\text{—}H_{\prime\prime}}{I\text{———}I}$$

Further investigation of this reaction indicated that such a mechanism could not be supported. Early molecular beam experiments using supersonically accelerated I_2 [see chapter 9] with H_2 target gas failed to show any evidence for the production of HI. Such negative evidence was not conclusive, of course, because a reaction with a small probability would not be observed in such experiments. But in 1967, Sullivan[11] studied the photochemical reaction between I_2 and H_2 at low temperatures (0–200°C), at which the thermal reaction studied by Bodenstein would be too slow to be measurable. In this experiment, white light is used to efficiently dissociate excited iodine molecules I_2^* via their A-X and B-X electronic transitions. This sets up a *photostationary* (steady-state, not thermal equilibrium) concentration of iodine atoms, according to the scheme:

$$I_2 + h\nu \longrightarrow I_2^*$$
$$I_2^* \longrightarrow 2I \qquad (2\text{-}168)$$
$$2I + M \longrightarrow I_2 + M$$

Figure 2–2 Plot of the composition of hydrogen-iodine mixtures with time. The left ordinate is the fraction of material present as HI; the right ordinate is the dissociated fraction, i.e., $H_2 + I_2$. The upper plot follows the dissociation of pure HI; the lower, the reaction of H_2 and I_2. The solid curves are the theoretical expression, equation (2–167). Both reactions reach a common equilibrium. The measurements shown were taken at a temperature of 448°C, maintained by a boiling sulfur bath (from Ref. 8).

Complex Reactions   Chap. 2

This is followed by the reaction of atomic iodine with hydrogen,

$$H_2 + 2I \longrightarrow 2HI \qquad (2-169)$$

The rate of production of HI is given by

$$\frac{d[HI]}{dt} = 2k_{atomic}[I]^2[H_2] \qquad (2-170)$$

Sullivan measured the rate of HI formation, as well as the steady-state concentration of I atoms. His essential finding was that k_{atomic} had an activation energy of approximately 21 kJ/mole, which, when added to the bond dissociation energy of I_2 (148.5 kJ/mole), gave just the overall activation energy observed by Bodenstein in his thermal experiments. Furthermore, when this process was combined with the thermal equilibrium concentration of iodine atoms at the higher temperatures, the observed production rate of HI was reproduced. In other words, if we postulate that the total formation rate of HI includes both atomic and bimolecular contributions, and we then find that the total rate is equal to the atomic contribution, then the bimolecular contribution must be zero. The rate law of equation (2-165) arises simply from combining equation (2-170) with the equilibrium expression for [I], i.e.,

$$[I]^2 = K_{dissoc}[I_2] \qquad (2-171)$$

to give

$$\frac{d[HI]}{dt} = 2k_{atomic}K_{dissoc}[H_2][I_2] \qquad (2-172)$$

so that the "observed" bimolecular rate coefficient is simply the product of the rate coefficient for the atomic process and the thermal dissociation equilibrium coefficient.

Subsequent work on this reaction indicated additional complications. The atomic reaction (2-169) most likely consists of two steps, formation of a weakly bound IH_2 complex (which may be assisted by a third body), followed by

$$I + IH_2 \longrightarrow 2HI \qquad (2-173)$$

A straightforward steady-state treatment of the IH_2 complex produces the rate law 2-170. Hammes and Widom[12] further suggested that vibrationally excited I_2 molecules lying just below the molecular dissociation limit may play an important role in the reaction as well.

All of these proposed mechanisms lead to the same rate law, equation (2-165) in the differential form and equation (2-167) in the integrated form which may be compared with experiment. This is a classic example of "Benson's Rule of Chemical Kinetics," which states that it is impossible to prove the correctness of a particular mechanism for a chemical reaction on the basis of experimental measurements; all that can be done is to disprove one or more of the postulated mechanisms.

We may summarize our state of knowledge about hydrogen-halogen kinetics as follows. The H_2-F_2, H_2-Br_2, and (with some qualification) the H_2-Cl_2 systems show rather complicated experimental rate laws, but these systems can be interpreted in terms of a rather straightforward free radical chain reaction mechanism. The H_2-I_2 system, on the other hand, has been studied longer than any of these other systems and possesses the simplest kinetic behavior; this system, however, is the one which is the least understood and is still the subject of investigation. The hydrogen + halogen systems thus illustrate another dictum, due to Holmes rather than Benson, to the effect that " . . . strange details, far from making the case more difficult, have really had the effect of making it less so"![13]

2.4 LAPLACE TRANSFORM METHOD

2.4.1 Introduction

The classical methods described in the preceding sections lead to the solutions of many kinetic linear differential equations, but as is plain, obtaining these solutions is not always straightforward. By contrast, using methods of *operational calculus* can lead directly to a solution satisfying given initial conditions without having to separately evaluate particular integrals or arbitrary constants. Through these methods, the calculation procedure can be both simplified and shortened considerably. The rules and operational procedures are introduced by means of a definite integral called a *transformation integral,* the general form of which is given by

$$F(p) = \int_a^b f(t)\mathcal{K}(p,t)\, dt \qquad (2\text{--}174)$$

where $F(p)$ is the *transform* of $F(t)$ with respect to the *kernel,* $\mathcal{K}(p,t)$. For chemical kinetic systems the most commonly used transform is the Laplace transform, which is represented by the kernel $\mathcal{K}(p,t) = e^{-pt}$ and integrated over the limits 0 to ∞. Some properties of this integral and its use in connection with solving chemical kinetic problems will be discussed in this section and also in Appendix 1 to this chapter.

2.4.2 The Laplace Transform

The transform $F(p)$ of a function $f(t)$ subjected to the Laplace transformation is defined by the integral:

$$F(p) = \mathcal{L}[f(t)] = \int_0^\infty e^{-pt}f(t)\, dt \qquad (2\text{--}175)$$

in which both factors, $f(t)$ and e^{-pt}, depend on t. After integration over the limits of $t = 0$ to $t = \infty$, the integral is found to be a function of p only.

The Laplace transform of a given function may be determined by direct integration of (2–175). For example, the transform $F(p)$ of $f(t) = e^{-at}$ where a is a constant, is computed as follows:

Complex Reactions   Chap. 2

$$F(p) = \mathcal{L}[e^{-at}] = \int_0^\infty e^{-at} e^{-pt}\, dt = \int_0^\infty e^{-(a+p)t}\, dt$$

$$= -\frac{1}{a+p}[e^{-(a+p)t}]_0^\infty \qquad\qquad (2\text{--}176)$$

$$= \frac{1}{p+a} \qquad (p > -a)$$

The condition $p > -a$ is required to make $e^{-(a+p)t}|_{t=\infty}$ equal to zero and to guarantee the convergence of the integral for $\mathcal{L}[e^{-at}]$.

2.4.3 Fundamental Properties of the Laplace Transform

The Laplace transform of a function has certain fundamental properties which are particularly useful in the evaluation of transforms and in the solution of chemical kinetic differential equations. Three of the most widely used properties will be mentioned here.

1. If $f(t)$ is a linear combination of known functions, i.e.,

$$f(t) = f_1(t) + f_2(t) + \cdots f_n(t) \qquad\qquad (2\text{--}177)$$

then its Laplace transform is

$$\mathcal{L}[f(t)] = \int_0^\infty [f_1(t) + f_2(t) + \cdots f_n(t)]e^{-pt}\, dt$$

$$= \int_0^\infty f_1(t)e^{-pt}\, dt + \int_0^\infty f_2(t)e^{-pt}\, dt + \cdots \int_0^\infty f_n(t)e^{-pt}\, dt$$

or

$$\mathcal{L}[f(t)] = F_1(p) + F_2(p) + \cdots F_n(p) \qquad\qquad (2\text{--}178)$$

This means that the Laplace transform of a linear combination of functions is the same linear combination of the transformed functions.

2. One of the most important features of the Laplace transform, especially in evaluating chemical kinetic equations, is that the Laplace transform of the derivative of $f(t)$, i.e., $f'(t)$, can be readily obtained. This is done by integrating equation (2–175):

$$\mathcal{L}[f'(t)] = \int_0^\infty f'(t)e^{-pt}\, dt$$

$$\mathcal{L}[f'(t)] = f(t)e^{-pt}\Big|_0^\infty + p\int_0^\infty f(t)e^{-pt}\, dt \qquad\qquad (2\text{--}179)$$

$$\mathcal{L}[f'(t)] = p\,\mathcal{L}[f(t)] - f(0)$$

For higher order derivatives, the Laplace transform can be found by applying equation (2–179) to the second derivative of $f(t)$, as follows:

$$\mathscr{L}[f''(t)] = p\mathscr{L}[f'(t)] - f'(0)$$

$$\mathscr{L}[f''(t)] = p(p\mathscr{L}[f(t)] - f(0)) - f'(0) \qquad (2-180)$$

$$\mathscr{L}[f''(t)] = p^2\mathscr{L}[f(t)] - pf(0) - f'(0)$$

Higher order derivatives, $f^{(n)}(t)$, can be obtained by repeating this operation n times, i.e.,

$$\mathscr{L}[f^{(n)}(t)] = p^n\mathscr{L}[f(t)] - \sum_{i=1}^{n} f^{(i-1)}(0)p^{n-i} \qquad (2-181)$$

3. The transform of an integral of a function $f(t)$ may be expressed as

$$\mathscr{L}\left[\int_0^t f(t)\, dt\right] = \frac{\mathscr{L}[f(t)]}{p} \qquad (2-182)$$

Additional properties of the Laplace transform are given in Appendix 1.A. A table of transforms of the functions most commonly encountered in kinetic problems is obtainable by direct integration and by using the preceding three properties. Extensive tables of Laplace transforms appear in Appendix 1.B. These tables may be used to find both the Laplace transform and the inverse Laplace transform symbolized by

$$\mathscr{L}^{-1}[F(p)] = f(t) \qquad (2-183)$$

2.4.4 Solving Chemical Kinetic Problems Using the Laplace Transform Method

As we can see from the general Laplace transform properties, the method allows differential and integral equations to be converted into linear algebraic equations that include initial conditions. The procedure for solving a set of kinetic equations is illustrated for the simplest example. Consider the opposing first-order reaction discussed in Sec. 2.1.2,

$$A_1 \underset{k_2}{\overset{k_1}{\rightleftharpoons}} A_2$$

The rate equations for this reaction are equations (2–3) and (2–4),

$$-\frac{d[A_1]}{dt} = k_1[A_1] - k_2[A_2]$$

and

$$-\frac{d[A_2]}{dt} = k_2[A_2] - k_1[A_1]$$

with initial conditions $[A_1] = A_0$ and $[A_2] = 0$ at $t = 0$. Laplace transformations can now be carried out on this set of differential equations to convert the equations into a set of algebraic equations.

$$(p + k_1)(\mathscr{L}[A_1]) - k_2\mathscr{L}[A_2] = [A_1]_0 \tag{2-184}$$

$$-k_1(\mathscr{L}[A_1]) + (p + k_2)\mathscr{L}[A_2] = [A_2]_0 \tag{2-185}$$

Solving for $\mathscr{L}[A_1]$ and $\mathscr{L}[A_2]$, and using Cramer's Rule for linear equations, we obtain

$$\mathscr{L}[A_1] = \frac{\begin{vmatrix} [A_1]_0 & -k_2 \\ [A_2]_0 & (p + k_2) \end{vmatrix}}{\begin{vmatrix} p + k_1 & -k_2 \\ -k_1 & p + k_2 \end{vmatrix}} \tag{2-186}$$

and

$$\mathscr{L}[A_2] = \frac{\begin{vmatrix} p + k_1 & [A_1]_0 \\ -k_1 & [A_2]_0 \end{vmatrix}}{\begin{vmatrix} p + k_1 & -k_2 \\ -k_1 & p + k_2 \end{vmatrix}} \tag{2-187}$$

The solution of this determinantal equation is

$$\mathscr{L}[A_1] = \frac{(p + k_2)[A_1]_0}{p(p + (k_1 + k_2))} + \frac{k_2[A_2]_0}{p(p + (k_1 + k_2))} \tag{2-188}$$

and

$$\mathscr{L}[A_2] = \frac{(p + k_1)[A_2]_0}{p(p + (k_1 + k_2))} + \frac{k_1[A_1]_0}{p(p + (k_1 + k_2))} \tag{2-189}$$

In order to find the time-dependent concentrations for A_1 and A_2, we carry out inverse Laplace transformations on equations (2–188) and (2–189). Using the Laplace tables we see that the solution to equation (2–188) is

$$[A_1] = \frac{[A_1]_0}{(k_1 + k_2)}(k_2 + k_1 e^{-(k_1+k_2)t}) + \frac{[A_2]_0 k_2}{(k_1 + k_2)}(1 - e^{-(k_1+k_2)t}) \tag{2-190}$$

and similarly for equation (2–189). Then, since $[A_2]_0 = 0$, we find the previous solution, equation (2–13):

$$[A_1] = \frac{[A_1]_0}{(k_1 + k_2)}(k_2 + k_1 e^{-(k_1+k_2)t})$$

Similarly for $[A_2]$, we recover equation (2–14):

$$[A_2] = \frac{k_1[A_1]_0}{(k_1 + k_2)}(1 - e^{-(k_1+k_2)t})$$

As can be seen from this simple example, as long as the rate equations are linear with respect to the reactants, the Laplace method will yield solutions. Nonlinear rate equations can be made linear by making the concentrations of certain reactants large relative to others so that pseudo-first-order equations, which are linear, can be obtained.

A somewhat more complex example of this method is the infrared multiphoton dissociation (IRMPD, see chapter 14) of vinylcyanide ($CH_2\!=\!CHCN$) which yields electronically excited and ground state C_2 according to the reaction

$$CH_2\!=\!CHCN \xrightarrow{nh\nu_{IR}} C_2(\tilde{X}^1\Sigma_g^+)$$
$$\longrightarrow C_2^*(\tilde{a}^3\Pi_u) \tag{2-191}$$

In the presence of a reactive scavenger such as O_2 the subsequent reactions are

$$C_2(\tilde{X}^1\Sigma_g^+) + O_2 \xrightarrow{k_1} \text{products} \tag{2-192}$$

$$C_2(\tilde{X}^1\Sigma_g^+) + O_2 \underset{k_e}{\overset{k_{e'}}{\rightleftharpoons}} C_2^*(\tilde{a}^3\Pi) + O_2 \tag{2-193}$$

and

$$C_2^*(\tilde{a}^3\Pi) + O_2 \xrightarrow{k_3} \text{products} \tag{2-194}$$

The reaction illustrated in equation (2–193) proceeds through collision induced intersystem crossing (see chapter 8). We can write equations (2–192) through (2–194) schematically as:

$$A + B \xrightarrow{k_1} D$$

$$A + B \underset{k_e}{\overset{k'_e}{\rightleftharpoons}} C + B \tag{2-195}$$

$$C + B \xrightarrow{k_3} D$$

This mechanism consists of a reversible and two concurrent reactions; the reversible step couples to the concurrent reactions. The rate equations for $[A] = [C_2(\tilde{X}^1\Sigma_g^+)]$ and $[C] = [C_2^*(\tilde{a}^3\Pi)]$ can be written as

$$\frac{d[A]}{dt} = -(k_1[B] + k'_e[B])[A] + k_e[B][C] \tag{2-196}$$

$$\frac{d[C]}{dt} = -(k_3[B] + k_e[B])[C] + k'_e[B][A] \tag{2-197}$$

Using a large excess of O_2, so that $[B]$ is approximately constant, we can rewrite these equations as

$$\frac{d[A]}{dt} = -K'[A] + K_e[C] \tag{2-198}$$

and

$$\frac{d[C]}{dt} = K'_e[A] - K[C] \tag{2-199}$$

where $K' = k_1[B] + k'_e[B]$, $K_e = k_e[B]$, $K'_e = k'_e[B]$ and $K = k_3[B] + k_e[B]$. The initial values of $[A]_0 = A_0$ and $[C]_0 = C_0$. The actual concentrations of $C_2(\tilde{X}^1\Sigma_g^+)$

Complex Reactions   Chap. 2

and $C_2^*(\tilde{a}^3\Pi)$ may be determined by laser-induced fluorescence (see chapter 3). The $C_2^*(\tilde{a}^3\Pi)$ radicals are monitored by exciting the $d^3\Pi \leftarrow \tilde{a}^3\Pi(0,0)$ transition and detecting LIF in the $(0,1)$ vibronic band, and $C_2(\tilde{X}^1\Sigma_g^+)$ radicals are monitored by exciting the $\tilde{a}^1\Pi - \tilde{X}^1\Sigma_g^+(0,0)$ transition and detecting the $(4,1)$ vibronic band fluorescence.[14]

Taking the Laplace transforms of equations (2–198) and (2–199), we find

$$(p + K')\mathscr{L}[A] - K_e\mathscr{L}[C] = A_0 \qquad (2-200)$$

$$-K_e'\mathscr{L}[A] + (p + K)\mathscr{L}[C] = C_0 \qquad (2-201)$$

Solving equations (2–200) and (2–201) by using Cramer's Rule, we obtain

$$\mathscr{L}[A] = \frac{A_0(p + K)}{(p + K')(p + K) - K_e K_e'} + \frac{K_e C_0}{(p + K')(p + K) - K_e K_e'} \qquad (2-202)$$

$$\mathscr{L}[C] = \frac{C_0(P + K')}{(p + K')(p + K) - K_e K_e'} + \frac{A_0 K_e'}{(p + K')(p + K) - K_e K_e'} \qquad (2-203)$$

The roots of $(p + K')(p + K) - K_e K_e' = 0$, found by using the quadratic formula, are:

$$s_1 = 1/2\{-(K + K') + [(K + K')^2 - 4(KK' - K_e K_e')]^{1/2}\} \qquad (2-204)$$

and

$$s_2 = 1/2\{-(K + K') - [(K + K')^2 - 4(KK' - K_e K_e')]^{1/2}\} \qquad (2-205)$$

Hence, we may split $\mathscr{L}[A]$ and $\mathscr{L}[C]$ into the partial fractions:

$$\mathscr{L}[A] = \frac{A_0(p + K)}{(p - s_1)(p - s_2)} + \frac{C_0 K_e}{(p - s_1)(p - s_2)} \qquad (2-206)$$

and

$$\mathscr{L}[C] = \frac{C_0(p + K')}{(p - s_1)(p - s_2)} + \frac{A_0 K_e'}{(p - s_1)(p - s_2)} \qquad (2-207)$$

Taking the inverse Laplace Transform of equations (2–202) and (2–203) using nos. 14 and 57 of Appendix 1.B, we obtain

$$[A(t)] = [C_2(\tilde{X}^1\Sigma_g^+)]_t = \frac{1}{s_1 - s_2}\{(A_0 s_1 + A_0 K + C_0 K_e)e^{s_1 t}$$
$$-(A_0 s_2 + A_0 K + C_0 K_e)e^{s_2 t}\} \qquad (2-208)$$

Similarly,

$$[C(t)] = [C_2^*(\tilde{a}^3\Pi)]_t = \frac{1}{s_1 - s_2}\{(C_0 s_1 + C_0 K' + A_0 K_e')e^{s_1 t}$$
$$- (C_0 s_2 + C_0 K' + A_0 K_e')e^{s_2 t}\} \qquad (2-209)$$

2.5 DETERMINANT (MATRIX) METHODS

Another method of solving kinetic equations uses matrix algebra. We apply this to the chemical reaction system $A_1 \rightleftarrows A_2$, for which the differential equations are as before

$$-\frac{d[A_1]}{dt} = k_1[A_1] - k_2[A_2]$$

and

$$-\frac{d[A_2]}{dt} = -k_1[A_1] + k_2[A_2]$$

These equations can be expressed in matrix form as

$$\begin{pmatrix} \dfrac{d[A_1]}{dt} \\ \dfrac{d[A_2]}{dt} \end{pmatrix} = \begin{pmatrix} -k_1 & k_2 \\ k_1 & -k_2 \end{pmatrix} \begin{pmatrix} [A_1] \\ [A_2] \end{pmatrix} \tag{2-210}$$

Thus, the matrix of rate coefficients is given by

$$\mathsf{K} = \begin{pmatrix} -k_1 & k_2 \\ k_1 & -k_2 \end{pmatrix} \tag{2-211}$$

The matrix form of the set of differential equations may be expressed as

$$\frac{d\mathbf{A}}{dt} = \mathsf{K}\mathbf{A} \tag{2-212}$$

where K is the rate coefficient matrix and $\mathbf{A}$ is the composition vector; because composition must be real and positive, $\mathbf{A}$ is a vector with real and positive components. K is an $n \times n$ matrix where n is the number of components, A_n. For this problem, K is diagonalizable and it has n linearly independent eigenvectors.[15]
If we define an orthogonal matrix P which is invertible such that

$$\mathbf{A} = \mathsf{P}\mathbf{B} \tag{2-213}$$

and $\mathsf{P}^{-1}\mathsf{K}\mathsf{P} = \Lambda$, where Λ is the matrix of negative eigenvalues with $-\lambda_1, -\lambda_2, \cdots$ $-\lambda_n$ as its successive diagonal elements and no off-diagonal elements, then equation (2-212) can be rewritten as

$$\frac{d\mathbf{B}}{dt} = \mathsf{P}^{-1}\mathsf{K}\mathsf{P}\mathbf{B} = \Lambda\mathbf{B} \tag{2-214}$$

Since equation (2-214) is uncoupled, its solution is

$$\mathbf{B} = e^{\Lambda t}\mathbf{B}_i \tag{2-215}$$

where $\mathbf{B}_i$ is the vector of initial values of $\mathbf{B}$ and is related to the initial conditions for $\mathbf{A}_i$ at $t = 0$ by

$$\mathbf{A}_i = \mathsf{P}\mathbf{B}_i \tag{2-216}$$

If we multiply equation (2–213) by the inverse matrix of P, from the left side we get

$$P^{-1}A = B \qquad (2\text{–}217)$$

Substituting **B** into equation (2–215) yields

$$P^{-1}A = e^{\Lambda t}B_i = e^{\Lambda t}P^{-1}A_i \qquad (2\text{–}218)$$

Multiplying through by P from the left we obtain the solution vector

$$A = Pe^{\Lambda t}P^{-1}A_i \qquad (2\text{–}219)$$

In order to evaluate the solution vector **A**, the matrix P that diagonalizes Kt must be found. To find the eigenvalues Λ, we must solve the characteristic equation of K satisfying the condition,

$$\det(K - \Lambda I) = 0 \qquad (2\text{–}220)$$

The characteristic equation for the reaction $A_1 \rightleftarrows A_2$ can be written as

$$(-k_1 - \lambda)(-k_2 - \lambda) - k_1 k_2 = 0$$

and the corresponding eigenvalues, which are roots of this equation, are

$$\lambda_1 = 0 \text{ and } \lambda_2 = -(k_1 + k_2)$$

Thus the eigenvectors are

$$P_1 = \begin{pmatrix} k_2 \\ k_1 \end{pmatrix} \text{ and } P_2 = \begin{pmatrix} 1 \\ -1 \end{pmatrix}$$

and they form a basis for the eigenspace corresponding to $\lambda_1 = 0$ and $\lambda_2 = -(k_1 + k_2)$. It is easy to check that P_1 and P_2 are linearly independent so that

$$P = \begin{pmatrix} k_2 & 1 \\ k_1 & -1 \end{pmatrix} \qquad (2\text{–}222)$$

diagonalizes K. The reader should verify that $P^{-1}KP = \Lambda$.

To find our solutions we use equation (2–219) and multiply the matrices. A_i is the initial value vector for the components given by

$$A = \begin{pmatrix} [A_1] \\ [A_2] \end{pmatrix} = \begin{pmatrix} A_0 \\ 0 \end{pmatrix} \text{ at } t = 0 \qquad (2\text{–}223)$$

The inverse matrix, P^{-1}, can be determined using standard matrix algebra.[15]

$$P^{-1} = \begin{pmatrix} \dfrac{1}{(k_1 + k_2)} & \dfrac{1}{(k_1 + k_2)} \\ \dfrac{k_1}{(k_1 + k_2)} & \dfrac{-k_2}{(k_1 + k_2)} \end{pmatrix} \qquad (2\text{–}224)$$

Therefore the solutions are given by

$$A = \begin{pmatrix} [A_1] \\ [A_2] \end{pmatrix} = \frac{A_0}{(k_1 + k_2)} \begin{pmatrix} k_2 + k_1 e^{-(k_1 + k_2)t} \\ k_1 - k_1 e^{-(k_1 + k_2)t} \end{pmatrix} \qquad (2\text{–}225)$$

and hence we recover equations (2–13) and (2–14),

$$[A_1] = \frac{A_0}{(k_1 + k_2)}(k_2 + k_1 e^{-(k_1+k_2)t})$$

$$[A_2] = \frac{k_1 A_0}{(k_1 + k_2)}(1 - e^{-(k_1+k_2)t})$$

The matrix method can be used to solve a set of first-order equations or pseudo-first-order equations. As is evident from the generalized treatment this procedure can also be used computationally to find the solution for n components (See Problem 2.5). One need only compute the eigenvalues and eigenvectors of **K** and assume a solution of the form given by equation (2–219). Computing eigenvalues for the rate coefficient matrix may be done without too much difficulty; calculating the eigenvectors can be done using matrix reduction algorithms such as the QR method.[16-18]

2.6 NUMERICAL METHODS

2.6.1 Introduction

It is usually best to solve kinetic equations exactly, provided that an exact solution does exist and also has a reasonable analytical form. As we have seen, for rather simple kinetic systems one can usually find an analytic solution, which makes it easy to obtain information about the concentration profiles and their time dependences. However, the search for an analytic solution has at least two limitations. The principal one is that such a solution may not be derivable using matrix or Laplace methods, or any other method for that matter. Secondly, even if an analytic solution does exist, it may be so complex that it may not yield the desired information easily nor in a useful form. An example of such a case is the reaction

$$A \xrightarrow{k_1} B + C \xrightarrow{k_2} D$$
$$\uparrow\!\!\!\underline{\hspace{3cm}} k_3 \underline{\hspace{3cm}}\!\!\!\rfloor$$

The solution of this complex reaction system consists of Bessel and Henkel functions.

When analytic solutions are impossible to obtain directly, approximation methods may be employed. One such method, which has historically played a useful role in the modeling of chemical systems, is the steady-state approximation. Although it provides analytic solutions, the question of their validity is open. Questions of validity of the steady state assumption have been raised in areas of kinetics that involve large sets of coupled differential equations, because it is difficult to verify experimentally the existence of steady state transient species.[19-21]

In treating the kinetics of many-component systems, such as those typically encountered in atmospheric and combustion chemistry (see Chapter 15), a numerical solution is often required. In this section, we present some useful computational tools for solving differential equation models that arise in chemical kinetics.

2.6.2 Analyzing Complex Reaction Systems Numerically

2.6.2.1 Categorizing kinetic differential equations. Before we can consider numerical methods that are suited to kinetic problems, we must be able to categorize the type of differential equation we have at hand. There are two distinct categories that are most commonly encountered. Equations involving derivatives of only one independent variable are called *ordinary differential equations*. If the equation contains more than one independent variable, then the equation becomes a *partial differential equation*. For most chemical kinetics problems, the equations are of the ordinary type, such as $y' = ky$. These may be further classified as either initial-value problems, in which the dependent variable and/or its derivative are required to have some specified value at the zero of the independent variable, i.e., $y(0)$ and $y'(0)$ are constants; or boundary-value problems, in which $y(x_1)$ and $y'(x_2)$ are specified constants, but x_1 and x_2 may be any values in the range of the independent variable.

Most coupled differential equations relevant to chemical kinetic systems are of the initial-value type and the problem is to solve a system of m equations involving n chemical species, each of which may be coupled to every other species by chemical reaction. In general, solving initial-value problems requires the problem to be in the form

$$\frac{dy}{dx} = f'(x,y) \tag{2-226}$$

with initial condition $y(x_0) = y_0$. Assuming that $\partial f/\partial y$ is continuous over the interval $x_0 \le x \le x_1$ guarantees that equation (2–226) possesses a solution; $x_1 = x_0 + h$, where h is a small increment. Note that the numerical solution is only a set of points determined by the slope of the curve calculated from equation (2–226).

In the remainder of this section we describe various methods for obtaining a numerical solution that are applicable to systems of chemical kinetic equations. These methods fall into two categories: one-step methods and predictor-corrector methods. One-step methods are ones that use information about a single point or a previous step to find the next point. Some common methods employing this basic algorithm are Euler's,[22] the Modified Euler technique[22,23] and Runge-Kutta Methods.[24] Predictor-corrector methods are multistep methods which make use of information about the solution and its derivative at more than one point in order to extrapolate to the next point. Methods that make use of this algorithm are Milne's[25] and the Adams-Bashforth[26] and Gear[26,27] methods.

2.6.2.2 One-step methods. Consider the first-order differential equation

$$y'(t) = f(t,y) \tag{2-227}$$

where y' represents dy/dt subject to the initial condition $y(t_0) = y_0$. The purpose of the one-step method is to provide a means of calculating a set of points $[(t_i,v_i)]$ where $v_0 = y_0$ and each point (t_i,v_i) is an approximation to the corresponding point $(t_i,y(t_i))$.

2.6.2.2.1 Euler Methods. Consider the first-order equation (2–227) and divide the interval $[t_0, t_N]$ into N equally spaced subintervals of step size,

$$h = \frac{t_N - t_0}{N} \tag{2-228}$$

(Note that the step size does not have to be constant over the interval, but for the moment let us assume it.) Each new increment in the independent variable can be approximated by

$$t_n = t_0 + nh \tag{2-229}$$

where $n = 0, 1, 2, \ldots N$. If $y(t)$ is continuous and has a unique solution, then it can be expanded in a Taylor's series about the point t_0 thus:

$$y(t_0 + h) = y(t_0) + f(t_0, y_0)h + \frac{h^2}{2!} f'(t_0, y_0) + \cdots \tag{2-230}$$

If the value of the step size is small, then higher order terms in h can be neglected. This is known as truncating and gives

$$y(t_0 + h) = y(t_0) + hy'(t_0) \tag{2-231}$$

This technique is called the Euler method, and the process can be continued for many steps:

$$y_{n+1} = y_n + hy'(x_n, y_n), \quad n = 0, 1, \ldots N - 1 \tag{2-232}$$

The Euler method is presented graphically in Fig. 2–3 and is the simplest of all methods for solving the initial-value problem.

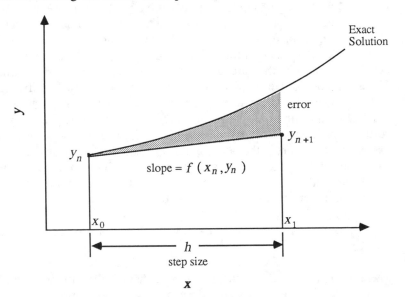

Figure 2–3 Graphical interpretation of the basic Euler method.

By truncating after the second term of the Taylor's expansion, we introduce errors in the value of y_{n+1}, on the order of h^2. The truncation error in each step can be made small provided the step size chosen is sufficiently small. The question that arises is whether the Euler method is able to provide a sufficiently accurate approximation to equation (2–227).

To answer this question, consider the decomposition of the trifluoromethoxy radical, CF_3O. The reaction rate for its decomposition is[28-30]

$$k = 10^{15.2} e^{-129.7 \text{ kJ/mole/RT}} \sec^{-1} \tag{2–233}$$

and the reaction is first order, i.e.,

$$-\frac{d[CF_3O]}{dt} = k[CF_3O] \tag{2–234}$$

At 1000K, $k = 2.66 \times 10^8$ $\sec^{-1}$, and at $t = 0$, $[CF_3O]_0 = 1$. The analytic solution is found by simple integration (Sec. 1-3-2) to be

$$[CF_3O]_t = [CF_3]_0 \exp (-2.66 \times 10^8 t) \tag{2–235}$$

Now, we solve equation (2–234) using the Euler method [equation (2–232)]

$$[CF_3O]_{n+1} = [CF_3O]_n - (k)h[CF_3O]_n \tag{2–236}$$

Table 2–1 shows the results. Notice that for $N = 10$, the numerical results from the Euler calculation oscillate and also give negative concentrations. This is an example where the Euler method is *unstable*. By increasing N to 20, the concentration of $[CF_3O]$ is still underestimated compared to the concentration given by the exact expression equation (2–236). Furthermore, it is clear that the concentration of CF_3O calculated by the Euler method decreases more rapidly, by about 2 orders of magnitude, than the concentration given by the exact expression. Consequently, the error grows with increasing values of the independent variable. We define the global error as the difference between the computed value of the solution and the exact solution. It accounts for the total accumulated error. For the CF_3O example, the global error is about 35% in the first step, $(N = 1)$. It continues to accumulate and by $N = 7$, the global error has reached 95%. Although the stability has been improved by increasing N, the method still remains unreliable. Results can be improved by increasing N further, which reduces the step size. This allows the function, which is rapidly decreasing, to be better estimated by the Euler method; but it increases the computation time since more steps are needed. To obtain more reliable and accurate numerical results for such problems, another method less susceptible to instability problems is required. Some higher order techniques, such as the Runge-Kutta methods, may be used.

2.6.2.2.2 Runge-Kutta Methods. Let us return to the Taylor expansion of the first-order equation (2–230). If the higher-order terms of the expansion are truncated at h^2, then

$$y(t_0 + h) = y(t_0) + hy'(t_0) + \frac{h^2}{2!} y''(t_0) + \cdots \tag{2–237}$$

TABLE 2–1 COMPARISON OF THE ACCURACY OF EULER INTEGRATION OF EQUATION (2–234) FOR 10- AND 20-STEP INTEGRATIONS.

$N = 10$	t, sec	$[CF_3O]_{Euler}$	$[CF_3O]_{exact}$	Deviation
1	5E-9	−.33	.2644772613	−.5944772613
2	1E-8	.1089	.06994822174	.03895177826
3	1.5E-8	−.035937	.01849971412	−.05443671412
4	2E-8	.01185921	4.892753725E-3	6.966456275E-3
5	2.5E-8	−3.9135393E-3	1.294022105E-3	−5.207561405E-3
6	3E-8	1.291467969E-3	3.422394225E-4	9.492285465E-4
7	3.5E-8	−4.261844298E-4	9.051454518E-5	−5.166989749E-4
8	4E-8	1.406408618E-4	2.393903902E-5	1.167018228E-4
9	4.5E-8	−4.64114844E-5	6.331331477E-6	−5.274281588E-5
10	5E-8	1.531578985E-5	1.674493209E-6	1.364129664E-5

$N = 20$				
1	2.5E-9	.335	.5142735277	−.1792735277
2	5E-9	.112225	.2644772613	−.1522522613
3	7.5E-9	.037595375	.1360136542	−.09841827917
4	1E-8	.01259445062	.06994822174	−.05735377112
5	1.25E-8	4.219140959E-3	.03597251875	−.03175337779
6	1.5E-8	1.413412221E-3	.01849971412	−.0170863019
7	1.75E-8	4.734930942E-4	9.513913242E-3	−9.040420148E-3
8	2E-8	1.586201865E-4	4.892753725E-3	−4.734133539E-3
9	2.25E-8	5.313776249E-5	2.516213718E-3	−2.463075956E-3
10	2.5E-8	1.780115044E-5	1.294022105E-3	−1.276220955E-3
11	2.75E-8	5.963385396E-6	6.654813131E-4	−6.595179277E-4
12	3E-8	1.997734108E-6	3.422394225E-4	−3.402416884E-4
13	3.25E-8	6.69240926E-7	1.760046751E-4	−1.753354342E-4
14	3.5E-8	2.241957102E-7	9.051454518E-5	−9.029034947E-5
15	3.75E-8	7.510556292E-8	4.654923446E-5	−4.647412889E-5
16	4E-8	2.516036358E-8	2.393903902E-5	−2.391387865E-5
17	4.25E-8	8.428721799E-9	1.231121404E-5	−1.230278532E-5
18	4.5E-8	2.823621803E-9	6.331331477E-6	−6.328507855E-6
19	4.75E-8	9.459133039E-10	3.256036174E-6	−3.25509026E-6
20	5E-8	3.168809568E-10	1.674493209E-6	−1.674176328E-6

In order to calculate the new point, the second derivative $y''(t_0)$ must be known. This can be estimated as the finite difference

$$y''(t_0) = \frac{\Delta y'}{\Delta t} = \frac{y'(t_0 + h) - y'(t_0)}{h} \tag{2–238}$$

Substituting this expression into the truncated Taylor expression results in

$$y(t_0 + h) = y(t_0) + hy'(t_0) + \frac{h^2}{2!}\left(\frac{y'(t_0 + h) - y'(t_0)}{h}\right) \tag{2–239}$$

and combining terms in like powers of h gives

$$y(t_0 + h) = y(t_0) + \frac{h^2}{2}(y'(t_0 + h) + y'(t_0)) \tag{2–240}$$

This is the second-order Runge-Kutta method because it utilizes the second term of the expansion, h^2. In order to derive other nth order Runge-Kutta methods, the nth order term of the Taylor expansion must be kept, and the nth order derivative calculated. This implies that more points are needed in order to evaluate the derivatives. The Runge-Kutta method thus provides a set of formulas, derived using the same procedure as for derivation of the second order method, for calculating the solutions of the first-order equation. The equations for the third-order method are

$$y_{n+1} = y_n + (\Delta Y_1 + 4\Delta Y_3 + \Delta Y_4)/6 \qquad (2-241)$$

where

$$\Delta Y_1 = hf(t_n, y_n)$$

$$\Delta Y_2 = hf(t_n + h, y_n + \Delta Y_1)$$

$$\Delta Y_3 = hf(t_n + h/2, y_n + \Delta Y_1/2)$$

$$\Delta Y_4 = hf(t_n + h, y_n + \Delta Y_2)$$

The fourth order Runge-Kutta method has found its widest application in chemical kinetics. This is because it is based on retaining all terms up through h^4 and consequently the error associated with any step is on the order of h^5. The equations for the fourth order method are

$$y_{n+1} = y_n + (\Delta Y_1 + 2\Delta Y_2 + 2\Delta Y_3 + \Delta Y_4)/6 \qquad (2-242)$$

where

$$\Delta Y_1 = hf(t_n, y_n)$$

$$\Delta Y_2 = hf(t_n + h/2, y_n + \Delta Y_1/2)$$

$$\Delta Y_3 = hf(t_n + h/2, y_n + \Delta Y_2/2)$$

$$\Delta Y_4 = hf(t_n + h, y_n + \Delta Y_3)$$

To illustrate the relative accuracy of the various one-step methods, let us model the reaction solved exactly in section 2.2.1 [equations (2–3) and (2–4)], viz.:

$$A \underset{k_2}{\overset{k_1}{\rightleftarrows}} B$$

$$\frac{d[A]}{dt} = -k_1[A] + k_2[B]$$

$$\frac{d[B]}{dt} = k_1[A] - k_2[B]$$

where $[A] = [A]_0$ and $[B] = 0$ at $t = 0$. The exact solution is given by equations (2–13) and (2–14), namely,

$$[A] = \frac{[A]_0}{(k_1 + k_2)}[k_2 + k_1 e^{-(k_1+k_2)t}]$$

and from mass conservation

TABLE 2-2 COMPARISON OF THE ACCURACY OF VARIOUS ONE-STEP METHODS FOR INTEGRATING THE KINETIC EQUATIONS FOR THE A $\rightleftarrows$ B REACTION

Comparison of one-step methods for [A]

Time	Euler	Modified Euler	Runge-Kutta-4	Exact
.00000	1000.00000	1000.00000	1000.00000	1000.00000
.00050	590.57145	607.16681	606.62113	606.62085
.00100	348.94227	368.80586	368.14394	368.14360
.00150	206.34191	224.17463	223.57245	223.57214
.00200	122.18458	136.41617	135.92920	135.92895
.00250	72.51813	83.16661	82.79743	82.79725
.00300	43.20687	50.85617	50.58748	50.58734
.00350	25.90849	31.25103	31.06091	31.06082
.00400	15.69963	19.35515	19.22337	19.22330
.00450	9.67476	12.13703	12.04712	12.04707
.00500	6.11910	7.75727	7.69668	7.69664
.00550	4.02069	5.09974	5.05932	5.05930
.00600	2.78229	3.48722	3.46048	3.46046
.00650	2.05143	2.50879	2.49122	2.49121
.00700	1.62010	1.91510	1.90362	1.90362
.00750	1.36555	1.55487	1.54741	1.54740
.00800	1.21533	1.33629	1.33146	1.33146
.00850	1.12667	1.20366	1.20055	1.20055
.00900	1.07434	1.12318	1.12118	1.12118
.00950	1.04347	1.07435	1.07307	1.07307
.01000	1.02524	1.04472	1.04390	1.04390

Comparison of one step methods for [B]

Time	Euler	Modified Euler	Runge-Kutta-4	Exact
.00000	.00000	.00000	.00000	.00000
.00050	409.42855	392.83319	393.37887	393.37915
.00100	651.05773	631.19414	631.85606	631.85640
.00150	793.65809	775.82537	776.42755	776.42786
.00200	877.81542	863.58383	864.07080	864.07105
.00250	927.48187	916.83339	917.20257	917.20275
.00300	956.79313	949.14383	949.41252	949.41266
.00350	974.09151	968.74897	968.93909	968.93918
.00400	984.30037	980.64485	980.77663	980.77670
.00450	990.32524	987.86297	987.95288	987.95293
.00500	993.88090	992.24273	992.30332	992.30336
.00550	995.97931	994.90026	994.94068	994.94070
.00600	997.21771	996.51278	996.53952	996.53954
.00650	997.94857	997.49121	997.50878	997.50879
.00700	998.37990	998.08490	998.09638	998.09638
.00750	998.63445	998.44513	998.45259	998.45260
.00800	998.78467	998.66371	998.66854	998.66854
.00850	998.87333	998.79634	998.79945	998.79945
.00900	998.92566	998.87682	998.87882	998.87882
.00950	998.95653	998.92565	998.92693	998.92693
.01000	998.97476	998.95528	998.95610	998.95610

$$[B] = \frac{k_1[A]_0}{(k_1 + k_2)} [1 - e^{-(k_1+k_2)t}]$$

We can use these equations to compare the relative accuracy of the various one-step methods for $k_1 = 1000$, $k_2 = 1$ and $[A]_0 = 1000$. Such a comparison is presented in Table 2–2. Clearly, the fourth-order Runge-Kutta method is better than the Euler method or the second-order method as shown also by the plot of the deviations given in Fig. 2–4. Notice that at large values of t the Euler deviation tends to zero. To explore why this is so, recall that the Euler method is based on the Taylor expansion. If we use only the first two terms we get equation (2–231) and the deviation of the Euler method from the exact value is

$$\Delta y = y(t + \Delta t) - [y(t) + y'(t)\Delta t] \qquad (2\text{–}243)$$

Thus,

$$\Delta y = \tfrac{1}{2} y''(t)\Delta t^2 + \tfrac{1}{6} y'''(t)\Delta t^3 + \cdots \qquad (2\text{–}244)$$

In the case of equation (2–14) the higher derivatives are

$$\frac{d^n[A]}{dt^n} = (-1)^n \frac{k_1[A]_0}{(k_1 + k_2)^n} e^{-(k_1+k_2)t} \qquad (2\text{–}245)$$

Consequently, as $t \to \infty$, the nth differential becomes close to zero, as does the deviation Δy, explaining why the Euler deviation tends to zero. Another way of looking at this physically (see Fig. 2–4) is that at large t, the curve becomes smoother and flatter, so that over the interval t_i and t_{i+1}, a linear extrapolation becomes a reasonable approximation.

A computer program for the fourth-order Runge-Kutta integration of simultaneous differential equations representing the reaction $A \rightleftarrows B$ is presented in Appendix 2.2. The equations given in the program are for the differential equations for [A] and [B] on page 61. A general program for handling rate problems by numerical integration is also available.

2.6.2.3 Multistep Methods: Predictor-Corrector Method.

The one-step methods discussed in the previous section all approximate the solution of the first-order differential equation (equation 2–227) with initial condition that $y_0 = y(x_0)$. Using the Euler method, the solution is represented by the sum across a series of single intervals,

$$y_1 = y_0 + \int_{x_0}^{x_1} f(x_0, y_0) \, dx$$

$$y_2 = y_1 + \int_{x_1}^{x_2} f(x_1, y_1) \, dx \qquad (2\text{–}246)$$

$$\vdots \qquad \vdots$$

$$y_{i+1} = y_i + \int_{x_i}^{x_{i+1}} f(x_i, y_i) \, dx$$

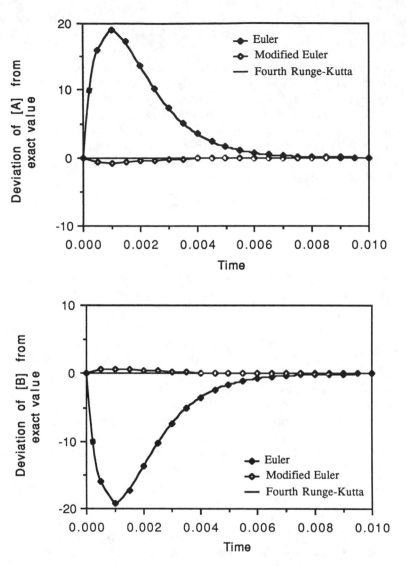

Figure 2–4 Plot of deviation of exact and numerical solution for the reaction A $\rightleftarrows$ B for various one-step numerical methods. (a) Deviation for species A. (b) Deviation for species B.

If the function $f(x_i, y_i)$ can be approximated by a polynomial that interpolates it at m points, where $x_i, x_{i-1}, \ldots, x_{i-m+1}$, and if the point at x_{i+1} is excluded, the method is known as an *explicit* type of formula. The *implicit* type is derived by basing the interpolating polynomial over the points x_i through x_{i-m+1}. All common multistep methods described by equation (2–246) that use the implicit and explicit formulas are known as *predictor-corrector* methods. In these methods, an approximate solution is obtained from the explicit formula predictor equation and refined iteratively by the implicit formula corrector equation. A flow chart of the predictor-corrector algorithm is shown in Fig. 2–5. The predictor-corrector procedure begins with a

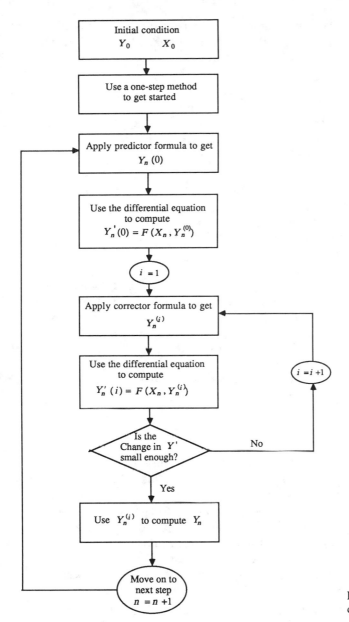

Figure 2–5 Flow chart of the predictor-corrector algorithm.

The flowchart contains the following boxes:

Initial condition
Y_0 X_0

Use a one-step method to get started

Apply predictor formula to get $Y_n(0)$

Use the differential equation to compute
$Y_n'(0) = F(X_n, Y_n^{(0)})$

$i = 1$

Apply corrector formula to get $Y_n^{(i)}$

Use the differential equation to compute
$Y_n'(i) = F(X_n, Y_n^{(i)})$

$i = i + 1$

Is the Change in Y' small enough?

No

Yes

Use $Y_n^{(i)}$ to compute Y_n

Move on to next step $n = n + 1$

one-step method to start the iterative procedure from an initial value. Then, using the predictor equation, an approximation to $y_n = y(x_n)$ is obtained which uses the function $f(x,y)$ to evaluate an approximation to y_n'. From there, a corrector equation is employed, one evaluates again, corrects again, and so on until one decides to move to the next point x_{n+1}. The different predictor-corrector methods differ primarily in the choice of method used to approximate the integral equation (2–246). Those that use Simpson's rule of integration form the basis of the Milne-Simpson predictor-corrector method.[25,31] The Adams-Bashforth method[25] uses the Newton backward interpolating formula of degree $k - 1$, where k refers to the number of previous steps used to interpolate $f(x,y)$. The order of the method is determined by the degree. The predictor equation for the fourth-order method is

$$y_{n+1} = y + \frac{h}{24}[55y'_n - 59y'_{n-1} + 37y'_{n-2} - 9y'_{n-3}] \qquad (2\text{--}247)$$

with an error $e_{i+1} = (251/720)h^5 y^{(v)}(\xi)$ where $x_{i-3} < \xi$. The corrector equation is

$$y_{n+1} = y_n + \frac{h}{24}[9y'_{n+1} + 19y'_n - 5y'_{n-1} + y'_{n-2}] \qquad (2\text{--}248)$$

with an error $e^*_{i+1} = -(19/720)h^5 y^{(v)}(\xi^*)$ and $\xi^* < x_{i+1}$.

2.6.3 Considerations in Selecting an Algorithm for Solving Differential Equations for Chemical Kinetic Systems

The first step in selecting a suitable algorithm for solving kinetic equations for a specific chemical application is to identify mathematically the type of differential equations, $y' = f(x,y)$ that describe the chemical system. If the differential equations are complicated, involving second- and/or third-order terms, a predictor-corrector method may be preferred since it requires only two evaluations of the function rather than the four required by the Runge-Kutta method. In specifying the problem the tolerance, τ, must be considered. Usually the local error is bounded by the tolerance, that is, the local error per unit step

$$\frac{|y_i^{calc} - y_i^{exact}|}{h} \leq \tau \qquad (2\text{--}249)$$

This specifies the accuracy required in the results of a differential equation, but it will depend on the step size. If the step size is too small, the computational process will consume vast amounts of computer time integrating through a large number of steps, which means that the number of operations increases and thus the global round-off errors increase. On the other hand, if the step size is too large, the local error due to truncation may be large, resulting in an accumulation of errors which will cause inaccurate results. Generally for a method of order n the local error will be on the order of a constant times the step size raised to the $n + 1$ power. Usually a differential equation of nth order will have n time constants. If two of them are of drastically different magnitudes, the numerical integration technique may become unstable. For systems of equations where the largest step-size is governed by the largest eigenvalue, the integration time for full evaluation of the solution is governed by the smallest eigenvalue. The system of kinetic equations which result in eigenvalues that differ in orders of magnitude is commonly referred to as *stiff equations*. These systems require that a step size be chosen small enough to account for the smallest eigenvalue. The difficulty with such a requirement is that the use of extremely small step sizes will cause the method to consume a large amount of time in reaching the end of the time interval. Also, round-off and truncation errors can accumulate through the large number of calculations that must be performed. However, these problems are not insurmountable, and the Gear routine is now commonly used for stiff chemical kinetic systems.[20]

2.7 STOCHASTIC METHOD

2.7.1 Introduction

Up to this point in the text, we have considered a number of complex kinetic processes and have gone through the mathematical theories used to predict the concentration of the reacting species as a function of time. The mathematical approach we have used thus far is *deterministic;* that is, once the initial values of the concentrations are specified (the initial state of the system), we can express them at any later time in some mathematical form determined by the postulated mechanism. The deterministic approach regards the evolution of the system in time as a continuous process governed by a set of coupled differential equations. However, this approach has no theoretical justification, and its use is based solely on empiricism. The *stochastic approach,* on the other hand, is based on the probability that a single reaction occurs in an ensemble of molecules.[32] It is similar in spirit to the statistical approach to thermodynamics. The motivation for considering a probabilistic approach is that a chemically reacting system is not deterministic when its state is specified merely by the number of molecules of each chemical species. If the position and velocity of each molecule are specified, and the dynamics is governed by classical mechanics, then the system is indeed deterministic; this is the molecular-dynamic approach taken in Chapters 6, 8, and 9. But knowing *only* the average species concentrations removes essential information about the system; hence, molecular reactions are essentially random occurrences. By making some plausible assumptions, we can regard the stochastic process as a Markovian one, described by ensemble averages and fluctuations. If the fluctuations turn out to be negligibly small compared to the average value, then we can say that the reactive process is effectively deterministic. For most simple reactions, the mean concentration will be essentially the same as the deterministic solution and the statistical fluctuations about the mean may be so small that they are negligible. However, a number of reactions are exceptions to this and are discussed in the review by McQuarrie.[33]

2.7.2 The Stochastic Approach

The stochastic approach to reaction kinetics stems from basic probability theory.[34] The probability of finding a number j of reactant molecules in the system at time t is represented by $P_j(t)$. The aim is to find the probability that a later time $t + \Delta t$, the number of reactant molecules is equal to some other number n. The relationship between the two probabilities is given by

$$P_n(t + \Delta t) = \sum_j P_j(t) W_{j,n}(\Delta t) \qquad (2\text{–}250)$$

where $W_{j,n}(\Delta t)$ is the conditional probability that n reactant molecules exist in the system at time $t + \Delta t$, given that there were j reactant molecules present at time t.

The product of $P_j(t)$ and $W_{j,n}(\Delta t)$ is summed over all values of j to find the total probability P_n at $t + \Delta t$. Equation (2–250) is the fundamental equation for the application of the stochastic approach to chemical reactions.

Consider the irreversible reaction A $\rightarrow$ B.[35,36] As the reaction takes place, then some A molecules will change with time to B. At any given time then there should be some A and some B molecules in the system, but at $t = 0$ only A molecules are present. The probability that any one of the n reactant (or A) molecules will undergo a reaction during the time interval t and $t + \Delta t$ is equal to

$$W_{n,n-1}(\Delta t) = kn\,\Delta t + O(\Delta t) \qquad (2\text{--}251)$$

where k is a constant. Note that k is defined so that for sufficiently small Δt, $k\,\Delta t$ is the probability that any A molecule will react to yield a B molecule in the next interval Δt. If there are n A molecules, and Δt is so small that we can ignore the probability of more than one A molecule reacting in the next Δt, we may deduce that the term $kn\,\Delta t$ is the probability that any one of the A molecules will react in the next Δt. The term $O(\Delta t)$ denotes the probability that more than one reaction occurs but is negligibly small by the preceding assumption. The probability that one of the $n + 1$ A molecules becomes a B molecule during Δt is given by

$$W_{n+1,n}(\Delta t) = k(n + 1)\Delta t + O(\Delta t) \qquad (2\text{--}252)$$

and the probability that the reactant stays in the same state and hence that an A molecule will remain an A molecule is

$$W_{n,n}(\Delta t) = 1 - W_{n,n-1}(\Delta t)$$

or $\qquad (2\text{--}253)$

$$W_{n,n}(\Delta t) = 1 - (kn\,\Delta t + O(\Delta t))$$

According to equation (2–250), the probability that n molecules are present at the end of the time interval Δt is

$$P_n(t + \Delta t) = P_{n+1}(t)W_{n+1,n}(\Delta t) + P_n(t)W_{n,n}(\Delta t)$$

or $\qquad (2\text{--}254)$

$$P_n(t + \Delta t) = k(n + 1)\Delta t P_{n+1}(t) + (1 - kn\,\Delta t)P_n(t) + O(\Delta t)$$

The probability of finding n A molecules in the system at time $t + \Delta t$ is given by a sum of two possible ways of arriving at that state from the earlier configuration. The first term on the right-hand side of equation (2–254) is the probability that, given $n + 1$ A molecules at time t, one of those $n + 1$ A molecules will react in the next interval Δt. The second term is the probability that, given n A molecules at time t, none of those molecules will react in the next Δt. The third term is an error term, representing the probability that two or more A molecules might react in the interval Δt. Transposing $P_n(t)$ to the left-hand side and dividing by Δt, we get

$$\frac{P_n(t + \Delta t) - P_n(t)}{\Delta t} = k(n + 1)P_{n+1}(t) - knP_n(t) + O(\Delta t) \qquad (2\text{--}255)$$

If we now take the limit as $\Delta t \rightarrow 0$, we get the differential-difference equation,

$$\frac{dP_n}{dt} = k(n + 1)P_{n+1}(t) - knP_n(t) \quad (n = 1, \ldots, n_0 - 1) \qquad (2\text{–}256)$$

which is a form of the master equation (see also chapter 14). At $t = 0$, $n = n_0$; therefore, $P_{n_0 + 1}(0) = 0$, and we obtain

$$\frac{dP_n}{dt} = -knP_n(t) \qquad (2\text{–}257)$$

The initial condition for equation (2–256) and (2–257) is:

$$P_n(0) = \begin{cases} 1, \text{ if } n = n_0 \\ 0, \text{ if } n \neq n_0 \end{cases}$$

Thus, the solution to equation (2–257) can easily be found. We first consider the solution that satisfies $P_{n_0}(0) = 1$, which is just

$$P_{n_0}(t) = e^{-kn_0 t} \qquad (2\text{–}258)$$

Using this expression equation (2–256) becomes a relatively simple linear differential equation for $P_{n_0 - 1}(t)$:

$$\frac{dP_{n_0 - 1}}{dt} = k(n_0 - 1 + 1)e^{-kn_0 t} - k(n_0 - 1)P_{n_0 - 1} \qquad (2\text{–}259)$$

The solution of this equation is subject to the initial condition $P_{n_0 - 1} = 0$, and is easily found. Iterating the procedure for decreasing n, to find $P_{n_0}(t)$, $P_{n_0 - 1}(t)$, . . . , $P_0(t)$, we find the general result by induction to be

$$P_n(t) = \binom{n_0}{n}(e^{-kt})^n(1 - e^{-kt})^{n_0 - n} \qquad (2\text{–}260)$$

Equation (2 260) has a simple interpretation. The first factor is the number of ways of picking n specific A molecules from a total of n_0 molecules; the second factor is the probability that a specific set of n A molecules will not have decayed by time t, and the third factor is the probability that the other $n_0 - n$ A molecules will have decayed by time t. We can determine the average number of reactant molecules at time t by

$$\langle n \rangle_t = \sum_{n=0}^{n_0} nP_n(t) \qquad (2\text{–}261)$$

The variance, which provides information on the fluctuations, is given by

$$\sigma_t^2 = \sum_{n=0}^{n_0} (n - \langle n \rangle_t)^2 P_n(t) \qquad (2\text{–}262)$$

or

$$\sigma_t^2 = \sum_{n=0}^{n_0} n^2 P_n(t) - \langle n \rangle_t^2 \qquad (2\text{–}263)$$

In the simple case of A $\rightarrow$ B the average number of reactant molecules at time t is given by

$$\langle n \rangle_t = n_0 e^{-kt} \tag{2-264}$$

and its variance is

$$\sigma_t^2 = n_0 e^{-kt}(1 - e^{-kt}) \tag{2-265}$$

Now consider the more complex reaction

$$A \underset{k_2}{\overset{k_1}{\rightleftharpoons}} B$$

discussed earlier. We set up the master equation by first defining

n_0 = the number of molecules of A at time zero;

n = the number of molecules of A at time t;

k_1 = the forward rate constant;

k_2 = the backward rate constant.

At some time t, there are n A molecules and $(n_0 - n)$ B molecules. If we assume that there is no other type of reaction except A $\rightleftharpoons$ B, then the probability that one of the $(n + 1)$ A molecules will become a B molecule over the time interval $t + \Delta t$ is

$$W_{n+1,n}(\Delta t) = k_1(n + 1)\Delta t \tag{2-266}$$

and the probability that one of the n A molecules will become a B molecule over the same time interval is

$$W_{n,n-1}(\Delta t) = k_1 n \Delta t \tag{2-267}$$

Similarly, the probability that one of the $[n_0 - (n - 1)]$ B molecules will become an A molecule is

$$W_{n_0-n+1,n_0-n}(\Delta t) = k_2(n_0 - n + 1)\Delta t \tag{2-268}$$

and the probability for one of the $(n_0 - n)$ B molecules to become an A molecule would be

$$W_{n_0-n,n_0-n-1}(\Delta t) = k_2(n_0 - n)\Delta t \tag{2-269}$$

Finally, the probability that one of the n A molecules will remain an A molecule is

$$W_{n,n}(\Delta t) = 1 - W_{n,n-1}(\Delta t) - W_{n_0-n,n_0-n-1}(\Delta t) \tag{2-270}$$

or

$$W_{n,n}(\Delta t) = 1 - k_1 n \Delta t - k_2(n_0 - n)\Delta t \tag{2-271}$$

From the theorem of total probability, we have

$$P_n(t + \Delta t) = \sum_j P_j(t)W_{j,n}(\Delta t) \tag{2-272}$$

where $P_j(t)$ is the probability of finding j reactant molecules at time t. Therefore,

$$P_n(t + \Delta t) = P_n(t)W_{n,n}(\Delta t) + P_{n+1}(t)W_{n+1,n}(\Delta t) + P_{n-1}(t)W_{n-1,n}(\Delta t) \qquad (2\text{--}273)$$

and substituting equations (2–266) through (2–271) into (2–273), we get

$$P_n(t + \Delta t) = P_n(t) - P_n(t)[k_1 n + k_2(n_0 - n)]\Delta t$$
$$+ P_{n+1}(t)k_1(n + 1)\Delta t + P_{n-1}(t)k_2(n_0 - n + 1)\Delta t \qquad (2\text{--}274)$$

Dividing both sides of the equation by Δt and taking the limit as $t \to \infty$, we obtain the master equation

$$\frac{dP_n}{dt} = k_2(n_0 - n + 1)P_{n-1}(t) + k_1(n + 1)P_{n+1}(t)$$
$$- [k_1 n + k_2(n_0 - n)]P_n(t) \qquad (2\text{--}275)$$

To solve this equation for $P_n(t)$, we follow the approach previously used for equations (2–258) through (2–260). The average number of reactant molecules at time t is found to be

$$\langle n \rangle_t = \frac{n_0}{k_1 + k_2}[k_2 + k_1 e^{-(k_1 + k_2)t}] \qquad (2\text{--}276)$$

which is just equivalent to the solution found by the deterministic approach, given by equation (2–14). A more generalized approach to setting up the master equation may be found in treatments by Gillespie.[37]

Since the stochastic and deterministic solutions are the same, why bother with a stochastic treatment? The beauty behind the formalism is that it provides a theoretical basis for the use of differential equations to model complex reaction systems. While solving the master equation for most complex reactions can be extremely difficult, there are still several good reasons for using the stochastic approach. First, its use may be appropriate either if σ_t is comparable to $\langle n \rangle_t$, or if the correlations present render $\langle n \rangle_t$ different from what you would get by solving the deterministic rate equations. One would never know for sure whether either of these conditions exist, unless the results of a stochastic analysis are examined. Second, one may want to try a stochastic rather than a deterministic approach, if one is forced to find a numerical solution, since the stochastic simulation is frequently less difficult to implement numerically than the integration of a set of reaction rate equations. Finally, the stochastic approach allows one to study fluctuations, and for systems involving a small number of molecules, fluctuations may be important. The stochastic approach can provide a measure of the fluctuations from the calculated variance.

It is clear from the simple examples considered in this section, the stochastic formulation can become mathematically intractable to solve analytically for more complex reaction systems. However, Gillespie[38-41] has developed a numerical method which simulates the stochastic evolution of any given chemical system in time. The basic stochastic numerical simulation algorithm and program are discussed in Appendix 2.3. The method is used to analyze the problem we have just considered, i.e., solving the kinetic equations for the reaction

$$A \; \rightleftarrows \; B$$

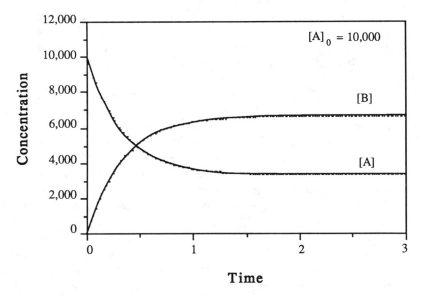

Figure 2–6 Plot of two stochastic simulation runs for the reaction A ⇌ B. (a) Plot of the reaction with $[A]_0 = 100$ molecules. (b) Plot of the reaction with $[A]_0 = 10,000$ molecules. Dotted lines correspond to deterministic mean value, solid lines to stochastic values. In plot (b), the dotted line is coincident with the solid line.

Complex Reactions   Chap. 2

As Fig. 2–6 shows, the evolution of the simple first-order reversible reaction shows significant fluctuations for the case where the initial reactant molecules are present in small numbers. However, when the system has a large number of molecules, e.g., $n > 10,000$, the fluctuations in the system are averaged out. The solution one then obtains is no different from the deterministic solution. For reactions occurring in the gas or solution phase where the number of molecules typically ranges from 10^{10} to 10^{23} molecules per unit volume, it becomes quite clear why a deterministic approach considered in earlier sections is acceptable. The stochastic approach is particularly useful in analyzing *in vivo* biochemical reactions taking place in the individual cells of organisms.[35] Similarly, the study of enzyme amplifier kinetics, in which a few activated molecules initiate an avalanche type of reaction, can be studied in detail with the stochastic approach. In this case, the number of molecules participating in the activation process is too small to model deterministically.[42,43]

REFERENCES

1. P. M. Jeffers, *J. Phys. Chem. 76*, 2829 (1972).
2. S. W. Benson, *Foundations of Chemical Kinetics* (New York: McGraw-Hill, 1960).
3. J-Y. Chien, *J. Am. Chem. Soc. 70*, 2258 (1948).
4. M. Bodenstein, *Z. physik. Chem. 85*, 329 (1913).
5. N. N. Semenov, *Zhur. Fiz. Khim. 17*, 187 (1943).
6. L. Volk, W. Richardson, K. H. Lau, M. Hall, and S. H. Lin, *J. Chem. Educ. 54*, 95 (1977).
7. J. C. Gliddings and H. Y. Shin, *J. Chem. Phys. 36*, 640 (1962).
8. M. Bodenstein, *Z. physik. Chem. 13*, 56 (1894).
9. M. Bodenstein, *Z. physik. Chem. 22*, 1 (1897).
10. M. Bodenstein, *Z. physik. Chem. 29*, 295 (1898).
11. J. H. Sullivan, *J. Chem. Phys. 46*, 73 (1967).
12. G. Hammes and B. Widom, *J. Chem. Phys. 96*, 7621 (1974).
13. A. C. Doyle, "A Study in Scarlet," in *The Complete Sherlock Holmes,* edited by C. Morley (New York: Doubleday, 1930), p. 50.
14. M. S. Mangir, H. Reisler, and C. Wittig, *J. Chem. Phys. 73*, 829 (1980).
15. G. Strang, *Linear Algebra and Its Applications* (New York: Academic Press, 1976).
16. J. G. F. Francis, *Computer J. 4*, 265, 332 (1961).
17. R. S. Martin, P. Peters, and J. H. Wilkinson, *Numerische Math. 4*, 219 (1970).
18. J. H. Wilkinson, *The Algebraic Eigenvalue Problem* (Oxford: University Press, 1965).
19. L. A. Fanow and D. Edelson, *Int. J. Chem. Kinet. 6*, 787 (1974).
20. D. Edelson, *J. Comp. Phys. 11*, 455 (1973).
21. D. L. Allara and D. Edelson, *Int. J. Chem. Kinet. 4*, 345 (1972).
22. D. F. deTar and C. E. deTar, *J. Phys. Chem. 70*, 3842 (1966).
23. S. D. Conte and C. deBoor, *Elementary Numerical Methods,* 2d ed. (New York: McGraw-Hill, 1972).

24. T. E. Shoup, *Applied Numerical Methods for the Microcomputer* (Englewood Cliffs, NJ: Prentice-Hall, 1984).

25. R. W. Hamming, *J. Assoc. Comp. Mach. 6,* 37 (1959).

26. C. W. Gear, *Commun. Am. Comput. Mach. 14,* 176 (1971).

27. C. W. Gear, *Numerical Initial Value Problems in Ordinary Differential Equations* (Englewood Cliffs, NJ: Prentice-Hall, 1971).

28. B. Descamps and W. Forst, *J. Phys. Chem. 80,* 933 (1976).

29. R. C. Kennedy and J. B. Levy, *J. Phys. Chem. 76,* 3480 (1972).

30. J. S. Francisco, Z. Li, and I. H. Williams, *Chem. Phys. Lett. 140,* 531 (1987).

31. W. E. Milne, *Amer. Math. Month. 33,* 455 (1926).

32. I. Oppenheim, K. Shuler, and G. H. Weiss, *Stochastic Processes in Chemical Physics: The Master Equation* (Cambridge, MA: M.I.T. Press, 1977).

33. D. A. McQuarrie, *J. Appl. Prob. 4,* 413 (1967).

34. W. Feller, *An Introduction to Probability Theory and Its Applications* (New York: Wiley, 1950).

35. A. F. Bartholomay, *Bull. Math. Biophys. 20,* 175 (1958).

36. E. A. Boucher, *J. Chem. Educ. 51,* 580 (1974).

37. D. T. Gillespie, *J. Chem. Phys. 72,* 5363 (1980).

38. D. T. Gillespie, *J. Phys. Chem. 81,* 2340 (1977).

39. D. T. Gillespie, *J. Stat. Phys. 16,* 311 (1977).

40. D. T. Gillespie, *J. Comp. Phys. 22,* 403 (1976).

41. D. T. Gillespie, *J. Chem. Phys. 75,* 704 (1981).

42. S. N. Levine, *Science 152,* 651 (1966).

43. R. G. MacFarlane, *Nature 202,* 498 (1964).

BIBLIOGRAPHY

BOWMAN, F. *Introduction to Bessel Functions.* New York: Dover, 1958.

BRADBURY, T. C. *Mathematical Methods with Applications to Problems in the Physical Sciences.* New York: Wiley, 1984.

CHURCHILL, R. V. *Operational Mathematics.* New York: McGraw-Hill, 1971.

COURANT, R., and HILBERT, D. *Methods of Mathematical Physics.* Vols. 1 and 2.

KAPLAN, W. *Ordinary Differential Equations.* Reading MA: Addison-Wesley, 1958.

MARGENAU, H., and MURPHY, G. M. *The Mathematics of Physics and Chemistry.* Vols. 1 and 2. New York: Van Nostrand, 1964.

MATHEWS, J., and WALKER, R. L. *Mathematical Methods of Physics.* 2d ed. New York: W. A. Benjamin, 1964.

RAINVILLE, E. D., and BEDIENT, P. E. *Elementary Differential Equations.* New York: Macmillan, 1981.

WILLIAMSON, R. E. *Introduction to Differential Equations.* Englewood Cliffs, NJ: Prentice-Hall, 1986.

APPENDIX 2.1 THE LAPLACE TRANSFORM

As discussed in section 2.4, the Laplace transform can be used to solve linear differential equations that describe kinetic processes occurring in reacting chemical systems. In section 2.4.3 we noted some properties which are particularly useful in carrying out Laplace transformations. Here, a few more useful properties are presented.

TABLE 2.1–1 LAPLACE TRANSFORM GENERAL PROPERTIES

No.	Name	$F(p)$	$f(t)$
1	Direct Laplace Transform	$\int_0^\infty f(t)\epsilon^{-pt}dt$	$f(t)$
2	Inverse Laplace Transformation	$F(p)$	$\dfrac{1}{2\pi l}\int_{c-i\infty}^{c+i\infty} F(p)e^{pt}dp$
3	Linearity	$F_1(p) \pm F_2(p)$	$f_1(t) \pm f_2(t)$
4	Linearity	$aF(p)$ where a is independent of p and t	$af(t)$
5	Real Differentiation	$PF(p) - f(0+)$	$\dfrac{df(t)}{dt}$
6	Real Differentiation	$P^n F(p) - \sum_{\ell=1}^{n} p^{(\ell-1)} \cdot f^{(n-\ell)}(0+)$ where $f^{(n-\ell)}(0+) \equiv \left.\dfrac{d^{(n-\ell)}f}{dt^{(n-\ell)}}\right|_{t=0+}$	$\dfrac{d^n f(t)}{dt^n}$
7	Real Indefinite Integration	$\dfrac{1}{p}F(p) + \dfrac{1}{p}\int^{o+} f(t)dt$	$\int f(t)dt$
8	Real Definite Integration	$\dfrac{1}{p}F(p)$	$\int_0^t f(t)dt$
9	Real Definite Integration	$\dfrac{1}{p^2}F(p)$	$\int_0^t \int_0^\gamma f(\lambda)d\lambda d\gamma$
10	Real Definite Integration	$\dfrac{1}{p}\int_0^p F(r)dr$	$\int_0^\infty \dfrac{f(\gamma)}{\gamma}d\gamma$
11	Real Translation	$e^{-ap}F(p)$ a is a nonnegative real number	$f(t-a)\cdot U(t-a)$ where $U(t-a)$ is a unit step at $t=a$.

TABLE 2.1–1 *(cont.)*

No.	Name	$F(p)$	$f(t)$
12	Real Translation	$\epsilon^{-ap}F(p)$ a is a nonnegative real number	$f(t+a)$ if $f(t+a)=0$ when $-a<t<0$
13	Multiplication by p	$PF(p)$	$\dfrac{df(t)}{dt}$ if $f(0+)=0$
14	Division by p	$\dfrac{1}{p}F(p)$	$\displaystyle\int_0^t f(t)\,dt$
15	Complex Multiplication	$F(p)\cdot G(p)$	$\displaystyle\int_0^t f(t-\tau)\cdot g(\tau)\,d\tau$
16	Complex Differentiation	$\dfrac{dF(p)}{dp}$	$-tf(t)$
17	Complex Differentiation	$\dfrac{d^nF(p)}{dp^n}$	$(-1)^n t^n f(t)$ $n=1,2,\ldots$
18	Complex Integration	$\displaystyle\int_p^\infty F(p)\,dp$	$\dfrac{1}{t}f(t)$
19	Complex Translation	$F(p+a)$ a complex with nonnegative real part.	$e^{-at}f(t)$
20	Complex Translation	$F(p-a)$ a complex with nonnegative real part.	$e^{at}f(t)$
21	Scale Change	$F(p)$ a positive and independent of p and t.	$\dfrac{1}{a}f\left(\dfrac{t}{a}\right)$
22	Differentiation with respect to a second independent variable.	$\dfrac{\partial F(p,r)}{\partial r}$ r independent of p and t.	$\dfrac{\partial f(t,r)}{\partial r}$
23	Integration with respect to a second independent variable.	$\displaystyle\int_{r_0}^r F(p,r)\,dr$ where r is independent of p and t.	$\displaystyle\int_{r_0}^r f(t,r)\,dr$
24	Commutativity of Laplace transformation with real transformation.	$\mathrm{RE}[F(p)]$	$\mathrm{RE}[f(t)]$ $f(t)$ is a complex function.
25	Commutativity of Laplace transformation with imaginary transformation.	$\mathrm{IM}[F(p)]$	$\mathrm{IM}[f(t)]$ $f(t)$ is a complex function.
26	Final Value	$\displaystyle\lim_{p\to 0} pF(p) = \lim_{t\to\infty} f(t)$ $pF(p)$ analytic on the axis of imaginaries and in the right half-plane.	
27	Initial Value	$\displaystyle\lim_{p\to\infty} pF(p) = \lim_{t\to 0} f(t)$	

TABLE 2.1-2 GENERAL TABLE OF LAPLACE TRANSFORM PAIRS

No.	$F(p)$	$f(t)$
1	p^n	$\delta^n(t)$ An impulse of order $(N+1)$ at $t = 0$ where N is a positive integer.
2	p	$\delta'(t)$ A doublet impulse at $t = 0$.
3	1	$\delta(t)$ A unit impulse at $t = 0$. $\delta(t) = \infty$ when $t = 0$; $\delta(t) = 0$ otherwise.
4	$\dfrac{1}{p}$	$1 = u_{-1}(t)$ A unit step of $t = 0$. $u_{-1}(t) = 1$ when $t \geq 0$; $u_{-1}(t) = 0$ when $t < 0$
5	$\dfrac{1}{p^2}$	t
6	$\dfrac{1}{p^N}$	$\dfrac{t^{(N-1)}}{(N-1)!}$ Where N is a positive integer
7	$\dfrac{1}{p^N}$	$\dfrac{t^{(N-1)}}{\Gamma(N)}$ $\Gamma(N)$ is the gamma function
8	$\dfrac{1}{\sqrt{p}}$	$\dfrac{1}{\sqrt{\pi t}}$
9	$\dfrac{1}{p^{3/2}}$	$2\sqrt{\dfrac{t}{\pi}}$
10	$\dfrac{1}{p+a}$	e^{-at} Valid for complex a.

TABLE 2.1–2. (cont.)

No.	$F(p)$	$f(t)$
11	$\dfrac{1}{p(p+a)}$	$\dfrac{1}{a}(1 - e^{-at})$
12	$\dfrac{1}{p^2(p+a)}$	$\dfrac{1}{a^2}(e^{-at} + at - 1)$
13	$\dfrac{1}{p^3(p+a)}$	$\dfrac{1}{a^2}\left[\dfrac{1}{a} - t + \dfrac{at^2}{2} - \dfrac{1}{a}e^{-at}\right]$
14	$\dfrac{1}{(p+a)(p+b)}$	$\dfrac{1}{(b-a)}(e^{-at} - e^{-bt})$
15	$\dfrac{1}{p(p+a)(p+b)}$	$\dfrac{1}{ab}\left[1 + \dfrac{1}{(a-b)}(be^{-at} - ae^{-bt})\right]$
16	$\dfrac{1}{p^2(p+a)(p+b)}$	$\dfrac{1}{(ab)^2}\left[\dfrac{1}{(a-b)}(a^2e^{-bt} - b^2e^{-at}) + abt - a - b\right]$
17	$\dfrac{1}{p^3(p+a)(p+b)}$	$\dfrac{1}{(ab)}\left[\dfrac{a^3 - b^3}{(ab)^2(a-b)} + \dfrac{1}{2}t^2 - \dfrac{(a+b)}{ab}t + \dfrac{1}{(a-b)}\left(\dfrac{b}{a^2}e^{-at} - \dfrac{a}{b^2}e^{-bt}\right)\right]$
18	$\dfrac{1}{(p+a)(p+b)(p+c)}$	$\dfrac{1}{(b-a)(c-a)}e^{-at} + \dfrac{1}{(a-b)(c-b)}e^{-bt} + \dfrac{1}{(a-c)(b-c)}e^{-ct}$
19	$\dfrac{1}{p(p+a)(p+b)(p+c)}$	$\dfrac{1}{abc} - \dfrac{1}{a(b-a)(c-a)}e^{-at} - \dfrac{1}{b(a-b)(c-b)}e^{-bt} - \dfrac{1}{c(a-c)(b-c)}e^{-ct}$
20	$\dfrac{1}{p^2(p+a)(p+b)(p+c)}$	$\dfrac{ab(ct-1) - ac - bc}{(abc)^2} + \dfrac{1}{a^2(b-a)(c-a)}e^{-at} + \dfrac{1}{b^2(a-b)(c-b)}e^{-bt}$ $+ \dfrac{1}{c^2(a-c)(b-c)}e^{-ct}$

TABLE 2.1–2. (cont.)

No.	$F(p)$	$f(t)$
21	$\dfrac{1}{p^3(p+a)(p+b)(p+c)}$	$\dfrac{1}{(abc)^3}[(ab+ac+bc)^2 - abc(a+b+c)] - \dfrac{ab+ac+bc}{(abc)^2}t + \dfrac{1}{2abc}t^2$ $- \dfrac{1}{a^3(b-a)(c-a)}e^{-at} - \dfrac{1}{b^3(a-b)(c-b)}e^{-bt} - \dfrac{1}{c^3(a-c)(b-c)}e^{-ct}$
22	$\dfrac{1}{(p+a)^2}$	te^{-at}
23	$\dfrac{1}{(p+a)^N}$	$\dfrac{t^{(N-1)}e^{-at}}{(N-1)!}$ where N is a positive integer
24	$\dfrac{1}{p(p+a)^2}$	$\dfrac{1}{a^2}[1 - e^{-at} - ate^{-at}]$
25	$\dfrac{1}{p^2(p+a)^2}$	$\dfrac{1}{a^3}[at - 2 + ate^{-at} - 2e^{-at}]$
26	$\dfrac{1}{p(p+a)^3}$	$\dfrac{1}{a^3}\left[1 - \left(\dfrac{1}{2}a^2t^2 + at + 1\right)e^{-at}\right]$
27	$\dfrac{1}{(p+a)(p+b)^2}$	$\dfrac{1}{(a-b)^2}\{e^{-at} + [(a-b)t - 1]e^{-bt}\}$
28	$\dfrac{1}{p(p+a)(p+b)^2}$	$\dfrac{1}{ab^2} - \dfrac{1}{a(a-b)^2}e^{-at} - \left[\dfrac{1}{b(a-b)}t + \dfrac{a-2b}{b^2(a-b)^2}\right]e^{-bt}$
29	$\dfrac{1}{p^2(p+a)(p+b)^2}$	$\dfrac{1}{a^2(b-a)^2}e^{-at} + \dfrac{1}{ab^2}\left(t - \dfrac{1}{a} - \dfrac{2}{b}\right) + \left[\dfrac{1}{b^2(a-b)}t + \dfrac{2(a-b)-b}{b^3(b-a)^2}\right]e^{-bt}$
30	$\dfrac{1}{(p+a)(p+b)(p+c)^2}$	$\left[\dfrac{1}{(c-b)(c-a)}t + \dfrac{2c-a-b}{(c-a)^2(c-b)^2}\right]e^{-ct} + \dfrac{1}{(b-a)(c-a)^2}e^{-at}$ $+ \dfrac{1}{(a-b)(c-b)^2}e^{-bt}$

TABLE 2.1–2 (cont.)

No.	$F(p)$	$f(t)$
31	$\dfrac{1}{p(p+a)(p+b)(p+c)^2}$	$\dfrac{1}{abc^2} + \dfrac{1}{a(a-b)(c-a)^2}e^{-at} + \dfrac{1}{b(b-a)(c-b)^2}e^{-bt}$ $+ \left[\dfrac{1}{c(c-b)(a-c)}t - \dfrac{(c-b)(c-a)+c(2c-b-a)}{c^2(c-b)^2(c-a)^2}\right]e^{-ct}$
32	$\dfrac{1}{p^2(p+a)(p+b)(p+c)^2}$	$\left[\dfrac{1}{c^2(c-b)(c-a)}t + \dfrac{2(c-b)(c-a)+c(2c-b-a)}{c^3(c-b)^2(c-a)^2}\right]e^{-ct}$ $+ \dfrac{1}{a^2(b-a)(c-a)^2}e^{-at} + \dfrac{1}{abc^2}\left(t - \dfrac{2}{c} - \dfrac{1}{a} - \dfrac{1}{b}\right) + \dfrac{1}{b^2(a-b)(c-b)^2}e^{-bt}$
33	$\dfrac{1}{(p+a)^2(p+b)^2}$	$\dfrac{1}{(b-a)^2}te^{-at} - \dfrac{2}{(b-a)^3}e^{-at} + \dfrac{1}{(a-b)^2}te^{-bt} - \dfrac{2}{(a-b)^3}e^{-bt}$
34	$\dfrac{1}{p(p+a)^2(p+b)^2}$	$\dfrac{(3a-b)}{a^2(b-a)^3}e^{-at} - \dfrac{1}{a(b-a)^2}te^{-at} + \dfrac{(3b-a)}{b^2(a-b)^3}e^{-bt} - \dfrac{1}{b(a-b)^2}te^{-bt} + \dfrac{1}{a^2b^2}$
35	$\dfrac{1}{p^2(p+a)^2(p+b)^2}$	$\dfrac{2(b-2a)}{a^3(b-a)^3}e^{-at} + \dfrac{1}{a^2(b-a)^2}te^{-at} + \dfrac{2(a-2b)}{b^3(a-b)^3}e^{-bt} + \dfrac{1}{b^2(a-b)^2}te^{-bt}$ $+ \dfrac{1}{a^2b^2}t - \dfrac{2(a+b)}{a^3b^3}$
36	$\dfrac{1}{(p+a)(p+b)^2(p+c)^2}$	$\dfrac{1}{(b-a)^2(c-a)^2}e^{-at} + \left[\dfrac{3c-(b+2a)}{(b-c)^3(a-c)^2}e^{-ct} + \dfrac{1}{(a-c)(b-c)^2}te^{-ct}\right.$ $\left.+ \dfrac{[3b-(c+2a)]}{(a-b)^2(c-b)^3}e^{-bt} + \dfrac{1}{(a-b)(c-b)^2}te^{-bt}\right]$
37	$\dfrac{1}{p^2+\omega^2}$	$\dfrac{1}{\omega}\sin \omega t$
38	$\dfrac{1}{p^2-a^2}$	$\dfrac{1}{a}\sinh at$

TABLE 2.1–2 (cont.)

No.	$F(p)$	$f(t)$
39	$\dfrac{1}{p^2(p^2 - a^2)}$	$\dfrac{1}{a^3}\sinh at - \dfrac{1}{a^2}t$
40	$\dfrac{1}{p^3(p^2 - a^2)}$	$\dfrac{1}{a^4}(\cosh at - 1) - \dfrac{1}{2a^2}t^2$
41	$\dfrac{1}{p^3 + a^3}$	$\dfrac{1}{3a^2}\left[e^{-at} - e^{(a/2)t}\left(\cos\dfrac{\sqrt{3}}{2}at - \sqrt{3}\sin\dfrac{\sqrt{3}}{2}at\right)\right]$
42	$\dfrac{1}{p^4 + 4a^4}$	$\dfrac{1}{4a^3}(\sin at \cosh at - \cos at \sin at)$
43	$\dfrac{1}{p^4 - a^4}$	$\dfrac{1}{2a^3}(\sinh at - \sin at)$
44	$\dfrac{1}{[(p + a)^2 - \omega^2]}$	$\dfrac{1}{\omega}e^{-at}\sinh \omega t$
45	$\dfrac{1}{p\sqrt{p+1}}$	$\operatorname{erf}(\sqrt{t});\ \operatorname{erf}(y) \equiv \text{the error function} = \dfrac{2}{\sqrt{\pi}}\displaystyle\int_0^y e^{-u^2}\,du$
46	$\dfrac{1}{\sqrt{p^2 + a^2}}$	$J_0(at)$; Bessel function of 1st kind, zero order
47	$\dfrac{1}{\sqrt{p^2 + a^2} + p}$	$\dfrac{J_1(at)}{at}$; J_1 is the Bessel function of 1st kind, 1st order
48	$\dfrac{1}{[\sqrt{p^2 + a^2} + p]^N}$	$\dfrac{N}{a^N}\dfrac{J_N(at)}{t}$; $N = 1,\,2,\,3,\,\cdots J_N$ is the Bessel function of 1st kind, Nth order
49	$\dfrac{1}{p[\sqrt{p^2 + a^2} + p]^N}$	$\dfrac{N}{a^N}\displaystyle\int_0^t \dfrac{J_N(au)}{u}\,du$; $N = 1,\,2,\,3,\,\cdots J_N$ is the Bessel function of 1st kind, Nth order

TABLE 2.1–2 (cont.)

No.	$F(p)$	$f(t)$
50	$\dfrac{1}{\sqrt{p^2+a^2}\,(\sqrt{p^2+a^2}+p)}$	$\dfrac{1}{a}J_1(at)$; J_1 is the Bessel function of 1st kind, 1st order
51	$\dfrac{1}{\sqrt{p^2+a^2}\,[\sqrt{p^2+a^2}+p]^N}$	$\dfrac{1}{a^N}J_N(at)$; $N = 1, 2, 3$. J_N is the Bessel function of 1st kind, Nth order
52	$\dfrac{1}{\sqrt{(p^2-a^2)}}$	$I_0(at)$; I_0 is modified Bessel function of 1st kind, zero order
53	$\dfrac{1}{\sqrt{p^2-1}}$	$I_0(t)$; I_0 is modified Bessel function of 1st kind, zero order
54	$\dfrac{1}{\sqrt{p}\,(p-1)}$	$e^t\,\mathrm{erf}(\sqrt{t})$; where $\mathrm{erf}(y) = \dfrac{2}{\sqrt{\pi}}\displaystyle\int_0^y e^{-u^2}\,du$
55	$\dfrac{1}{1+\sqrt{p}}$	$\dfrac{1}{\sqrt{\pi t}} - e^t + e^t\,\mathrm{erf}(\sqrt{t})$; where $\mathrm{erf}(y) = \dfrac{2}{\sqrt{\pi}}\displaystyle\int_0^y e^{-u^2}\,du$
56	$\dfrac{1}{\sqrt{p^2-a^2}\,(p+\sqrt{p^2-a^2})^N}$	$\dfrac{1}{a^N}I_N(at)$; I_N is modified Bessel function of 1st kind, Nth order
57	$\dfrac{p}{(p+a)(p+b)}$	$\dfrac{1}{(a-b)}\big[ae^{-at} - be^{-bt}\big]$
58	$\dfrac{p}{(p-a)(p-b)}$	$\dfrac{1}{(a-b)}\big[ae^{at} - be^{bt}\big]$
59	$\dfrac{p}{(p+a)(p+b)(p+c)}$	$\dfrac{-a}{(b-a)(c-a)}e^{-at} - \dfrac{b}{(a-b)(c-b)}e^{-bt} - \dfrac{c}{(a-c)(b-c)}e^{-ct}$
60	$\dfrac{p}{(p+a)^2}$	$(1-at)e^{-at}$

TABLE 2.1-2 (cont.)

No.	$F(p)$	$f(t)$
61	$\dfrac{p}{(p-a)^3}$	$\left(1 - \dfrac{a}{2}t\right)te^{-at}$
62	$\dfrac{p}{(p+a)(p+b)^2}$	$\dfrac{a}{(b-a)^2}\left[e^{-bt} - e^{-at} + \dfrac{b}{a}(b-a)te^{-bt}\right]$
63	$\dfrac{p}{(p+a)^2(p+b)^2}$	$\dfrac{1}{(a-b)^2}\left[\left(\dfrac{a+b}{a-b}\right)(e^{-bt} - e^{-at}) - t(ae^{-at} + be^{-bt})\right]$
64	$\dfrac{p}{p^2+\omega^2}$	$\cos\omega t$
65	$\dfrac{p}{(p+a)(p^2+\omega^2)}$	$\dfrac{1}{a^2+\omega^2}[\sqrt{a^2+\omega^2}\sin(\omega t + \phi) - ae^{-at}];\ \phi = \dfrac{\pi}{2} - \tan^{-1}\left(\dfrac{\omega}{a}\right)$
66	$\dfrac{p}{(p+a)^2(p^2+\omega^2)}$	$\dfrac{\omega^2-a^2}{(a^2+\omega^2)^2}e^{-at} - \dfrac{a}{a^2+\omega^2}te^{-at} + \dfrac{1}{a^2+\omega^2}\sin(\omega t + \phi);\ \phi = \tan^{-1}\left(\dfrac{a^2-\omega^2}{2a\omega}\right)$
67	$\dfrac{p}{(p+a)(p+b)(p^2+\omega^2)}$	$\dfrac{a}{(a^2+\omega^2)(a-b)}e^{-at} - \dfrac{b}{(b^2+\omega^2)(a-b)}e^{-bt}$ $+ \dfrac{1}{\sqrt{(ab-\omega^2)^2 + (a+b)^2\omega^2}}\sin(\omega t + \phi)$ $\phi = \tan^{-1}\dfrac{ab-\omega^2}{\omega(a+b)}$
68	$\dfrac{p}{(p^2+a^2)(p^2+\omega^2)}$	$\dfrac{1}{\omega^2-a^2}[\cos at - \cos\omega t]$
69	$\dfrac{p}{(p^2+\omega^2)^2}$	$\dfrac{1}{2\omega}t\sin\omega t$

TABLE 2.1-2 (cont.)

No.	$F(p)$	$f(t)$
70	$\dfrac{p}{p^2 + 2ap + b^2}$	$\dfrac{1}{2\sqrt{a^2 - b^2}}\left[(a + \sqrt{a^2 - b^2})e^{-(a+\sqrt{a^2-b^2})t} - (a - \sqrt{a^2 - b^2})e^{-(a-\sqrt{a^2-b^2})t}\right]$
71	$\dfrac{p}{p^3 + a^3}$	$\dfrac{1}{3a}\left[e^{-at} + e^{(1/2)at}\left(\cos\dfrac{\sqrt{3}}{2}at + \sqrt{3}\sin\dfrac{\sqrt{3}}{2}at\right)\right]$
72	$\dfrac{p}{p^4 - \omega^4}$	$\dfrac{1}{2\omega^2}\left[\cosh\omega t - \cos\omega t\right]$
73	$\dfrac{p}{p^4 + 4a^4}$	$\dfrac{1}{2a^2}\left[\sin(at)\,\sinh(at)\right]$
74	$\dfrac{p}{(p - a)^{3/2}}$	$\dfrac{(1 + 2at)}{\sqrt{\pi t}}e^{at}$
75	$\dfrac{p + a}{p(p + b)}$	$\dfrac{a}{b} - \dfrac{(a - b)}{b}e^{-bt}$
76	$\dfrac{p + a}{p^2(p + b)}$	$\dfrac{a}{b}t + \dfrac{b - a}{b^2}(1 - e^{-bt})$
77	$\dfrac{p + a}{p^3(p + b)}$	$\dfrac{a}{2b}t^2 + \left(\dfrac{b - a}{b^2}\right)\left[t + \dfrac{1}{b}e^{-bt} - \dfrac{1}{b}\right]$
78	$\dfrac{p + a}{(p + b)(p + c)}$	$\dfrac{1}{(c - b)}\left[(a - b)e^{-bt} - (a - c)e^{-ct}\right]$
79	$\dfrac{p + a}{p(p + b)(p + c)}$	$\dfrac{a}{bc} + \dfrac{a - b}{b(b - c)}e^{-bt} + \dfrac{a - c}{c(c - b)}e^{-ct}$
80	$\dfrac{p + a}{p^2(p + b)(p + c)}$	$\dfrac{1}{bc}(1 + at) - \dfrac{(b + c)a}{b^2c^2} + \dfrac{1}{c - b}\left[\dfrac{(a - b)}{b^2}e^{-bt} - \dfrac{(a - c)}{c^2}e^{-ct}\right]$

TABLE 2.1–2 (cont.)

No.	$F(p)$	$f(t)$
81	$\dfrac{p+a}{(p+b)(p+c)(p+d)}$	$\dfrac{(a-b)}{(c-b)(d-b)}e^{-bt} + \dfrac{(a-c)}{(b-c)(d-c)}e^{-ct} + \dfrac{(a-d)}{(b-d)(c-d)}e^{-dt}$
82	$\dfrac{p+a}{p(p+b)(p+c)(p+d)}$	$\dfrac{a}{bcd} + \dfrac{(b-a)}{b(c-b)(d-b)}e^{-bt} + \dfrac{(c-a)}{c(b-c)(d-c)}e^{-ct} + \dfrac{(d-a)}{d(b-d)(c-d)}e^{-dt}$
83	$\dfrac{p+a}{p^2(p+b)(p+c)(p+d)}$	$\dfrac{1}{bcd}(1+at) - \dfrac{a(bc+cd+db)}{b^2c^2d^2} + \dfrac{(a-b)}{b^2(c-b)(d-b)}e^{-bt} + \dfrac{(a-c)}{c^2(b-c)(d-c)}e^{-ct} + \dfrac{(a-d)}{d^2(b-d)(c-d)}e^{-dt}$
84	$\dfrac{p+a}{(p+b)^2}$	$[1+(a-b)t]e^{-bt}$
85	$\dfrac{p+a}{p(p+b)^2}$	$\dfrac{a}{b^2} + \left[\left(1-\dfrac{a}{b}\right)t - \dfrac{a}{b^2}\right]e^{-bt}$
86	$\dfrac{p+a}{p^2(p+b)^2}$	$\dfrac{1}{b^2}\left\{\left(1+at-\dfrac{2a}{b}\right) - \left(1-\dfrac{2a}{b}+[b-a]t\right)e^{-bt}\right\}$
87	$\dfrac{p+a}{(p+b)(p+c)^2}$	$\dfrac{(a-b)}{(b-c)^2}e^{-bt} + \left[\dfrac{a-c}{b-c}t + \dfrac{b-a}{(b-c)^2}\right]e^{-ct}$
88	$\dfrac{p+a}{p(p+b)(p+c)^2}$	$\dfrac{a}{bc^2} + \dfrac{(b-a)}{b(c-b)^2}e^{-bt} + \left[\dfrac{(a-c)}{c(c-b)}t - \dfrac{c^2-a(2c-b)}{c^2(c-b)^2}\right]e^{-ct}$
89	$\dfrac{p+a}{p^2(p+b)(p+c)^2}$	$\dfrac{a}{bc^2}\left[\left(\dfrac{1}{a}-\dfrac{1}{b}-\dfrac{2}{c}\right)+t\right] - \dfrac{(b-a)}{b^2(c-b)^2}e^{-bt} - \left[\dfrac{c-a}{c^2(b-c)}t + \dfrac{(2a-c)(2c-b)-ac}{c^3(c-b)^2}\right]e^{-ct}$
90	$\dfrac{p+a}{(p+b)^2(p+c)^2}$	$\left[\dfrac{(a-c)}{(b-c)^2}t + \dfrac{b+c-2a}{(b-c)^3}\right]e^{-ct} + \left[\dfrac{(a-b)}{(c-b)^2}t + \dfrac{b+c-2a}{(c-b)^3}\right]e^{-bt}$

No.	$F(p)$	$f(t)$
91	$\dfrac{p+a}{p(p+b)^2(p+c)^2}$	$\dfrac{a}{b^2c^2} + \left[\dfrac{(c-a)}{c(b-c)^2}t + \dfrac{3ac-ab-2c^2}{c^2(b-c)^3}\right]e^{-ct} + \left[\dfrac{(b-a)}{b(c-b)^2}t + \dfrac{3ab-ac-2b^2}{b^2(c-b)^3}\right]e^{-bt}$
92	$\dfrac{p+a}{p^2(p+b)^2(p+c)^2}$	$\left[\dfrac{1}{b^2c^2}(1+at) - \dfrac{2a(b+c)}{b^3c^3}\right] + \left[\dfrac{(a-b)}{b^2(c-b)^2}t + \dfrac{2a(c-2b)+b(3b-c)}{b^3(c-b)^3}\right]e^{-bt} + \left[\dfrac{(a-c)}{c^2(b-c)^2}t + \dfrac{2a(b-2c)+c(3c-b)}{c^3(b-c)^3}\right]e^{-ct}$
93	$\dfrac{p+a}{(p+b)(p+c)(p+d)^2}$	$\dfrac{(a-b)}{(c-b)(d-b)^2}e^{-bt} + \dfrac{(a-c)}{(b-c)(d-c)^2}e^{-ct} + \left[\dfrac{(a-d)}{(d-c)(d-b)}t + \dfrac{a(2d-b-c)-d^2+bc}{(d-b)^2(d-c)^2}\right]e^{-dt}$
94	$\dfrac{p+a}{p(p+b)(p+c)(p+d)^2}$	$\dfrac{a}{bcd^2} - \dfrac{(a-c)}{c(b-c)(d-c)^2}e^{-ct} - \dfrac{(a-b)}{b(c-b)(d-b)^2}e^{-bt} + \left[\dfrac{(d-a)}{d(d-c)(d-b)}t - \dfrac{a(d-c)(d-b)+d(a-d)(2d-b-c)}{d^2(d-c)^2(d-b)^2}\right]e^{-dt}$
95	$\dfrac{p+a}{(p+b)(p+c)^2(p+d)^2}$	$\dfrac{(a-b)}{(b-c)^2(b-d)^2}e^{-bt}$ $+ \left[\dfrac{(c-2a)(b-c)+(db-c^2)-a(d-c)}{(c-b)^2(d-c)^3} + \dfrac{(a-c)}{(b-c)(c-d)^2}t\right]e^{-ct}$ $+ \left[\dfrac{(d-2a)(b-d)+(bc-d^2)-a(c-d)}{(d-b)^2(c-d)^3} + \dfrac{(a-d)}{(b-d)(c-d)^2}t\right]e^{-dt}$
96	$\dfrac{p+a}{p^2+\omega^2}$	$\dfrac{\sqrt{a^2+\omega^2}}{\omega}\sin(\omega t+\phi); \ \phi = \tan^{-1}\left(\dfrac{\omega}{a}\right)$
97	$\dfrac{p+a}{p(p^2+\omega^2)}$	$\dfrac{a}{\omega^2} - \dfrac{\sqrt{a^2+\omega^2}}{\omega^2}\cos(\omega t+\phi); \ \phi = \tan^{-1}\left(\dfrac{\omega}{a}\right)$

TABLE 2.1-2 (cont.)

No.	$F(p)$	$f(t)$
98	$\dfrac{p \sin \phi + \omega \cos \phi}{(p^2 + \omega^2)}$	$\sin(\omega t + \phi)$
99	$\dfrac{p \cos \phi - \omega \sin \phi}{(p^2 + \omega^2)}$	$\cos(\omega t + \phi)$
100	$\dfrac{p + a}{(p + b)^2 + \omega^2}$	$\dfrac{1}{\omega}\sqrt{(a-b)^2 + \omega^2}\; e^{-bt} \sin(\omega t + \phi); \; \phi = \tan^{-1}\left(\dfrac{\omega}{a - b}\right)$
101	$\dfrac{(p + b) \cos \phi \pm \omega \sin \phi}{(p + b)^2 + \omega^2}$	$e^{-bt} \cos(\omega t \pm \phi)$
102	$\dfrac{\omega \cos \phi \pm (p + b) \sin \phi}{(p + b)^2 + \omega^2}$	$e^{-bt} \sin(\omega t \pm \phi)$
103	$\dfrac{p + a}{(p + a)^2 + \omega^2}$	$e^{-at} \cos \omega t$
104	$\dfrac{p + a}{[(p + a)^2 - \omega^2]}$	$e^{-at} \cosh \omega t$
105	$\dfrac{p + a}{(p + b)^3}$	$\left[t + \left(\dfrac{a - b}{2}\right)t^2\right]e^{-bt}$
106	$\dfrac{p + a}{p(p + b)^3}$	$\dfrac{1}{b^3}\left\{a + \left[\dfrac{b^2(b - a)}{2}t^2 - a - abt\right]e^{-bt}\right\}$
107	$\dfrac{p + a}{p^2(p + b)^3}$	$\dfrac{a}{b^3}t + \dfrac{b - 3a}{b^4} + \left[\dfrac{3a - b}{b^4} + \dfrac{a - b}{2b^2}t^2 + \dfrac{2a - b}{b^3}t\right]e^{-bt}$
108	$\dfrac{p + a}{(p + c)(p + b)^3}$	$\dfrac{a - c}{(b - c)^3}e^{-ct} + \left[\dfrac{a - b}{2(c - b)}t^2 + \dfrac{c - a}{(c - b)^2}t + \dfrac{a - c}{(c - b)^3}\right]e^{-bt}$

TABLE 2.1–2 (cont.)

No.	$F(p)$	$f(t)$
109	$\dfrac{p^2}{(p+a)(p+b)(p+c)}$	$\dfrac{a^2}{(b-a)(c-a)}e^{-at} + \dfrac{b^2}{(a-b)(c-b)}e^{-bt} + \dfrac{c^2}{(a-c)(b-c)}e^{-ct}$
110	$\dfrac{p^2}{(p+a)(p+b)^2}$	$\dfrac{a^2}{(b-a)^2}e^{-at} + \left[\dfrac{b^2}{(a-b)}t + \dfrac{b^2-2ab}{(a-b)^2}\right]e^{-bt}$
111	$\dfrac{p^2}{(p+a)^3}$	$\left[2 - 2at + \dfrac{a^2}{2}t^2\right]e^{-at}$
112	$\dfrac{p^2+a}{p^2(p+b)}$	$\dfrac{b^2+a}{b^2}e^{-bt} + \dfrac{a}{b}t - \dfrac{a}{b^2}$
113	$\dfrac{p^2+a}{p^3(p+b)}$	$\dfrac{a}{2b}t^2 - \dfrac{a}{b^2}t + \dfrac{1}{b^3}[b^2 + a - (a+b^2)e^{-bt}]$
114	$\dfrac{p^2+a}{p(p+b)(p+c)}$	$\dfrac{a}{bc} + \dfrac{(b^2+a)}{b(b-c)}e^{-bt} - \dfrac{(c^2+a)}{c(b-c)}e^{-ct}$
115	$\dfrac{p^2+a}{p^2(p+b)(p+c)}$	$\dfrac{b^2+a}{b^2(c-b)}e^{-bt} + \dfrac{c^2+a}{c^2(b-c)}e^{-ct} + \dfrac{a}{bc}t - \dfrac{a(b+c)}{b^2c^2}$
116	$\dfrac{p^2+a}{(p+b)(p+c)(p+d)}$	$\dfrac{b^2+a}{(c-b)(d-b)}e^{-bt} + \dfrac{c^2+a}{(b-c)(d-c)}e^{-ct} + \dfrac{d^2+a}{(b-d)(c-d)}e^{-dt}$
117	$\dfrac{p^2+a}{p(p+b)(p+c)(p+d)}$	$\dfrac{a}{bcd} + \dfrac{b^2+a}{b(b-c)(d-b)}e^{-bt} + \dfrac{c^2+a}{c(b-c)(c-d)}e^{-ct} + \dfrac{d^2+a}{d(b-d)(d-c)}e^{-dt}$
118	$\dfrac{p^2+a}{p^2(p+b)(p+c)(p+d)}$	$\dfrac{a}{bcd}t - \dfrac{a}{b^2c^2d^2}(bc + cd + db) + \dfrac{b^2+a}{b^2(b-c)(b-d)}e^{-bt} + \dfrac{c^2+a}{c^2(c-b)(c-d)}e^{-ct}$ $+ \dfrac{d^2+a}{d^2(d-b)(d-c)}e^{-dt}$

TABLE 2.1–2 (cont.)

No.	$F(p)$	$f(t)$
119	$\dfrac{p^2+a}{(p+b)(p+c)(p+d)(p+f)}$	$\dfrac{b^2+a}{(c-b)(d-b)(f-b)}e^{-bt} + \dfrac{c^2+a}{(b-c)(d-c)(f-c)}e^{-ct}$ $+ \dfrac{d^2+a}{(b-d)(c-d)(f-d)}e^{-dt} + \dfrac{f^2+a}{(b-f)(c-f)(d-f)}e^{-ft}$
120	$\dfrac{p^2+a}{p(p+b)(p+c)(p+d)(p+f)}$	$\dfrac{a}{bcdf} - \dfrac{b^2+a}{b(c-b)(d-b)(f-b)}e^{-bt} - \dfrac{c^2+a}{c(b-c)(d-c)(f-c)}e^{-ct}$ $- \dfrac{d^2+a}{d(b-d)(c-d)(f-d)}e^{-dt} - \dfrac{f^2+a}{f(b-f)(c-f)(d-f)}e^{-ft}$
121	$\dfrac{p^2+a}{(p+b)(p+c)(p+d)(p+f)(p+g)}$	$\dfrac{b^2+a}{(c-b)(d-b)(f-b)(g-b)}e^{-bt} + \dfrac{c^2+a}{(b-c)(d-c)(f-c)(g-c)}e^{-ct}$ $+ \dfrac{d^2+a}{(b-d)(c-d)(f-d)(g-d)}e^{-dt} + \dfrac{f^2+a}{(b-f)(c-f)(d-f)(g-f)}e^{-ft}$ $+ \dfrac{g^2+a}{(b-g)(c-g)(d-g)(f-g)}e^{-gt}$
122	$\dfrac{p^2+a}{(p+b)(p+c)(p+d)^2}$	$\dfrac{b^2+a}{(c-b)(d-b)^2}e^{-bt} + \dfrac{c^2+a}{(b-c)(d-c)^2}e^{-ct} + \dfrac{d^2+a}{(b-d)(c-d)}te^{-dt}$ $+ \dfrac{a(2d-b-c)+d[d(b+c)-2bc]}{(b-d)^2(c-d)^2}e^{-dt}$
123	$\dfrac{p^2+a}{p(p+b)(p+c)(p+d)^2}$	$\dfrac{a}{bcd^2} - \dfrac{d^2+a}{d(c-d)(b-d)}te^{-dt} + \dfrac{2}{(c-d)(b-d)}e^{-dt} + \dfrac{b^2+a}{b(d-b)^2(c-b)}e^{-bt}$ $- \dfrac{[(b-d)(c-d)-d(b-d)-d(c-d)](d^2+a)}{d^2(b-d)^2(c-a)}e^{-dt}$ $- \dfrac{c^2+a}{c(b-c)(d-c)^2}e^{-ct}$

TABLE 2.1–2 (cont.)

No.	$F(p)$	$f(t)$
124	$\dfrac{p^2+a}{p(p+b)^2}$	$\dfrac{a}{b^2} - \dfrac{(a+b^2)}{b}te^{-bt} + \dfrac{(b^2-a)}{b^2}e^{-bt}$
125	$\dfrac{p^2+a}{p^2(p+b)^2}$	$\dfrac{a}{b^2}t - \dfrac{2a}{b^3} + \dfrac{1}{b^2}\left[(b^2+a)t + \dfrac{2a}{b}\right]e^{-bt}$
126	$\dfrac{p^2+a}{(p+b)(p+c)^2}$	$\dfrac{b^2+a}{(c-b)}e^{-bt} + \left[\dfrac{c^2+a}{(b-c)}t + \dfrac{c^2-2bc-a}{(b-c)^2}\right]e^{-ct}$
127	$\dfrac{p^2+a}{p(p+b)(p+c)^2}$	$\dfrac{a}{bc^2} - \dfrac{(b^2+a)}{b(c-b)^2}e^{-bt} + \left[\dfrac{c^2+a}{c(c-b)}t + \dfrac{bc^2+(2c-b)a}{c^2(c-b)^2}\right]e^{-ct}$
128	$\dfrac{p^2+a}{p^2(p+b)(p+c)^2}$	$\dfrac{a}{bc^2}t - \dfrac{a(c+2b)}{c^3b^2} + \dfrac{b^2+a}{b^2(c-b)^2}e^{-bt}$ $+ \left[\dfrac{c^2+a}{c^2(b-c)}t - \dfrac{2}{c(b-c)} + \dfrac{(2b-3c)(c^2+a)}{c^3(b-c)^2}\right]e^{-ct}$
129	$\dfrac{p^2+a}{p(p+b)^2(p+c)^2}$	$\left[\dfrac{b^2+a}{(c-b)^2}t - \dfrac{2(bc+a)}{(c-b)^3}\right]e^{-bt} + \left[\dfrac{c^2+a}{(c-b)^2}t + \dfrac{2(bc+a)}{(c-b)^3}\right]e^{-ct}$
130	$\dfrac{p^2+a}{p(p+b)^2(p+c)^2}$	$\dfrac{a}{b^2c^2} - \left[\dfrac{b^2+a}{b(c-b)^2}t - \dfrac{2}{(c-b)^2} + \dfrac{(b^2+a)(c-3b)}{b^2(c-b)^3}\right]e^{-bt}$ $- \left[\dfrac{c^2+a}{c(b-c)^2}t - \dfrac{2}{(b-c)^2} + \dfrac{(c^2+a)(b-3c)}{c^2(b-c)^3}\right]e^{-ct}$
131	$\dfrac{p^2+a}{p^2(p+b)^2(p+c)^2}$	$\dfrac{a}{b^2c^2}t - \dfrac{2a(b+c)}{b^3c^3} + \left[\dfrac{(b^2+a)}{b^2(c-b)^2}t + \dfrac{2a(c-b)-2b(a+b^2)}{b^3(c-b)^3}\right]e^{-bt}$ $+ \left[\dfrac{(c^2+a)}{c^2(c-b)^2}t + \dfrac{2a(b-c)-2c(a+c^2)}{c^3(b-c)^3}\right]e^{-ct}$

TABLE 2.1–2 (cont.)

No.	$F(p)$	$f(t)$
132	$\dfrac{p^2+a}{(p+d)(p+b)^2(p+c)^2}$	$\left[\dfrac{b^2+a}{(c-b)^2(d-b)}t - \dfrac{2b}{(c-b)^2(d-b)} - \dfrac{(b^2+a)[(c-b)+2(d-b)]}{(c-b)^3(d-b)^2}\right]e^{-bt}$ $+\left[\dfrac{c^2+a}{(b-c)^2(d-c)}t - \dfrac{2c}{(b-c)^2(d-c)} - \dfrac{(c^2+a)[(b-c)+2(d-c)]}{(b-c)^3(d-c)^2}\right]e^{-ct}$ $+\dfrac{d^2+a}{(b-d)^2(c-d)^2}e^{-dt}$
133	$\dfrac{p^2+a}{p^2(p+b)^3}$	$\left[\dfrac{b^2+a}{2b^2}t^2 + \dfrac{2a}{b^3}t + \dfrac{3a}{b^4}\right]e^{-bt} + \dfrac{a}{b^3}t + \dfrac{3a}{b^4}$
134	$\dfrac{p^2+a}{p(p^2+\omega^2)}$	$\dfrac{a}{\omega^2} - \dfrac{(a-\omega^2)}{\omega^2}\cos\omega t$
135	$\dfrac{p^2+2a^2}{p(p^2+4a^2)}$	$\cos^2 at$
136	$\dfrac{p^2+a}{p^2(p^2+\omega^2)}$	$\dfrac{a}{\omega^2}t - \dfrac{(a-\omega^2)}{\omega^3}\sin\omega t$

The table of Laplace transforms presented on pages 77–91 is extensive, but by no means complete. For more comprehensive tables of Laplace transforms, see the following texts:

1. A. Erdelyi, W. Magnus, F. Oberhettinger, and F. Tricomi, *Tables of Integral Transform,* vols. 1 and 2 (New York: McGraw-Hill, 1954).
2. F. Oberhettinger, *Tabellen zur Fourier Transformation* (Berlin: Springer-Verlag, 1957).
3. P. A. McCollum and B. F. Brown, *Laplace Transform Tables and Theorems* (New York: Holt, Rinehart and Winston, 1965).

APPENDIX 2.2 NUMERICAL ALGORITHMS FOR DIFFERENTIAL EQUATIONS

This appendix contains a sample FORTRAN computer program which simulates the rate equations for a set of chemical reactions. One can choose between the Euler, modified Euler, or fourth-order Runge-Kutta methods by specifying the appropriate subroutine option. The differential equation specifying the kinetic system is incorporated in the external subroutine DERY. The name of the DERY must be declared in the calling program by an EXTERNAL statement. The subroutine EXACT provides the solution for the chemical reaction $A_1 \rightleftarrows A_2$ discussed in section 2.1.2; for other solutions, a different subroutine must be provided.

```
C       Runge-Kutta program
C       ******************************************************************
C       This program contains the Euler method,
C       the modified Euler method (or second-order Runge-Kutta),
C       the fourth-order Runge-Kutta, and the exact solution
C       for a system of any number of differential equations.
C       The step size and the number of outputs can be changed.
C       ******************************************************************
C       SUBROUTINE LIST:
C       SUB-1 INICON             INITIAL CONDITION
C       SUB-2 DERY               DEFINE THE DIFFERENTIAL EQUATIONS
C       SUB-3 EXACT              EXACT SOLUTIONS
C       SUB-4 FOURRK             THE FOURTH-ORDER RUNGE-KUTTA METHOD
C       SUB-5 ONERK              EULER METHOD
C       SUB-6 TWORK              MODIFIED EULER OR SECOND-ORDER
C                                RUNGE-KUTTA METHOD
C       SUB-7 OUTPUT             WRITE OUTPUT FILE TO DISK
C
C       ******************************************************************
C       Execution Begins
C       Main Program
C
```

```
        PARAMETER(M=20)
        IMPLICIT DOUBLE PRECISION(A-H,O-Z)
        CHARACTER*30 NDATA
        DIMENSION Y(M),DY(M),B(M),C(M),EA(M)
        DIMENSION YF(M),DYF(M),YT(M),DYT(M),BT(M),CT(M)
C
        EXTERNAL DERY
C
C       Create an output file
100     PRINT *,'The name of output file is ?'
        READ(*,80) NDATA
80      FORMAT(A30)
        OPEN(2,FILE=NDATA)
C
C       NEQS is the number of equations
        NEQS=2
C
C       Add the first equation, which is the independent variable
        N=NEQS+1
C
C       Initial conditions
        CALL INICON(N,Y)
C       where Y corresponds to the fourth order method
        CALL INICON(N,YF)
        where YF corresponds to the Euler method
        CALL INICON(N,YT)
C       where YT corresponds to the second order method
C
C       H is the step size
        PRINT *,'STEP SIZE H=?'
        READ *,H
C       H2 and H3 also are the same step size.
        H2=H
        H3=H
C
C       Control the number of outputs
        PRINT *,'BEGINNING,FINAL,AND STEP FOR OUTPUT(?,?,?)'
        READ *,XMI,XMAX,XSTEP
C
C       Select the output for concentration [A] or [B].
        PRINT *,'[A]=1,[B]=2, ?'
        READ *,NN
        NN=NN+1
C
C       Calculation
        CALL FOURRK(N,H,0,Y,DY,B,C)
        CALL EXACT(N,Y(1),EA)
        CALL ONERK(N,H2,0,YF,DYF)
        CALL TWORK(N,H3,0,YT,DYT,BT,CT)
```

```
            CALL OUTPUT(N,Y,EA,YF,YT,NN)
            XMI=XMI+XSTEP
    50      CALL FOURRK(N,H,1,Y,DY,B,C)
            CALL EXACT(N,Y(1),EA)
            CALL ONERK(N,H2,1,YF,DYF)
            CALL TWORK(N,H3,1,YT,DYT,BT,CT)
            XY1=Y(1)
            IF (Y(1) .GE. 10.*H-H) XY1=Y(1)+H
            IF (XY1 .LT. XMI) GOTO 40
            XMI=XMI+XSTEP
            CALL OUTPUT(N,Y,EA,YF,YT,NN)
    40      IF (Y(1).LT.XMAX) GOTO 50
    C
    C       Ask for stopping or recalculation.
            PRINT *,'DO AGAIN?(Y=1)'
            READ *,NS
            IF (NS.EQ.1) GOTO 100
            STOP
            END
    C
    C       ****************************************************************
            SUBROUTINE INICON(N,YO)
    C       ****************************************************************
    C       (SUB1-INICON)
    C       Initial condition
    C
            IMPLICIT DOUBLE PRECISION(A-H,O-Z)
            DIMENSION YO(N)
    C
            YO(1)=0.D0
            YO(2)=1000.D0
            YO(3)=0.D0
            RETURN
            END
    C
    C       ****************************************************************
            SUBROUTINE DERY(N,FY,FDY)
    C       ****************************************************************
    C       (SUB-2 DERY)
    C       Define differential equations.
    C
            IMPLICIT DOUBLE PRECISION(A-H,O-Z)
            DIMENSION FY(N),FDY(N)
    C
    C       C1 and C2 are the rate constants.
    C
            C1=1000.D0
            C2=1.D0
    C
```

```
         FDY(1)=1.D0
         FDY(2)=-C1*FY(2)+C2*FY(3)
         FDY(3)=C1*FY(2)-C2*FY(3)
         RETURN
         END
C
C        ******************************************************************
         SUBROUTINE EXACT(N,T,EA)
C        ******************************************************************
C        (SUB-3 EXACT)
C        Exact Solutions
C
         IMPLICIT DOUBLE PRECISION(A-H,O-Z)
         DIMENSION EA(N)
C
C        EAO is the initial concentration of A.
C        C1 and C2 are the rate constants.
C
         EA0=1000.D0
         C1=1000.D0
         C2=1.D0
         EA(1)=(EA0/(C1+C2))*(C2+C1*EXP(-(C1+C2)*T))
         EA(2)=(EA0*C1/(C1+C2))*(1.D0-EXP(-(C1+C2)*T))
         RETURN
         END
C
C        ******************************************************************
         SUBROUTINE FOURRK(N,H,L,Y,DY,B,C)
C        ******************************************************************
C        (SUB-4 FOURRK)
C        Subroutine for the fourth-order Runge-Kutta method
C
         IMPLICIT DOUBLE PRECISION(A-H,O-Z)
         DIMENSION A(4),Y(N),DY(N),B(N),C(N)
C
         A(1)=H/2.0
         A(2)=A(1)
         A(3)=H
         A(4)=H
         IF(L) 2,2,4
2        DO 3 I=1,N
3        B(I)=Y(I)
         CALL DERY(N,Y,DY)
         RETURN
4        DO 5 K=1,3
         DO 6 I=1,N
         C(I)=B(I)+A(K)*DY(I)
6        Y(I)=Y(I)+A(K+1)*DY(I)/3.0
5        CALL DERY(N,C,DY)
```

```
          DO 7 I=1,N
7         Y(I)=Y(I)+A(1)*DY(I)/3.0
          GOTO 2
          END
C
C         ***************************************************************
          SUBROUTINE ONERK(N,H2,L,YF,DYF)
C         ***************************************************************
C         (SUB-5 ONERK)
C         Subroutine for the Euler Method
C
          IMPLICIT DOUBLE PRECISION(A-H,O-Z)
          DIMENSION YF(N),DYF(N)
C
          IF(L) 11,11,13
11        CALL DERY(N,YF,DYF)
          YF(1)=YF(1)+H2
          RETURN
13        DO 15 I=2,N
15        YF(I)=YF(I)+H2*DYF(I)
          GOTO 11
          END
C
C         ***************************************************************
          SUBROUTINE TWORK(N,H3,L,YT,DYT,BT,CT)
C         ***************************************************************
C         (SUB-6 TWORK)
C         Subroutine for the modified Euler or second-order Runge-Kutta method.
          IMPLICIT DOUBLE PRECISION(A-H,O-Z)
          DIMENSION YT(N),DYT(N),BT(N),CT(N)
C
          IF(L) 21,21,23
21        DO 22 J=1, N
22        BT(J)=YT(J)
          CALL DERY(N,YT,DYT)
          YT(1)=YT(1)+H3
          RETURN
23        DO 25 I=1, N
25        BT(I)=BT(I)+H3*DYT(I)
          CALL DERY(N,BT,CT)
          DO 27 I=2, N
27        YT(I)=YT(I)+0.5*H3*(CT(I)+DYT(I))
          GOTO 21
          END
C
C         ***************************************************************
          SUBROUTINE OUTPUT(N,Y,EA,YF,YT,NN)
C         ***************************************************************
C         (SUB-7 OUTPUT)
C         Write output file to disk
```

```
C
          IMPLICIT DOUBLE PRECISION(A-H,O-Z)
          DIMENSION Y(N),EA(N),YF(N),YT(N)
C
          WRITE(*,30) Y(1),YF(NN),YT(NN),Y(NN),EA(NN-1)
          WRITE(2,30) Y(1),YF(NN),YT(NN),Y(NN),EA(NN-1)
30        FORMAT(' ',2X,5(F13.5))
          RETURN
          END
```

APPENDIX 2.3 STOCHASTIC NUMERICAL SIMULATION OF CHEMICAL REACTIONS

Often the master equation turns out to be intractable to solve analytically, and even numerically, for some cases, if attempted by finite-difference methods. However, Gillespie[38–40] has developed a numerical method which simulates the stochastic evolution in time of a given chemical system. To illustrate the method let us numerically analyze the problem previously considered in section 2.7.2 within the framework of the stochastic formulation by using Gillespie's stochastic simulation algorithm. The reaction is

$$A \underset{k_2}{\overset{k_1}{\rightleftharpoons}} B$$

We define a set of reaction events by R_i, so when A $\rightarrow$ B we say that an R_1 reaction has occurred, and when B $\rightarrow$ A we say that an R_2 reaction has occurred. We can write the set of reaction events as

$$R_i = \{R_1, R_2\}$$

We also define reaction rate coefficients for each reaction as

$$k_i = \{k_1, k_2\}$$

where k_1 is the forward rate constant and k_2 is the backward rate constant. Now, as mentioned earlier, k_1 is defined so that $k_1 \Delta t$ gives the probability that any given A molecule will turn into a B molecule in the next infinitesimal time interval Δt. With this definition, one can in principle calculate k_1 from a knowledge of the mechanism responsible for the reaction A $\rightarrow$ B. Such a probabilistic formulation reflects either the random nature of collisions responsible for bimolecular reactions, or the random decay of molecules in unimolecular processes.

We consider a chemical system containing N molecular species denoted by the concentration variables $C_i = \{C_1, C_2\}$, where C_1 is the number of A molecules and C_2 is the number of B molecules. These species can interact through M chemical channels $R_i(i = 1, \ldots, M)$. We note that the number of chemical components need not be equal to the number of reaction channels. This particular problem can be regarded as a single-variable problem since $C_2 = N - C_1$.

The Gillespie algorithm is based on a reaction probability density function, $P(\tau, i)$, defined by

$P(\tau,i)d\tau$ = Probability at time t that the next reaction will occur in the differential time interval $(t + \tau, t + \tau + d\tau)$ and will be an R_i reaction

More formally,

$$P(\tau,i) = k_i C_i \exp\{-\sum_{i=1}^{2} k_i C_i \tau\} \tag{A3-1}$$

where C_i is the number of molecules for species i and k_i is the set of rate constants. If we let

$$a_i = k_i C_i \quad (i = 1,2)$$

and

$$a_t = \sum_{i=1}^{2} a_i = \sum_{i=1}^{2} k_i C_i$$

then equation (A3–1) becomes

$$P(\tau,i) = k_i C_i \exp[-(k_1 C_1 + k_2 C_2)\tau] \quad (i = 1,2) \tag{A3-2}$$

The probability density function, $P(\tau,i)$, is a function of two variables and can be written as the product of two probability density functions of one variable. In other words, $P(\tau,i)$ can be expressed as the product of the probability of a reaction occurring in time interval $d\tau$ and the probability of that reaction being an i reaction:

$$P(\tau,i) = P(\tau) \cdot P(i) \tag{A3-3}$$

Here, $P(\tau)d\tau$ represents the probability that the next reaction will occur in the time interval $(t + \tau, t + \tau + d\tau)$ regardless of what the reaction is, and $P(i)$ is the probability that the next reaction will be an R_i reaction. Now by the addition theorem for probability, $P(\tau)$ is obtained from

$$P(\tau) = \sum_{i=1}^{2} P(\tau,i) \tag{A3-4}$$

and $P(i)$ is obtained by substitution of the above expression into eq. (A3–3),

$$P(i) = P(\tau,i)/\sum_{i=1}^{2} P(\tau,i) \tag{A3-5}$$

Now, if we substitute Eq. (A3–2) into Eq. (A3–4)

$$P(\tau) = a_t \exp(-a_t \tau) \quad \text{where } (0 \le \tau < \infty); \tag{A3-6}$$

and if we substitute Eq. (A3–2) into Eq. (A3–5) we obtain

$$P(i) = a_i/a_t \quad (i = 1,2) \tag{A3-7}$$

Notice that $P(\tau)$ and $P(i)$ are properly normalized in that they satisfy

$$\int_0^{\infty} P(\tau)d\tau = 1$$

and

$$\sum_{i=1}^{2} P(i) = 1 \tag{A3-8}$$

The problem now is to generate a random number τ according to the probability function $P(\tau)$ in Eq. (A3-6), and a random number i according to the probability function $P(i)$ in Eq. (A3-7). The function $P(\tau)$ is known as *an exponential random variable with parameter a_t,* and it can be shown that one way to generate a sample value of such a variable is to first generate a random number r_1 on the uniform interval between 0 and 1, and then take

$$\tau = (1/a_t) \ln(1/r_1) \tag{A3-9}$$

One can generate a sample value for the random number i according to eq. (A3-7) by first generating another random number r_2 on the uniform interval between 0 and 1, and then taking

$$i = \begin{cases} 1, & \text{if } r_2 < a_1/a_t \\ 2, & \text{if } r_2 > a_1/a_t \end{cases} \tag{A3-10}$$

The simulation time variable is then incremented by τ (i.e., we advance to the next occurrence of a reaction), and the concentration variables are changed as follows:

$$C_1 \longrightarrow C_1 - 1, C_2 \longrightarrow C_2 + 1, \text{ if } i = 1$$

$$C_1 \longrightarrow C_1 + 1, C_2 \longrightarrow C_2 - 1, \text{ if } i = 2$$

Summary of the Stochastic Algorithm Procedure

In general, Gillespie's procedure uses Monte Carlo techniques to simulate the stochastic process described by $P(\tau,i)$. The simulation algorithm may be outlined as follows:

1. Set the time variable $t = 0$. Then specify the initial values A(0) and B(0), the rate constants k_1 and k_2, and the time step size Δt and the "stopping time" t_{stop}.
2. Calculate the m equations $a_i = k_i C_i$ which collectively determine the reaction probability density function $P(\tau,i)$.
3. Using suitable Monte Carlo techniques generate one random pair (τ,i) according to the joint probability density function $P(\tau,i)$.
4. Using the numbers τ and i generated in step (c), advance t by τ, and change the C_i value of those species involved in reaction R_i to reflect the occurrence of one R_i whose reactant C_i values have just been changed.
5. If t has just been advanced through Δt, print out the current molecular population values. If $t > t_{stop}$, or if no more reactants remain (all $C_i = 0$), terminate the calculation; otherwise return to step (b).

A flow chart of this simulation algorithm is given in Fig. 2A-1, followed by a simple BASIC program that solves the reversible first-order reaction problem stochastically. A version of this program is also available on the distributed software package.

A plot of typical results using this simulation has been given in Fig. 2-6.

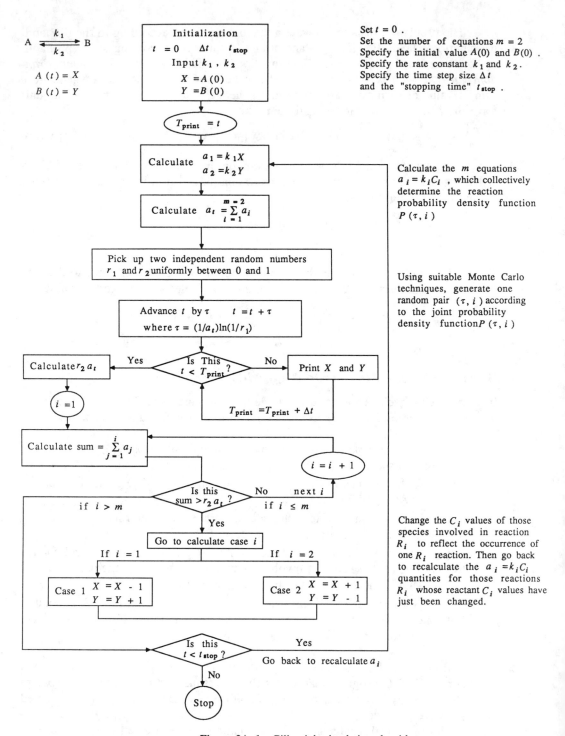

Figure 2A–1 Gillespie's simulation algorithm.

```
! MONTE CARLO SIMULATION
! PROGRAM TO SIMULATE PARALLEL FIRST-ORDER REACTIONS
!**********************************************************
!       C(MU):    REACTION PARAMETERS c1,c2 ......
!       T:        INITIAL TIME t1
!       T2:       STOPPING TIME t2
!       TINT:     TIME INTERVAL!
        X,Y:      INITIAL NUMBERS OF MOLECULES AT THE INITIAL TIME t1
!**********************************************************
clearwindow
set output toscreen
RANDOMIZE
        DIM C(4),A(4)
        PRINT "Give a name of the output file"
        INPUT NDATA$
!       NDATA$="MKOUT"
        OPEN #2:NDATA$,outin
!contant
5       PRINT "C(1)=? C(2)=? C(3)=? C(4)=? Xo=?"
        INPUT C1,C2,C3,C4,Xo
        C(1)=C1
        C(2)=C2
        C(3)=C3
        C(4)=C4

        Y=0
        Z=0
        X=Xo
        EPS=1.E-5
        PRINT C(1),C(2),C(3),C(4),X,Y,Z
        PRINT #2: C(1),C(2),C(3),C(4),X,Y,Z

        PRINT "INITIAL To? STOPPING T2? & INTERVAL TINT?"
        INPUT T,T2,TINT
        PRINT "T=";T," T2=";T2," TINT=";TINT
        PRINT "TO RUN? (Y=1,N=0)"
        INPUT NST
        IF NST = 0 THEN GOTO 50

        TPRT=T
10      A(1)=C(1)*X
        A(2)=C(2)*X
        A(3)=C(3)*Y
        A(4)=C(4)*Z
        AO=A(1)+A(2)+A(3)+A(4)
        IF AO < EPS THEN GOTO 50
        R1=RND(1)
        R2=RND(1)
21      T=T+LOG(1./R1)/AO
22      IF T < TPRT THEN GOTO 25
```

```
!          Call LOGXYZ
!          Call Exact(TPRT)
           F$="#######.##"
           PRINT TPRT;FORMAT$(F$;X,Y,Z)
           PRINT #2: TPRT;FORMAT$(F$;X,Y,Z)
           TPRT=TPRT+TINT
           GOTO 22
25         R2AO=R2*AO
           SUM=0.
           FOR NU=1 TO 4
           MU=NU
           SUM=SUM+A(NU)
           IF SUM > R2AO  THEN EXIT
           NEXT NU
           SELECT CASE MU
           CASE=1
                   X=X-1
                   Y=Y+1
           CASE=2
                   X=X-1
                   Z=Z+1
           CASE=3
                   Y=Y-1
                   Z=Z+1
           CASE=4
                   Z=Z-1
                   Y=Y+1

           END SELECT
40         IF T <= T2 THEN GOTO 10
50         PRINT "STOP? (Y=1,N=0)"
           INPUT NSTOP
           IF NSTOP=0  THEN GOTO 5
           STOP
           END


! ***********************************************
SUB LOGXYZ
! ***********************************************
           LX=Log(X)
           LY=Log(Y)
           LZ=Log(Z)
End sub


! ***********************************************
SUB EXACT(tim)
! ***********************************************
           YA=Xo*(C2+C1*exp(-(C1+C2)*tim))/(C1+C2)
           YB=Xo-YA
End sub
```

PROBLEMS

2.1 The reaction $A + B \rightarrow C$ takes place in two steps by the mechanism $2A \rightleftarrows D$ followed by $B + D \xrightarrow{k_2} A + C$. The first step comes to a rapid equilibrium (constant K_1). Derive an expression for the rate of formation of C in terms of K_1, k_2, [A], and [B].

2.2 Set up the rate expressions for the following mechanisms:

$$A \xrightarrow{k_1} B \qquad B \xrightarrow{k_{-1}} A$$

$$B + C \xrightarrow{k_3} D$$

If the concentration of B is small compared with the concentrations of A, C, and D, the steady-state approximation may be used to derive the rate law. Show that this reaction may follow first-order kinetics at high pressures and second-order kinetics at low pressures.

2.3 Consider the reaction

$$A_1 \xrightarrow{k_1} A_2 \xrightarrow{k_2} A_3 \cdots A_{n-2} \xrightarrow{k_{n-2}} A_{n-1} \xrightarrow{k_{n-1}} A_n$$

at $t = 0$, $[A_1] = [A_1]_0$, $[A_2] = [A_3] = \cdots = [A_{n-2}] = [A_{n-1}] = [A_n] = 0$

Derive an expression that will describe the concentration of the intermediate existing at any time t during the reaction. [E. Abel, *Z. Phys. Chem. A56*, 558 (1906)].

2.4 Consider the reaction mechanism

$$A + B \underset{k_{-1}}{\overset{k_1}{\rightleftarrows}} X + B$$

$$X \xrightarrow{k_2} C + D$$

(a) Write chemical rate equations for [A] and [X].

(b) Employing the steady-state approximation, show that an effective rate equation for [A] is

$$\frac{d[A]}{dt} = -k_{\text{eff}}[A][B].$$

(c) Give an expression for k_{eff} in terms of k_1, k_{-1}, k_2, and [B].

2.5 For each of the following reaction mechanisms, give the equilibrium concentration, if it exists, of all the chemical species for the initial concentrations stated.

(a)

$$A + B \underset{k_{-1}}{\overset{k_1}{\rightleftarrows}} X \xrightarrow{k_2} C + D$$

Initial concentrations

$$[A]_0 = 2, [B]_0 = 1, [X]_0 = [C]_0 = [D]_0 = 0.$$

(b) All rate coefficients equal; initial concentrations

$$[A]_0 = 1, [B]_0 = [C]_0 = 0.$$

(c) $X + X \xrightarrow{k_1} 3X$ Assume $k_1 = k_2$; initial concentrations $[X_0] = \frac{1}{2}$, $[C_0] = 0$.

$$X \xrightarrow{k_2} C$$

2.6 Consider a reaction with two intermediates I_1 and I_2 which both decompose to the same product P by competing parallel steps [S.I. Miller, *J. Chem. Education 62*, 490 (1985)]:

$$A + M \underset{k_{-1}}{\overset{k_1}{\rightleftharpoons}} I_1$$

$$I_1 \underset{k_{-3}}{\overset{k_3}{\rightleftharpoons}} I_2$$

$\searrow$ P $\nearrow$

(a) Find an expression for dP/dt by applying the steady state approximation with $[I_1]_{ss} \simeq [I_2]_{ss} \simeq 0$.

(b) Find an expression for dP/dt by applying the steady state approximation with $[A]_t = [A]_0 - [I_1] - [I_2] - [P]$.

(c) How do the expressions found in **(a)** and **(b)** differ, and what conclusions can be drawn from this difference?

2.7 Consider the first-order cyclic reversible reaction

A

B $\rightleftharpoons$ C

Assume that initially, at $t = 0$ $[A] = A_0$, $[B] = [C] = 0$, and that the amounts of A, B, and C which have reacted at a later time satisfy the following relation

$$[A]_0 = [A] + [B] + [C].$$

Using the detailed balance method, solve for $[A]$, $[B]$, and $[C]$.

2.8 (a) The reaction $2NO + O_2 \rightarrow 2NO_2$ is third order. Assuming that a small amount of NO_3 exists in rapid reversible equilibrium with NO and O_2 and that the rate-determining step is the slow bimolecular reaction $NO_3 + NO \rightarrow 2NO_2$, derive the rate equation for this mechanism.

(b) Another possible mechanism for the reaction $2NO + O_2 \rightarrow 2NO_2$ is

(1) $NO + NO \rightarrow N_2O_2$ k_1
(2) $N_2O_2 \rightarrow 2NO$ k_2
(3) $N_2O_2 + O_2 \rightarrow 2NO_2$ k_3

Apply the steady state approximation to $[N_2O_2]$ to obtain the rate law for $d(NO_2)/dt$.

    If only a very small fraction of the N_2O_2 formed in (1) goes to form products in reaction (3), while most of the N_2O_2 reverts to NO in reaction (2), and if the activation energies are $E_1 = 79.5$ kJ/mole, $E_2 = 205$ kJ/mole, and $E_3 = 84$ kJ/mole, what is the overall activation energy?

(c) How would you distinguish experimentally between the mechanisms suggested in parts **(a)** and **(b)**?

2.9 The mechanism for the decomposition of ozone into oxygen,

$$2 O_3 = 3 O_2$$

is stated to be as follows:

Step 1, $O_3 \rightarrow O_2 + O$,   rate constant, k_1

Step 2, $O + O_2 \rightarrow O_3$,   rate constant, k_2

Step 3, $O + O_3 \rightarrow 2 O_2$,   rate constant, k_3

The activation energies, in kilojoules per mole, for each of the steps are as follows:

$$\text{Step 1, } E_{act} = 103 \text{ kJ/mole}$$

$$\text{Step 2, } E_{act} = 0$$

$$\text{Step 3, } E_{act} = 21 \text{ kJ/mole}$$

(a) Obtain the differential equation for the steady state rate of decomposition of ozone, $-d[O_3]/dt$, in terms of the constants, k_1, k_2 and k_3, the concentration of ozone, $[O_3]$, and the concentration of oxygen, $[O_2]$.

(b) On the basis of the values of E_{act} given above, simplify the expression for $-d[O_3]/dt$ obtained in (a) by eliminating any terms which can become negligible. State clearly the basis for the simplification.

(c) Calculate the energy of activation for the overall reaction.

2.10 As a final variation on the theme of Problems 8 and 9, let us consider the reaction between NO and O_3. In this experiment [E. Bar-Ziv, J. Moy, and R. J. Gordon, *J. Chem. Phys.* **68**, 1013 (1978)], it is necessary to distinguish between ozone in its ground (000), excited stretching (001), and excited bending (010) vibrational states.

A CO_2 laser is used to excite ozone from the (000) to the (001) state. The following reactions then ensue in the presence of nitric oxide:

(1) V-V equilibration of ozone:

$$O_3(001) + NO \underset{k_{-1}}{\overset{k_1}{\rightleftharpoons}} O_3(010) + NO$$

(2-4) reaction of ozone in specified vibrational states with NO to form NO_2:

$$O_3(000) + NO \xrightarrow{k_2} NO_2 + O_2$$

$$O_3(001) + NO \xrightarrow{k_3} NO_2 + O_2$$

$$O_3(010) + NO \xrightarrow{k_4} NO_2 + O_2$$

(5,6) V-T relaxation of ozone:

$$O_3(001) + NO \xrightarrow{k_5} O_3(000) + NO$$

$$O_3(010) + NO \xrightarrow{k_6} O_3(000) + NO$$

The ozone (000) could be followed by its Hartley-band absorption at $\lambda = 254$ nm, and the ozone (001) by its infrared emission at $\lambda = 9 \mu m$.

(a) Write the kinetic equations, from the above (simplified) mechanism, for the time derivatives of $[O_3(000)]$, $[O_3(010)]$, $[O_3(001)]$, and $[NO_2]$. If the experiment is carried out in a large excess of NO, then pseudo-first-order kinetics may be obtained; rewrite these four expressions in pseudo-first-order, using $K_i = k_i[NO]$.

(b) Find the Laplace transform of the four pseudo-first-order rate equations. (A table of Laplace transforms is included in the Appendix to Chapter 2.) Set up this set of transformed equations in determinantal form.

(c) Find $[O_3(000)]_t$ and $[O_3(010)]_t$ in terms of pseudo-first-order rate coefficients derived from k_1-k_6 and the initial concentrations of $O_3(000)$ and $O_3(001)$ following the laser pulse. Since the experiment is carried out at 300K, and $\nu_2 \approx 700$ cm^{-1} for ozone, the initial concentration of $[O_3(010)]$ cannot be neglected.

2.11 Consider the following system of first order reversible reactions.

$$A_1 \underset{k_1'}{\overset{k_1}{\rightleftharpoons}} A_2 \underset{k_2'}{\overset{k_2}{\rightleftharpoons}} A_3$$

Assume that initially, at $t = 0$ $[A_1] = [A_1]_0$, $[A_2]_0 = [A_3]_0 = 0$, and that amounts of A_1, A_2, and A_3, which have reacted at a later time, satisfy the following relation:

$$[A_1]_0 = [A_1] + [A_2] + [A_3]$$

(a) Solve for A_1, A_2, and A_3 using the detailed balance method.
(b) Solve for A_1, A_2, and A_3 using the Laplace transform method.
(c) Show that the two methods are equivalent as $t \rightarrow \infty$.
(d) Assuming that $k_1 = k_1' = k_2 = k_2' = 1$ sec^{-1}, and $[A_1]_0 = 0.01$ M, provide a plot for A_1, A_2, and A_3 from $t = 0$ to $t = 6$ sec. Which of the two preceding methods is valid at shorter time periods? Indicate the time after the method which is invalid at shorter times becomes valid, and explain why it then becomes valid.

2.12 The free radical decomposition of methane occurs via

$$CH_4 \rightleftharpoons CH_3 + H \tag{1}$$

$$H + CH_4 \rightleftharpoons CH_3 + H_2 \tag{2}$$

$$2CH_3 \rightleftharpoons C_2H_6 \tag{3}$$

where $k_1 = 14$ s^{-1}, k_{-1} 1.2×10^{10} mole^{-1} 1 s^{-1}
$k_2 = 1.5 \times 10^9$ mole^{-1} 1 s^{-1}, $k_{-2} = 2.9 \times 10^7$ mole^{-1} 1 s^{-1}
$k_3 = 2.0 \times 10^{10}$ mole^{-1} 1 s^{-1}, $k_{-3} = 4.5 \times 10^4$ mole^{-1} 1 s^{-1}

with an initial methane concentration of $[CH_4]_0 = 1.0 \times 10^{-3}$ mole 1^{-1}
(a) Write the kinetic equations for this (simplified) mechanism for the time derivatives of the intermediates, reactant, and products.
(b) Assuming steady state concentrations for the free radicals, what are the steady state expressions for $[H]_{ss}$ and $[CH_3]_{ss}$; derive an expression for the time dependence of $[C_2H_6]$ in terms of the concentration of $[CH_4]$.
(c) Plot the time dependence of $[CH_3]$ radicals and $[C_2H_6]$ over 30 μsec.
(d) Using the Euler algorithm which follows shortly, numerically integrate the differential equations over 30 μsec using an integration step size of $\Delta t = 0.1$ μsec and 1.0 μsec. Plot the time dependence of $[CH_3]$ and $[C_2H_6]$.
(e) Compare the steady-state and numerical solutions. Over what times is it appropriate to use the steady-state assumption?
(f) What percentage increase in the concentration of C_2H_6 is observed if k_2 is increased by a factor of 5?

Euler algorithm for Problem 2–12

1. Assume an initial concentration for each chemical species. Input step size and rate constants.
2. For each species calculate its incremental concentration change starting with the initial conditions for the time interval Δt.

3. Using the incremental concentration change, calculate the concentration of each species for the next interval.
4. Repeat until the desired time is reached.

Example: For the reaction

$$A \xrightarrow{k_1} \text{products}$$

The differential equation for the decay of A is

$$d[A]/dt = -k_1[A]$$

The incremental change in the concentration of A over the time interval Δt is

$$\Delta[A] = -k_1*[A]_i*\Delta t$$

$$[A]_{i+1} = [A]_i + \Delta[A]$$

2.13 Consider the reaction system

$$A_1 + A_2 + A_3 + \cdots + A_{n-1} + A_n \underset{k_2}{\overset{k_1}{\rightleftharpoons}}$$

$$B_1 + B_2 + B_3 + \cdots + B_{n-1} + B_n$$

Let $A_1(t), A_2(t), A_3(t), \ldots, A_{n-1}(t), A_n(t)$ and $B_1(t), B_2(t), B_3(t), \ldots, B_{n-1}(t), B_n(t)$ be random variables which represent the number of molecules of species $A_1, A_2, A_3, \ldots, A_{n-1}, A_n$ and $B_1, B_2, B_3, \ldots, B_{n-1}, B_n$, respectively, present at time t. We let $a_1, a_2, a_3, \ldots, a_{n-1}, a_n$ and $b_1, b_2, b_3, \ldots, b_n$ be the respective (integer) values which these random variables can achieve. The possible states of the system at time t which could lead to the state specified by $a_1, a_2, a_3, \ldots, a_{n-1}, a_n$ and $b_1, b_2, b_3, \ldots, b_{n-1}, b_n$ at time $t + \Delta t$ involving not more than one molecular transformation in the interval Δt are

$$\{a_1 + 1, a_2 + 1, a_3 + 1, \ldots, a_{n-1} + 1,$$

$$a_n + 1, b_1 - 1, b_2 - 1, b_3 - 1, \ldots, b_{n-1} - 1, b_n - 1\},$$

$$\{a_1 - 1, a_2 - 1, a_3 - 1, \ldots, a_{n-1} - 1,$$

$$a_n - 1, b_1 + 1, b_2 + 1, b_3 + 1, \ldots, b_{n-1} + 1, b_n + 1\},$$

and

$$\{a_1, a_2, a_3, \ldots, a_{n-1}, a_n, b_1, b_2, b_3, \ldots, b_{n-1}, b_n\}$$

(a) Write an expression $P(a_1, a_2, a_3, \ldots, a_{n-1}, a_n; b_1, b_2, b_3, \ldots, b_{n-1}, b_n; t + \Delta t)$ for probability vs. time during the course of this reaction. At time t, $A_1(t) = a_1$, $A_2(t) = a_2, \ldots, A_n(t) = a_n, B_1(t) = b_1, B_2(t) = b_2, \ldots, B_n(t) = b_n$.

(b) Since the system is conserved, the random variables $a_1, a_2, a_3, \ldots, a_{n-1}, a_n$ and $b_1, b_2, b_3, \ldots, b_{n-1}, b_n$ can be related through the initial concentrations which are taken to be $A_1(0) = \alpha_1, A_1(0) = \alpha_2, \ldots, A_n(0) = \alpha_n, B_1(0) = \beta_1, B_2(0) = \beta_2, \ldots, B_n(0) = \beta_n$. Clearly $\alpha_1 - a_1 = \alpha_2 - a_2 = \cdots = \alpha_n - a_n = b_1 - \beta_1 = b_2 - \beta_2 = \cdots = b_n - \beta_n$ Write an expression for probability vs. time during the course of the reaction for a single random variable $A_1(t)$.

14. In the reaction

$$X + Y \xrightarrow{k_1} 2Y$$

$$2Y \xrightarrow{k_2} Z$$

there are two steady-state solutions [M. Malek-Mansour and G. Nicolis, *J. Stat. Phys.* *13,* 197 (1975)].

(a) Using the steady-state approximation, find these two solutions.
(b) Using the stochastic method, set up the master equation for this problem.
(c) Simulate the reaction using the numerical stochastic method described in Appendix 2–3. Can you obtain the two steady-state solutions?
(d) What does the simulation imply about the physical significance of one of the steady-state solutions which you found?

CHAPTER 3
Kinetic Measurements

3.1 INTRODUCTION

Experiments in macroscopic chemical kinetics have two broadly defined goals. The first aim is to characterize and confirm details of the mechanism involved in complex chemical processes. In these studies, the experimenter attempts to measure as many properties as possible of the reaction, such as the dependence of the reaction rate on concentration and temperature, the identity of all species present in the system and their concentrations and time dependences, or the extent of progress of the reaction. Such measurements provide the criteria for testing the proposed model of the reaction mechanism. With the experimental data obtained, the validity of the proposed mechanism can be tested by transforming the postulated mechanism into a mathematical model from which predictions can be made to see whether the model is reasonable and consistent with prior and new experimental data.

The next stage involves fitting the proposed mechanism to the experimental data. Once new rate parameters are obtained, the adequacy of the mechanism can be tested by investigating the deviations of the predictions from the experimental results. One procedure is to measure parameters such as rate constants at a variety of initial concentrations. If the model is correct, the values of the rate constants will be invariant under such changes. If the model is inadequate, a new mechanism must be proposed which is consistent with new and existing experimental data. Changes are usually made in species that are involved in the mechanisms, for which evidence is provided by the experimental detection and identification of such species in the reacting system. Then the cycle is repeated until the model explains the set of experimental data within the limits of experimental error. (But, as noted in chapter 2, sec. 2.3.2, no mechanism can be proven to be correct; it can only by disproved.)

The basic strategy for carrying out the preceding process is shown in Figure 3–1 and is the essence of applied macroscopic chemical kinetics.

A second goal of experiments in macroscopic chemical kinetics is to study the kinetics of individual elementary reactions. These studies provide rate data for ele-

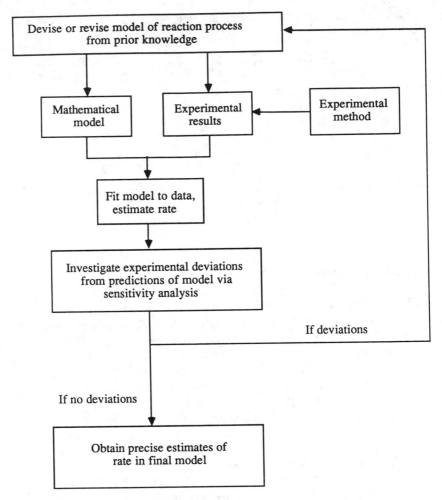

Figure 3–1. The basic approach to applied macroscopic chemical kinetics.

mentary steps that are needed in assessing complex reaction mechanisms as well as examine the relation between the measured macroscopic rate constant and details of the molecular dynamics (i.e., the microscopic kinetics), allowing us to test various microscopic rate theories. Thus, the experimental design is one that allows a single reaction step to be studied in complex reaction mechanisms. Such studies are now possible largely because of the advances that have been made in direct time-resolved detection of short-lived intermediates by spectroscopic means.

It is our aim in this chapter to introduce the basic concepts of kinetic measurement techniques and show how information on rate constants may be determined from such measurements. In the last portion of the chapter, we consider how a postulated mechanism is fitted to a set of experimental data. In discussing the evaluation of rate constants for elementary steps by fitting observations on complex systems, one cannot overemphasize the importance of sensitivity analysis, because knowing the experimental error allows a realistic appraisal of the data. This information is needed not only by experimentalists to assess the sensitivities of calculations from models, e.g., in atmospheric chemistry, combustion, or pollution studies, but also by theorists who need to know the reliability of results from such measurements in guiding and assessing various microscopic theories of reaction kinetics. Many of the techniques discussed in this chapter can be used for either gas phase or solution kinetics studies. However, the major emphasis in the chapter is on gas phase kinetics. Techniques specific to solution kinetics are described in chapter 4.

3.2 TECHNIQUES FOR KINETIC MEASUREMENTS

3.2.1 General Features

Experimental methods in chemical kinetics are characterized by three features: a method of initiation, a suitable medium in which a reaction may take place, and a method of detection. Let us first consider some methods which are used to initiate chemical reactions.

3.2.1.1 Initiation methods. Before a reaction can take place, the reactant molecules must be activated in some way to a minimum total energy necessary to induce the reaction. Various methods can be used to initiate reactions, including thermal activation (heating), electrical discharge, chemical activation, and photoactivation. Perhaps the most conventional means of starting a reaction is by thermal activation. Here, a reactant molecule is activated by collision with another molecule or with the walls of a reaction vessel heated by a furnace. In the collision, energy is transferred to the reactant molecule and distributed internally throughout it. (We shall go into further details of this process in subsequent chapters.) If the energy content of the activated molecule is above its minimum energy for reaction, reaction may occur. Typically the system is characterized by a single temperature, which for most experiments ranges from 300 to 3,000 K.

Electrical discharges are also effective in initiating reactions. The two types most often used in kinetics studies are microwave ($\approx$2450 MHz) and radio-frequency ($\approx$20 MHz) discharges. In both cases, free electrons which are present in the gaseous reaction mixture are accelerated to high kinetic energies by the rF or microwave fields. These electrons in turn produce additional charged species by electron-impact ionization, as well as by molecular fragmentation and electronic excitation. Because the light electrons are inefficient at transferring momentum to heavier molecular particles, the kinetic temperature of the gas remains relatively cool. Unlike thermal excitation, which tends to break only the weakest bond in the molecule, electrical discharges cause extensive fragmentation and produce atoms, radicals, and ions. Consequently, such discharges are not clean sources of free radicals but are an excellent source of atoms.

Photoactivation sources usually initiate a reaction by photodissociating the reactant molecule following its absorption of light. The typical wavelengths used to initiate photochemical reaction range from 120 nm to 700 nm. If the light is of low intensity, then usually one photon of light will be absorbed by the molecule, in accordance with Einstein's law of photochemical equivalence. If the light is ultraviolet, say, 300 nm in wavelength, a photon will have an energy of about 400 kJ/mole. [This figure is obtained by converting the wavelength λ(nm) to energy units E(kJ/mol) by the conversion expression $E = 1.1962 \times 10^5/\lambda$ kJ/mol.] With this amount of energy, the molecule may be promoted to electronically excited states and dissociate. One conventional light source that can provide these energies is the mercury lamp, which is basically a high-intensity lamp with output between 3 and 1,000 watts covering a broad range of wavelengths that extend from the ultraviolet to the visible region. In order to do accurate and quantitative kinetic studies using photoactivation, it is necessary to use monochromatic light. Unfortunately, atomic lamp sources frequently have a broad wavelength distribution, and the output is not tunable over the range of their operation. Some degree of selectivity in wavelength can be obtained by using a monochromator, but this reduces the available intensity. These limitations may be overcome through the use of lasers, which have the unique advantage of being monochromatic, tunable, and highly intense. The energy available in laser pulses, with pulse widths that can range from femtoseconds to milliseconds, makes the laser an extremely useful initiation source, since species can be produced for kinetic study over a broad range of time scales. This expands the scope of kinetic problems that can be studied in both solution and state-to-state chemistry. Furthermore, with high-intensity laser sources it is also possible to induce dissociation by absorption of multiple photons rather than just one photon. One difficulty with inducing dissociation by absorption of a single ultraviolet photon is that the process can leave the resulting photofragments highly excited. Dissociation by multiple infrared photons can overcome this problem. Details of the principles of the process are discussed in chapter 14, sec. 14.4.1. Lasers have become major tools for chemical kinetic studies and are sure to increase in importance as their frequency, pulse length, and power characteristics continue to improve.

3.2.1.2 Kinetic systems (measurement media)

3.2.1.2.1 Static Systems. Chemical reactions are generally observed taking place in some container—a gas bulb, test tube, living cell, etc.—which holds very large numbers of each of the reactive species. Such a closed, constant-volume container, which serves as the measurement medium, constitutes a *static system*. For gas phase reaction, the reaction vessel, often cylindrically or spherically shaped and made of glass, is situated in a temperature-controlled furnace and connected to a vacuum system. After evacuation, the vessel is filled to some measured pressure with an appropriate reactant or mixture of reactants and possibly an inert buffer gas, and is then closed off from the system. If a mixture of reagents is used, the reaction vessel in the static system must be designed so that mixing can take place during admission of the reagents into the vessel. In a static system there is usually a finite time required for the reactant to fill the vessel, mix with the other reagents, and reach an equilibrated temperature. This time often limits the rates of some reactions that can be measured using conventional measurement techniques. For a cylindrical reaction vessel, the mixing time can be estimated[1] as

$$t = \frac{3\ell^2}{4\mathcal{D}} \tag{3-1}$$

where ℓ is the length of the vessel in cm and $\mathcal{D}$ is the diffusion coefficient in $(cm^2 \ sec^{-1})$. The diffusion coefficient[2] depends on the total pressure in the reaction vessel according to

$$\mathcal{D} = \frac{3\bar{v}}{32n_t\sigma^2} \tag{3-2}$$

where n_t is the total density of molecules (cm^{-3}), σ is the collision diameter for the molecular pair, and the mean relative velocity

$$\bar{v} = \left(\frac{8k_BT}{\pi\mu}\right)^{1/2} \tag{3-3}$$

where, in turn, k_B is the Boltzmann constant $(1.38 \times 10^{-16} \ erg \ K^{-1})$, T is the absolute (Kelvin) temperature, and μ is the reduced mass of the collision pair. As a typical example, $\ell = 25$ cm, $T = 343$ K, $\sigma = 3 \times 10^{-8}$ cm, and $\bar{v} = 5 \times 10^4$ cm sec^{-1}, giving $\mathcal{D} \approx 3 \times 10^{-1}$ cm^2 sec^{-1} at a pressure of 5 Torr, so that the diffusion time $t \approx 15$ sec. Clearly, the reaction time to be measured must be longer than a few seconds, or serious errors in rate measurements will result.

A way to overcome this limitation of mixing time is to initiate the reaction by electric discharge or pulsed light and to observe the products immediately following initiation by time-resolved spectroscopic detection. It is then possible to study many reactions in premixed, static systems. Consequently, kinetic experiments which would otherwise be limited to intermittent measurements in static systems can be monitored continuously.

3.2.1.2.2 Flow Systems. Although laser initiation techniques have greatly expanded the range of reactant species and time scales that can be studied in static systems, there are a number of reactions that cannot be conveniently studied in this way. These include reactions of atoms and ions formed in a continuous discharge. For such reactions, *flow systems* are convenient measurement media. Unlike the closed, static systems discussed thus far, a flow system is an *open system,* in that reactants are constantly entering and products leaving the reaction zone. A typical *discharge-flow* apparatus is shown in Figure 3-2.

The analysis of a flow system is particularly simple if mixing is complete in a direction transverse to the flow but does not take place at all along the direction of flow. The conditions required for this to hold true will be discussed later in this section. In this case, the net change in the number N of species i in some volume dV a distance z down the flow is given by

$$dN_i = r_i dV - u dc_i$$

On the right, the first term represents the chemical rate of change r_i taking place in volume element $dV = Adz$, where A is the cross-sectional area of the flow tube, while the second term represents the transport of species i through volume element

Figure 3–2. Schematic of typical flow tube apparatus. The experiment shown is the measurement of the rate of the N + O_2 → NO + O reaction. N is produced by microwave discharge dissociation of N_2 and is mixed with O_2 via the movable "shower-head" injector. The detection system illustrated works through multiphoton ionization of the product NO molecules (see section 3.2.2.6 and chapter 9) [From I. C. Winkler, R. A. Stachnik, J. I. Steinfeld, and S. M. Miller, *J. Chem. Phys.* **85**, 890 (1986)].

dV with volume flow rate u. At steady-state flow conditions $dN_i = 0$ at a fixed z; therefore,

$$r_i dV = u dc_i$$

Integration gives

$$\frac{1}{u} \int dV = \frac{A}{u} \int dz = \frac{A}{u} z = \int_{c(0)}^{c(z)} \frac{dc_i}{r_i} \qquad (3\text{–}4)$$

For a simple first-order reaction X → products with rate constant k_1,

$$r_x = \frac{dc_x}{dt} = -k_1 c_x$$

Substituting this expression into equation 3–4 gives

$$\frac{Az}{u} = -\int_{c(0)}^{c(z)} \frac{dc_x}{k_1 c_x} = \frac{1}{k_1} \ln \frac{c_0}{c_2} \qquad (3\text{–}5)$$

If the concentration c of species X is measured at intervals along the flow tube, and if $(u/A)\ln(c/c(0))$ is plotted against z, the slope of the plot will be $-1/k_1$. The same result would be obtained by measuring the concentration decay with time and plotting $\ln(c(t)/c(0))$ vs. t; therefore, the flow method is equivalent to replacing the reaction time t by the flow distance $z = (u/A)t$. For a second-order reaction we may simply use a large excess of one of the reactants to establish *pseudo-first-order* kinetics, as discussed in chapter 2, sec. 2.2.2.

The principal advantage of the flow system is that processes that proceed at a moderately fast rate can be measured. For example, linear flow velocities of several thousand cm sec^{-1} can be achieved with Roots blowers or other high-speed pumps, and, if steady flow conditions are maintained, so that the concentration-distance profile is stable, a series of measurements may be made along the flow tube with sufficient time to afford a good signal-to-noise ratio. Of course, many of the detection techniques employed in flow-tube measurement, such as mass-spectrometric sampling, laser-induced fluorescence, and multiphoton ionization, are not simple to move along the tube; but in such cases the distance z is varied by moving the reactant injector along the flow direction.

The principal requirement for successful flow measurements is nonturbulent flow, the condition for which is that the Reynolds number

$$Re = \frac{vd\rho}{\eta} \tag{3-6}$$

be less than 2,300. In equation 3–6 v is the volume flow rate u/A, d is the tube diameter, ρ is the gas density, and η is the gas specific viscosity. In typical experiments a 100-scfm pump is used, giving $u \approx 5 \times 10^4$ cm^3 sec^{-1}; in a tube of 10 cm^2 cross section, $v = u/A \approx 5 \times 10^3$ cm sec^{-1}. The tube diameter $d = (4A/\pi)^{1/2} = 3.6$ cm, the density of atmospheric gases at $p = 1$ torr is about 2×10^{-6} g cm^{-3}, and η(air) ≈ 200 μpoise $= 2 \times 10^{-4}$ dyne sec cm^{-2}. Inserting these values in equation 3–6 gives Re ≈ 180, safely under 2,300. Since measurements of $\ln(c/c(0))$ over one or two decades can be made over a distance of 1 to 10 cm, this means that rate constants on the order of 10^3 sec^{-1} can be measured using flow techniques. Faster reactions can be measured by using larger pumps to get faster flow speeds and also lower partial pressures (for bimolecular reactions); but the limitation then becomes the mixing time following injection of the reagent into the flow, which may be several msec. Other factors which limit flow system measurements, such as concentration gradients and nonlaminar flow corrections, are discussed in detail in the excellent treatments by Howard,[3] Hoyermann,[4] and Kaufman.[5] Flow systems have also proven very useful in studies of ion-molecule reactions over a wide temperature range.[6]

3.2.2 Description of Techniques

As noted earlier, the typical apparatus for kinetic studies includes a suitable initiation source, measurement medium, and detection method. Since the essence of kinetic studies is measuring changes in composition with time, the detection method used must be sensitive to the magnitude of the expected concentration changes on the time scale at which they are changing. Analytical techniques used in classical kinetic studies to make such measurements include pressure measurements, nuclear magnetic resonance spectroscopy, electron paramagnetic resonance spectroscopy, ultraviolet and visible spectroscopy, gas titration, chemical trapping, ultrasonic measurement, polarography, gas chromatography, and mass spectroscopy. Most of these techniques are incapable of monitoring reactions that occur on fast time scales. A general requirement for any technique is that the analysis time must be short compared to the time during which the reaction occurs. There are several ways of achieving this requirement, one of which is simply to slow the reaction down. Since the reaction rate can be expressed as $k = A\exp(-E_{act}/RT)$ (see chapter 1, sec. 1.4), lowering the temperature will reduce the reaction rate to a point at which the reaction can be monitored. For example, a reaction with a temperature-independent activation energy of 84 kJ/mol will be about 10^7 times slower at 157 K than at room temperature. This method has found good use in low-temperature matrix isolation studies, but it is hardly an adequate solution, of course, for very fast reactions.

A limitation of any detection system is electronic processing of the signal; the best state-of-the-art devices have 10-GHz bandwidths (about 0.1 nsec response time), although new instruments based on superconducting and Josephson devices, operating at liquid helium temperatures, may ultimately provide one or two decades faster response. The most important technological advance in detection is the laser, for many of the same reasons that make it a superb device for initiating reactions (see section 3.2.1.1). Its narrow-bandwidth, tunable output, together with its short pulse duration and high peak power, make it possible to monitor specific species on short time scales. For example, mode-locked laser systems are capable of producing femtosecond (10^{-15} sec) pulses, which allow details of molecular motion in condensed phases to be monitored in real time.

Figure 3–3 indicates the time scales which can be accessed by a number of

Figure 3–3. Summary of various experimental techniques and their time ranges [adapted from G. G. Hammes, *Principles of Chemical Kinetics* (New York: Academic Press, 1978)].

fast-reaction techniques. In the sections that follow, we introduce several of these techniques, with an emphasis on methods that employ lasers for detection.

3.2.2.1 Flash photolysis and laser flash photolysis.

One of the early techniques which overcame many of the limitations of classical techniques is flash photolysis. Originally developed by Norrish and Porter in the 1950s,[7,8] flash photolysis has established itself as a powerful technique for directly studying the rates of elementary reactions both in the gas phase and in solution. In flash photolysis, a reactant is irradiated with an intense flash of ultraviolet or visible light. The intensity of the light is sufficiently high to initiate a chemical reaction in the mixture, and the duration of the flash is short compared with the time scale of the reaction to be studied. This allows the unstable intermediates that are generated to be followed in real time using a suitable monitoring technique, such as absorption spectroscopy. Consequently, the concentration of the intermediates can be measured as a function of time, and their physical and chemical properties can be determined.

The typical flash photolysis apparatus is shown in Figure 3–4. The reaction vessel is comprised of a quartz cylinder up to 1 m in length with plane windows at each end. One or more photoflash tubes are situated parallel to the cell, and the complete assembly is surrounded by suitable reflectors, usually coated with magnesium oxide.

The photoflash lamp consists of a long quartz tube fitted at each end with metal electrodes, usually made of magnesium. The tube contains an inert gas at low pressure. The light flash is generated by discharging a large capacitor, previously charged to a high potential, across the two electrodes. If the inductance of the circuit is kept low and the conductivity of the leads is high, the duration of the flash (the fall time to half the maximum intensity) is about 20 μs. In general, the duration depends on the energy dissipated and varies from nanoseconds for microjoule flashes to milliseconds for flashes of several hundred joules. An empirical relation that exists

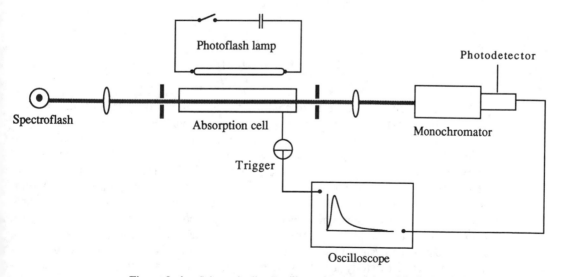

Figure 3–4. Schematic diagram illustrating principles of flash photolysis.

between the duration of the flash and the dissipated electrical energy is that a tenfold reduction in duration requires a hundredfold reduction in energy. Since scattered light from the flash interferes with spectroscopic measurements, the time resolution of the technique is limited by the duration of the flash. Although the technique is necessarily restricted to molecules with suitable absorption characteristics, the light output is essentially a continuum down to the quartz cutoff, so that the range of reactions which can be investigated is clearly very large.

Ensuing reactions must be followed in real time. Ideally, one would like to scan the absorption spectrum repeatedly after the flash, but this is ruled out by signal-to-noise problems. The usual procedure is either to monitor a narrow spectral band continuously or to obtain a complete spectrum at a predetermined interval after the flash. The latter has proved the more profitable, at least until the reaction mechanism has been fully analyzed. The experiment is repeated with different time delays until the complete pattern of events has been resolved.

The absorption spectrum is obtained by utilizing a second flash lamp situated at one end of the reaction vessel. The spectroflash energy and duration are chosen such that the detector situated at the opposite end of the reaction vessel receives light that is adequate for detection but insufficient to interfere with ongoing photochemistry.

Once the transient spectrum has been established, kinetic data may be obtained by measuring time-resolved absorption at a fixed wavelength. A typical flash photolysis kinetic curve is shown in Figure 3–5. For an atomic recombination reaction such as $2A + M \rightarrow A_2 + M$, the instantaneous concentration of atoms can be found[9] from such a curve to be

$$[A]_t = 2[A_2]_\infty \frac{\ln(x_t/x_\infty)}{\ln(x_0/x_\infty)}$$

Using $x_t = x_\infty + \Delta x$, $\ln(x_t/x_\infty) = \ln(1 + \Delta x/x_\infty)$. For optically thin samples, this is approximately equal to $\Delta x/x_\infty$, so that

$$[A]_t = \gamma \Delta x$$

where

$$\gamma = \frac{2[A_2]_\infty}{x_\infty \ln(x_0/x_\infty)}$$

One difficulty with the flash photolysis technique has been the inability to generate a concentration of transient species sufficiently high to permit the recording of an absorption spectrum from conventional flash lamps of short duration and low energy. However, pulsed lasers have provided a new source of excitation which is not subject to the energy-duration characteristics of flash lamps. The peak power of these lasers can be enormous (10^9 to 10^{12} watts), and the emitted light is coherent, so the beam can be focused on very small areas. The beam can also be monochromatic and of very short (nanosecond to femtosecond) pulse duration. A typical short-pulse laser pump-probe experiment is described in reference 10.

Flash photolysis and its laser variant are extremely versatile methods. They have been used to study a variety of reactions: radical-radical recombination, unimolecular decay, and radical-molecule, atom-atom, and atom-molecule bimolecular

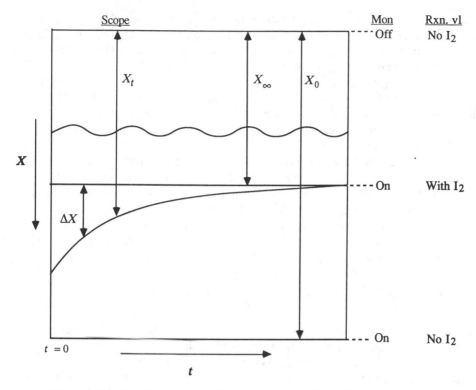

Figure 3–5. Plot of a typical oscilloscope flash photolysis curve. Shown are the scope deflections x_o, x_∞, x_t, and $\Delta x = x_t - x_\infty$. Mon = monitor light, Rxn.vl = reaction vessel. The flash was fired at $t = 0$. [Adapted from Yamanashi and Nowak, *J. Chem. Ed. 45*, 705 (1968).]

reactions. They also have proved particularly useful in the identification of transient species such as radicals, atoms, and molecules in excited states. A similar technique, which uses transient infrared absorption spectroscopy, is described in the next section.

3.2.2.2 Absorption spectroscopy. Absorption spectroscopy encompasses all those techniques that measure a change in the intensity of some light source due to absorption that occurs as the light passes through a gas along some specified path length. The absorption measurement approach is in general quite versatile, and the technique can use light sources which range from broadband to monochromatic laser; in principle, it can cover the entire electromagnetic spectral region. Absorption spectroscopy is used in kinetic measurements as a means of quantitatively detecting species whose reaction rate is of interest. Consequently, a requirement for using this technique for kinetic studies is that there exists a unique spectral signature associated with the species of interest which allows it to be monitored and is such that the isolated absorption line or feature must be free from interference from other species produced in the reaction.

The infrared region has traditionally been considered the "fingerprint" for polyatomic molecules, since most such molecules have vibrational-rotational bands in that region. A new technique which is very versatile as well as highly species-specific is the time-resolved diode laser absorption technique, abbreviated TRDLA. Because of its great specificity and selectivity, this technique is used to study gas phase reactions involving transient species.[11-13]

The selectivity of TRDLA is based on the fact that diode lasers have a very high spectral resolution, usually 10^{-3} cm^{-1}, so that the real spectral resolution for molecules in the gas phase can easily achieve the Doppler limit and the individual vibrational-rotational absorption lines of small molecules with fewer than ten atoms can be resolved. Because of the high spectral resolution a particular species of interest can be easily identified, even in a very complicated mixture, provided that the spectroscopy of the species is well established. The high sensitivity of TRDLA is related to its high selectivity: since individual vibrational-rotational lines are usually resolved, the background of the signal is the intensity profile of the diode laser beam instead of contours of unresolved clusters of lines as in low-resolution spectroscopy. The sensitivity of the diode laser absorption can be characterized[14] by the minimum detectable concentration n_{min}. For a path length of 100 cm,

$$n_{min} = \frac{6.3 \times 10^{14} \gamma_D}{s(n_j/n)} \text{ molecule cm}^{-3} \tag{3-7}$$

where s is the absorption band strength in cm^{-2} atm^{-1}, γ_D is the Doppler half-width at half maximum, and n_j/n is the fraction of molecules in a particular vibrational-rotational state. For most small molecules in gas phase, $n_{min} \approx 10^{12}$ molecule cm^{-3}, which corresponds to a partial pressure of about 10^{-3} torr at room temperature.

The principle of TRDLA is that the absorbance at the diode laser frequency is approximately proportional to the concentration of the absorbing species. The measured absorption signal is therefore a direct measure of the species of interest, and the evolution with time of its concentration can be monitored in real time. Because of the direct detection of the species under investigation, kinetic quantities such as reaction rate constants are likely to be more reliable than those obtained with indirect methods such as gas chromatography or pressure measurements, and the errors involved in the experiment can be more easily analyzed and estimated.

This is not to say that there are no limitations to the TRDLA technique. First, the technique is limited to species which have reasonable absorption band strengths within the operating region of the diode laser. Second, the spectral appearance may be very complicated due to the high resolution of the diode laser. Therefore, the spectroscopy of the species to be monitored must be well known for positive identification of that species. Furthermore, calibrations in the infrared region are inherently more difficult than in the visible region. This is because in a frequency region where calibration lines are sparse, one may have to transfer a calibration frequency from a known line several wave numbers away from the wanted line, and the practice will introduce errors in frequency measurements of the line. In that case it will be very difficult to identify the wanted line from a group of closely spaced lines of different species. Finally, even though the sensitivity of the diode laser absorption is high, it is in some cases not high enough to be used in kinetic measurements if the absorption of the species is too small.

In principle, for kinetic measurements, any reaction involving a transient species which absorbs infrared laser radiation can be investigated by TRDLA. To illustrate how the technique is used in chemical kinetic measurement, consider the rate constant for methyl (CH_3) recombination, i.e.,

$$CH_3 + CH_3 \xrightarrow{\ k_R\ } C_2H_6$$

A diagram of the basic components of the TRDLA system for these measurements is shown in Figure 3–6. First, a KrF laser is used to photolyze CH_3I to generate CH_3 radicals, the concentration of which is measured by the transient absorption of the diode laser output. The KrF laser and the diode laser beams are made colinear by a combination of lithium fluoride plates which serve as 90% reflectors to the diode laser while passing the KrF laser beam. Absorption of CH_3 radicals from the diode radiation is detected by a HgCdTe infrared detector, and the signal from this detector is recorded and averaged by a transient digitizer controlled by a computer. An example of the signal from methyl radical recombination is shown in Figure 3–7.

From chapter 2, the kinetic rate equation for the bimolecular recombination of methyl is

$$\frac{d(CH_3)}{dt} = -2k_R[CH_3]^2 \tag{3–8}$$

which yields

$$\frac{[CH_3]_0}{[CH_3]_t} = 2k_R[CH_3]_0(t - t_0) \tag{3–9}$$

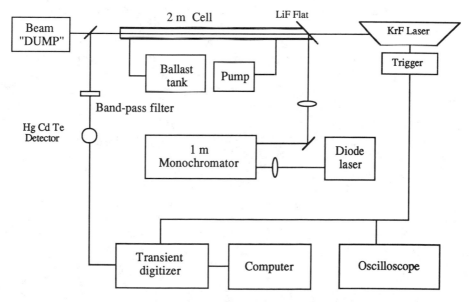

Figure 3–6. Schematic arrangement for detection of species using the time-resolved diode laser absorption technique.

Figure 3–7. Absorption signal of methyl radicals as a function of the time after the photolysis pulse. The radicals are monitored by the time-resolved diode laser absorption technique. [Adapted from G. A. Laguna and S. L. Baughcum, *Chem. Phys. Lett. 88,* 568 (1982).]

A least-squares fit of the signal shown in Figure 3–6 to equation 3–9 is made to obtain the recombination rate constant k_R using the initial concentration of CH_3 calculated from the CH_3I pressure, photolysis laser power, and absorption of the gas. We shall examine this reaction in more detail in a later section.

3.2.2.3 Laser photolysis/chemiluminescence.

The chemiluminescence technique uses the emission from excited species as a means for detecting and following the kinetics of these species.[15] The emission from the excited species is one of the several paths by which excess energy may be lost by the molecule, the general phenomenon of light emission from an excited state being known as luminescence. Emission following excitation by chemical reaction is referred to as chemiluminescence. For example, the reaction between NO and O_3 yields the reaction product NO_2, some of which is generated in an excited electronic state ($NO + O_3 \rightarrow NO_2^*$) and is subsequently relaxed by photon emission. Detection of the NO_2^* provides a measure of the reaction rate of $NO + O_3$.

In the laser photolysis/chemiluminescence technique, the role of the laser is primarily that of initiating the reaction. For example, in studying the reaction of Cl atoms with hydrogen halides (HX), the laser is used to photolyze molecular chlorine (Cl_2) to yield $Cl(^2P_{3/2})$ atoms according to the reaction[16]

$$Cl_2 \xrightarrow{h\nu (347.1 \text{ nm})} 2Cl(^2P_{3/2})$$

which is followed by the reactions

$$Cl + HX \xrightarrow{k_1} HCl^* + X \qquad (X = Br, I)$$

and

$$HCl^* + M \xrightarrow{k_2} HCl + M$$

The rate of the reaction under investigation is followed by monitoring the time-resolved infrared chemiluminescence of the excited product HCl molecules (see Figure 3–8). If the reaction conditions are such that $[HX] \gg [Cl]_0$, where $[Cl]_0$ is the initial concentration of chlorine atoms produced by the photolyzing laser (pseudo-first-order conditions), then the chemiluminescence intensity signal if related to the rate by the expression

$$I = \exp(-k_2[M]t) - \exp(-k_1[HX]_0 t) \qquad (3\text{--}10)$$

By fitting the signal to a functional form

$$S(t) = A[\exp(-t/t_d) - \exp(-t/t_r)] \qquad (3\text{--}11)$$

where A is a constant and t_d and t_r are the decay and rise times, respectively, we can obtain the rise time, which is related to the rate constant of interest by

$$\frac{1}{t_r} = k_1[HX]_0 \qquad (3\text{--}12)$$

Figure 3–8. Schematic diagram of the apparatus used for measurement of the Cl + HX rate constant [adapted from F. J. Wodarczyk and C. B. Moore, *Chem. Phys. Lett. 26*, 484 (1974).]

Thus, we need know only the initial concentration of HX and the rise time in order to determine rate constants using this technique.

The detection limit of the chemiluminescence technique is determined by the background against which the measurements are made. Nevertheless, the sensitivity of the technique is about 10^{12} molecules cm^{-3}. For greater sensitivity, the fluorescence technique (see next) may be used.

3.2.2.4 Laser-induced fluorescence.

The basic laser-induced fluorescence (LIF) approach typically employs one or more pulsed lasers to selectively produce an excited state of the species being monitored.[17, 18] This is done by varying the laser frequency until the laser output overlaps an absorption line of the molecule. If the lifetime of the excited state is short compared to the collisional deactivation process, the excited state will decay to a more stable state via fluorescence. This fluorescence generally occurs at ultraviolet or visible wavelengths and can be detected using high-efficiency photomultiplier tubes and photon-counting techniques. To further enhance the signal-to-noise ratio of the detection process, gated photon-counting synchronized to the pulsed laser is generally employed. The basic experimental arrangement is shown in Figure 3–9. In order to use this technique for kinetic measurements, three major requirements must be fulfilled. First of all, the molecule must fluoresce. Although this requirement is well satisfied in general for small molecules, particularly diatomic molecules, there are many polyatomic molecules whose excited states are known to undergo radiationless transitions instead of fluorescence. The second requirement is that the band system must be well characterized spectroscopically. The third requirement is that the molecular band system should be accessible with available lasers; unfortunately, this restricts the type of molecules that can be studied to those having low-lying excited electronic states.

Despite its limitation to molecules which satisfy these requirements, LIF does have an advantage over typical absorption techniques such as time-resolved diode laser absorption spectroscopy. In the absorption technique, the sensitivity of the technique is limited by the smallest $\Delta I/I_0$ that can be measured. In LIF, the fluorescence is measured against a zero background, and the intensity of the fluorescence is proportional to the intensity of the irradiating light. In principle, 10^3 atoms can be detected by LIF, which makes it far more sensitive than the TRDLA technique, but in practice this limit is rarely reached because of scattered light and saturation problems. For practical purposes, LIF can detect species at about 10^{10} particles/cm^3. However, with great effort, OH radicals, for example, have been detected at concentration levels of 10^6 molecules/cm^3. With these sensitivity limits, a variety of reactions can be studied with the LIF method: atom-molecule, radical-molecule, and unimolecular. Zare and co-workers[19,20] have studied the family of reactions between Ba atoms and hydrogen halides, i.e.,

$$\text{Ba} + \text{HX}(\text{X} = \text{F,Cl,Br,I}) \longrightarrow \text{BaX} + \text{H}$$

using the LIF technique. Other examples of species detection by LIF include NO, CN, NO_2, CF_2, NO_3, and CH_3O. Since the intensity of the fluorescence is proportional to the population of the state or states from which fluorescence occurs, by

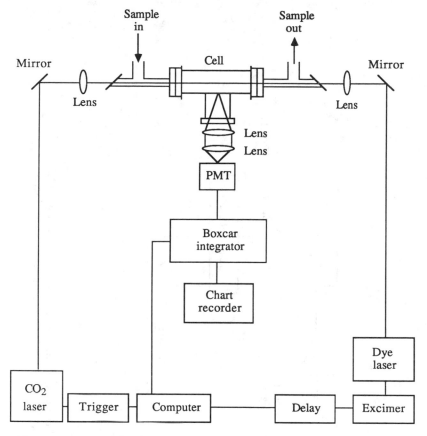

Figure 3–9. Schematic diagram of a typical laser photolysis using a CO_2 laser followed by laser-induced fluorescence detection.

measuring the intensity of the fluorescence during the reaction and plotting intensity versus time, a rate constant can be obtained.

3.2.2.5 Molecular modulation. The modulation technique[21] allows direct measurement of concentrations of transient species. This is done by modulating the concentration of intermediates produced by periodic interruptions of the photolysis. The technique is based upon the principles that (a) phase-sensitive detection discriminates against unwanted signals introduced after the initial chopping or modulation of the photolysis lamp, and (b) measurement of the phase and amplitude of the output signal as a function of modulation frequency can provide information about the rates of processes which occur subsequently to the initial modulation step. Figure 3–10 shows a block diagram of a chemical modulation experiment. The phase sensitive detector (PSD) plays a very important role. As implied by its name, it responds to the phase of the input signal. A signal which is of the same frequency but differs in phase from the output chopper gives an output whose magnitude depends upon the

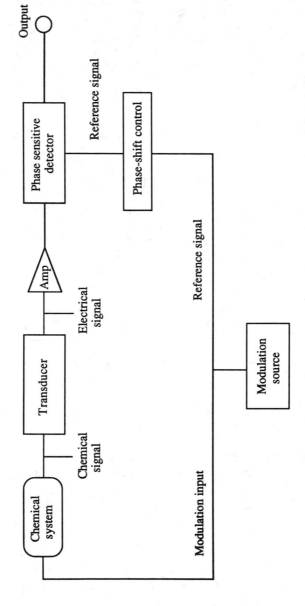

Figure 3–10. Molecular modulation experiment, schematic diagram (adapted from L. F. Phillips, *Progr. Reaction Kinetics*, Vol. 7, ed. K. R. Jennings and R. B. Cundall (Oxford: Pergamon Press, 1973)].

phase difference. If the phase difference is exactly 90° or 270° the output is zero, while if the difference is 180° a maximum negative output results. Noise is of random phase and therefore gives an output which averages to zero. A signal which is not quite of the correct frequency gives an output whose value, averaged over a number of chopper cycles, is also zero. Thus, the PSD circuit provides an effective means of discriminating against noise signals which are introduced subsequent to the initial modulation step. Light from a suitable source, usually a mercury arc, is brought to a focus, and a rotating sector is inserted at this point. The slots occupy about one-quarter of the total, so that the light periods have one-third the duration of the dark periods. The sector is driven by a synchronous motor to give close control over the speed of rotation. The light is made parallel by a simple lens system and is allowed to fall on the reaction cell.

To learn how modulation affects a chemical reaction, consider a very simple case. Suppose we have a pair of first-order consecutive reactions with rate constants k_1 and k_2:

$$A \xrightarrow{k_1} B$$

$$B \xrightarrow{k_2} Products$$

The steady-state concentration of B is given by $[B]_{ss} = k_1[A]/k_2$, and the mean lifetime of species B is $t_B = 1/k_2$.

The differential equation describing the evolution over time of the concentration of B is

$$\frac{d[B]}{dt} = k_1[A] - k_2[B] \tag{3-13}$$

which can be integrated at constant $[A] = a$ to give

$$[B] = b(1 - e^{-k_2 t}) \tag{3-14}$$

for the case $[B] = 0$ at $t = 0$, where b is equal to $a(k_1/k_2)$. If the concentration of A follows a square wave as in Figure 3–11(a), the concentration of B will follow a waveform like that in Figure 3–11(b). If k_2 is large, i.e., if the lifetime t_B is very short compared with the modulation period, the concentration of B will be able to attain its steady-state value very rapidly, as in Figure 3–11(c), and will follow the periodic variations in the concentration of A with a high degree of accuracy. If, however, t_B is long compared with the modulation period, the concentration of B will fluctuate around some mean value, as in Figure 3–11(d). The amplitude of the signal corresponding to the modulation of [B] is seen to be reduced at high frequency. If, in addition, the removal of B is not first order, it can be shown that the average value of [B] at high frequency is also different from the the low-frequency value. This is the basis of the rotating sector method for determining the lifetimes of reaction intermediates. A plot of typical results is shown in Figure 3–12. The inflection point t' on the curve is close, although not exactly equal, to the mean lifetime t_B. For this specific case, $k_2 = 1/t_B$. For more complicated circumstances, the results have been tabulated[21] so that the measured value of t' can be transformed to the corresponding lifetime t_1 and hence to the required rate constant.

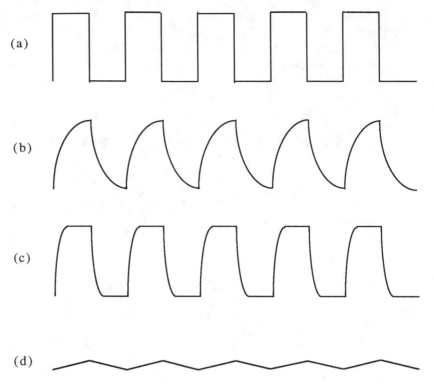

(a)

(b)

(c)

(d)

Figure 3–11. Waveforms from rotating sector experiments. (a) Square wave, concentration of species A. (b) Waveform for [B], time constant τ comparable with period of modulation. (c) Waveform for [B], τ less than modulation period. (d) Waveform for [B], τ much greater than modulation period.

To calculate the effect of the rate constant k_2 on the phase of a signal corresponding to [B], we expand the time-varying concentration of A and B in terms of their component harmonics as

$$[A] = a_0 + a_1 \sin \omega t + a_2 \sin 2\omega t + \cdots \qquad (3\text{–}15)$$

and

$$[B] = b_0 + b_1 \sin(\omega t - \delta) + b_2 \sin(2\omega t - \delta') + \cdots \qquad (3\text{–}16)$$

where δ, δ', etc., are the phase shifts associated with the fundamental and successive harmonics of [B], and $\omega = 2\pi f$. Differentiating equation 3–16, we obtain

$$\frac{d[B]}{dt} = b_1\omega \cos(\omega t - \delta) + 2b_2\omega \cos(2\omega t - \delta') + \cdots \qquad (3\text{–}17)$$

Substituting equations 3–15, 3–16, and 3–17 into equation 3–13 and neglecting everything but the fundamental component of the signals at frequency ω, equation 3–13 becomes

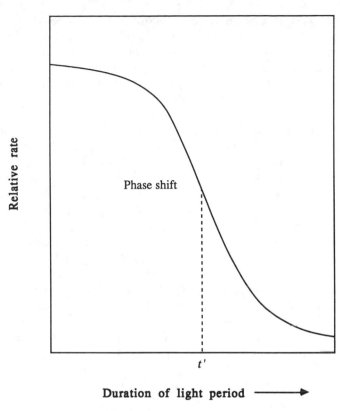

Figure 3–12. Typical modulation curve.

$$b_1 \cos(\omega t - \delta) = k_1 a_1 \sin \omega t - k_2 b_1 \sin(\omega t - \delta) \qquad (3\text{–}18)$$

The terms involving cos and $\sin(\omega t - \delta)$ can be expanded using standard formulas, and we may equate the coefficients of cos ωt on both sides of the resulting equation to obtain

$$\frac{\omega}{k_2} = \tan \delta \qquad (3\text{–}19)$$

or

$$\tan \delta = 2\pi f t_B \qquad (3\text{–}20)$$

Equation 3–20 is the basis of the phase-sensitive method of measuring the lifetime t_1 of short-lived species, especially in the excited state. Since $k_2 = 1/t_B$, using this equation, we can relate the phase signal to the measured rate by

$$\tan \delta = \frac{2\pi f}{k_2} \qquad (3\text{–}21)$$

The modulation method is used mostly to study radical-radical recombinations and radical-molecule reactions.[22,23]

3.2.2.6 Photoionization Techniques. Photoionization spectroscopy is a comparatively new and sensitive laser technique which is well suited for kinetic measurements. Its main advantage is that the ionization process is highly selective, so that one can ionize and identify transient or stable species at low concentrations in the presence of a large excess of other species. Furthermore, molecules that cannot be detected using LIF because fluorescence quantum yields are too low can be studied using photoionization spectroscopy. The basic concept of the technique utilizes one or more pulsed visible- or ultraviolet-light lasers focused between two biased ion collection plates housed inside a sample cell. If the photon energy absorbed by the molecule exceeds its ionization threshold, ions will be generated and collected by collection plates. The resulting signal is a measure of the number of molecules in the cell. Molecules can be ionized using the multiple photon ionization process (MPI). This is achieved by first exciting the molecule from its ground electronic state to an intermediate state, and then ionizing the molecule from the intermediate state by one photon excitation. The technique is referred to as the $n + 1$ MPI process, where n photons are used to reach the intermediate state and one additional photon is used to ionize that excited state.[24] The $2 + 1$ MPI process is illustrated[25] for Br atoms in Figure 3–13. In general, the MPI process has high sensitivity. For example, detection limits of Br atoms using $2 + 1$ MPI is reported to be about 10^{10} atoms/cm^3. Not only is the technique highly sensitive, but also it is highly specific. The net result of the absorption of $n + 1$ photons by the molecule results in the selective production of the ionized molecule in a complex product mixture with the advantage of not ion-

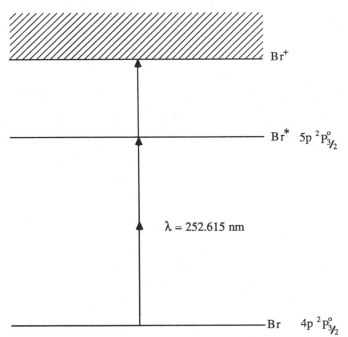

Figure 3–13. Multiple-photon ionization of bromine atoms. Two-photon excitation at 252.615 nm excites the $5p\,{}^2\mathrm{P}^{\mathrm{o}}_{3/2}$—$4p\,{}^2\mathrm{p}^{\mathrm{o}}_{3/2}$ transition, and an additional photon of the same wavelength ionizes the atom.

izing interfering compounds of the reaction mixture. The sensitivity and selectivity of the method make MPI well suited for kinetic studies.

Coupling photoionization spectroscopy with mass spectrometry produces a powerful technique which allows one to identify molecules not only by their mass but also by their characteristic optical ionization spectrum. A typical experimental setup is shown in Figure 3–14. Since the ions produced in the photoionization process are generated in a very short time interval, usually a few nanoseconds during the laser pulse, and in a small volume of about 10^{-5} cm^3 (the focal volume of the laser), it is most convenient to use time-of-flight mass spectroscopic detection. Slagle et al.[26] provide a good illustration of the use of the photoionization mass spectroscopic technique for kinetic rate constant measurement for the reaction of oxygen atoms with ethylene, i.e.,

$$O + C_2H_4 \xrightarrow{k_1} CH_3 + HCO$$

The complete mechanism includes the additional reactions

$$O + CH_3 \xrightarrow{k_2} H_2CO + H$$
$$C_2H_4 + CH_3 \xrightarrow{k_3} C_3H_7 \qquad\qquad (3\text{–}22)$$
$$CH_3 + O_2 \xrightarrow{k_4} CH_3O_2$$

and

$$2CH_3 \xrightarrow{k_5} C_2H_6$$

The desired conditions of the model are achieved by mixing O atoms and C_2H_4 in a flow system in which the concentrations of the reactants are held constant to ensure a constant rate of CH_3 production. Under these conditions, $[CH_3]$ remains constant

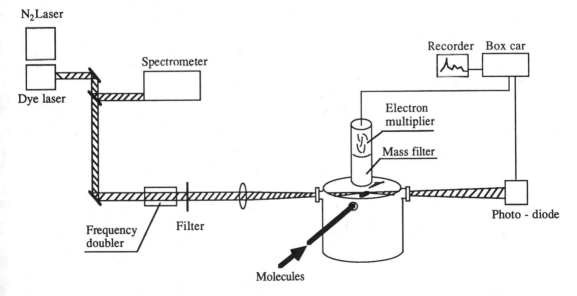

Figure 3–14. Experimental setup for mass-selective detection of ions produced in a multiple-photon ionization process.

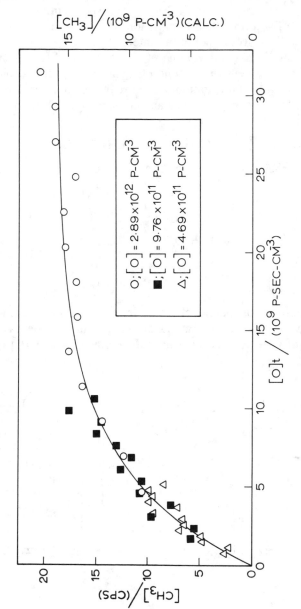

Figure 3–15. Plot of CH_3 ion signal versus $[O]_t$ in a kinetic experiment employing multi-photon ionization with mass spectrometric detection [From I. R. Slagle, F. J. Pruss, and D. Gutman, *Int. J. Chem. Kinet.* **6**, 111 (1974)].

and equal to the steady-state concentration $[CH_3]_{ss}$. The second and third reactions in the mechanism given by equations 3–22 are the only significant reactions consuming CH_3. Since the CH_3 concentration is proportional to the CH_3 ion signal, k_1 can be determined by fitting the half-life of the CH_3 ion-count growth to its steady-state value, as shown in Figure 3–15.

3.3 TREATMENT OF KINETIC DATA

3.3.1 General Analysis of Kinetic Data

3.3.1.1 Introduction. Two questions usually arise when rate data are obtained experimentally. The first deals with the inherent errors in the data as contributed by instrumental factors and experimental procedures; the second pertains to how these errors determine the uncertainty in the fitted rate constants. Correct estimates of kinetic parameters and their uncertainties are fundamentally important for chemical kinetics, because such data are needed for assessing kinetic models and microscopic rate theories in chemical kinetics. These two questions are also important in assessing just how well correlated a postulated reaction mechanism is to the raw data. Since errors are inherent in any measurement correlations are never exact, and a method for judging whether a particular correlation is significant is needed to assess the model. This section is given over to considering briefly some of these points.

3.3.1.2 Types of measurement errors. No measurement can be made perfectly and with zero error; thus, it becomes important to know the uncertainty of a measurement and how different errors that arise from the experiment or the instrument enter into it. Furthermore, once sources of errors are identified, ways of reducing them may be undertaken. It is the duty of every experimental kineticist to take every precaution to minimize errors and to report the accuracy of the final result. The accuracy of kinetic results involves estimating the systematic errors in each observation in the experimental apparatus or procedure. Although this step is difficult to quantify, it is possible to correct for the errors incurred. The precision of the result depends on random or statistical errors, which are fluctuations in repeated measurements and are usually beyond the control of the experimentalist. To estimate such errors, we treat the experimental data statistically using standard methods to obtain the standard deviation in the results from the mean value. Knowing these uncertainties in the experimental data permits the uncertainty in the measured rate constants to be estimated and also enables us to further understand how these sources of errors contribute to the overall uncertainty in the measured rate. In this section our purpose is not to provide the reader with an extensive review of statistical methods of treating data (indeed, this has already been done by others[27-29]), but rather to acquaint the would-be kineticist with the procedures for analyzing and reporting kinetic data together with appropriate error limits.

3.3.1.2.1 Systematic Errors. Systematic errors in kinetic measurements have been discussed in some detail by Cvetanovic et al.[30-33] In kinetic measurements there are four kinds of systematic errors: (1) *instrumental* (shortcomings of the instruments or the effects of the environment on the instrument or its user); (2) *operational* (the personal judgment of the kineticist, which enters into readings of the instrument, and errors in instrument calibration); (3) *methodological* (inaccuracies in the modeling equations, which are imperfect approximations to the true solutions); and (4) *mechanistic* (errors introduced by inaccurate representations of the underlying chemical mechanism or some unrecognized chemical interference from secondary reactions or impurities in reagents). There are several ways of searching for systematic errors in measured kinetic parameters:

1. *Comparison of measured kinetic parameters under a variety of experimental conditions.* Inconsistencies between measured rate coefficients greater than the cumulative random error indicate the presence of systematic errors. Caution must be exercised since apparently consistent results can conceal systematic errors.
2. *Comparison of the absolute value of rate parameters obtained by different techniques.* Here, the importance of determining the same kinetic parameters by vastly different techniques cannot be overemphasized.
3. *Through insight and guidance from theoretical and semi-empirical methods such as those to be presented in later chapters.*

In chemical kinetics, an inadequate mechanism is frequently the largest source of systematic errors. If a mechanistic complication is the cause of systematic error, its magnitude and sign may be assessed by computer simulation. In this case, the reaction or reactions that may be significant are included. Once upper and lower bounds of the systematic error are estimated, it is possible to evaluate their contribution to the overall uncertainty in the results of the measurements. When upper (e_s^u) and lower (e_s^L) bounds of the systematic error are given, a correction for the error is given by the mean,

$$\bar{e}_s = \frac{(e_s^u + e_s^L)}{2} \tag{3–23}$$

In this case the remaining error becomes

$$\pm(e_s)_{\text{max}} = \frac{\pm(e_s^u + e_s^L)}{2} \tag{3–24}$$

An excellent discussion of other ways of estimating systematic errors is given by Eisenhart.[34,35]

3.3.1.2.2 Random Errors.
Random errors are usually due to unknown causes and may occur even when all systematic errors have been accounted for. They also contribute to the uncertainty of the measured value. Because of their randomness, there is no known method of controlling or correcting these errors. The only way to offset them is to increase the number of repetitive measurements and use statistical means to obtain the estimate of the true value of the parameter under measurement. For a large number of observations, the random errors have a Gaussian distribution;[27,36] that is, the probability of their occurrence is given by

$$f(\epsilon) = \left(\frac{1}{\sigma\sqrt{2\pi}}\right)\exp\left(\frac{-(x-\mu)^2}{2\sigma^2}\right) \qquad (3\text{--}25)$$

where ϵ is the magnitude of the random error given by $x - \mu$, and μ and σ are moments of the distribution. The quantity μ is the *mean* of the population, and σ is the *standard deviation* of the population—it measures true spread of the individual observations about the mean. The term $x - \mu$ represents the magnitude of the random error in an individual measurement.

3.3.1.2.3 Propagation of Errors.

Rate constants of chemical reactions are calculated from experimental measurements of quantities such as gas pressure, intensity alteration, intensity of chemiluminescence or fluorescence, etc. Consequently, the uncertainty in the rate constant is dependent on the individual uncertainties in these measured parameters. For example, consider a resonance fluorescence experiment done in a flow system. The critical parameters affecting the results of measurement are (1) gas flow, (2) pressure, (3) temperature, (4) distance, and (5) light intensity. Table 3–1 provides a list of typical uncertainty ranges for these parameters.[37] Uncertainties in these values together with scatter in the kinetic data ultimately determine the precision of reported results. Using the propagation-of-errors method,[27,28,36] the overall uncertainty in the measured parameter is given by

$$\sigma_k = \left[\sum_i \left(\frac{\partial f}{\partial\phi_i}\sigma_i\right)^2\right]^{1/2} \qquad (3\text{--}26)$$

where the σ_i are the uncertainties in the measurement parameters and the partial derivatives $(\partial f/\partial\phi_i)$ are measured at the sample mean of the measurement parameters.

TABLE 3–1 RANDOM ERROR SOURCES AND MAGNITUDES IN EXPERIMENTAL FLOW-TUBE MEASUREMENTS (FROM REF. 37)

Measurement (units)	Device/method used	Range used	Uncertainty range[a]
Gas flow, u	Volume displacement	0.05–250	0.01–15
(mliter (STP)sec^{-1})	Critical flow orifices	5–500	1.0–5.0
Pressure, P (torr)	Closed-end manometers	1–1000	0.1–5.0
	Bourdon gauges	1–50	0.1–2.5
	U-tube manometer	0.5–20	0.05–1.0
Temperature, T (°C)	Chromel-alumel thermocouple	-80–200	0.1
	Pt vs. Pt-Rh thermocouple	300–2000	10–60
Distance, z (cm)	Centimeter scale	0–100	0.05
Light intensity			
I (mV)	Absorption-monochromator/PMT	10–100	0.5–2.0
F (mV)	Fluorescence-light filter/PMT	1–20	0.1–1.0
I_{CL} (pA)	Chemiluminescence-calibrated monochromator/PMT	0.1–500	0.02–50[b]

[a]Values correspond to uncertainties at extremes of range; units are the same as measurement units.

[b]Includes contribution of scatter in detection system calibration ($\approx\pm5\%$) used to obtain absolute intensities.

Let us determine the precision of a simple first-order rate constant from a flow reaction system (see section 3.2.1.2.2). The specific rate constant k is given by

$$k = \frac{-v(\ln[A]_1 - \ln[A]_2)}{(z_1 - z_2)} \qquad (3-27)$$

where $[A]_1$ and $[A]_2$ are the concentrations of A present at the distances z_1 and z_2, respectively, and $v = u/A$ is the linear gas flow velocity. If we assume that errors in the measured quantities are independent of each other, then we can use equation 3–26 to estimate the uncertainty Δk in k as follows:

$$(\Delta k)^2 = \left(\frac{\partial k}{\partial z_1}\right)^2 (\Delta z_1)^2 + \left(\frac{\partial k}{\partial z_2}\right)^2 (\Delta z_2)^2 + \left(\frac{\partial k}{\partial \ln[A]_1}\right)^2 (\Delta \ln[A]_1)^2$$

$$+ \left(\frac{\partial k}{\partial \ln[A]_2}\right)^2 (\Delta \ln[A]_2)^2 + \left(\frac{\partial k}{\partial v}\right)^2 (\Delta v)^2 \qquad (3-28)$$

The relative error can be found by dividing through by k^2 and then multiplying and dividing each term by the square of the independent variable:

$$\left(\frac{\Delta k}{k}\right)^2 = \left(\frac{z_1}{k}\frac{\partial k}{\partial z_1}\right)^2 \left(\frac{\Delta z_1}{z_1}\right)^2 + \left(\frac{z_2}{k}\frac{\partial k}{\partial z_2}\right)^2 \left(\frac{\Delta z_z}{z_z}\right)^2 + \left(\frac{\ln[A]_1}{k}\frac{\partial k}{\partial \ln[A]_1}\right)^2 \left(\frac{\Delta \ln[A]_1}{\ln[A]_1}\right)^2$$

$$+ \left(\frac{\ln[A]_2}{k}\frac{\partial k}{\partial \ln[A]_2}\right)^2 \left(\frac{\Delta \ln[A]_2}{\ln[A]_2}\right)^2 + \left(\frac{v}{k}\frac{\partial k}{\partial v}\right)^2 \left(\frac{\Delta v}{v}\right)^2 \qquad (3-29)$$

We use equation 3–27 to calculate the partial derivatives, and if we substitute the values in equation 3–29, we obtain

$$\left(\frac{\Delta k}{k}\right)^2 = \left(\frac{z_1}{z_1 - z_2}\right)^2 \left(\frac{\Delta z_1}{z_1}\right)^2 + \left(\frac{z_2}{z_1 - z_2}\right)^2 \left(\frac{\Delta z_2}{z_2}\right)^2 + \left(\frac{\ln[A]_1}{\ln[A]_1 - \ln[A]_2}\right)^2 \left(\frac{\Delta \ln[A]_1}{\ln[A]_1}\right)^2$$

$$+ \left(\frac{\ln[A]_2}{\ln[A]_1 - \ln[A]_2}\right)^2 \left(\frac{\Delta \ln[A]_2}{\ln[A]_2}\right)^2 + \left(\frac{\Delta v}{v}\right)^2 \qquad (3-30)$$

To show how we use equation 3–30, suppose that for a particular reaction $z_1 - z_2 = 100$ cm, $[A]_2 = 0.80[A]_1$, the relative error in determining $[A]_1$ and $[A]_2$ is about $\pm1\%$, the flow velocity is 400 cm/sec, its uncertainty is ±5 cm/sec, and the uncertainty in the distance is ±0.1 cm. Substituting these values into equation 3–30, we obtain, for the precision in k,

$$\frac{\Delta k}{k} = \pm\{(0.005)^2 + (0.001)^2 + (0.0448)^2 + (0.0448)^2 + (0.0125)^2\}^{0.5}$$

$$= \pm0.0645 \qquad (3-31)$$

so the expected error is $\pm6.5\%$.

3.3.1.3 Evaluation of rate constants

3.3.1.3.1 Least-Squares Analysis. Once the uncertainties in each measurement are known, how does one determine the uncertainties in the rate constant? Here, we turn to the question of relating the mechanistic model expressed in some mathematical form to the actual experimental rate data. Since there are errors in the measurements and perhaps some inaccuracies in the model, it is often difficult to obtain an exact fit. Consequently, it becomes difficult to judge the appropriateness of the model to the experimental data by inspection. Assuming that the errors, excluding systematic errors, are random, a suitable technique typically used in kinetic data treatment is the least-squares method.[27,36,38] To estimate values of parameters in the model, we must minimize some measure of the errors. In the least-squares method the measure of error is defined as the sum of squares of the errors of the individual measurements. If we have some model to be fitted to the form

$$y = f(\mathbf{p}, t) \tag{3-32}$$

where $\mathbf{y}$ is the set of experimental data points and $\mathbf{p}$ are parameters we wish to find, the least-squares criterion gives

$$Q = \sum_{j=1}^{M} \omega_j [y_j - f(\mathbf{p}, t_j)]^2 = \sum_{j=1}^{M} \omega_j \epsilon_j^2 \tag{3-33}$$

In this equation, $\epsilon_j = y_j - f(\mathbf{p}, t_j)$ represents the residual error given by the difference between the observed value y_j and the calculated value $f(\mathbf{p}, t_j)$, which is then weighted by ω_j for each observation. It is the quantity Q that is minimized with respect to estimated parameters.

There are two cases associated with the least-squares procedure. In the linear least-squares case, the functional form of the values of $\mathbf{y}$ are presumed to follow some general linear equation of the form

$$y = p_1 + p_2 t \tag{3-34}$$

In the second case, the function y has a nonlinear form. For many cases in kinetics, the general linear functional form is more commonly used. If the y values have unequal weights, equation 3–33 can be rewritten as

$$Q = \sum_{j=1}^{M} \omega_j [y_j - (p_1 + p_2 t_j)]^2 \tag{3-35}$$

The derivatives of Q with respect to both p_1 and p_2 must be taken and equated to zero in order to find the optimal values. Taking the derivative of equation 3–33 with respect to p_1 yields

$$\frac{\partial Q}{\partial p_1} = -2 \sum_{j=1}^{M} \omega_j (y_j - p_1 - p_2 t_j) = 0 \tag{3-36}$$

so that

$$p_1 = \frac{\sum_{j=1}^{M} \omega_j y_j - p_2 \sum_{j=1}^{M} \omega_j t_j}{\sum_{j=1}^{M} \omega_j} \tag{3-37}$$

Thus, we now have an equation which relates the optimal value of p_1 to the optimal value of p_2 and the means of the y and t values. Taking the derivative of Q with respect to p_2 yields

$$\frac{\partial Q}{\partial p_2} = -2 \sum_{j=1}^{M} t_j \omega_j (y_j - p_1 - p_2 t_j) = 0 \tag{3-38}$$

Rewriting this expression, we obtain

$$\sum_{j=1}^{M} \omega_j t_j y_j - p_1 \sum_{j=1}^{M} \omega_j t_j - p_2 \sum_{j=1}^{M} \omega_j t_j^2 = 0 \tag{3-39}$$

Substituting equation 3–36 into equation 3–38 yields

$$p_2 = \frac{(\Sigma \omega_j)(\Sigma \omega_j t_j y_j) - (\Sigma \omega_j t_j)(\Sigma \omega_j y_j)}{(\Sigma \omega_j)(\Sigma \omega_j t_j^2) - (\Sigma \omega_j t_j)^2} \tag{3-40a}$$

and substituting this into equation 3–37 yields

$$p_1 = \frac{\Sigma \omega_j t_j^2 \, \Sigma \omega_j y_j - \Sigma \omega_j t_j \Sigma \omega_j t_j y_j}{(\Sigma \omega_j)(\Sigma \omega_j t_j^2) - (\Sigma \omega_j t_j)^2} \tag{3-40b}$$

The standard deviations of p_1 and p_2 are useful indicators of the goodness of fit. They are given by

$$\sigma(p_1) = \sigma \left(\frac{\Sigma \omega_j t_j^2}{(\Sigma \omega_j)(\Sigma \omega_j t_j^2) - (\Sigma \omega_j t_j)^2} \right)^{1/2} \tag{3-41}$$

and

$$\sigma(p_2) = \sigma \left(\frac{\Sigma \omega_j t_j}{(\Sigma \omega_j)(\Sigma \omega_j t_j^2) - (\Sigma \omega_j t_j)^2} \right)^{1/2} \tag{3-42}$$

where

$$\sigma = \left(\frac{\sum_{j=1}^{M} \omega_j (y_j - y_j^*)^2}{(M - 1)} \right)^{1/2} \tag{3-43}$$

In this expression, y_j^* are the computed values for y_j from $y = p_1 + p_2 t$ using the computed values of p_1 and p_2. A more general derivation of the linear least-squares expression from matrix methods is given in Appendix 3–1.

There are other measures of the quality of fit of the data. The simplest is the root mean square (rms) deviation, which is

$$\text{rms error} = \left(\frac{\sum_{j=1}^{M} (\omega_j \epsilon_j)^2}{n} \right)^{1/2} \tag{3-44}$$

where n is the number of data points. Another commonly used quantity is the correlation coefficient, which allows the significance of correlation to be tested. This quantity is given by

$$r = \frac{\Sigma \omega_j t_j \, \Sigma \omega_j y_j - n \Sigma \omega_j t_j y_j}{[(\Sigma \omega_j y_j)^2 - n(\Sigma \omega_j y_j^2)][(\Sigma \omega_j t_j)^2 - n(\Sigma \omega_j t_j^2)]} \tag{3-45}$$

When $r = 1$, there is a perfect correlation between the calculated line and experimental points; when $r = 0$, the data cannot be correlated linearly. If kinetic data are reasonably fitted, the correlation is generally $r > 0.95$, while poor correlations have $r < 0.9$.

Although the linear least-squares method is widely used in the treatment of kinetic data, it does have its limitations. One such limitation is that the method applies only to cases that can be expressed in linear form. Thus, for most complex kinetic models, the linear least-squares method is not applicable. Sometimes the equation can be linearized with respect to the constants by simple rearrangement. For example, the expression for the concentration-vs.-time dependence of the second-order reaction, $A + A \xrightarrow{\kappa}$ products, is

$$[A] = \frac{[A]_0}{1 + \kappa [A]_0 t} \tag{3–46}$$

This form is nonlinear, but it can be rewritten linearly as

$$\frac{1}{[A]} = \frac{1}{[A]_0} + \kappa t \tag{3–47}$$

This form is identical to equation 3–34 with $1/[A] = y$, $1/[A]_0 = p_1$, and $\kappa = p_2$. When an equation is linearized by either rearrangement or transformation, the original errors are often not preserved. It may then be the case that statistical weights have to be adjusted appropriately in order to obtain the correct estimates of p_1 and p_2 by the least-squares method.[30–33] Furthermore, for models that involve several dependent variables, linearization is not feasible. In such cases we turn to the nonlinear least-squares method,[39] which does not require that the model equations be linear in the unknown parameters. Nonlinear least-square methods are beyond the scope of this treatment.

3.3.1.3.2 Evaluation of Rate Constants at a Given Temperature from Direct Measurements.
A simple illustration of the use of the least-squares method to evaluate rate constants from experimental determinations is illuminating. Let us consider how the rate constant for the recombination of methyl radicals is measured using the laser photolysis time-resolved diode laser absorption technique (see section 3.2.2.2). The rate equation for the recombination reaction

$$CH_3 + CH_3 \xrightarrow{\kappa_1} C_2H_6$$

is given by

$$\frac{d[CH_3]}{dt} = -2\kappa_1 [CH_3]^2 \tag{3–48}$$

which yields, upon integration,

$$\frac{[CH_3]_0}{[CH_3]} = 1 + 2\kappa_1 [CH_3]_0 (t - t_0) \tag{3–49}$$

In the experiment the measured quantity is the voltage (V) on the voltage meter corresponding to the signal from the HgCdTe infrared detector. According to Beer's law

$$I = I_0 \exp(-\epsilon c \ell) \qquad (3\text{--}50)$$

where ϵ is the molar extinction, c the molar concentration, and ℓ the absorption path length. If $\epsilon c \ell$ is very small (weak absorption), we can expand equation 3–50 via a Taylor series and truncate the expansion beyond the linear term to obtain

$$I = I_0(1 - \epsilon c \ell) \qquad (3\text{--}51)$$

Consequently, the concentration c is given as

$$c = \frac{(\epsilon \ell)^{-1}(I_0 - I)}{I_0} \qquad (3\text{--}52)$$

From equation 3–52, we see that the concentration of the absorbing molecule is proportional to the relative change in the transmittance I. Since the transmittance is proportional to the voltage, we have

$$c \propto \frac{(V_0 - V)}{V_0} = \frac{S}{V_0} \qquad (3\text{--}53)$$

where S is the strength of the signal which is the difference between the voltage for $c = 0$ and $c \neq 0$. Now, we can rewrite equation 3–49 as

$$\frac{S(t_0)}{S(t)} = 1 + 2\kappa_1[CH_3]_0(t - t_0) \qquad (3\text{--}54)$$

and if we let

$$y = \frac{S(t_0)}{S(t)} - 1 \qquad (3\text{--}55)$$

$\kappa' = 2\kappa_1[CH_3]_0$, and $x = t - t_0$, we get a simple expression of the form

$$y = \kappa'x \qquad (3\text{--}56)$$

Using the formula of linear least-squares fitting described in section 3.3.1.3.1, we get

$$\sigma_{\kappa'} = \frac{\sigma}{\left[\displaystyle\sum_i \omega_i x_i^2\right]^{1/2}} \qquad (3\text{--}57)$$

where

$$\sigma = \frac{(\Sigma \omega_j(y_j - \kappa_1 x_i)^2)^{1/2}}{n - 1} \qquad (3\text{--}58)$$

$$\omega_i = \frac{c^2}{\sigma_i^2} \qquad (3\text{--}59)$$

and where n is the number of y_i measurements and c is an arbitrary proportionality constant. To evaluate σ_i, we apply the propagation-of-errors formula, equation 3–26, as follows:

$$\sigma_i = \left[\left(\frac{\partial y}{\partial S(t_0)}\sigma_S(t_0)\right)^2 + \left(\frac{\partial y}{\partial S(t_i)}\sigma_S(t_i)\right)^2\right]^{1/2} \tag{3–60}$$

This is no more than

$$\sigma_i = \left[\left(\frac{\sigma_S(t_0)}{S(t_0)}\right)^2 + \left(\frac{\sigma_S(t_i)}{S(t_i)}\right)^2\right]^{1/2}\frac{S(t_0)}{S(t_i)} \tag{3–61}$$

where $\sigma_S(t_0)$ and $\sigma_S(t_i)$ are the standard deviations of measured signal at $t = t_0$ and $t = t_i$, respectively. To evaluate the standard deviation of κ', we substitute equations 3–60 and 3–61 into equations 3–58 and 3–59. To evaluate the standard deviation of κ_1, we again apply equation 3–26 to the expression

$$\kappa_1 = \frac{\kappa'}{2[CH_3]_0} \tag{3–62}$$

to obtain

$$\sigma_{\kappa_1} = \left[\left(\frac{\partial \kappa_1}{\partial \kappa'}\sigma_{\kappa'}\right)^2 + \left(\left(\frac{\partial \kappa_1}{\partial [CH_3]_0}\right)\sigma_{[CH_3]_0}\right)^2\right]^{1/2} \tag{3–63}$$

giving

$$\sigma_{\kappa_1} = \kappa_1\left[\left(\frac{\sigma_{\kappa'}}{\kappa_1}\right)^2 + \left(\frac{\sigma_{[CH_3]_0}}{[CH_3]_0}\right)^2\right]^{1/2} \tag{3–64}$$

Now, $[CH_3]_0$ is proportional to the pressure P_0 of the initial reactant (CH_3I) and the photolysis laser fluence Φ, i.e.,

$$[CH_3]_0 \propto P_0(CH_3)\Phi \tag{3–65}$$

So again, using the equation 3–26 with equation 3–65, we obtain

$$\sigma[CH_3]_0 = [CH_3]_0\left\{\left(\frac{\sigma_{P_0}}{P_0}\right)^2 + \left(\frac{\sigma_\Phi}{\Phi}\right)^2\right\}^{1/2} \tag{3–66}$$

Finally,

$$\sigma_{\kappa_1} = \kappa_1\left[\left(\frac{\sigma_{\kappa'}}{\kappa'}\right)^2 + \left(\frac{\sigma_{P_0}}{P_0}\right)^2 + \left(\frac{\sigma_\Phi}{\Phi}\right)^2\right\}^{1/2} \tag{3–67}$$

The quantity σ'_κ is the standard deviation of the sample mean value $\kappa' = 2\kappa_1[CH_3]_0$ obtained by evaluating equations 3–57 through 3–61, σ_{p_0} is the standard deviation of the sample mean value of the reactant $(CH_3I)_0$ pressure, and σ_Φ is the standard deviation of the sample mean value of photolysis laser fluence.

3.3.1.3.3 Evaluation of Temperature-Dependent Rate Constants.

Perhaps the simplest illustration of the use of the least-squares method to analyze kinetic data is in evaluating Arrhenius parameters from temperature-dependent rate data. In the form

$$\kappa(T) = A \exp\left(\frac{-E_{\text{act}}}{RT}\right) \tag{3-68}$$

the temperature rate expression is nonlinear. Accordingly, if the linear least-square method is to be used to evaluate the coefficients, this expression must be linearized. The common practice is to transform equation 3-68 to the logarithmic form

$$\ln \kappa(T) = \ln A - \frac{B}{T} \tag{3-69}$$

where $B = E_{\text{act}}/R$. Then, to obtain E_{act} and A (the preexponential factor), a linear least-squares fit can be performed on equation 3-69 rather than using a nonlinear fit to equation 3-68. However, when equation 3-68 is thus linearized and a least-square fit performed, the weights of the $\kappa(T)$ values must be adjusted to the weight of $\ln \kappa(T)$ to ensure that the least-squares procedure does indeed minimize the residuals of the measured rate constant $\kappa(T)$.

3.3.1.3.4 Reporting Kinetic Data.

Results of kinetic measurements have great value for anyone who uses them, particularly for those performing atmospheric and combustion modeling. However, the results are useful only when they are accompanied by realistic estimates of their overall uncertainty. In reporting kinetic data, it is recommended[34] that information regarding both random and systematic errors be given. In particular, one should report known and even suspected systematic errors and, when available, estimates of their bounds and how the estimates were obtained. The value of the random error of the sample mean should be reported at a confidence level of at least $\pm 2\sigma$. In reporting kinetic data, the form $a \pm b$ is used, but only when the exact meaning of the uncertainty $\pm b$ is explicitly and clearly stated.

3.3.2 Sensitivity Analysis

In chemical kinetics it is often the case that the concentrations of chemical species are known more precisely than the rate constants of the kinetic mechanism. Consequently, in analyzing results from kinetic experiments, it is important to assess the effects of such parameter uncertainties on the predicted concentrations of the various intermediate and product species. Such information can be useful in identifying those rate constants which need to be determined more precisely, and may also help the experimentalist decide which concentration to monitor in order to improve the estimate of a specific rate constant. An analysis of this type is known as a *sensitivity analysis*.[40-48] Its purpose is to assess the sensitivity of the system's solutions to changes in input parameters, which in turn can be of enormous help in specifying the reaction mechanism and in designing experimental measurements.

How does one carry out a sensitivity analysis? Suppose we have a kinetic mechanism described by a set of coupled first-order differential equations having the following general form:

$$\frac{dy_i}{dt} = f_i(y_1, y_2, \ldots, y_N, t, \kappa_1, \kappa_2, \ldots, \kappa_M) \tag{3-70}$$

where there are N chemical components with concentrations y_i and M reactions in the mechanism with rate parameters κ_i. The solution of equation 3–70 will be a function of the two variables κ and t, that is $y_i(\kappa,t)$. A rate constant is called sensitive if a small change in its value produces large changes in the solution. Likewise, if changes in rate constants produce little or no change in the solution, then the rate constant is said to be insensitive. Therefore, to measure the sensitivity of the system of equations given by equation 3–70 with respect to a particular rate constant parameter, we need to know what small change in y_i is produced as a result of a small change in the rate constant κ_j, i.e.,

$$\Delta y_i = y_i(\ldots, \kappa_j + \Delta\kappa_j, \ldots) - y_i(\ldots, \kappa_j, \ldots) \qquad (3-71)$$

If this expression has a limiting value as $\Delta\kappa_j$ approaches zero, i.e., if

$$\lim_{\Delta\kappa_j \to 0} \frac{y_i(y_i, t, \kappa_j + \Delta\kappa_j) - y_i(y_i, t, \kappa_j)}{\Delta\kappa_j} = \frac{\partial y_i}{\partial \kappa_j} \qquad (3-72)$$

then we can define the sensitivity matrix element Z_{ij} as

$$\frac{\partial y_i}{\partial \kappa_j} = Z_{ij} \qquad (3-73)$$

The coefficients of the sensitivity matrix are referred to as *sensitivity coefficients*. In order to calculate Z_{ij} we must relate it to the set of differential equations of the system, i.e., equation 3–70. We do this by interchanging the order of differentiation

$$\frac{d(Z_{ij})}{dt} = \frac{d}{dt}\frac{\partial y_i}{\partial \kappa_j} = \frac{\partial}{\partial \kappa_j}\frac{dy_i}{dt} \qquad (3-74)$$

Since y_i is a function of κ_j the right-hand expression is expanded to give

$$\frac{d(Z_{ij})}{dt} = \frac{\partial f_i}{\partial \kappa_j} + \sum_{\ell=1}^{N} \frac{\partial f_i}{\partial y_\ell}\frac{\partial y_\ell}{\partial \kappa_j} \qquad (3-75)$$

where

$$i = 1, 2, \ldots, N$$

and

$$j = 1, 2, \ldots, M$$

Finally, substituting equation 3–73 into equation 3–75, we obtain

$$\frac{d(Z_{ij})}{dt} = \frac{\partial f_i}{\partial \kappa_j} + \sum_{\ell=1}^{N} \frac{\partial f_i}{\partial y_\ell} Z_{\ell j} \qquad (3-76)$$

Equation 3–76 is the fundamental sensitivity equation.

In deriving the sensitivity equation 3–76, we did not speak of the initial condition. If in equation 3–70 the initial condition is

$$f_i(y_i, 0, \kappa_j) = 0 \qquad (3-77)$$

then there is no deviation between the two systems. The initial condition for equation 3–76 is

$$Z_{ij}(0) = 0 \qquad (3-78)$$

This is sometimes referred to as the constant parameter case. Indeed, for most chemical kinetic systems one needs to consider only this case. The solution of equation 3–76, then, provides us with information about the sensitivity of the system to variations in each rate constant parameter. To obtain the sensitivity coefficient we must solve the differential equation 3–76 along with equation 3–70 since both of these two sets of equations are usually coupled. First, equation 3–70 must be solved to yield $y_i(\kappa,t)$ in order to evaluate $\partial f_i/\partial \kappa_i$ and $\partial f_i/\partial y_i$; then equation 3–76 must be solved. To measure the total sensitivity of the system we use the sensitivity coefficients to calculate the total change contributed by each of the n parameters from the expression

$$\Delta y_i = \sum_{j=1}^{N} Z_{ij} \Delta \kappa_j \qquad (3\text{–}79)$$

We shall illustrate the procedure for a hypothetical first-order decay reaction

$$A_1 \xrightarrow{\ k\ } A_2 \qquad (3\text{–}80)$$

with rate constant $k = \kappa_1 \exp(-\kappa_2/T)$. We wish to examine the sensitivity of the concentration of A_1 to variations in the Arrhenius parameters κ_1 and κ_2 at $T = 298$ K. The rate equation for the decay of A_1 is

$$\frac{d[A_1]}{dt} = -k[A_1] = -[A_1]\kappa_1 e^{-\kappa_2/T} \qquad (3\text{–}81)$$

subject to the initial condition that

$$[A_1]_t = [A_1]_0 \text{ at } t = 0.$$

The terms in the sensitivity equations are

$$\frac{\partial f_1}{\partial \kappa_1} = -[A_1]e^{-\kappa_2/T} \qquad (3\text{–}82)$$

$$\frac{\partial f_1}{\partial \kappa_2} = \frac{\kappa_1[A_1]}{T}e^{-\kappa_2/T} \qquad (3\text{–}83)$$

and

$$\frac{\partial f_1}{\partial [A_1]} = -\kappa_1 e^{-\kappa_2/T} \qquad (3\text{–}84)$$

With these terms we can write out the sensitivity differential equations as

$$\frac{dZ_{11}}{dt} = -[A_1]e^{-\kappa_2/T} - \kappa_1 e^{-\kappa_2/T}Z_{11} \qquad (3\text{–}85)$$

$$\frac{dZ_{12}}{dt} = +\frac{\kappa_1[A_1]e^{-\kappa_2/T}}{T} - \kappa_1 e^{-\kappa_2/T}Z_{12} \qquad (3\text{–}86)$$

Even in this simple single reaction case one can clearly see that equations 3–85 and 3–86 are coupled to equation 3–81. Solving these equations simultaneously yields

$$[A_1] = [A_1]_0 \exp(-\kappa_1 t e^{-\kappa_2/T}) \tag{3-87}$$

The sensitivity coefficients are

$$\frac{\partial[A_1]}{\partial \kappa_1} = Z_{11} = -[A_1]_0 t e^{-\kappa_2/T}[\exp(-\kappa_1 t e^{-\kappa_2/T})] \tag{3-88}$$

and

$$\frac{\partial[A_1]}{\partial \kappa_2} = Z_{12} = \frac{\kappa_1}{T}[A_1]_0 t e^{-\kappa_2/T}[\exp(-\kappa_1 t e^{-\kappa_2/T})] \tag{3-89}$$

To see what the sensitivity coefficients tell us about the effect of variations in κ_1 and κ_2 on the concentrations of A_1, let us choose $\kappa_1 = 1.79 \times 10^6$ sec^{-1} and $\kappa_2 = 500$ K. Also, let us suppose that $[A_1]_0 = 1$ mole ℓ^{-1} at $T = 298$ K and that the uncertainties in κ_1 and κ_2 are

$$8.97 \times 10^5 \leq \kappa_1 \leq 3.59 \times 10^6 \text{ sec}^{-1} \text{ and } 0 \leq \kappa_2 \leq 1000 \text{ K}$$

Figure 3–16 shows the sensitivity function at various times for these values of the constants; it can be seen that the most sensitive parameter is κ_2. Note that the sensitivity function typically attains a maximum. Comparing the relative sensitivities in Figure 3–17, we see trends similar to those in Figure 3–16, but for the parameter range we have studied, we conclude that all the sensitivity measures, including the

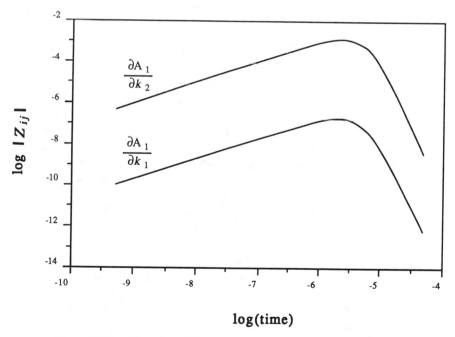

Figure 3–16. Plot of sensitivity coefficients $|Z_{ij}|$ for the reaction $A_1 \rightarrow A_2$.

$$\frac{\partial A_1}{\partial k_2} \Delta k_2$$

$$\frac{\partial A_1}{\partial k_1} \Delta k_1$$

Figure 3–17. Plot of relative sensitivities for the reaction $A_1 \rightarrow A_2$.

analytical sensitivity coefficients indicate that the concentration of A_1 is more sensitive to changes in κ_2 than to changes in κ_1. If we increase the uncertainty in κ_1 by two more orders of magnitude, then the concentration of A_1 becomes more sensitive to κ_1 than to κ_2.

The method we have just illustrated for the simple case above is often referred to as the direct method of sensitivity analysis.[40–42] If we wish to explore reactions of larger systems that include more parameters, we have to resort to numerical means of solving equations 3–70 and 3–76. First, these expressions must be reformulated in matrix terms. Then either of two strategies are used[41,42]: (1) solving 3–70 and 3–76 simultaneously which amounts to solving $2N$ differential equations M times; or (2) solving equation 3–70 to produce $y_i(\kappa,t)$, then interpolating two values of $y_i(\kappa,t)$ to evaluate the terms in equation 3–76, to solve the N differential equations m times. The latter technique amounts to solving N differential equations once, followed by N differential equations m times. Other approaches make use of Green's functions.[43–45] With these methods, it becomes computationally feasible to explore the importance of parameter variations in large systems, such as those pertinent to atmospheric and combustion systems.

To demonstrate the usefulness of sensitivity analysis in identifying elementary reaction steps in a mechanism consisting of many steps, consider a moderate size model for the combustion of CH_4 in O_2-Ar mixtures[46] (see chapter 15). In this system there are 13 species present and 23 rate constants; therefore, there is a total of 24 input parameters, 23 rate constants and one initial condition. Of interest is an analysis of the system near ignition, i.e., when $T \cong 2000°K$. The reactions and their rates used in this study are shown in Table 3-2. Integrating the system of kinetic rate equations for the set of reactions given in Table 3-2 up to 10^{-2} sec, we obtain the concentration profiles shown in Figure 3-18 for the initial concentrations

$$[CH_4] = 2.265 \times 10^{18} \text{ molecules/cm}^3$$

$$[O_2] = 4.530 \times 10^{18} \text{ molecules/cm}^3$$

$$[Ar] = 1.790 \times 10^{19} \text{ atoms/cm}^3$$

$$[CO_2] = [H_2] = 1.0 \text{ molecule/cm}^3$$

The concentrations of all other intermediates and products are set to zero. As in the previous example, the sensitivity coefficients vary with time, so in this example we look at sensitivity information at $t = 2.5 \times 10^{-5}$ sec; this corresponds to the time at which most species in the reaction have reached their maximum values (see Figure 3-18). Using the Green's Function Method[43-45] and solving the simultaneous differential equations (equations 3-70 and 3-76), we find the reactions which show

TABLE 3-2 METHANE OXIDATION REACTION KINETICS [REFERENCE 46].

Reaction	Forward rate constant[a]	Forward rate constant at 2,000 K	Reverse rate constant at 2,000 K
(1) $CH_4 + M \rightarrow CH_3 + H + M$	$3.32 \times 10^{-7} \exp(-44,500/T)$	7.2×10^{-17}	3.05×10^{-31}
(2) $CH_4 + OH \rightarrow CH_3 + H_2O$	$9.96 \times 10^{-10} \exp(-6,290/T)$	4.3×10^{-11}	1.74×10^{-13}
(3) $CH_4 + H \rightarrow CH_3 + H_2$	$3.72 \times 10^{-20}T^2 \exp(-4,400/T)$	3.3×10^{-11}	1.34×10^{-12}
(4) $CH_4 + O \rightarrow CH_3 + OH$	$3.49 \times 10^{-11} \exp(-4,560/T)$	3.6×10^{-12}	1.1×10^{-13}
(5) $CH_3 + O \rightarrow CH_2O + H$	1.66×10^{-10}	1.66×10^{-10}	8.8×10^{-17}
(6) $CH_3 + O_2 \rightarrow CH_2O + OH$	3.32×10^{-14}	3.32×10^{-14}	7.2×10^{-20}
(7) $CH_2O + O \rightarrow CHO + OH$	$8.30 \times 10^{-11} \exp(-2,300/T)$	2.63×10^{-11}	1.5×10^{-14}
(8) $CH_2O + OH \rightarrow CHO + H_2O$	$8.97 \times 10^{-10} \exp(-3,170/T)$	1.84×10^{-10}	1.42×10^{-14}
(9) $CH_2O + H \rightarrow CHO + H_2$	$2.24 \times 10^{-11} \exp(-1,890/T)$	8.7×10^{-12}	7.0×10^{-15}
(10) $CH_2O + M \rightarrow CHO + H + M$	$6.64 \times 10^{-12} \exp(-18,500/T)$	6.38×10^{-16}	5.4×10^{-32}
(11) $CHO + O \rightarrow CO + OH$	1.66×10^{-10}	1.66×10^{-10}	1.9×10^{-19}
(12) $CHO + OH \rightarrow CO + H_2O$	1.66×10^{-10}	1.66×10^{-10}	2.1×10^{-20}
(13) $CHO + H \rightarrow CO + H_2$	3.32×10^{-10}	3.32×10^{-10}	4.6×10^{-19}
(14) $CHO + M \rightarrow CO + H + M$	$8.30 \times 10^{-12} \exp(-9,570/T)$	7.0×10^{-14}	1.0×10^{-35}
(15) $CO + OH \rightarrow CO_2 + H$	$6.64 \times 10^{-12} \exp(-4,030/T)$	8.9×10^{-13}	3.8×10^{-13}
(16) $H_2 + OH \rightarrow H + H_2O$	$4.82 \times 10^{-10} \exp(-5,530/T)$	3.0×10^{-11}	3.0×10^{-12}
(17) $H_2 + O \rightarrow H + OH$	$5.31 \times 10^{-10} \exp(-7,540/T)$	1.2×10^{-11}	1.1×10^{-11}
(18) $H + O_2 \rightarrow O + OH$	$3.65 \times 10^{-10} \exp(-8,450/T)$	5.34×10^{-12}	2.3×10^{-11}
(19) $H + OH + Ar \rightarrow H_2O + Ar$	$2.31 \times 10^{-26}T^{-2}$	5.8×10^{-33}	5.14×10^{-21}
(20) $H + OH + H_2O \rightarrow H_2O + H_2O$	$3.86 \times 10^{-25}T^{-2}$	9.65×10^{-32}	8.56×10^{-20}
(21) $H + HO_2 \rightarrow OH + OH$	$4.15 \times 10^{-10} \exp(-950/T)$	2.58×10^{-10}	1.0×10^{-15}
(22) $H + O_2 + M \rightarrow HO_2 + M$	$4.13 \times 10^{-33} \exp(500/T)$	5.32×10^{-33}	3.6×10^{-14}
(23) $OH + OH \rightarrow H_2O + O$	$9.13 \times 10^{-11} \exp(-3,520/T)$	1.57×10^{-11}	2.0×10^{-12}

[a]Units: no. particles, cm^3, sec, K.

significant sensitivities (see Table 3–3). The conclusion that can be drawn is that reactions 1, 6, 14, and 18 are the four most important reactions for all the species. Reaction 15 is important for the production of CO_2, and some reactions of lesser significance are 10, 3, 8, 13, 2 and 5; all other reactions are insensitive. As is plain from this example, the number of reactions that are needed to model the process can be reduced considerably. This becomes very useful in studying atmospheric and combustion processes, which may involve over 200 elementary reactions.

Figure 3–18. Calculated concentration profile for CH_4-O_2-Ar system at 2,000 K and 1 atm total pressure (a) Profiles for CH_4, H_2O, CO, and hydrocarbon species.

Figure 3–18. (b) Profiles for CH_4, CO, and HO_x species. [From *Combusion Science and Technology 15*, 99 (1976).]

In this section, we have introduced a technique which can be of great utility in analyzing complex kinetic models. Although the emphasis here has been on analysis of phenomenological kinetic equations, the same principles can be applied to elucidating the relationships between microscopic quantities such as details of the potential energy surface, discussed in chapter 7, and properties such as scattering cross sections (chapter 8) and statistical rate coefficients (chapters 10 and 11). This use of sensitivity techniques[47,48] can answer such basic questions as which characteristic molecular parameters are most relevant for determining the rates and products of chemical reactions.

Sec. 3.3   Treatment of Kinetic Data   **149**

TABLE 3–3 METHANE OXIDATION SENSITIVITY ANALYSIS, TIME $= 2.5 \times 10^{-5}$ SEC [REFERENCE 46]

Component	Most sensitive rate constants[a]									
CH_4	1	1	18	14	10	3	13	4	8	2
CH_3	6	1	14	5	8	2	13	4	18	10
H	1	6	14	5	8	2	13	4	18	10
OH	1	18	6	14	2	8	10	13	8	7
H_2O	1	6	18	3	10	14	4	5	13	7
H_2	1	6	14	10	8	18	2	5	13	12
CO	1	6	18	8	10	14	3	2	8	13
O	18	1	6	5	14	10	8	3	4	2
CO_2	1	6	18	15	2	10	14	3	8	5
O_2	1	6	18	10	3	14	13	4	5	12
CH_2O	6	8	1	2	5	18	7	14	10	13
CHO	6	1	18	14	13	8	10	12	3	4
HO_2	1	6	14	22	10	21	13	3	18	8

[a]Reaction numbers are given in Table 3–2. Reactions are listed in order of decreasing sensitivity from left to right.

REFERENCES

1. L. Batt and F. R. Cruickshank, *J. Chem. Soc. A*, 261 (1967).

2. S. Chapman, *Phil. Trans. Roy. Soc. (London), A217*, 115 (1918).

3. C. J. Howard, *J. Phys. Chem. 83*, 3 (1976).

4. K. H. Hoyermann, "Interaction of Chemical Reactions, Transport Processes, and Flow" in *Physical Chemistry, An Advanced Treatise: Kinetics of Gas Reactions*, edited by H. Eyring, D. Henderson, and W. Jost (New York: Academic Press, 1975). Vol. 6B, pp. 931–1006.

5. F. Kaufman, *Prog. React. Kinet. 1*, 3 (1961).

6. V. M. Bierbaum, G. B. Ellison, and S. R. Leone, in *Gas Phase Ion Chemistry*, Vol. 3, edited by M. T. Bowers, (New York: Academic Press, 1984), p. 1.

7. G. Porter, *Proc. Roy. Soc. (London) A200*, 284 (1950).

8. M. I. Christie, R. G. W. Norrish, and G. Porter, *Proc. Roy. Soc. (London) A216*, 152 (1952).

9. B. S. Yamanashi and A. V. Nowak, *J. Chem. Ed. 45*, 705 (1968).

10. A. Declemy and C. Rulliere, *Rev. Sci. Instrum. 57*, 2733 (1986).

11. G. A. Laguna and S. L. Baughcum, *Chem. Phys. Lett. 88*, 568 (1982).

12. A. Wahner and C. Zetzsch, *Berg. Bunsenges. Phys. Chem. 89*, 323 (1985).

13. R. J. Balla and L. Pasternack, *J. Phys. Chem. 91*, 73 (1987).

14. J. Wormhoudt, A. C. Stanton, and J. A. Silver, *Proc. SPIE 452*, 88 (1984).

15. P. L. Houston, *Adv. Chem. Phys. 47*, 615 (1981).

16. F. J. Wodarczyk and C. B. Moore, *Chem. Phys. Lett. 26*, 484 (1974).

17. R. N. Zare and P. J. Dagdigian, *Science 185,* 739 (1974).

18. J. L. Kinsey, *Ann. Rev. Phys. Chem. 28,* 349 (1977).

19. A. Schultz, H. W. Cruse, and R. N. Zare, *J. Chem. Phys. 57,* 1354 (1972).

20. H. W. Cruse, P. Dagdigian, and R. N. Zare, *Farad. Disc. Chem. Soc. 55,* 277 (1973).

21. L. F. Phillips, *Progr. Reaction Kinetics,* vol. 7, edited by K. R. Jennings and R. B. Cundall (Oxford: Pergamon Press, 1973).

22. H. S. Johnston, S. E. McGraw, T. T. Paukert, L. W. Richards, and J. van den Bogaerde, *Proc. Nat. Acad. Sci. USA 57,* 1146 (1967).

23. D. A. Parkes, D. M. Paul, C. P. Quinn, and R. C. Robson, *Chem. Phys. Lett. 23,* 425 (1973).

24. P. M. Johnson and C. E. Otis, *Ann. Rev. Phys. Chem. 32,* 139 (1981).

25. S. Arepalli, N. Presser, D. Robie, and R. J. Gordon, *Chem. Phys. Lett. 117,* 64 (1985).

26. I. R. Slagle, F. J. Pruss, and D. Gutman, *Int. J. Chem. Kinet. 6,* 111 (1974).

27. P. R. Bevington, *Data Reduction and Error Analysis for the Physical Sciences* (New York: McGraw-Hill, 1969).

28. H. D. Young, *Statistical Treatment of Experimental Data* (New York: McGraw-Hill, 1962).

29. N. R. Draper and H. Smith, *Applied Regression Analysis* (New York: Wiley, 1966).

30. R. J. Cvetanovic, R. P. Overend, and G. Paraskevopoulos, *Int. J. Chem. Kinet. S1,* 249 (1975).

31. R.J. Cvetanovic and D. L. Singleton, *Int. J. Chem. Kinet. 9,* 481 (1977).

32. R. J. Cvetanovic and D. L. Singleton, *Int. J. Chem. Kinet. 9,* 1007 (1977).

33. R. J. Cvetanovic, R. P. Overend, and G. Paraskevopoulos, *J. Phys. Chem. 83,* 50 (1979).

34. C. Eisenhart, *Science 160,* 1201 (1968).

35. C. Eisenhart, *J. Res. Natl. Bur. Stand., Sec. C 67,* 161 (1963).

36. E. M. Pugh and G. H. Winslow, *The Analysis of Physical Measurements* (Reading: Addison-Wesley, 1966).

37. A. Fontijn and W. Felder in *Reactive Intermediates in the Gas Phase: Generation and Monitoring,* edited by D. W. Setser (New York: Academic Press, 1979), pp. 59–151.

38. C. Daniel and F. S. Wood, *Fitting Equations to Data,* 2d ed. (New York: Wiley, 1980).

39. D. W. Marquardt, *J. Soc. Indust. Appl. Math. 11,* 431 (1963).

40. R. W. Atherton, R. B. Schainker, and E. R. Ducot, *AIChE J. 21,* 441 (1975).

41. R. P. Dickinson and R. J. Gelinas, *J. Comput. Phys. 21,* 123 (1976).

42. J. W. Tilden, V. Costanza, G. J. McRae, and J. H. Seinfeld in *Modelling of Chemical Reaction Systems,* edited by K. H. Ebert, P. Deuflhard, and W. Jüger (Berlin: Springer-Verlag, 1981) pp. 69–91.

43. E. P. Dougherty, J.-T. Hwang, and H. Rabitz, *J. Chem. Phys. 71,* 1794 (1979).

44. J.-T. Hwang, E. P. Dougherty, S. Rabitz, and H. Rabitz, *J. Chem. Phys. 69,* 5180 (1978).

45. E. P. Dougherty and H. Rabitz, *Int. J. Chem. Kinet. 11,* 1237 (1979).

46. A. A. Boni and R. C. Penner, *Combust. Sci. Tech. 15,* 99 (1976).

47. H. Rabitz, *Comput. Chem. 5,* 167 (1981).

48. L. Eno and H. Rabitz, *Adv. Chem. Phys. 51,* 177 (1982).

APPENDIX 3.1 LEAST-SQUARES METHOD IN MATRIX FORM

In section 3.3.1.3, the basic concepts of the least-squares method were introduced. When large sets of data need to be analyzed it becomes convenient to use a matrix formalism of this method. Accordingly, we develop such a formalism here. Recall that the basic principle of the least-squares method is to minimize the sum of the squares of the errors of the individual measurements, given by the difference between the measured and calculated values, for a function

$$\mathbf{y} = f(\mathbf{p}, t) \tag{3A–1}$$

which is linear with respect to the parameters to be determined. The residual, which is the weighted sum of squares of the errors, i.e.,

$$Q = \sum_j \omega_j (y_j - f(\mathbf{p}, t))^2 \tag{3A–2}$$

is maximized with respect to each parameter p_j by

$$\frac{\partial Q}{\partial p_j} = 0 \qquad j = 1, 2, \ldots, M \tag{3A–3}$$

where M is the number of parameters.

Equation 3A–3 can be written in matrix form. To illustrate, suppose that

$$\mathbf{y} = f(\mathbf{p}, t) = p_1 + p_2 t \tag{3A–4}$$

Then we can write the residual as

$$Q = \sum_j^M \omega_j (y_j - (p_1 + p_2 t))^2 \tag{3A–5}$$

The derivatives with respect to the parameters p_1 and p_2 are given by

$$\frac{\partial Q}{\partial p_2} = -2 \sum_{j=1}^M \omega_j (y_j - (p_1 + p_2 t_j)) = 0 \tag{3A–6}$$

and

$$\frac{\partial Q}{\partial p_1} = -2 \sum_{j=1}^M \omega_j t_j (y_j - (p_1 + p_2 t_j)) = 0 \tag{3A–7}$$

These equations can be rearranged as

$$+ \sum_{j=1}^M \omega_j p_1 + \sum_{j=1}^M \omega_j p_2 t_j = \sum_{j=1}^M \omega_j y_j \tag{3A–8}$$

and

$$+ \sum_{j=1}^M \omega_j t_j p_1 + \sum_{j=1}^M \omega_j t_j p_2 t_j = \sum_{j=1}^M \omega_j t_j y_j \tag{3A–9}$$

which in turn can be formulated in matrix form as

$$\begin{pmatrix} \sum\limits_j \omega_j & \sum\limits_j \omega_j t_j \\ \sum\limits_j \omega_j t_j & \sum\limits_j \omega_j t_j^2 \end{pmatrix} \begin{pmatrix} p_1 \\ p_2 \end{pmatrix} = \begin{pmatrix} \sum\limits_j \omega_j y_j \\ \sum\limits_j \omega_j t_j \end{pmatrix} \qquad (3A-10)$$

We now define the matrix A as

$$A = \begin{pmatrix} \sum\limits_j \omega_j & \sum\limits_j \omega_j t_j \\ \sum\limits_j \omega_j t_j & \sum\limits_j \omega_j t_j^2 \end{pmatrix} \qquad (3A-11)$$

the column matrix $\mathbf{p}$ as

$$\mathbf{p} = \begin{pmatrix} p_1 \\ p_2 \end{pmatrix} \qquad (3A-12)$$

and the column matrix C as

$$C = \begin{pmatrix} \sum\limits_j \omega_j y_j \\ \sum\limits_j \omega_j t_j y_j \end{pmatrix} \qquad (3A-13)$$

Equation 3A–10 can then be expressed in matrix form as

$$A \cdot \mathbf{p} = C \qquad (3A-14)$$

To find the set of parameters given by $\mathbf{p}$, we multiply equation 3A–14 by the inverse of matrix A from the left, i.e.,

$$A^{-1}A \cdot \mathbf{p} = A^{-1}C \qquad (3A-15)$$

Consequently, the best-fit parameter vector is given by

$$\mathbf{p} = A^{-1}C \qquad (3A-16)$$

The inverse matrix A^{-1} is calculated by standard procedures of linear algebra,[1,2] but is given here for convenience:

$$A^{-1} = \frac{\begin{pmatrix} \sum\limits_j \omega_j t_j^2 & \sum\limits_j \omega_j t_j \\ \sum\limits_j \omega_j t_j & \sum\limits_j \omega_j \end{pmatrix}}{\mathrm{DET} \begin{vmatrix} \sum\limits_j \omega_j & \sum\limits_j \omega_j t_j \\ \sum\limits_j \omega_j t_j & \sum\limits_j \omega_j t_j^2 \end{vmatrix}} \qquad (3A-17)$$

If one performs the matrix multiplication of equation 3A–16 using equations 3A–13 and 3A–17, the result is the vector of parameters, equation 3–43.

The standard deviation is

$$\sigma_{p_j} = \sigma (A^{-1})_{jj}^{1/2} \qquad (3A-18)$$

where $\sigma = [\sum_j \omega_j (y_j - f(\mathbf{p}, t))^2 / (M - 1)]^{1/2}$. When there is a large number of parameters to be determined, the matrix could become large, in which case it is not easily treated. In general, the matrix formalism can readily be used on the computer to find the vector **p** of the best-fit parameters and the standard deviation σp_j in each parameter.

REFERENCES

1. E. D. Nering, *Linear Algebra and Matrix Theory* (New York: Wiley, 1963).
2. K. Hoffman and R. Kunze, *Linear Algebra* (Englewood Cliffs, N.J.: Prentice-Hall, 1981).

PROBLEMS

3.1 Glucose is formed by the reaction of cellulose in the presence of HCl at various temperatures. The rate is found to be first order in the cellulose concentration, but depends on the HCl concentration [E. E. Harris and A. A. Kline, *J. Phys. Colloid Chem. 53*, 344 (1949)]. The following data are measured for k_1, the rate of hydrolysis of Douglas-fir wood cellulose at various temperatures:

[HCl] moles/ℓ	$T = 160°C$	k_1, sec^{-1} $T = 180°C$	$T = 190°C$
0.055	0.00203	0.0190	0.0627
0.11	0.00486	0.0455	0.149
0.22	0.01075	0.011	0.357
0.44	0.0261	0.284	—
0.88	0.0672	—	—

Find the best least-squares values of k_1' and n that fit the relationship

$$k_1 = k_1' [HC1]^n$$

at each temperature.

3.2 An experiment is designed in which the reaction rate between gaseous H_2 and Br_2 will be measured in a static photolysis vessel of 50 cm diameter at a total pressure of 1 torr. How much time must be allowed for the sample to be well mixed before the reaction is initiated?

3.3 This problem examines the sensitivity of the concentration of a substance A to variations in Arrhenius parameters at $T = 208$ K.

(a) Calculate analytic sensitivity coefficients for the second-order reaction

$$A + A \quad \xrightarrow{K} \quad products$$

where $K = k_1 \exp(-k_2/T)$. Suppose that

$$[A]_0 = 2.0 \text{ mole } \ell^{-1}$$

$$k_1 = 1.79 \times 10^{10} \ell \text{ mole}^{-1} \text{ sec}^{-1}$$

$$k_2 = 500 \text{ K}$$

and the uncertainty range for the parameters is

$$8.9 \times 10^9 < k_1 < 3.59 \times 10^{10}$$

and

$$0 < k_2 < 1,000$$

(b) Plot the sensitivity coefficients as a function of time. What can you conclude from your plot?

(c) Plot the relative sensitivities $\Delta k_i \partial A_j / \partial k_i$ as a function of time. What can you conclude about the importance of the effects of parameter uncertainties on the concentration of A?

(d) Suppose that the range of uncertainty for the parameter k_1 is

$$0 < k_1 < 3.59 \times 10^{10}$$

Plot the sensitivity coefficients and the relative sensitivities as a function of time. What can you conclude?

3.4 Consider the following set of four reactions in three components.

Numerical values are assigned to the rate constants and initial conditions as follows:

$k_1 = 1$	$[C_1]_0 = 2$
$k_2 = 499.5$	$[C_2]_0 = 1$
$k_3 = 499.5$	$[C_3]_0 = 0$
$k_4 = 1$	

(a) Using the Laplace transform method or the matrix method solve for $[C_1]$, $[C_2]$, and $[C_3]$. Plot the evolution with time of these concentrations.

(b) Calculate analytic sensitivity coefficients for this reaction system.

(c) Plot the sensitivity coefficients as a function of time. When will the sensitivity function attain a maximum?

(d) Suppose that the uncertainty range for the parameters is

$$0.1k_i < k_i < 10k_i \qquad\qquad (i = 1, \ldots, 4)$$

Plot the relative sensitivity as a function of time.

CHAPTER 4

Reactions in Solution

Although the principal emphasis in this text is on reactions that take place in the gas phase, many important chemical transformations occur in liquid solutions. The synthesis of organic chemicals, the commercial production of many chemical commodities, and the metabolic processes which enable a person to be reading this book all occur in liquids, often in aqueous solutions. It is therefore essential to have an understanding of the nature of chemical reactions in solutions.

4.1 GENERAL PROPERTIES OF REACTIONS IN SOLUTION

There are significant differences between reactions in solution and reactions in the gas phase. In solution, the concept of isolated collisions between reactant molecules is practically meaningless. The reactant molecules interact continuously with solvent molecules. The distribution of collisions is also quite different in the two cases. In the gas phase, there is a random distribution of collisions between reactant molecules, so that the occurrence of a collision does not affect the likelihood of a subsequent collision. Such processes are said to be *Markovian* in nature. By contrast, the mean collision frequency per unit density may be the same in liquids as in gases, but the distribution of the collisions in time is quite different. If two solute molecules collide in solution, it is highly probable that they will undergo numerous successive collisions before they separate. In effect, the liquid structure forms a "cage" of solvent molecules around the reactants, holding them in close proximity.[1] Such a set of consecutive collisions in solution is referred to as an *encounter*. The different distributions of collisions for the gas phase and in solution are illustrated in Figure 4–1.

The effects of solvent on reaction rate may be either chemical or physical, i.e., environmental. If the solvent molecules participate directly in the reaction mechanism, the effect is chemical. In this case, the solvent will appear explicitly in the stoichiometric equation as a reactant or product. If the effect of the solvent is merely to modify the interaction between reacting species, as in the alteration of the collision distribution just described, its effect may be termed physical. Solvent molecules may also play a catalytic role (see chapter 5), in which case they appear in the explicit kinetic equations but not in the stoichiometric equation for the reaction.

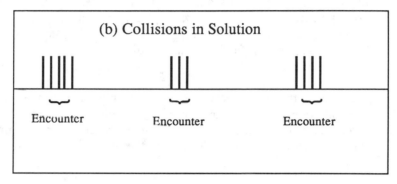

Figure 4–1. Distribution of collisions between solute molecules. Record (a) shows schematically the gas-phase collisions, while record (b) shows the distribution of collisions when a solvent is present.

The solvent may play a variety of roles during the course of the reaction. For example, solvation stabilizes the formation of ions: $NaCl \rightarrow Na^+ + Cl^-$ is highly endothermic in the gas phase, but exothermic in aqueous solution. Thus, ionic reactions are quite common in solution mechanisms, but relatively rare in the gas phase. Rapid energy transfer between reactant and solvent molecules maintains thermal equilibrium in solution reactions, while thermal disequilibrium is often important in gas-phase reactions. Also, as discussed in the sections that follow, interactions between reacting ions are influenced by the dielectric character of the solvent and by the presence of other ions in solution.

4.2 PHENOMENOLOGICAL THEORY OF REACTION RATES

There is an upper limit to the rate of a bimolecular reaction in solution which is analogous to the hard-sphere encounter frequency in the kinetic theory of gases (see Sec. 8.1). The rate with which two reactant molecules A and B can come together is governed by the phenomenological laws of diffusion and conduction. Once the reactants have come together, they do not simply collide and separate; instead, they are constrained by the solvent molecules, and typically remain trapped for approximately 10^{-11} sec undergoing numerous collisions with each other, as shown in Figure 4–1(b). If the activation energy for the chemical reaction is low, so that the reaction between A and B is facile, the encounter frequency sets an upper bound to the bimolecular reaction rate. This frequency, in turn, is controlled by the processes of diffusion and conduction, which determine the rate at which the reactive molecules can get into the same solvent cage.

4.2.1 Diffusion and Conduction

The two factors which control the movement of reactant molecules in solution are random diffusive motion and any conductive forces that are present if a potential energy gradient exists in the solution. The former process describes the tendency of solute molecules to move from regions of high to low concentration in the solution. This is expressed by Fick's first law of diffusion,

$$\phi_i = -D_i \nabla c_i \qquad (4-1)$$

where ϕ_i is the flux density of solute i in units of molecules $cm^{-2}sec^{-1}$, ∇c_i is the concentration gradient in units of molecules $cm^{-3}cm^{-1}$, and D_i is the diffusion coefficient for solute i relative to the solvent. If the concentration gradient is spherically symmetric, so that it depends only on the distance r from an origin, equation 4–1 becomes

$$\phi_i = -D_i \left(\frac{dc_i}{dr} \right) \qquad (4-2)$$

A potential energy gradient can also cause a net motion of the solute in a definite direction. The solute flux arising from a potential gradient is written as

$$\phi_i = v_i c_i \qquad (4-3)$$

where v_i and c_i are the velocity and concentration respectively, of the solute i. Potential energy gradients, arising from an electric field, are particularly important for ions in solution. An ion i with a charge $z_i e$ moving in an electric potential gradient has a mobility μ_i given by

$$\mu_i = \frac{v_i}{\nabla E} \qquad (4-4)$$

where ∇E is the electric potential gradient. For spherically symmetric concentration and potential gradients, we have, from equations 4–3 and 4–4,

$$\phi_i = \frac{-z_i}{|z_i|} \mu_i c_i \frac{dE}{dr} \qquad (4-5)$$

where the factor $-z_i/|z_i|$ is introduced to fix the sign. A negative ion has a positive flux in the direction of increasing E. Standard units for dE/dr and μ_i are volts cm^{-1} and cm^2 volt$^{-1}sec^{-1}$.

The total flux density of a solute is the sum of the diffusion term and the term for the potential energy gradient. Thus, for an ion, we have

$$\phi_i = -\left(D_i \frac{dc_i}{dr} + \frac{z_i}{|z_i|}\mu_i c_i \frac{dE}{dr}\right) \qquad (4\text{–}6)$$

by combining equations 4–2 and 4–5. The first term in equation 4–6 represents diffusion, the second conduction. In a chemical reaction both terms are important. For example, in a reaction between two positively charged ions conduction will cause the ions to move apart. Diffusion must then bring the reactants together.

At equilibrium the flux density must be zero everywhere in the solution, so that

$$\phi_i = 0 = -D_i \frac{dc_i}{dr} - \frac{z_i}{|z_i|}\mu_i c_i \frac{dE}{dr}$$

and

$$\frac{dc_i}{c_i} = \frac{-z_i}{|z_i|}\frac{\mu_i}{D_i} dE \qquad (4\text{–}7)$$

Integrating equation 4–7 gives

$$\ln\frac{c_i}{c_i^0} = \frac{-z_i}{|z_i|}\frac{\mu_i}{D_i} E \qquad (4\text{–}8)$$

where c_i^0 is the concentration of ions at a reference value of r for which $E = 0$. It can be assumed that at equilibrium the concentration must obey the Boltzmann law

$$c_i = c_i^0 \exp\left[\frac{-V(r)}{k_B T}\right] \qquad (4\text{–}9)$$

with the potential energy $V(r)$ of the ions in the electric field given by

$$V(r) = z_i e E \qquad (4\text{–}10)$$

Equations 4–8, 4–9, and 4–10 can be combined to obtain the relation between the ion's mobility and its diffusion coefficient, viz.,

$$\frac{\mu_i}{D_i} = \frac{|z_i|e}{k_B T} \qquad (4\text{–}11)$$

This equation is important for determining an ion's diffusion coefficient, since its mobility can be determined directly from a conductance measurement. Finally, if equations 4–10 and 4–11 are inserted into equation 4–6, we have

$$\phi_i = -D_i\left[\frac{dc_i}{dr} + \frac{1}{k_B T}c_i \frac{dV(r)}{dr}\right] \qquad (4\text{–}12)$$

4.2.2 Solution of the Conduction-Diffusion Equation

The relative motion of A and B molecules in a solvent is depicted in Figure 4–2. At a microscopic level this motion is related to local concentration and potential energy gradients. It will thus be assumed that on the average these gradients are described

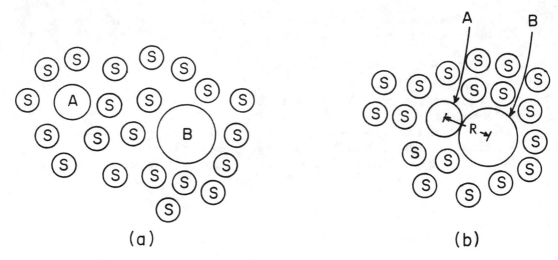

Figure 4–2. Model for AB relative diffusion through solvent. (a) A and B separated by solvent sheath. (b) A and B have diffused into contact when reaction occurs; the contact distance is R. [Adapted from P. C. Jordan, *Chemical Kinetics and Transport* (New York: Plenum, 1979).]

by the preceding equations. Consequently, from equation 4–12, the flux density of B across a sphere of radius r around each A is given by

$$\phi_B = -(D_A + D_B)\left[\frac{d[B]_r}{dr} + \frac{1}{k_B T}[B]_r \frac{dV(r)}{dr}\right] \tag{4–13}$$

where the diffusion coefficient is $D_A + D_B$, since A and B are diffusing simultaneously, and $[B]_r$ is the local concentration of B at a distance r from A. Reaction occurs whenever the A and B molecules come into contact, which is stipulated by the contact radius R. Thus, the total flux (flux density times area) of B across a sphere of radius R about A equals the reaction rate:

$$\frac{-d[B]}{dt} = k[A][B] = -[A][4\pi R^2 \phi_B] \tag{4–14}$$

At steady state the total flux through a sphere of radius r about B is constant for all values of r. Thus, R in equation 4–14 can be replaced by r and ϕ_B can be replaced with the right-hand side of equation 4–13 to give

$$k[B] = 4\pi r^2[D_A + D_B]\left[\frac{d[B]_r}{dr} + \frac{[B]_r}{k_B T}\frac{dV(r)}{dr}\right] \tag{4–15}$$

since k and $[B]$ on the left-hand side are independent of r. To integrate equation 4–15, write it in the form

$$k[B] = 4\pi r^2[D_A + D_B]e^{-V(r)/k_B T}\frac{d}{dr}\{[B]_r e^{V(r)/k_B T}\} \tag{4–16}$$

Dividing both sides of the equation by $r^2 \exp[-V(r)/k_B T]$ and integrating over r gives

$$k[B] \int_{R}^{\infty} \frac{e^{V(r)/k_B T}}{r^2} dr = 4\pi (D_A + D_B) \int_{R}^{\infty} \left(\frac{d}{dr} \{ [B]_r e^{V(r)/k_B T} \} \right) dr \qquad (4\text{--}17)$$

To complete the integration first define

$$\beta^{-1} \equiv \int_{R}^{\infty} \frac{e^{V(r)/k_B T}}{r^2} dr \qquad (4\text{--}18)$$

and then change the limits of integration from distance to concentration. At the limits $r = R$ and $r = \infty$, the concentration of B is $[B]_R$ and the bulk concentration, $[B]$, respectively. Thus equation 4–17 becomes

$$k[B] = 4\pi (D_A + D_B)\beta \int_{[B]_R}^{[B]} d\{ [B]_r e^{V(r)/k_B T} \} \qquad (4\text{--}19)$$

As $r \to \infty$, $V(r) \approx 0$, so one has, from equation 4–19,

$$k = \frac{4\pi (D_A + D_B)\beta}{[B]} \{ [B] - [B]_R e^{V(R)/k_B T} \} \qquad (4\text{--}20)$$

For the general case there is a finite rate constant k_R for reaction during the encounter, where

$$\frac{-d[B]}{dt} = k[A][B] = k_R[A][B]_R \qquad (4\text{--}21)$$

Substituting $[B]_R = (k/k_R)[B]$ into equation 4–20 and rearranging gives

$$k = \frac{4\pi (D_A + D_B)\beta}{1 + [4\pi (D_A + D_B)\beta/k_R \exp\{-V(R)/k_B T\}]} \qquad (4\text{--}22)$$

This equation is the general phenomenological rate constant for a bimolecular reaction in solution. Before illustrating its uses, it is worthwhile reviewing the model assumed in deriving the equation. Equation 4–12, which describes bulk diffusion and conduction, is used to describe the relative motion between molecules of A and molecules of B at the microscopic level. Note that to integrate equation 4–16 the variation of the diffusion coefficient with concentration is neglected. However, if the reactant concentrations are low and if a constant solution environment is maintained, the diffusion coefficient does remain approximately constant during the reaction. To obtain the general diffusion equation 4–12, a steady state has been assumed; this assumption breaks down for reactions with short half-lives. Finally, spherical symmetry has been assumed for the reactants, but it holds only for simple reactions. Introducing nonspherical symmetry results in steric and alignment effects for reaction and reduces the solid angle in equation 4–22 from 4π to a smaller value. (See discussion preceding equation 4–14).

4.3 DIFFUSION-LIMITED RATE CONSTANT

The diffusion-limited rate constant results when each encounter between A and B molecules leads to reaction. There are two approaches for finding this rate constant from the foregoing equations. If A and B do indeed react at each encounter, $[B]_R$ is zero since no B molecules exist at the critical radius. Thus, from equation 4–20, the diffusion-limited rate constant is

$$k_D = 4\pi (D_A + D_B)\beta \qquad (4\text{–}23)$$

Alternatively, for the rate constant to be diffusion-limited, the rate constant k_R during the encounter is very large, so that the denominator of equation 4–22 becomes unity.

For uncharged reactants $V(r) \approx 0$, so that equation 4–18 gives $\beta = R$. The diffusion-limited rate constant is then

$$k_D = 4\pi (D_A + D_B)R \qquad (4\text{–}24)$$

For the spherical particle model used in the derivation of equation 4–22, the diffusion coefficient is related to the viscosity via the Einstein-Stokes relationship

$$D = \frac{k_B T}{6\pi r \eta(T)} \qquad (4\text{–}25)$$

where $\eta(T)$ is the temperature-dependent viscosity of the solution, and r is the radius of the particle. Writing the critical radius R as $(r_A + r_B)$ and substituting equation 4–25 into equation 4–24 gives

$$k_D = \frac{2k_B T (r_A + r_B)^2}{3\eta(T) r_A r_B} \qquad (4\text{–}26)$$

Thus, the rate constant is proportional to temperature and inversely proportional to viscosity.

It is of interest to estimate k_D for neutral reactants. A representative value for r_A and r_B is 2×10^{-8} cm, so that the critical radius R is 4×10^{-8} cm. A typical aqueous solution diffusion coefficient at 300 K is 1×10^{-5} cm^2/sec. Thus, from equation 4–24, the value for k_D in liters/mole-sec is

$$k_D = \frac{4\pi (4 \times 10^{-8})(2 \times 10^{-5})(6 \times 10^{23})}{10^3}$$

$$= 6 \times 10^9$$

This is compared with 3×10^{11} liters/mole-sec for a normal binary gas collision rate constant. Accordingly, liquid-phase reactions are slowed down by a factor of about 50 due to diffusion, with the rate dependent on the viscosity of the solvent.

A diffusion-limited rate constant can also be evaluated for reactions between ions. The potential energy between two ions in solution is (see chapter 7, section 7.1)

$$V(r) = \frac{z_A z_B e^2}{\epsilon r} \qquad (4\text{–}27)$$

where ϵ is the dielectric constant. Inserting this expression for $V(r)$ into equation 4–18 yields

$$\beta = \frac{-z_A z_B r_0}{1 - \exp(z_A z_B r_0 / R)} \qquad (4\text{–}28)$$

The parameter r_0 has dimensions of length and is given by

$$r_0 = \frac{e^2}{\epsilon k_B T} \tag{4-29}$$

For water at 25°C, r_0 equals 7 Å. The term β is positive for any combination of ion charges. For ions of the same charge the exponential in equation 4–28 is larger than unity, since r_0 is ordinarily greater than R. If the ions are oppositely charged, the exponential is small and β is approximately $-z_A z_B r_0$. The diffusion-limited rate constant for reactions between ions is found by inserting the expression for β in equation 4–28 into equation 4–23. The rate constant here differs from that for reactions between neutrals, given by equation 4–24, by the ratio β/R. Representative values of this ratio are listed in Table 4–1 for a water solution with R equal to 4×10^{-8} cm. As expected, ions of opposite charge react more rapidly than neutral molecules, while ions of the same charge react more slowly.

TABLE 4–1 EFFECT OF REACTANT CHARGE ON THE DIFFUSION-LIMITED RATE CONSTANT[a]

$z_A z_B$	β/R
0	1
−1	2.14
−2	3.65
−3	5.30
1	0.362
2	0.105
3	0.029

[a] β/R is for an aqueous solution at 25°C; r_0 in equation (4–29) equals 7.1×10^{-8} cm and R equals 4.0×10^{-8} cm.

4.4 SLOW REACTIONS

The preceding solution reactions may all be depicted by the mechanism

$$A + B \underset{k_{uni}}{\overset{k_D}{\rightleftharpoons}} AB \overset{k_2}{\longrightarrow} \text{products} \tag{4-30}$$

where AB is the encounter complex and k_{uni} is the unimolecular rate constant for dissociation of the encounter complex (i.e., diffusion of the reactants apart from the solvent cage). A steady-state treatment of this mechanism yields the rate law

$$\frac{-d[B]}{dt} = k[A][B] = \frac{k_2 k_D}{k_{uni} + k_2}[A][B] \tag{4-31}$$

Only if k_2 is much larger than k_{uni} is the overall reaction rate diffusion-limited. For a slow reaction where $k_{uni} \gg k_2$, the first step in equation 4–30 is in equilibrium and

$$k = Kk_2 \tag{4-32}$$

where $K = k_D/k_{uni}$.

The approach used in section 4.2 for calculating the bimolecular rate constant may also be used to calculate the maximum rate constant for dissociation of the encounter complex. The only difference is in the boundary conditions which become [AB] = 0 and $V = 0$ at $r = \infty$, and [AB] = $1/\Delta V$ and $V = V(R)$ at $r = R$. In other words, at infinite separation the complex does not exist. On the other hand, the volume ΔV contains the encounter complex AB and is approximately $4/3\pi R^3$. Beginning with equation 4–13, and using the same reasoning as before, we obtain

$$k_{\text{uni}} = 4\pi (D_A + D_B) \frac{\beta \exp[V(R)/k_B T]}{\Delta V} \qquad (4\text{--}33)$$

The equilibrium constant in equation 4–32 is then

$$K = \frac{k_D}{k_{\text{uni}}} = \frac{4\pi R^3}{3} 10^{-3} N_0 \exp[-V(R)/k_B T] \qquad (4\text{--}34)$$

for concentration in units of M^{-1}.

The phenomenological treatment developed in section 4.2 can be used to interpret slow reactions. In these cases there is a substantial barrier for the encounter complex to form reaction products, so that $k_D \gg k_R \exp[-V(R)/k_B T]$. Thus, equation 4–22 reduces to

$$k = k_R \exp[-V(R)/k_B T] \qquad (4\text{--}35)$$

This equation can be combined with equations 4–32 through 4–34 to find the relationship between k_2 in equation 4–30 and k_R.

Equation 4–35 has been used most frequently for interpreting reactions between ions. Substituting the potential given by equation 4–27 gives a rate constant dependent on the dielectric constant and contact radius; i.e.,

$$k = k_R \exp(-z_A z_B e^2 / \epsilon k_B T R) \qquad (4\text{--}36)$$

By varying the composition of the solvent, the dielectric constant for the reaction can be adjusted. A plot of the logarithm of k versus $1/\epsilon$ should be linear with the slope giving the value of R. Such an analysis has been performed for slow ionic reactions, and physically realistic values of R have been obtained.

4.5 EFFECT OF IONIC STRENGTH ON REACTIONS BETWEEN IONS

If the ionic strength of a solution is sufficiently high ($> 10^{-4}$ M) so that the ion-atmosphere affects conductance and equilibrium properties, the derivation of the diffusion-controlled rate constant using equations 4–27 through 4–29 is no longer valid. Instead, the simple coulombic potential given by equation 4–27 must be modified to account for the shielding which results from the ion atmosphere. The approach taken here is to use a form for the potential developed by Debye and Hückel.[2] Using that approach, a general expression for the phenomenological rate constant can be derived from equation 4–22.

To calculate the potential $V(r)$ between two ions in an ionic atmosphere, we begin with the Poisson differential equation. In the case of a spherically symmetric distribution of charges about a central ion, the electrical potential $E(r)$ is a function only of r, the distance from the central ion. For this situation the Poisson equation is

$$\frac{1}{r^2}\frac{d}{dr}\left[r^2\frac{dE(r)}{dr}\right] = -\frac{4\pi\rho}{\epsilon} \tag{4-37}$$

where ρ is the charge density at radius r. The charge density is positive if positive charges predominate, negative if negative charges predominate, and zero if they balance. The electrostatic potential $E(r)$ is due to the central ion and its surrounding atmosphere. The solutions to equation 4–37 give this potential at any r as a function of ρ. To solve Equation (4–37) requires evaluating ρ for the distribution of ions.

In an ionic solution ρ will not be zero in the region surrounding an ion A, since there is a tendency for ions of opposite sign to concentrate around A. The Boltzmann law, equation (4–9) gives the concentration of a particular ion i at a radius r from A. Since $V(r)$ is related to the electric potential via equation 4–10, we have

$$c_i = c_i^0 \exp[-z_i e E(r)/k_B T] \tag{4-38}$$

The charge density in a region of the solution is the summation of $c_i z_i e$ over the different kinds of ions (positive and negative) that may be present; i.e.,

$$\rho = \sum_i c_i z_i e$$

or

$$\rho = \sum_i c_i^0 z_i e \exp[-z_i e E(r)/k_B T] \tag{4-39}$$

The Debye-Hückel approximation, based on the assumption that the solution is so dilute that ions will rarely be close together, is now made. For this situation $z_i e E(r) \ll k_B T$, since the interionic potential is much less than the average thermal energy. The exponential factor in equation 4–39 may then be expanded as

$$\exp[-z_i e E(r)/k_B T] = 1 - \frac{z_i e E(r)}{k_B T} + \cdots. \tag{4-40}$$

where the higher-order terms are negligible. With this expansion equation 4–39 becomes

$$\rho = \sum_i c_i^0 z_i e - \frac{E(r)e^2}{k_B T}\sum_i c_i^0 z_i^2 \tag{4-41}$$

The first term is zero because there is electrical neutrality at large r, where $V(r)$ equals zero. Thus,

$$\rho = \frac{-e^2 E(r)}{k_B T}\sum_i c_i^0 z_i^2 \tag{4-42}$$

Substituting equation 4–42 into equation 4–37 gives the Poisson-Boltzmann equation

$$\frac{1}{r^2}\frac{d}{dr}\left[r^2\frac{dE(r)}{dr}\right] = \frac{4\pi e^2 E(r)}{\epsilon k_B T}\sum_i c_i^0 z_i^2$$

or

$$\frac{d}{dr}\left[r^2\frac{dE(r)}{dr}\right] = b^2 r^2 E(r) \tag{4–43}$$

where

$$b^2 = \frac{4\pi e^2}{\epsilon k_B T}\sum_i c_i^0 z_i^2 \tag{4–44}$$

The quantity $1/b$ has the dimensions of length and is often called the *Debye screening length*. It may be regarded as an approximate measure of the thickness of the ionic atmosphere surrounding an ion.

The term $\sum c_i^0 z_i^2$ is closely related to the ionic strength, defined as $I = \frac{1}{2}\sum m_i z_i^2$, where m_i is the molality of ionic species i in the solution. Since c_i^0 is the number of ions per cm³, we have

$$\sum_i c_i^0 z_i^2 = \frac{2N_0 d}{1,000} I \tag{4–45}$$

where N_0 is Avogadro's number and d is the density of the solution. Thus b^2 can be expressed as

$$b^2 = \frac{8\pi\, e^2 N_0 d}{1,000\epsilon k_B\, T} I \tag{4–46}$$

The solution to equation 4–43 is obtained by substituting $x = rE$ from which $d^2x/dr^2 = b^2 x$, and therefore,

$$x = C_1 e^{-br} + C_2 e^{br}$$

or

$$E(r) = \frac{C_1}{r}e^{-br} + \frac{C_2}{r}e^{br} \tag{4–47}$$

The integration constants C_1 and C_2 are determined by the boundary conditions. Clearly, C_2 is zero, since $E(r)$ approaches zero as r goes to infinity. C_1 can be determined by noting that when $b = 0$, i.e., the ionic strength is zero, the electrical potential is simply that of a single ion A in the absence of any other ions and is given by $E(r) = z_A e/\epsilon r$. Thus, the final solution is

$$E(r) = \frac{z_A e}{\epsilon r}\exp(-br) \tag{4–48}$$

Since, as previously discussed, a dilute ionic solution with small b is assumed in deriving this expression, the exponential may be expanded without further approximation to give

$$E(r) = \frac{z_A e}{\epsilon r} - \frac{z_A e}{\epsilon} b \qquad (4\text{--}49)$$

The first term is simply the electrical potential of an ion in a medium with dielectric constant ϵ. The second term is the potential resulting from the ions which form the ionic atmosphere about the ion A.

The interaction potential $V(r)$ between two ions A and B in an ionic solution can now be written. For B with charge z_B in the electric field of A, the potential energy is

$$V(r) = z_B e E(r) = \frac{z_A z_B e^2}{\epsilon r} - \frac{z_A z_B e^2}{\epsilon} b \qquad (4\text{--}50)$$

The first term is the potential at infinite dilution, denoted by $V^0(r)$. Dividing by $k_B T$ gives

$$\frac{V(r)}{k_B T} = \frac{V^0(r)}{k_B T} - z_A z_B r_0 b \qquad (4\text{--}51)$$

where r_0 is as defined in equation 4–29. To obtain the rate constant, β in equation 4–18 is evaluated using equation 4–51. The result is

$$\beta = \beta^0 \exp(z_A z_B r_0 b) \qquad (4\text{--}52)$$

where β^0, given by equation 4–28, is the value of β at infinite dilution. Inserting this expression for β and equation 4–51 for $V(r)$ into equation 4–22 yields

$$k = \frac{k_D^0 \exp(z_A z_B r_0 b)}{1 + k_D^0/k_R \exp[-V^0(R)/k_B T]} \qquad (4\text{--}53)$$

which may be written as

$$k = k^0 \exp(z_A z_B r_0 b) \qquad (4\text{--}54)$$

The rate constants k_D^0 and k^0 are for infinite dilution.

Equation 4–54 becomes

$$\log_{10} k = \log_{10} k^0 + 1.02 z_A z_B I^{1/2} \qquad (4\text{--}55)$$

at 25°C in water. Thus, a plot of $\log k$ vs. $I^{1/2}$ is linear with a slope of $1.02 z_A z_B$. A few examples are illustrated in Figure 4–3. The effect of the ionic strength is to increase the rate for ions with charges of the same sign and to decrease it when the charges are opposite in sign; the increased ionic atmosphere counteracts the simple electrostatic interaction between A and B. This variation of rate constants with ionic strength is often called the *primary salt effect*.

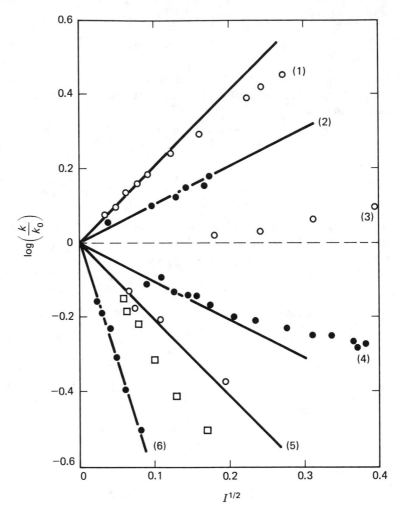

Figure 4–3. Plots of $\log_{10}(k/k^0)$ versus the square root of the ionic strength for ionic reactions of various types. The reactants are:

(1) $BrCH_2COO^- + S_2O_3{}^{2-}$

(2) $e_{aq} + NO_2^-$

(3) $H_3O^+ + C_{12}H_{22}O_{11}$ (inversion of sucrose)

(4) $H_3O^+ + Br^- + H_2O_2$

(5) $OH^- + Co(NH_3)_5Br^{2+}$ [ionic strength varies with NaBr (circles) and Na_2SO_4 (squares)]

(6) $Fe(H_2O)_6^{2+} + Co(C_2O_4)_3^{3-}$

[Adapted from R. E. Weston, Jr., and H. A. Schwarz, *Chemical Kinetics* (Englewood Cliffs, NJ: Prentice-Hall, Inc., 1972).]

4.6 LINEAR FREE-ENERGY RELATIONSHIPS

Empirical correlations have been very useful in systematizing a large number of rate coefficients and equilibrium constants for organic reactions in solution. They provide a predictive method for estimating rate data when no such data are available. Many of these correlations are of a general type called *linear free-energy relationships*.

Consider a general solution reaction denoted by

$$RX + A \longrightarrow \text{products} \tag{4-56}$$

where R contains the reactive site and X is a substituent not directly involved in the reaction. The empirical correlation which is made is to plot $\log k$ vs. $\log K$ for reactions of this type which simply differ by the substituent X, where k is the rate constant and K the equilibrium constant for the particular reaction. What is observed for many cases is that the plot is linear, i.e., $\log k = \alpha \log K + \text{constant}$, so that

$$\frac{k'}{k} = \left(\frac{K'}{K}\right)^{\alpha} \tag{4-57}$$

where k' and K' are for the reaction with the substituent X'.

Such a relationship between k and K is characteristic of an overall reaction mechanism controlled by one elementary reaction, which is apparently the case for many solution reactions. To derive equation 4–57, two assumptions must be made. The first is that the elementary rate constant is proportional to $\exp(-\Delta G^a/RT)$, where ΔG^a is the free energy of activation. In chapter 10 it is shown that this proportionality is predicted by transition state theory. The second assumption is that, for a specific class of reactions like those in equation 4–56, ΔG^a is related to the free energy of the overall reaction ΔG^0 according to

$$\Delta G^a = \alpha \, \Delta G^0 + \text{constant} \tag{4-58}$$

This is the linear free-energy relationship.

One of the best known linear free-energy relationships is the Hammett equation,[3]

$$\log_{10} K = \log_{10} K_0 + \sigma\rho \tag{4-59}$$

which correlates the effects of meta- or para-substituents on equilibrium constants for reactions involving aromatic compounds. K is the equilibrium constant for the reaction involving the substituted compound, K_0 is for the same reaction of the unsubstituted compound, σ is an empirical parameter that varies only with the substituent and its position, and ρ is an empirical parameter that is characteristic of the particular reaction series, including the solvent, that is being studied.

The ionization of benzoic acid in water at 25°C is the reference reaction used to establish values for σ and ρ. The definitions are $\rho = 1$ and $\sigma = \log(K_i/K_{i0})$, where K_i and K_{i0} are the ionization constants for the substituted and unsubstituted benzoic acid, respectively. Representative values of σ are listed in Table 4–2. These values can be inserted into equation 4–59 to determine values of ρ for benzoic acid in other solvents, or other sets of equilibria in various solvents.

TABLE 4–2 σ-VALUES OF COMMON SUBSTITUENTS

Substituent	σ_{meta}	σ_{para}
H	0	0
NH$_2$	−0.16	−0.66
CH$_3$	−0.07	−0.17
C$_6$H$_5$	0.06	−0.01
OH	0.12	−0.37
OCH$_3$	0.12	−0.27
F	0.34	0.06
Cl	0.37	0.23
Br	0.39	0.23
I	0.35	0.18

By combining equations 4–57 and 4–59, a free-energy relationship is found for correlating rates of substituted aromatic compounds. This relationship is given by

$$\log_{10}k = \log_{10}k_0 + \sigma\rho^a \qquad (4\text{–}60)$$

The σ values are those for substituted benzoic acids given in the previous paragraph. As stated above, the parameter ρ depends on the particular reaction.

Table 4–3 lists some values of ρ for acid dissociation equilibria. There are also extensive lists of ρ-values for rates of various reactions in different solvents. Essentially, the parameter ρ gives the sensitivity of the reaction (or equilibrium) to a change in substituent. A positive value of ρ means that a reaction (or particular equilibrium) responds to the substituent in the same ways as does benzoic acid ionization equilibrium. If $\rho > 1$ the reaction (equilibrium) is more sensitive to the substituent, while the reaction is less sensitive if $0 < \rho < 1$. If ρ is negative, the substituents affect the reaction (equilibrium) in a manner opposite to what is found for benzoic acid dissociation.

There are several ways to identify the Hammett equations 4–59 and 4–60 as linear free-energy relationships. Since the equilibrium constant is related to the free energy via $\Delta G^0 = -RT \ln K$, equation 4–59 can be written as

$$\Delta G^0 = \Delta G_0^0 - 2.303RT\sigma\rho \qquad (4\text{–}61)$$

TABLE 4–3 VALUES OF THE REACTION CONSTANT ρ FOR ACID DISSOCIATION EQUILIBRIA

Acid[a]	Solvent	ρ
Ar-COOH	H$_2$O	1.00
Ar-COOH	C$_2$H$_5$OH	1.96
Ar-OH	H$_2$O	2.26
Ar-CH$_2$COOH	H$_2$O	0.56
Ar-NH$_3$	H$_2$O	2.94
Ar(NO$_2$)-COOH	H$_2$O	0.91

[a] Ar = substituted phenyl group

The same equation holds for another class of reactions, except that ΔG^0, ΔG_0^0, and ρ are different and identified by a ('). Dividing the two equations by ρ and ρ', respectively, and subtracting gives

$$\frac{\Delta G^0}{\rho} - \frac{\Delta G^{0\prime}}{\rho'} = \frac{\Delta G_0^0}{\rho} - \frac{\Delta G_0^{0\prime}}{\rho'} \qquad (4\text{--}62)$$

where it is assumed that ΔG^0 and $\Delta G^{0\prime}$ are for the same substituent and position. This equation can be rewritten as

$$\Delta G_0^{0\prime} - \Delta G^{0\prime} = \frac{\rho'}{\rho}(\Delta G_0^0 - \Delta G^0) \qquad (4\text{--}63)$$

Thus, in the new reaction series the difference between the free-energy changes for the unsubstituted and substituted reactions is proportional to the analogous difference for the initial reaction class. An equation similar to equation 4–63 can be derived for the free energies of activation.

A relationship can also be written between the free energy of activation ΔG^a and the overall free energy change ΔG^0 for a chemical reaction. From equation 4–60 and the relation $k \propto \exp(-\Delta G^a/RT)$, we have

$$\Delta G^a = \Delta G_0^a - 2.303RT\sigma\rho^a \qquad (4\text{--}64)$$

where ρ^a is used to differentiate it from ρ in the equilibrium free-energy equation 4–61. Combining equations 4–61 and 4–64 yields

$$\Delta G^a = \frac{\rho^a}{\rho}\Delta G^0 + \left[\Delta G_0^a - \frac{\rho^a}{\rho}\Delta G_0^0\right] \qquad (4\text{--}65)$$

This is the relationship between ΔG^a and ΔG^0 predicted by the Hammett equations and may be compared with equation 4–58. Although the Hammett correlations are extremely successful, they do not apply well to the reactions of aliphatic and some aromatic compounds. Related correlations have been suggested for these cases. Another widely used free-energy relationship is the Brönsted correlation of acid-base catalysis.[4]

4.7 RELAXATION METHODS FOR FAST REACTIONS

Most solution reactions involve a complex multistep mechanism. In analyzing the mechanism, it is often necessary to determine rate constants for the individual elementary reactions. The time scales for these reactions may range from minutes to 10^{-12} sec. Different experimental techniques have been developed for measuring rate constants over such a broad time range. Flow techniques may be used if the reaction time is less than 10^{-4} sec, while ESR and NMR spectral line shape methods are applicable for reaction times between 10^{-1} and 10^{-10} sec. One of the most successful techniques, which spans the $1-10^{-11}$-sec time scale, is the *relaxation method*.

As a typical example, consider the formation of transition-metal complexes given by the reactions

$$M^{+2}(H_2O)_6 + L \underset{k_{-1}}{\overset{k_1}{\rightleftharpoons}} ML^{+2}(H_2O)_5 + H_2O$$

$$ML^{+2}(H_2O)_5 + L \underset{k_{-2}}{\overset{k_2}{\rightleftharpoons}} ML_2^{+2}(H_2O)_4 + H_2O$$

$$ML_2^{+2}(H_2O)_4 + L \underset{k_{-3}}{\overset{k_3}{\rightleftharpoons}} ML_3^{+2}(H_2O)_3 + H_2O \qquad (4\text{--}66)$$

$$\vdots \qquad\qquad \vdots \qquad\qquad \vdots$$

where M^{+2} is a cation such as Ni^{+2}, Co^{+2}, or Cu^{+2} and L is a unidentate ligand like NH_3. The mechanism can be easily generalized to include charged ligands or a multidentate ligand such as $NH_2(CH_2)_3NH_2$. If all six reactions are occurring at once, analysis of the system is quite complex. However, by choosing the proper relative concentrations, it is often possible to isolate one or two of the reactions. For example, with $[M^{+2}] \gg [L]$, there is mostly M^{+2} and ML^{+2}. At a higher concentration of L there may be predominantly ML_2^{+2} and ML_3^{+2}. Thus, for the correct choice of initial concentrations the complex mechanism may be dominated by only one equilibrium, denoted by

$$ML_{n-1} + L \underset{k_{-n}}{\overset{k_n}{\rightleftharpoons}} ML_n$$

with equilibrium constant $K_n = k_n/k_{-n}$. The generic reaction under investigation is then

$$A + B \underset{k_{-1}}{\overset{k_1}{\rightleftharpoons}} C \qquad (4\text{--}67)$$

with concentrations [A], [B], and [C]. The rate constants can be measured by the relaxation method developed by Eigen[5] in the 1950s. In this technique, the position of the equilibrium is shifted and the time required for equilibrium to be reestablished is measured. If the displacement from equilibrium is small, the relaxation to the new equilibrium will be first order.

Equation 4–67 is described by the usual rate equation,

$$\frac{-d[A]}{dt} = \frac{-d[B]}{dt} = \frac{d[C]}{dt} = k_1[A][B] - k_{-1}[C] \qquad (4\text{--}68)$$

If a, b, and c are the new equilibrium conditions resulting from the perturbation of the system, we have

$$x = a - [A] = b - [B] = [C] - c \qquad (4\text{--}69)$$

where x is the difference between the instantaneous concentrations and the concentrations for the new equilibrium. Equation 4–68 may then be written as

$$\frac{dx}{dt} = k_1(a - x)(b - x) - k_{-1}(c + x)$$

or

$$\frac{dx}{dt} = k_1ab - k_1(a + b)x + k_1x^2 - k_{-1}c - k_{-1}x \qquad (4\text{--}70)$$

By the definition of equilibrium, $k_1ab - k_{-1}c = 0$. The equation can then be *linearized* by making the assumption that the displacement is small, so that $x^2 \ll x$. When we do so, we have

$$\frac{dx}{dt} = -[k_1(a + b) + k_{-1}]x$$

$$= \frac{-x}{\tau} \tag{4-71}$$

where τ is the relaxation time. By measuring τ as a function of $a + b$, and knowing the equilibrium constant, both k_n and k_{-n} can be obtained. Note that in deriving equation 4-71 we have identified a general principle, viz., that "all rate equations near equilibrium turn into first-order differential equations which have simple exponential solutions."

Often it is impossible to isolate one reaction in a complex mechanism. In that case, a set of coupled reactions as depicted in equation 4-66 must be solved. By using the procedure just described, we can derive, for the coupled reactions, a system of linear differential equations which may be solved by the matrix method described in chapter 2, section 2-5. To illustrate this method, consider the two coupled equations

$$A + B \underset{k_{-1}}{\overset{k_1}{\rightleftharpoons}} C$$

and

$$C + B \underset{k_{-2}}{\overset{k_2}{\rightleftharpoons}} D \tag{4-72}$$

If the rate equations are linearized and the material balance and equilibrium relations used, one obtains the linear differential equations

$$\frac{-d\Delta[A]}{dt} = \{k_1(a + b) + k_{-1}\}\Delta[A] + (k_{-1} - k_1a)\Delta[D]$$

$$= c_{11}\Delta[A] + c_{12}\Delta[D] \tag{4-73}$$

and

$$\frac{-d\Delta[D]}{dt} = k_2(b - c)\Delta[A] + \{k_2(b + c) + k_{-2}\}\Delta[D]$$

$$= c_{21}\Delta[A] + c_{22}\Delta[D] \tag{4-74}$$

where the Δ's are the displacement in the concentrations from equilibrium; i.e., $[A] = a + \Delta[A]$.

These equations are solved as outlined in chapter 2, section 2-5, and the relaxation times are found from the secular determinantal equation

$$\begin{vmatrix} c_{11} - 1/\tau & c_{12} \\ c_{21} & c_{22} - 1/\tau \end{vmatrix} = 0 \tag{4-75}$$

The two solutions to the resulting quadratic equation are given by

$$\frac{1}{\tau_{1,2}} = \frac{1}{2}(c_{11} + c_{22}) \pm [\{c_{11} + c_{22}\}^2 - 4\{c_{11}c_{22} - c_{12}c_{21}\}]^{1/2} \qquad (4\text{--}76)$$

and the displacements $\Delta[A]$ and $\Delta[D]$ from equilibrium go as sums of exponentials; e.g.,

$$\Delta[A] = a_1 \exp(-t/\tau_1) + a_2 \exp(-t/\tau_2) \qquad (4\text{--}77)$$

The coefficients a_1 and a_2 are found from the secular equations as described in chapter 2, section 2.5.

Several experimental methods are available for performing relaxation kinetics measurements on microsecond time scales. One of the most versatile, pioneered by Eigen and co-workers, is the *temperature-jump* method. In this method, the electrical charge on a storage capacitor is released through a small volume of electrolytic solution containing the reactants. The energy heats the solution, which has a small but finite electrical resistance. In a typical apparatus, 45 J of electrical energy may be discharged in 1 μsec through 10 cm^3 of solution. The specific heat of aqueous solutions is essentially that of water, viz., 4.18 J/g. The rise in temperature of the solution is thus on the order of 1°C, which is enough to shift the equilibrium by van't Hoff's law,

$$\left(\frac{\partial \log_{10} K}{\partial T}\right)_P = \frac{\Delta H^\circ}{2.303 R T^2} \qquad (4\text{--}78)$$

A variation of this technique is the *pressure-jump* method, in which a volume of compressed gas is suddenly released into a chamber containing the solution. In this case, the equilibrium shift is given by

$$\left(\frac{\partial \log_{10} K}{\partial P}\right)_T = -\frac{\Delta \tilde{V}}{2.303 R T} \qquad (4\text{--}79)$$

Properties of excited electronic states can be exploited in this technique. Since acid dissociation constants of organic species are often quite different in the ground and excited electronic states, excitation may release protons into the solution, resulting in a *pH-jump*. The excitation may be carried out by a pulsed laser, thus affording nanosecond or even picosecond time resolution.

Of course, the method used for following the adjustment of the reaction to the new equilibrium must be fast enough to detect whatever changes occur. The two most commonly used methods are optical absorption (frequently with a pH indicator added to amplify the absorbance changes) and electrical conductance for ionic reactions. In addition to acid-base and metal-ligand reactions, the kinetics of many enzyme-catalyzed reactions have been elucidated by the use of relaxation techniques.

REFERENCES

1. E. Rabinowitch and W. C. Wood, *Trans. Faraday Soc. 32*, 1381 (1936).

2. P. Debye and E. Hückel, *Physik. Z. 24*, 185 (1923.)

3. L. P. Hammett, *Physical Organic Chemistry* (New York: McGraw-Hill, 1940), chapter 7.

4. J. N. Brönsted and K. Pederson, *Z. Phys. Chem.* (*Leipzig*) **108**, 185 (1924).

5. M. Eigen, *Disc. Faraday Soc.* **17**, 194 (1954).

BIBLIOGRAPHY

BERRY, R. S., RICE, S. A., and ROSS, J. *Physical Chemistry*. New York: Wiley, 1980.

GARDINER, W. C., JR., *Rates and Mechanisms of Chemical Reactions*. Menlo Park, CA: W. A. Benjamin, 1972.

HAMMES, G. G. *Principles of Chemical Kinetics*. New York: Academic Press, 1978.

JORDAN, P. C. *Chemical Kinetics and Transport*. New York: Plenum Press, 1979.

LAIDLER, K. J. *Chemical Kinetics*. New York: Harper and Row, 1987.

LOWRY, T. H., and RICHARDSON, K. S. *Mechanism and Theory in Organic Chemistry*. New York: Harper and Row, 1981.

MOORE, W. J. *Physical Chemistry*. Englewood Cliffs, NJ: Prentice-Hall, 1962.

WESTON, R. E., JR., and SCHWARZ, H. A. *Chemical Kinetics*. Englewood Cliffs, NJ: Prentice-Hall, 1972.

PROBLEMS

4.1 The results of the alkaline hydrolysis of ethyl nitrobenzoate at various times are reported below. Determine the order of the reaction and find the rate constant.

t (sec)	0	100	200	300	400	500	600	700	800
[A] $(10^{-2} \text{mol/dm}^3)$	5.00	3.55	2.75	2.25	1.85	1.60	1.48	1.40	1.38

4.2 Consider the following mechanism for a chemical reaction in solution.

$$A + B \underset{k_{-d}}{\overset{k_d}{\rightleftharpoons}} (AB) \overset{k}{\longrightarrow} \text{products, P.}$$

(a) Show that the overall rate law is second-order if the steady-state assumption is applied to AB.

(b) What is the expression for the second-order rate constant k_2 in terms of the mechanism?

(c) Show what relations between the rate constants give the diffusion-controlled limit and the activation-controlled limit.

4.3 The substitution constant for OCH_3 is $+0.12$ in the meta position and -0.27 in the para position. What is the rate constant ratio for the ionization of

meta   and   *para*

in H_2O at 25°C?

4.4 The bimolecular rate constant for a reaction in solution is

$$k = \frac{4\pi(D_A + D_B)\beta}{1 + [4\pi(D_A + D_B)\beta/k_R e^{-V(R)/k_B T}]}$$

(a) What are the units for the rate constant?
(b) What is the expression for the diffusion limited rate constant?
(c) Express the diffusion limited rate constant in units of liters/mole-sec.
(d) What is the rate constant expression for a slow reaction?

4.5 How does increasing the ionic strength affect the rates of the following reactions?
(a) $S_2O_8^{-2} + 2I^- \rightarrow I_2 + 2SO_4^{-2}$
(b) $H_2O_2 + 2H^+ + 2Br^- \rightarrow 2H_2O + Br_2$
(c) $Co(NH_3)_5 Br^{+2} + OH^- \rightarrow Co(NH_3)_5 OH^{+2} + Br^-$
(d) $O_2NNCOOC_2 H_5^- + OH^- \rightarrow N_2O + CO_3^{-2} + C_2 H_5OH$
For each of these cases, what ionic strength will cause a change in rate of 50%?

4.6 Following the linearizing of rate expressions presented in Section (4–7), determine the relationships between the relaxation time and the rate constants for the equilibria $A \rightleftharpoons B$, $2A \rightleftharpoons A_2$, and $A + B \rightleftharpoons C + D$.

4.7 What is the temperature dependence of a slow solution reaction? What is the temperature dependence of the rate constant for the diffusion-controlled reaction $A^+ + B^- \rightarrow AB$?

4.8 The complexation reaction between $Rh_2(OAc)_4 \cdot 2H_2O$ and $5'-AMP$ (adenosine monophosphate) has been investigated using temperature-jump techniques.[a] The reaction provides an interesting example of relaxation kinetics because two equilibria are maintained. Denoting the dirhodium complex by M and the $5'-AMP$ by L, we find that the important equilibria are

$$M \cdot 2H_2O + L \underset{k_{-1}}{\overset{k_1}{\rightleftharpoons}} M \cdot L(H_2O) + H_2O, \qquad K_1 = 1,893 \text{ M}^{-1}$$

$$M \cdot L(H_2O) + L \underset{k_{-2}}{\overset{k_2}{\rightleftharpoons}} M \cdot 2L + H_2O, \qquad K_2 = 202 \text{ M}^{-1}$$

where the equilibrium constants were determined at 25°C.[b] The approach to equilibrium was followed by a spectrophotometric technique which was sensitive to the concentration of the dirhodium complex $M \cdot 2H_2O$. The decay constant was determined at 25°C for a variety of *total* concentrations of both M and L, c_M, and c_L. The data are given in the following table:

DEPENDENCE OF THE DECAY CONSTANT IN THE $Rh_2 (OAc)_4 \cdot 2H_2O - (5'-AMP)$
SYSTEM ON CONCENTRATION[a]

c_M (mM)	c_L (mM)	$1/\tau \times 10^{-3}$ (sec^{-1})	[M] (mM)	[L] (mM)
1.94	1.96	5.88	0.79	0.67
1.92	2.91	6.25	0.50	1.21
1.92	3.38	7.14	0.41	1.51
1.91	3.85	7.69	0.33	1.84
1.90	4.31	8.70	0.27	2.19
1.89	4.76	9.09	0.23	2.54

[a] K. Das, E. L. Simmons, and J. L. Bear, *Inorg. Chem. 16,* 1268 (1977).
[b] L. Rainen, R. A. Howard, A. P. Kimball, and J. L. Bear, *Inorg. Chem. 14,* 2752 (1975).

(a) Show that if the reaction $ML + L \rightleftharpoons ML_2$ were being monitored for this system, the decay constant would be

$$\frac{1}{\tau} = k_2\left([ML] + [L] + \frac{1}{K_2}\right)$$

(b) Use the data in the table and the values of K_1 and K_2 to compute $[ML]$.

(c) Test the hypothesis of part (a) both graphically and via a least-squares procedure.

4.9 The following bimolecular rate coefficients relate to a substitution reaction of bromacetic acid. Deduce the product of the formal charges of the reactants.

Rate coef/$dm^3 mol^{-1} min^{-1}$:	0.288	0.301	0.334	0.371	0.429
$10^3 \times$ ionic strength/mol dm^{-3}	1	2	4	10	20

4.10 The diffusivity of I_2 in CCl_4 is $1.5 \times 10^{-5} cm^2 s^{-1}$ at 320 K. At the same temperature the rate coefficient for combination of atomic iodine in CCl_4 was recorded as 7×10^{12} $cm^3 mol^{-1} s^{-1}$ with a flash method. If the reaction is diffusion controlled, and the diffusivity of the atom is the same as for the diatomic molecule, do the data give an acceptable model? The equilibrium nuclear separation of I_2 is 2.66×10^{-8} cm.

4.11 Verify equation 4–76 for relaxation times τ_1 and τ_2 using equations 4–73 through 4–75. The material balance relations are

$$\Delta[A] = -\Delta[C] - \Delta[D] \quad \text{and} \quad \Delta[A] - \Delta[B] = \Delta[D]$$

CHAPTER 5

Catalysis

Many of the mechanisms discussed in the preceding chapters are characterized by a *rate-limiting step,* that is, one elementary reaction whose rate is significantly slower than that of the other steps with which it is coupled. The overall rate of the chemical transformation is constrained by the slow step through which the reactants must pass. (A toll booth across a busy highway is a familiar analogy!) Very often, in practical situations, the greatest possible net rate of chemical conversion, or *throughput,* is desirable. While the rate coefficient of any reaction possessing typical Arrhenius behavior (see chapter 1, section 1.5) can be increased by raising the temperature, it is often impractical to do so because the equilibrium concentration of the desired product may decrease with temperature, or undesirable side reactions which consume product or yield impurities may occur at the high temperatures. In such situations the use of a *catalyst* is usually helpful. A catalyst is defined as a chemical substance which increases the rate of a chemical reaction without itself being consumed in the reaction.

If the catalyst is uniformly dispersed in a gas or liquid phase solution, we may speak of *homogeneous* catalysis (see section 5.2). Enzyme-catalyzed biochemical reactions (see section 5.4) are a special case of homogeneous catalysis. Very often, however, the catalytic process occurs at the surface of a solid particle that is in contact with the gaseous or liquid solution. This is called *heterogeneous* catalysis (see section 5.5). Many important industrial catalysts are of the heterogeneous variety. Before dealing with these specific types of catalysis in detail, we consider, in the next section, some general principles of catalysis.

5.1 CATALYSIS AND EQUILIBRIUM

The effect of a catalyst C on a slow reaction A $\rightarrow$ B can be written formally as

$$A + C \longrightarrow B + C$$

If the overall reaction is reversible, i.e., if

$$A \underset{k_{-1}}{\overset{k_1}{\rightleftharpoons}} B$$

with

$$\frac{k_1}{k_{-1}} = K_{eq} = \frac{[B]_{eq}}{[A]_{eq}} \tag{5-1}$$

then the catalyzed path is

$$A + C \underset{k_{-2}}{\overset{k_2}{\rightleftharpoons}} B + C$$

with

$$\frac{k_2}{k_{-2}} = K'_{eq} = \frac{[B]_{eq}[C]}{[A]_{eq}[C]} = \frac{[B]_{eq}}{[A]_{eq}} = K_{eq} \tag{5-2}$$

That is, the equilibrium coefficient K'_{eq} for the catalyzed path must be equal to the equilibrium coefficient K_{eq} for the uncatalyzed path. The catalyst concentration [C] appears in the rate expression, but not in the equilibrium ratio. Another way of saying this is that for the catalyzed reaction, the forward rate coefficient k_2 must be ac-

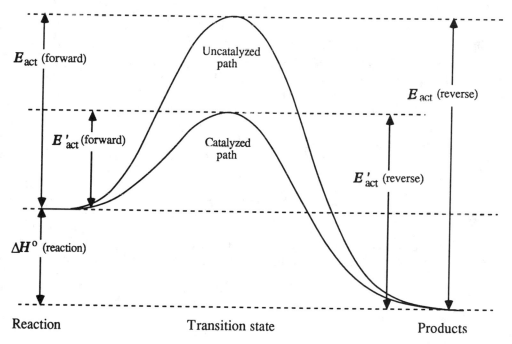

Figure 5–1. The enthalpy diagram shown here is based on the potential energy surface for a chemical reaction, which will be discussed in chapter 7. From this figure, we can see that the catalyst reduces the enthalpy of activation for the forward reaction by exactly the same amount as it reduces the enthalpy of activation for the reverse reaction. Thus, the net enthalpy change in the reaction, $\Delta H^\circ = \Delta H^\ddagger_{fwd} - \Delta H^\ddagger_{rev}$, remains unchanged. The equilibrium constant can be written as $K_{eq} = \exp(-\Delta G^\circ/RT)$, where $\Delta G^\circ = \Delta H^\circ - T\Delta S^\circ$; since the entropy change in the reaction is also unaffected by the presence of the catalyst, the position of equilibrium remains the same for the catalyzed and uncatalyzed processes.

celerated by exactly the same factor as the reverse rate coefficient k_{-2} with respect to the uncatalyzed rate coefficients k_1 and k_{-1}. This can be interpreted in terms of an enthalpy diagram similar to that discussed in chapter 1; such a diagram is shown in Figure 5–1.

5.2 HOMOGENEOUS CATALYSIS

Homogeneous catalysis occurs when the catalyst is uniformly dispersed in the reaction mixture, be it a gaseous or a liquid solution. An example of each kind will be given.

Consider the reaction

$$2SO_2 + O_2 = 2SO_3$$

at 300 K. The free energy change for this reaction, as calculated from standard free energies of formation, is

$$\Delta G^\circ = 2(-395) - (0) - 2(-297) = -196 \text{ kJ/mole}$$

so that the equilibrium coefficient for the reaction as written is

$$K = e^{-\Delta G^\circ/RT} = 2.3 \times 10^{34}$$

for the reactants in their standard states. Despite this enormously favorable equilibrium coefficient, mixtures of SO_2 and O_2 are stable at room temperature: the reaction does not proceed spontaneously.

As a step in the production of sulfuric acid, the oxidation of SO_2 is a very important industrial process toward which much effort has been expended in finding an efficient catalyst. A solid catalyst containing vanadium is most widely used. In the gas phase, the principle of catalysis can be illustrated by the effect of adding nitric oxide. The following reactions occur:

$$2NO + O_2 \longrightarrow 2NO_2 \tag{1}$$

$$NO_2 + SO_2 \longrightarrow NO + SO_3 \tag{2}$$

The second reaction proceeds much faster than the first. Taking the net reaction to be one of step 1 combined with two of step 2, we have the following conditions:

$$\frac{d[NO]}{dt} = 0 \tag{5–3a}$$

and

$$\frac{d[NO_2]}{dt} = 0 \tag{5–3b}$$

Equation 5–3a is just the necessary condition on a catalyst, viz., that its concentration does not change during the reaction. Equation 5–3b can be regarded as a steady-state condition on NO_2, which, since step 2 is stipulated to be much faster than step 1, remains at low concentration during the course of the reaction. The rate of formation of SO_3 is

$$\frac{d[SO_3]}{dt} = k_2[NO_2][SO_2]$$

Applying Equation (5–3b) to the kinetic equation for $[NO_2]$, we have

$$\frac{d[NO_2]}{dt} = 2k_1[NO]^2[O_2] - k_2[NO_2][SO_2] = 0$$

so that

$$[NO_2]_{ss} = \frac{2k_1[NO]^2[O_2]}{k_2[SO_2]}$$

and

$$\frac{d[SO_3]}{dt} = k_2[SO_2]\frac{2k_1[NO]^2[O_2]}{k_2[SO_2]} = 2k_1[NO]^2[O_2] \qquad (5\text{–}4)$$

Notice that the concentration of SO_2 does not appear in the expression for the rate of formation of SO_3 (equation 5–4). This is because the rate-limiting step (1) does not include SO_2 as a reactant. This will be true until the reaction proceeds so far that most of the SO_2 is used up. Step (1) will then no longer be rate-limiting, and equation 5–4 will no longer be valid.

Catalytic processes of a similar nature to the oxidation of SO_2 are thought to be responsible for the removal of atmospheric ozone; some of these are discussed in chapter 15.

A simple example of a catalyzed process in a liquid solution is the hydrolysis of an ester, such as ethyl acetate:

$$CH_3COOCH_2CH_3 + H_2O \longrightarrow CH_3COOH + C_2H_5OH$$

This reaction proceeds immeasurably slowly in a neutral solution, but quite readily in the presence of added acid or base. The mechanism of the acid-catalyzed hydrolysis is the following:

$$CH_3COOCH_2CH_3 + H_3O^+ \rightleftharpoons CH_3\overset{H}{\underset{\overset{\|}{O}}{C}}-\overset{+}{O}CH_2CH_3 + H_2O$$

$$+H_2O \qquad CH_3\overset{+}{C}OOH_2 + CH_3CH_2OH$$

$$CH_3COOH + H^+$$

In this mechanism, two rapid protonation equilibria are coupled to the rate-determining step which liberates the free alcohol. The overall rate of disappearance of ester, given by

$$-\frac{d[ester]}{dt} = k_{hyd}[H_3O^+][ester]$$

is proportional to the concentration of acid. The overall equilibrium coefficient,

$$K_{eq} = \frac{a_{acid}\, a_{alcohol}}{a_{ester}\, a_{water}}$$

is expressed in terms of molar activities rather than concentrations, as is appropriate for a concentrated solution. K_{eq} is independent of acidity, except as the pH may affect the activity coefficients.

In the absence of added acid, the only H^+ that would be present results from the self-hydrolysis of water, viz.,

$$2H_2O \quad \rightleftharpoons \quad H_3O + OH^-$$

with

$$K_W = \frac{a(H_3O^+)a(OH^-)}{a^2(H_2O)} = \frac{[H_3O^+][OH^-]}{[H_2O]^2} \cdot \frac{\gamma_{H^+}\, \gamma_{OH^-}}{\gamma_w^2} \qquad (5\text{--}5)$$

where the γ's are activity coefficients for H^+, OH^-, and water. The hydrolysis rate would thus be logarithmically proportional to the activity coefficient product in equation 5–5, which, according to the Debye-Hückel theory,[1] is in turn proportional to the ionic strength $I = \frac{1}{2}\Sigma_i\, c_i z_i^2$. This means that adding an inert electrolyte to a solution, which increases the value of $c_i z_i^2$, will increase the hydrolysis rate even though the reactants themselves are electrically neutral but the catalytic species is charged. This is often referred to as the *secondary salt effect*.

5.3 AUTOCATALYSIS AND OSCILLATING REACTIONS

An interesting new feature is introduced by the presence of *autocatalysis*, which is just the appearance of one of the products of the reaction as a reactant in the same or a coupled reaction. The simplest possible example is the reaction

$$A + B \quad \xrightarrow{k} \quad 2B$$

whose rate law is

$$-\frac{d[A]}{dt} = k[A][B]$$

To find a solution, let $[A]_0 - [A(t)] = [B(t)] - [B]_0$ and introduce a progress variable $x = [A(t)]$; we then have

$$[B(t)] = [A]_0 + [B]_0 - [A(t)] = A_0 + B_0 - x$$

Then

$$-\frac{dx}{dt} = kx(A_0 + B_0 - x)$$

which is simply integrated:

$$\int \frac{dx}{(A_0 + B_0)x - x^2} = -\int k\, dt$$

The solution is

$$\frac{1}{A_0 + B_0} \ln\left(\frac{A_0}{B_0} \cdot \frac{[B]}{[A]}\right) = -kt$$

or

$$[B(t)] = \frac{A_0 + B_0}{1 + (A_0/B_0)e^{-k(A_0 + B_0)t}} \tag{5-6}$$

The time dependence of [B] is shown in Figure 5–2. Note that at $t = 0$, [B] just has its initial value B_0. If $B_0 = 0$, then the only solution is [B] $= 0$ for all t. If any small amount $B_0 > 0$ is initially present, then there is first a slow increase in [B] called an *induction period;* the smaller B_0 is, the longer is this period. The rate of increase of [B] then increases (the autocatalytic effect), goes through an *inflection point* (maximum $d[B]/dt$) at some time t^*, and levels off at a value of [B] $= A_0 + B_0$ at long times.

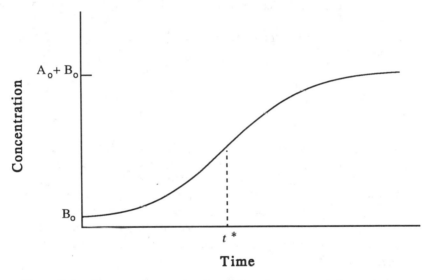

Figure 5–2. Concentration vs. time for product of an autocatalytic reaction, as given by equation 5–6. The inflection point occurs at t*.

The concentration-vs.-time dependence shown in Figure 5–2 is the classic "S" curve associated with increasing populations. For example, if A is food and B is bacteria (or, perhaps, people), the organisms eat the food and in turn produce more organisms. The reaction is intrinsically irreversible unless cannibalism is included in the mechanism. An actual population model for humans or other organisms that reproduce sexually would necessarily be more complex:

$$A + X_m \longrightarrow X_m^*$$

$$A + X_f \longrightarrow X_f^*$$

$$X_m^* + X_f^* \longrightarrow X_m + X_f + B$$

$$A + B \longrightarrow X, \text{ etc.}$$

In this model the organisms are divided into subsets of m (males) and f (females), an initial activation step is included, and the activated species then react to produce B (babies), which consume food and in turn become reproductive species. Another kind of population model is explored in problem 2 at the end of the chapter.

The presence of one or more autocatalytic steps in a complex chemical mechanism can lead to the phenomenon known as *chemical oscillation*. Normally, chemical systems approach equilibrium in a smooth, frequently exponential relaxation; under special circumstances, however, the concentrations can oscillate between several stationary states. By analogy with electrical or electro-optical systems that display oscillation or bistability, such systems are sometimes described as examples of *chemical feedback*. In addition to temporal oscillation, these systems can also produce nonuniform spatial structures which appear to arise spontaneously from a homogeneous medium. While chemical oscillation is rather unusual, the phenomenon has been the subject of a considerable amount of study, in part because of its possible significance in biological evolution. Further background may be found in Jordan[2] and Field.[3]

In order to illustrate how chemical oscillation may arise, consider the reaction system

$$A + X \xrightarrow{k_1} 2X$$
$$X + Y \xrightarrow{k_2} 2Y \qquad (5\text{–}7)$$
$$Y \xrightarrow{k_3} Z$$

This "Lotka" mechanism has been specifically invented to display oscillatory behavior; it does not correspond to any chemical system that is known. All its steps are irreversible, and steps (1) and (2) are autocatalytic, as in the reaction $A + B \rightarrow 2B$ considered previously. We also postulate that the system is open and that [A] is maintained equal to A_0 at all times. This can be realized, for example, in a flow reactor, in which species A is introduced and Z is drawn off continuously, with intermediates X and Y remaining at a steady state. The kinetic equations for these intermediates are then

$$\frac{d[X]}{dt} = +k_1 A_0[X] - k_2[X][Y] \qquad (5\text{–}8a)$$

and

$$\frac{d[Y]}{dt} = +k_2[X][Y] - k_3[Y] \qquad (5\text{–}8b)$$

A steady-state solution will occur when $d[X]/dt = d[Y]/dt = 0$, or

$$k_2[Y]_{ss} = k_2 Y_0 = k_1 A_0 \qquad (5\text{–}9a)$$

and

$$k_2[X]_{ss} = k_2 X_0 = k_3 \qquad (5\text{–}9b)$$

Note that, since [A] is to be maintained at a constant value A_0, the steady-state concentrations given by equations 5–9a and 5–9b are completely independent of time.

We now wish to see what will happen when the concentrations are temporarily displaced from their steady-state values. In order to treat the problem mathematically, we introduce *displacement variables* as in relaxation kinetics (see chapter 4, section 4.7). In this case these are

$$[X] = x(t) + X_0$$

and

$$[Y] = y(t) + Y_0$$

In terms of these displacement variables, equation 5–8a becomes

$$\frac{dx}{dt} = k_1 A_0(x + X_0) - k_2(x + X_0)(y + Y_0)$$

$$= k_1 A_0 X_0 + k_1 A_0 x - k_2 xy - k_2 X_0 y - k_2 Y_0 x - k_2 Y_0 X_0$$

Using equation 5–9a, $k_2 Y_0 = k_1 A_0$, to cancel terms, we have

$$\frac{dx}{dt} = -k_2 X_0 y - k_2 xy \qquad (5\text{–}10a)$$

Similarly, from equations 5–8b and 5–9b, we obtain

$$\frac{dy}{dt} = k_2(x + X_0)(y + Y_0) - k_3(y + Y_0) = k_2 Y_0 x + k_2 xy \qquad (5\text{–}10b)$$

If the displacements from the steady state are small, we can neglect the xy terms in equations 5–10 and obtain a simple set of coupled linear first-order differential equations:

$$\frac{dx}{dt} = -k_2 X_0 y$$

$$\frac{dy}{dt} = +k_2 Y_0 x \qquad (5\text{–}11)$$

By writing $\dot{x} = -ay$, $\dot{y} = +bx$, we can see that equations 5–11 are equivalent to a single second-order differential equation,

$$\ddot{x} = -a\dot{y} = -(ab)x \qquad (5\text{–}12)$$

The solution of equations 5–11 or equation 5–12 can be written down by inspection; but let us proceed systematically, as was done previously, and find an eigenvalue solution. We write the solution as

$$x(t) = [X]_t - X_0 = (\text{const})e^{-t/\tau}$$

The eigenvalue τ^{-1} is found from the determinant

$$\begin{vmatrix} +\dfrac{1}{\tau} & -k_2 X_0 \\[2mm] +k_2 Y_0 & +\dfrac{1}{\tau} \end{vmatrix} = 0$$

or

$$\left(\frac{1}{\tau}\right)^2 + k_2^2 X_0 Y_0 = 0$$

$$\left(\frac{1}{\tau}\right)^2 = -k_2^2(X_0 Y_0)$$

(5–13)

Since the rate coefficient k_2 and the concentrations X_0 and Y_0 are all positive quantities, the eigenvalues given by equation 5–13 must be purely imaginary and thus can be written as

$$\frac{1}{\tau} = \pm i k_2^2 (X_0 Y_0)^{1/2} = \pm i (k_1 k_3 A_0)^{1/2}$$

The time-varying displacement is then

$$x(t) = c_1 e^{i\omega t} + c_2 e^{-i\omega t}$$

(5–14)

The functional form of equation 5–14 is an oscillation with angular frequency $\omega = |1/\tau| = |k_1 k_3 A_0|^{1/2}$, since oscillating functions such as $\cos(\omega t)$ can be written as $e^{i\omega t} + e^{-i\omega t}$. Physically, this means that if the system is displaced from its steady state, rather than quietly returning to their previous values the species concentrations will oscillate and the displacements may even grow in magnitude. This can be represented by the *limit cycle* diagram shown in Figure 5–3.

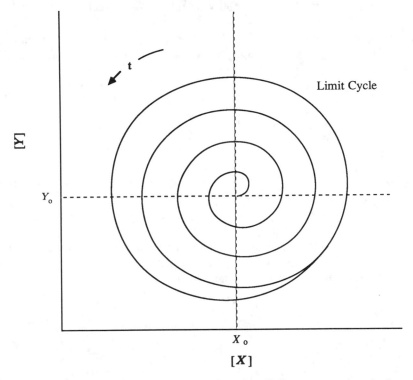

Figure 5–3. *X-Y* phase plane representation of oscillating concentrations in the Lotka mechanism.

Note that if all steps are reversible, that is, if the mechanism is actually

$$A + X \underset{k_{-1}}{\overset{k_1}{\rightleftharpoons}} 2X \qquad X + Y \underset{k_{-2}}{\overset{k_2}{\rightleftharpoons}} 2Y \qquad Y \underset{k_{-3}}{\overset{k_3}{\rightleftharpoons}} Z$$

then simple exponential relaxation, and not oscillation, occurs. This is explored further in problem 2.

While no chemical system is known which exactly follows the Lotka mechanism given in the reactions 5–7, several "chemical oscillators" have been discovered. The one which has been most extensively studied[2-5] is the Belousov-Zhabotinsky reaction, which is the oxidation of malonic acid by bromate, catalyzed by cerium ions. There are at least nine coupled reactions in this system. Two of these are

$$2H^+ + BrO_3^- + Br^- \xrightarrow{k_1} HBrO_2 + HOBr \tag{1}$$

and

$$H^+ + HBrO_2 + Br^- \xrightarrow{k_2} 2HOBr \tag{2}$$

These are coupled with the two rapid reactions.

$$Br^- + HOBr + H^+ \longrightarrow Br_2 + H_2O \tag{3}$$

and

$$Br_2 + CH_2(COOH)_2 \longrightarrow BrCH(COOH)_2 + H^+ + Br^- \tag{4}$$

An overall reaction α, leading to bromination of malonic acid, is obtained by adding to steps 1 and 2 three times each of reactions 3 and 4:

$$BrO_3^- + 2Br^- + 3CH_2(COOH)_2 + 3H^+ \longrightarrow 3BrCH(COOH)_2 + 3H_2O \tag{α}$$

Reactions leading to the oxidation of cerium ions are

$$H^+ + BrO_3^- + HBrO_2 \xrightarrow{k_5} 2BrO_2 + H_2O \tag{5}$$

$$H^+ + BrO_2 + Ce^{III} \xrightarrow{k_6} HBrO_2 + Ce^{IV} \tag{6}$$

and

$$2HBrO_2 \xrightarrow{k_7} BrO_3^- + HOBr + H^+ \tag{7}$$

Adding two of reaction 5 to four of reaction 6 and one of reaction 7 gives an overall reaction β, given by

$$BrO_3^- + 4Ce^{III} + 5H^+ \longrightarrow HOBr + 4Ce^{IV} + 2H_2O \tag{β}$$

Finally, the oxidation of bromomalonic acid by cerium is

$$BrCH(COOH)_2 + 4Ce^{IV} + 2H_2O \longrightarrow$$
$$Br^- + HCOOH + 2CO_2 + 4Ce^{III} + 6H^+ \tag{8}$$

coupled with a fast reaction

$$HOBr + HCOOH \longrightarrow Br^- + CO_2 + H^+ + H_2O \tag{9}$$

to give a net reaction γ of

$$BrCH(COOH)_2 + 4Ce^{IV} + HOBr + H_2O \longrightarrow$$

$$2Br^- + 3CO_2 + 4Ce^{III} + 6H^+ \qquad (\gamma)$$

The overall Belousov-Zhabotinsky reaction, obtained by adding together reactions α, β, and γ, is

$$2BrO_3^- + 3CH_2(COOH)_2 + 2H^+ \xrightarrow{Ce^{III},\ Br^-} 2BrCH(COOH)_2 + 3CO_2 + 4H_2O$$

Note that in this reaction the concentrations of cerium ion and bromide ion are canceled out, so that these species play a true catalytic role.

In order to analyze such a complicated set of reactions, simplifying assumptions are of course necessary. The most appropriate such assumptions are steady-state conditions on intermediate-valence and odd-electron bromine species, so that

$$\frac{d[HBrO_2]}{dt} = \frac{d[BrO_2]}{dt} = 0$$

From the kinetic equations, this gives

$$\frac{d[HBrO_2]}{dt} = k_1[H^+]^2[BrO_3^-][Br^-] - k_2[H^+][HBrO_2][Br^-]$$

$$- k_5[H^+][BrO_3^-][HBrO_2] + k_6[H^+][BrO_2][Ce^{III}] \qquad (5-15)$$

$$- 2k_7[HBrO_2]^2 = 0$$

and

$$\frac{d[BrO_2]}{dt} = 2k_5[H^+][BrO_3^-][HBrO_2] - k_6[H^+][BrO_2][Ce^{III}] = 0 \qquad (5-16)$$

Equation 5–16 allows us to replace $k_6[H^+][BrO_2][Ce^{III}]$, the fourth term in equation 5–15, with $2k_5[H^+][BrO_3^-][HBrO_2]$, to give the relatively simple steady-state condition,

$$k_1[H^+]^2[BrO_3^-][Br^-] - k_2[H^+][HBrO_2]_{ss}[Br^-]$$

$$+ k_5[H^+][BrO_3^-][HBrO_2]_{ss} - 2k_7[HBrO_2]_{ss}^2 = 0 \qquad (5-17)$$

Two limiting cases can be distinguished:

1. If the steady-state bromide ion concentration is much greater than the steady-state $HBrO_2$ concentration, we may neglect the last term in equation 5–17 and solve for

$$[HBrO_2]_{ss(1)} = \frac{k_1[BrO_3^-][Br^-][H^+]}{k_2[Br^-] - k_4[BrO_3^-]}$$

2. If, on the other hand, the steady-state bromide ion concentration is much less than the steady-state $HBrO_2$ concentration, the first and second terms in equation 5–17 may be neglected, and we obtain

$$[HBrO_2]_{ss(2)} = \frac{k_4[H^+][BrO_3^-]}{2k_7}$$

Figure 5–4. Oscillatory behavior of bromide ion concentration and ratio $[Ce^{III}]/[Ce^{IV}]$ in the Belousov-Zhabotinsky reaction. [From R. J. Field, *Am. Scientist 73*, 142 (1985).]

So there are two possible steady-state solutions for at least one of the key intermediates, and the system may be *nonstationary*, that is, it may switch back and forth between these two steady states. Another term for such switching behavior is *bistability*. We can see qualitatively how such bistability may arise. If initially the bromide ion concentration is small, then reaction cycle α is shut off, cerous ion is oxidized to the ceric form, and reaction cycle γ proceeds to generate Br^-. As the latter builds up, reaction cycle α turns back on, Ce^{IV} is no longer produced, and the Br^- is consumed. The concentrations of these species may oscillate back and forth between the two steady states, as shown in Figure 5–4.

A rather different type of instability, which is encountered far more frequently than the preceding oscillatory type of reaction, is an explosion. An example of explosion kinetics is presented in chapter 15, section 15.2.

5.4 ENZYME-CATALYZED REACTIONS

A very important class of catalysts is the *biological enzymes:* without them, you would not be reading this book, nor indeed would it ever have been written. Enzymes are protein molecules which, through a process of chemical evolution, have become exquisitely tailored to carry out highly selective and efficient biochemical reactions in living organisms. The biochemistry of enzymes is a subject of great depth and complexity; in this section we can only touch on the simplest aspects of the kinetics of these systems. For greater detail, the reader should consult Hammes[5] and Walsh.[6]

A generic enzyme-catalyzed mechanism may be illustrated as follows:

| S (Substrate) | E (Enzyme) | X or E · S (Enzyme–Substrate Complex) | P_1 P_2 (Products) | E |

The substrate may be anything from a protein to a simple sugar molecule to a strand of RNA or even a complete cell; the enzyme, typically a protein of molecular weight 10,000 to 100,000 atomic mass units, may in fact be much bigger than the substrate, or it may attach to a specific site such as a part of a cell membrane. The enzyme-substrate complex is a true chemical intermediate, but is typically present in such low concentrations that it cannot be isolated. The product may be separated molecules, or the protein with one bond cleaved or one site oxidized. A simple example of this mechanism is the hydrolysis of a complex sugar, sucrose, to two smaller and more

*The suffix "-ase" is a general designator for enzymes, with the remainder of the name suggesting the specific functionality. The yeast enzyme is named invertase because the hydrolysis converts a dextrorotary sugar ($[\alpha]_D^{20} = +66.53$) into an overall levorotary mixture of glucose ($[\alpha]_D^{20} = +52.7$) and fructose ($[\alpha]_D^{20} = -92.4$). The first measurements on this system were indeed performed using quantitative polarimetry, but the current technique employs a colorimetric assay.[7]

readily metabolized sugars, with the assistance of an enzyme, called invertase,* that is derived from yeast:

$$CH_2OH \quad HOCH_2 \qquad + H_2O \rightarrow \qquad CH_2OH \qquad + \qquad HOCH_2$$

Sucrose $\qquad\qquad\qquad\qquad$ Glucose $\qquad$ Fructose

Kinetic equations can readily be derived from the general mechanism described earlier:

$$E + S \underset{k_{-1}}{\overset{k_1}{\rightleftarrows}} X \underset{k_{-2}}{\overset{k_2}{\rightleftarrows}} E + P$$

$$-\frac{d[X]}{dt} = (k_2 + k_{-1})[X] - k_1[E][S] - k_{-2}[E][P] \qquad (5\text{--}18)$$

In a reaction such as the hydrolysis of sucrose, in which the products are separated molecules, the route from complex to products is essentially irreversible, so that k_{-2} may safely be set equal to zero. Also, because the enzyme-substrate complex is rapidly destroyed and thus is present at very low concentration, the steady-state condition $d[X]/dt = 0$ may be safely invoked. Further, we have conservation conditions on the total enzyme, which is present either as free enzyme or bound in the complex, thus

$$[E]_0 = [E] + [X] \qquad (5\text{--}19a)$$

and on the total substrate-cum-product concentration, viz.,

$$[S]_0 = [S] + [P] \qquad (5\text{--}19b)$$

since $[X] \approx [E] \ll [S]$. Hence,

$$-\frac{d[S]}{dt} = +\frac{d[P]}{dt} = k_1[E][S] - k_{-1}[X]$$

Setting equation 5–18 equal to zero, solving for $[X]_{ss}$, and substituting in the preceding expression gives

$$-\frac{d[S]}{dt} = +\frac{\{k_1 k_2 [S] - k_{-1} k_{-2} [P]\}[E]_0}{k_1[S] + k_{-2}[P] + k_{-1} + k_2} \qquad (5\text{--}20)$$

Measurements of the reaction velocity v are typically carried out at initial times, when $[S] \gg [P]$. This permits us to neglect the terms containing the concentration of product in equation 5–20, and so obtain

$$v = -\frac{d[S]}{dt}\bigg|_{t\approx0} = \frac{k_1 k_2 [S][E]_0}{k_1[S] + k_{-1} + k_2}$$

or

$$v = \frac{k_2[E]_0}{1 + \dfrac{k_{-1} + k_2}{k_1[S]}} \tag{5–21}$$

If we define $v_s = k_2[E]_0$ and $K_M = (k_{-1} + k_2)/k_1$, then equation 5–21 takes the form

$$v = \frac{v_s}{1 + K_M/[S]} \tag{5–22}$$

which is known as the *Michaelis-Menten* equation. The ratio K_M is known as the *Michaelis Constant*. Although a ratio of rate coefficients, it is not a true equilibrium constant, since its definition does not satisfy detailed balancing (see chapter 2). A plot of the Michaelis-Menten equation is shown in Figure 5–5(a).

Because of the curvature of the Michaelis-Menten plot, it is often difficult to determine the limiting value v_s, the reaction velocity when all of the enzyme is tied up in the enzyme-substrate complex. Similarly, K_M, which is numerically equal to the substrate concentration when v is exactly equal to one-half v_s, is not precisely determined. To obtain better precision, it is customary to recast the kinetic data into a linear form. Two such forms are commonly used, the *Lineweaver-Burk* plot,

$$\frac{1}{v} = \frac{K_M}{v_s} \frac{1}{[S]} + \frac{1}{v_s} \tag{5–23}$$

and the *Eadie-Hofstee* plot, one form of which is

$$\frac{v}{[E]_0[S]} = \frac{k_2}{K_M} - \frac{v}{K_M[E]_0} \tag{5–24}$$

Lineweaver-Burk (LB) and Eadie-Hofstee (EH) plots are shown in Figures 5–5(b) and (c), respectively. In the LB plot, $1/v_s$ is obtained as the intercept when $1/[S] \rightarrow 0$, and the slope of the straight line is equal to K_M/v_s. In the EH plot, the negative of the slope gives $1/K_M$ directly, and the intercept as $v/[E]_0 \rightarrow 0$ is k_2/K_M.

No matter what method of analysis is used, at best two parameters of the enzyme reaction can be extracted using initial-velocity data. If we know $[E]_0$, then we can find k_2 from v_s. In many cases, however, the molecular weight, and thus the exact concentration of the enzyme, is not precisely known, since the enzyme is obtained only as a complex mixture of proteins from some biological extraction. In such a case, only the *specific activity* (moles substrate converted per mg of enzyme preparation) can be reported. If k_2 can be determined, then combining that with the measurement of K_M allows us to find the ratio k_1/k_{-1}. To obtain all of the rate coefficients for a particular enzyme-catalyzed reaction, additional measurements are necessary. If the reaction is intrinsically reversible, then it can be run backwards, starting with products, and the corresponding v_p and $K_M(p)$ values, combined with those of v_s and $K_M(s)$, would permit determination of all of the k's. When this is not possible, as is often the case, then fast-reaction techniques such as stopped-flow or relaxation measurements (see chapter 4, section 4.7) must be used. Additional information may be found in Hammes[5] and Walsh.[6]

Catalysis   Chap. 5

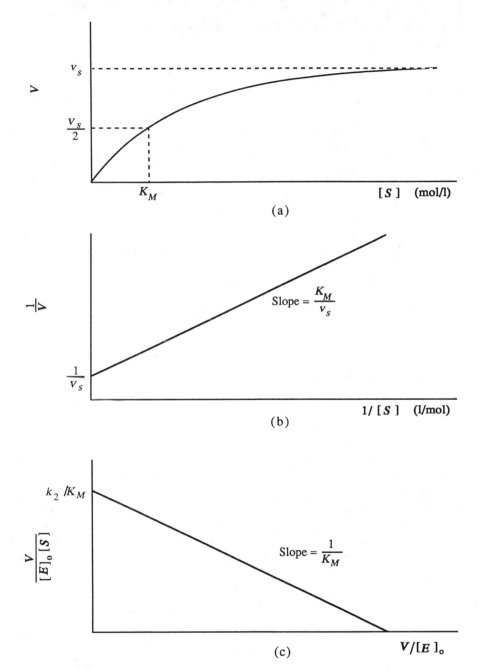

Figure 5–5. Kinetic plots for a simple enzyme-catalyzed reaction. (a) Michaelis-Menten plot, equation 5–22. (b) Lineweaver-Burk plot, equation 5–23. (c) Eadie-Hofste plot, equation 5–24.

5.5 HETEROGENEOUS CATALYSIS AND GAS-SURFACE REACTIONS

An extremely important aspect of chemical kinetics, from the points of view of both fundamental processes and practical applications, is *heterogeneous catalysis,* i.e., reactions occurring at the gas-solid or liquid-solid interface. The discussion here touches on only a few of the principal features of these processes; further details are available in the literature.[8–11]

The generalized surface-catalyzed reaction bears a formal similarity to an enzyme-catalyzed system:

$$A \xrightarrow[\text{on surface S}]{} P(\text{roducts})$$

As we shall see, the resemblance is more than superficial.

A prototypical example of heterogeneous catalysis is the formation of hydrocarbons from hydrogen and carbon monoxide over a suitable solid catalyst. The methanation reaction

$$3H_2 + CO \longrightarrow CH_4 + H_2O$$

is the simplest case of such a reaction. The free-energy change in this reaction at 500 K is

$$\Delta G^0 = -219.05 - 32.74 - (3(0) - 155.41) = -96.38 \text{ kJ/mole}$$

giving a very favorable equilibrium ratio

$$K_{eq} = \exp(-\Delta G^0/RT) = \exp(-(-96.38/0.0083 \times 500)$$

$$= 1.22 \times 10^{10} \text{ at } 500 \text{ K}$$

However, the homogeneous reaction rate between H_2 and CO in the gas phase is vanishingly small. The reaction is carried out efficiently at moderate temperatures (250^0C–450^0C) by contacting the gas mixture with finely divided transition metals, such as "Raney nickel" (60% Ni supported on Al_2O_3). Variation of the reaction conditions produces mixtures of higher molecular weight paraffins and olefins, a process known as *Fischer-Tropsch synthesis.* This process is of technical importance because it provides a route from the products of coal gasification (CO and H_2) to products which must otherwise be obtained by refining petroleum.[12,13] Use of a different catalyst, such as copper-zinc oxide on Al_2O_3, produces methanol and higher alcohols rather than hydrocarbons via the reaction

$$2H_2 + CO \longrightarrow CH_3OH$$

even though the equilibrium for this reaction is much less favorable than for methane production. We shall consider possible mechanisms for these reactions in greater detail later in this section.

The basic mechanism for a surface-catalyzed reaction begins with *adsorption* of the reactant at a localized site on the surface:

$$A + S \rightleftharpoons A \cdot S \qquad (5\text{--}25)$$

The magnitude of the adsorption coefficient $K_{ads}(A)$ may depend on the nature of the surface S and the ambient temperature, as well as on the identity of species A. This is because the interaction between the surface and the adsorbate can range from the rather weak binding of a valence-saturated species due to van der Waals attraction (physisorption) to the formation of a strong chemical bond between the adsorbate and a surface atom (chemisorption).

In order to express K_{ads} quantitatively, we must be able to define the concentration of the adsorbed species, $A \cdot S$. This is best done in terms of a dimensionless quantity, the *coverage* θ. For a given solid catalyst, there is a finite number S_0 of binding sites available. The maximum number is the number of surface atoms, on the order of 10^{14} to 10^{15} per cm^2 of surface area, but typically the number of active binding sites is a small fraction of that. The coverage is defined as

$$\theta = \frac{\text{number of } A \cdot S}{S_0} \tag{5-26}$$

The numerator and denominator of equation 5–26 are expressed in terms of either the total number of adsorbed species and binding sites for a given piece of catalyst or the site density per unit area, but must be consistent in either case.

The physical resemblance to enzyme catalysis may be apparent at this point. An enzyme is typically a large protein whose "surface" includes one or more active sites. The enzyme-catalyzed reaction begins by binding of the substrate to one of these active sites; the resulting enzyme-substrate complex is formally analogous to the adsorbed molecule. If we define the equilibrium coefficient for the adsorption process given by equation 5–25 as K_{ads}, the coverage is

$$\theta = \frac{K_{ads}[A]}{1 + K_{ads}[A]} \tag{5-27}$$

If the adsorption is from a low-pressure gas which obeys the ideal gas law $p = (N/V)k_B T$, equation 5–27 can be expressed as

$$\theta = \frac{bp}{1 + bp} \tag{5-28}$$

where $b = K_{ads}/k_B T$. Equation 5–28 is the familiar *Langmuir adsorption isotherm*; notice its similarity to the Michaelis-Menten expression, equation 5–22.

If the ensuing surface reaction is simply conversion of the adsorbed species to product, i.e.,

$$A \cdot S \xrightarrow{k_2} \text{products}$$

then the rate of conversion is given by

$$R = -\frac{d[A]}{dt} = k_2 \theta S_0 = \frac{k_2 K_{ads} S_0 [A]}{1 + K_{ads}[A]} \tag{5-29}$$

Two limiting cases of equation 5–29 should be noted. At low [A], $K_{ads}[A] \ll 1$, so that the rate is proportional to the product of the partial pressure of A and the total amount of catalyst. By contrast, at large [A], $K_{ads}[A] \gg 1$, and the rate becomes independent of the pressure of A. The catalyst is then said to be *saturated*.

For a surface-catalyzed bimolecular reaction, two mechanisms are possible. In the first, both species must be preadsorbed, and the reaction occurs between adsorbed A and B species at neighboring sites on the surface. This mechanism can be written as follows:

$$A + S \rightleftharpoons A \cdot S \qquad K_{ads}(A)$$
$$B + S \rightleftharpoons B \cdot S \qquad K_{ads}(B)$$
$$A \cdot S + B \cdot S \longrightarrow \text{products} \qquad k_2'$$

The rate of product formation associated with this *Langmuir-Hinshelwood mechanism* is

$$R = k_2' \theta_A \theta_B S_0^2 = \frac{k_2' K_{ads}^{(A)}[A] K_{ads}^{(B)}[B] S_0^2}{\{1 + K_{ads}^{(A)}[A] + K_{ads}^{(B)}[B]\}^2} \qquad (5\text{--}30)$$

Note that there is a competition for available surface sites between A and B molecules; one consequence of this is that the rate equation 5–30 at constant p_A may go through a maximum as p_B is varied.

The other possible mechanism, often referred to as an *Eley-Rideal* process, involves reaction of an impinging gas-phase species with an adsorbate. This can be written

$$A + S \rightleftharpoons A \cdot S \qquad K_{ads}^{(A)}$$
$$A \cdot S + B \longrightarrow \text{products} \qquad k_2''$$

The rate expression for this mechanism is

$$R = k_2'' \theta_A S_0 p_B = k_2'' \frac{K_{ads}^{(A)}[A] p_B S_0}{1 + K_{ads}^{(A)}[A]} \qquad (5\text{--}31)$$

This expression is linear in [B] and saturates at high [A]; it does not possess a maximum with variation of p_B. No clear instance of an Eley-Rideal mechanism has yet been found.

The actual mechanism for catalytic processes at surfaces can be much more complex than the preceding two simple models. For the steps of the Fischer-Tropsch synthesis mentioned earlier, over a dozen mechanisms have been proposed. These range from complete dissociation of the reactants on the surface, e.g.,

$$CO \longrightarrow C(ads) + O(ads)$$
$$H_2 \longrightarrow 2H(ads)$$
$$4H(ads) + C(ads) \longrightarrow CH_4(g)$$

to a concerted, super-Eley-Rideal mechanism in which CO and H_2 are conveniently bound at the same site, e.g.,

$$CO(g) + H_2(g) \longrightarrow H_2CO(ads)$$

$$H_2CO(ads) + 2H_2(ads) \longrightarrow CH_4(g) + H_2O(g)$$

(The surface S is implicit in all of these reactions.)

A mechanism which plausibly accounts for the variety of products observed includes the following steps:

1. Adsorption of CO on neighboring surface sites:

$$CO(g) + S \longrightarrow$$

2. Dissociative chemisorption of H_2:

$$H_2(g) + S \longrightarrow \qquad \text{and}$$

3. Reaction of chemisorbed H with chemisorbed CO:

4. Subsequent hydrogenation reactions on the surface, followed by desorption of stable product species:

$$
\underset{\displaystyle S}{\overset{\displaystyle O=CH_2}{|}} \quad + \quad \underset{\displaystyle S}{\overset{\displaystyle H}{|}} \quad \longrightarrow \quad \underset{\displaystyle S}{\overset{\displaystyle O-CH_3}{|}} \quad \xrightarrow{\;+\,H\,(ads)\;} \quad CH_3OH\;(g)
$$

$$\Big\downarrow +\,H\,(ads)$$

$$O=\!\!=\!\!S \qquad +\;CH_4\;(g)$$

$$\Big\downarrow +\,2H\,(ads)$$

$$H_2O\;(g)$$

In this mechanism, the higher molecular weight olefins and alcohols arise from polymerization of the adsorbed H_2CO species prior to desorption of the gaseous products. Rate laws for the production of hydrocarbons, derived from such mechanisms, are complex, to say the least. Typically, they involve irrational powers of the partial pressures of CO and hydrogen.

As a matter of fact, phenomenological rate laws are an insufficient basis for deciding among the various mechanisms postulated for surface-catalyzed reactions. Their proof (or, more likely, disproof) rests on experimental observations of the postulated surface-adsorbed species, using sensitive techniques such as photoelectron spectroscopy, Auger electron spectroscopy, electron energy-loss spectroscopy, and thermal desorption mass spectrometry. (These techniques are discussed in books on surface science, such as Gasser.[10]) However, such measurements must be carried out under ultra-high-vacuum conditions, typically 10^{-13} atm. Extrapolating from these conditions to those under which catalytic conversions are frequently carried out, which may be tens or hundreds of atmospheres, is by no means straightforward.

REFERENCES

1. R. S. Berry, S. A. Rice, and J. Ross, *Physical Chemistry* (New York: Wiley, 1980), section 26.5.
2. P. C. Jordan, *Chemical Kinetics and Transport* (New York: Plenum Press, 1979), section 7.4–7.6.

3. R. J. Field, *Amer. Scientist 73*, 142 (1985).

4. J. H. Espenson, *Chemical Kinetics and Reaction Mechanisms* (New York: McGraw-Hill, 1981), section 7.8.

5. G. G. Hammes, *Principles of Chemical Kinetics* (New York: Academic Press, 1978), chapter 9.

6. C. T. Walsh, *Enzymatic Reaction Mechanisms* (San Francisco: W. H. Freeman Co., 1979).

7. D. P. Shoemaker, C. W. Garland, J. I. Steinfeld, and J. W. Nibler, *Experiments for Physical Chemistry*, 4th ed. (New York: McGraw-Hill, 1981), pp. 271–281.

8. P. C. Jordan, *op. cit.*, pp. 147–152.

9. R. J. Madix, *Science 233*, 1159 (1986).

10. R. P. H. Gasser, *An Introduction to Chemisorption and Catalysis by Metals* (Oxford, England: Clarendon Press, 1985).

11. K. J. Laidler, *Chemical Kinetics*, 3rd ed. (New York: Harper and Row, 1987), chapter 7.

12. M. A. Vannice, *Catal. Rev. Sci. Eng. 14*, 153 (1976).

13. J. Haggin, *Chem. and Eng. News* (Oct. 26, 1981), pp. 22–32.

PROBLEMS

5.1 (a) Using integration by partial fractions, derive the autocatalytic rate law, equation 5–6. That is, integrate

$$\int \frac{dx}{(A_0 + B_0)x - x^2} = -k \int dt$$

(b) Find the time at which the curve of $[B(t)]$ undergoes a point of inflection.

5.2 Consider the hypothetical mechanism

$$A + X \xrightarrow{k_1} 2X, \quad X + Y \xrightarrow{k_2} 2Y, \quad Y \xrightarrow{k_3} Z$$

which does not describe any known chemical process, but has the following crude ecological analog. If A represents cabbage, X rabbits, Y foxes, and Z dead foxes (or fertilizer), then rabbits eat cabbage and reproduce, foxes eat rabbits and reproduce, and foxes die. In wilderness management it is well known that animal populations oscillate rather than settle into a steady state. As we have seen, the preceding mechanism can account for such behavior.

The related problem, where reverse reaction is possible, is

$$A + X \underset{k_{-1}}{\overset{k_1}{\rightleftharpoons}} 2X, \quad X + Y \underset{k_{-2}}{\overset{k_2}{\rightleftharpoons}} 2Y, \quad Y \underset{k_{-3}}{\overset{k_3}{\rightleftharpoons}} Z$$

For simplicity, assume that [A] has a constant value A_0 which may be arranged if A is initially in great excess or if the system is open and A is continually replenished. Since reverse reaction is possible, this mechanism admits of a true equilibrium state.

(a) Using the principle of detailed balance, find the equilibrium constraints on the system. Write the coupled rate equation for the system, i.e., $d[X]/dt$, $d[Y]/dt$, and $d[Z]/dt$.

(b) Define displacement variables $\xi = [X] - X_0$, $\eta = [Y] - Y_0$, and $\zeta = [Z] - Z_0$, and use the result of part (a) to show that, near equilibrium,

$$\dot{\xi} = -(k_{-1}X_0 + k_2 Y_0)\xi + k_{-2} Y_0 \eta$$

$$\dot{\eta} = k_2 Y_0 \xi - (k_{-2} Y_0 + k_3)\eta + k_{-3}\zeta$$

$$\dot{\zeta} = k_3 \eta - k_{-3}\zeta$$

(c) Assume that $[Z] = [Z_0]$, i.e., that product is always drawn off so that $\zeta = 0$. Show that the decay constants are solutions of the equation

$$\lambda^2 - [k_{-1}X_0 + (k_{-2} + k_2)Y_0 + k_3]\lambda + k_{-1}X_0(k_3 + k_{-2}Y_0) + k_2 k_3 Y_0 = 0$$

(d) Show that both roots of the preceding equation are positive and therefore that the system relaxes to equilibrium without oscillation. (The roots of the equation $\lambda^2 - B\lambda + C = 0$ satisfy the relation $\lambda_1 \lambda_2 = C$.)

5.3 Using the stochastic method described in chapter 2, section 2.7, and appendix 2.3,
  (a) Set up the stochastic master equation for problem 2.
  (b) Show by stochastic numerical simulation what happens to the fox and rabbit populations.
  (c) If there is a short-term fluctuation in the supply of cabbage, what effect does this have on the fox and rabbit populations? Choose reasonable numerical values for the rate constants and initial "concentrations". The following values give interesting results:

	Set I	Set II	Set III	Set IV
k_1	2×10^{-4}	2×10^{-3}	2×10^{-3}	1.5×10^{-3}
k_{-1}	0.0	0.0	4×10^{-3}	0
k_2	1×10^{-4}	1×10^{-3}	1×10^{-3}	1×10^{-3}
k_{-2}	0.0	0.0	2×10^{-3}	0
k_3	1×10^{-2}	1×10^{-1}	1×10^{-1}	1.1×10^{-1}
k_{-3}	0.0	0.0	2×10^{-1}	0

5.4 The reaction catalyzed by the enzyme hexokinase is

$$Mg\,ATP + G \rightleftharpoons Mg\,ADP + G6P$$

where $Mg\,ATP$ is magnesium adenosine $5'$-triphosphate, G is glucose, $Mg\,ADP$ is magnesium adenosine $5'$-diphosphate, and G6P is glucose 6-phosphate. What mechanisms are consistent with the initial-rate steady-state data given in the following table for yeast hexokinase at pH 8, 25°C, and 0.3 M $(CH_3)_4NCl$ [G. G. Hammes and D. Kochavi, *J. Am. Chem. Soc.* 84, 2069, 2073, 2076 (1962)].

[MgATP], M $\times$ 10^4	[G], M $\times$ 10^4	$E_0/v \times 10^3$, sec
4.73	10	2.55
	5	2.90
	2	4.54
	1	7.48
9.20	10	2.21
	5	2.51
	2	3.90
	1	6.10
19.4	10	1.83
	5	2.17
	2	3.17
	1	4.87
40.0	10	1.72
	5	2.03
	2	2.93
	1	4.44

CHAPTER 6

The Transition
from the Macroscopic
to the Microscopic Level

Thus far, we have considered chemical kinetics from the macroscopic, or molar, point of view. Although informed throughout by the microscopic, or molecular, viewpoint, concepts such as the rate coefficient, the reaction order, mechanism, and even activation energy could be interpreted without reference to the discrete, atomic nature of matter. We now turn to the deeper insight afforded by a description of chemical reactions in terms of ensembles of molecular collisions.[1,2]

6.1 RELATION BETWEEN CROSS SECTION AND RATE COEFFICIENT

Let us return to the prototypical bimolecular chemical reaction with which we began our discussion in chapter 1. Rather than dealing with molar concentrations in a bulk system, we shall instead consider a scattering process in laboratory coordinates; this will be an idealized version of the molecular-beam experiments discussed in chapter 9. In lieu of a thermodynamic temperature, we shall characterize the reacting A and B particles by their velocities, and we shall also take explicit account of the rotational, vibrational, and electronic states of the reactant and product molecules, which will be denoted by Γ. The reaction is then described as

$$A(v_A \Gamma_A) + B(v_B \Gamma_B) \longrightarrow C(v_C \Gamma_C) + D(v_D \Gamma_D)$$

As shown in Figure 6–1, we measure the number dN_c of C molecules collected per unit time and the solid angle $d\Omega$ emerging in a direction specified by angles θ and ϕ from the scattering center:

$$\frac{dN_C(\theta, \phi) d\Omega}{dt} = I_R(\theta, \phi; v\{\Gamma\}) v n_A f_A(v_A) n_B f_B(v_B) V d\Omega \qquad (6\text{–}1)$$

Let us take a closer look at the variables in this equation, paying particular attention to their dimensional units:

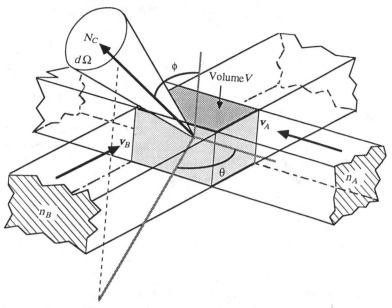

Figure 6–1. Definition of the scattering process $A + B \rightarrow C + D$ in laboratory coordinates, with collection of product C.

n_A and n_B are the densities of A and B particles, respectively, with units of number/cm³   (dimension ℓ^{-3})

dN_C/dt is the number of C molecules collected per second at scattering angles θ and ϕ (dimension t^{-1})

$v = |v_A - v_B|$ is the relative velocity between A and B particles   (dimension ℓt^{-1})

$d\Omega = \sin\theta \, d\theta d\phi$ is the element of solid angle   (dimensionless)

$f_i(v_i)$ is the dimensionless normalized probability density for species i having a velocity between v_i and $v_i + dv_i$

V is the volume defined by the intersection of the beams of A and B molecules   (dimension ℓ^3)

$\{\Gamma\}$ denotes the internal state specification of the reactant and product molecules; we shall see how to deal with these variables shortly.

A straightforward dimensional analysis allows us to determine the units of the coefficient $I_R(\theta, \phi; v\{\Gamma\})$:

$$(t^{-1}) = (?)(\ell t^{-1})(\ell^{-3})(\ell^{-3})(\ell^3) = (?)(\ell^{-2}t^{-1})$$

Therefore, the units of I_R must be ℓ^2 which corresponds to a *cross section*. I_R is termed a *differential* cross section, since it depends on the angular variables θ and ϕ in addition to the molecular variables v and $\{\Gamma\}$. We can also define a *total* cross section by integrating over the angular variables, thus:

$$\sigma_R(v\{\Gamma\}) = \int_0^{2\pi} \int_0^{\pi} I_R(\theta, \phi; v\{\Gamma\}) \sin\theta \, d\theta \, d\phi \qquad (6\text{--}2)$$

Now we wish to see how the cross section, as just defined, connects with the phenomenological rate coefficient which has formed the basis for our discussion thus far. We first have to address the specification of internal states $\{\Gamma\}$ of the reactants and $\{\Gamma'\}$ of the product molecules. We shall see in chapter 9 that these internal state variables can have a profound influence on the course of the reaction. Molecules may react preferentially from specified initial states (selective energy consumption), or a reaction may preferentially populate specific product states (specific energy release). For comparison with reactions at thermal equilibrium, it is necessary to average these variables to obtain total species concentrations. The procedure for doing so, sometimes referred to as the canon of molecular dynamics,[3] is to average over initial (reactant) states and sum over final (product) states. The result is

$$\frac{dN_C(\theta, \phi; \Gamma') \, d\Omega}{dt} = \sum_\Gamma I_R(\theta, \phi; v\Gamma\Gamma') v n_A f_A(v_A) f_A(\Gamma_A) n_B f_B(v_B) f_B(\Gamma_B) V \, d\Omega$$

$$(6\text{--}3)$$

This expression gives us the rate of appearance of product C in a specified state Γ' at scattering angles θ and ϕ. The distribution over reactant states $f(\Gamma)$ may be Maxwell-Boltzmann, or may be some nonequilibrium distribution prepared in the experiment, e.g., by laser pumping. Summing over all product states gives the total rate of appearance of species C:

$$\frac{dN_C(\theta, \phi) d\Omega}{dt} = \sum_{\Gamma'} I_R(\theta, \phi; v\Gamma') v n_A f_A(v_A) n_B f_B(v_B) V \, d\Omega \qquad (6\text{--}4)$$

We next expand the beam intersection volume V to be the entire reaction vessel containing A and B molecules and integrate over all velocities and angular variables:

$$\frac{dn_C}{dt} = \frac{d}{dt}\left(\frac{N_C}{\text{volume}}\right) = \frac{1}{V}\frac{dN_C}{dt}$$

$$= \int_0^{2\pi} \int_0^{\pi} \int_0^{\infty} \int_0^{\infty} v I_R(\theta, \phi; v) n_A f_A(v_A) n_B f_B(v_B) \times 4\pi v_A^2 \, dv_A \, 4\pi v_B^2 \, dv_B \sin\theta \, d\theta \, d\phi$$

or

$$\frac{dn_C}{dt} = \int\int_0^{\infty} v\sigma_R(v) n_A f_A(v_A) n_B f_B(v_B) 4\pi v_A^2 \, dv_A \, 4\pi v_B^2 \, dv_B \qquad (6\text{--}5)$$

Recalling the definition of the phenomenological rate coefficient given in chapter 1, i.e.,

$$\frac{dn_C}{dt} = -\frac{dn_A}{dt} = k n_A n_B$$

we can identify the right-hand side of this equation with that of equation 6–5 and take the species densities n_A and n_B outside of the integral to give an expression for the rate coefficient as the velocity-weighted, velocity-averaged reactive cross section:

$$k = \iint_0^\infty v\sigma_R(v) f_A(v_A) f_B(v_B) 4\pi v_A^2 \, dv_A 4\pi v_B^2 \, dv_B \qquad (6\text{--}6)$$

Corresponding expressions for state-resolved rate coefficients in terms of the appropriate cross sections are obtained if the averaging indicated by equation 6–3 or 6–4 is not carried out.

We now assume that the molecular velocities are described by a Maxwell-Boltzman distribution

$$f_A(v_A; T) = \left(\frac{m_A}{2\pi k_B T}\right)^{3/2} e^{-m_A v_A^2/2k_B T}$$

We shall use the symbol k_B for Boltzmann's constant, which has the value 1.381×10^{-23} J per degree K. This is the molecular analogue of the gas constant $R = k_B N = (1.381 \times 10^{-23})(6.022 \times 10^{23}) = 8.3145$ J per degree K per mole. Using the Maxwell-Boltzmann distribution in equation 6–6 gives the following expression for the thermal rate coefficient:

$$k(T) = \left(\frac{\mu}{2\pi k_B T}\right)^{3/2} \int_0^\infty v\sigma_R(v) e^{-\mu v^2/2k_B T} 4\pi v^2 \, dv \qquad (6\text{--}7a)$$

Substituting the relative kinetic energy $E = \frac{1}{2}\mu v^2$ results in the form

$$k(T) = \frac{1}{k_B T}\left(\frac{8}{\pi \mu k_B T}\right)^{1/2} \int_0^\infty E\sigma_R(E) e^{-E/k_B T} \, dE \qquad (6\text{--}7b)$$

In equations 6–7a and 6–7b, the reduced mass μ is $m_A m_B/(m_A + m_B)$.

6.2 MICROSCOPIC REVERSIBILITY AND DETAILED BALANCING

In addition to providing us with a formula for transforming from scattering cross section to rate coefficient, and thus forming the basis for subsequent developments of microscopic rate theory, equation 6–6 affords insight into the real origin of detailed balancing (see chapter 2, section 2.1.2). Consider the collision process

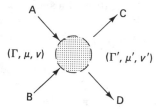

which takes A and B, having internal states Γ, reduced mass μ, and relative velocity v, into C and D having corresponding internal states Γ', μ', and v'. The process is described by the differential cross section $I_R(\Gamma \to \Gamma', v, \theta, \phi)$. The fundamental physical principle of invariance under time reversal gives us the relationship between the forward and the reverse processes, i.e., $C + D[\Gamma', \mu', -v'] \to A + B[\Gamma, \mu, -v]$. This relationship,[4] called *microscopic reversibility*, is

$$(\mu v)^2 g_\Gamma I_R(\Gamma \to \Gamma'; v\theta\phi) = (\mu' v')^2 g_{\Gamma'} I_R(\Gamma' \to \Gamma; v'\theta\phi) \qquad (6\text{-}8)$$

The ratio $(\mu v)^2/(\mu' v')^2$ is just the ratio of the translational states available to the reactants and the products; from conservation of energy, v' and v are related by

$$\frac{1}{2}\mu' v'^2 = \frac{1}{2}\mu v^2 - \Delta E_{int}(\text{chemical}, \Gamma) \qquad (6\text{-}9)$$

The ratio $g_\Gamma/g_{\Gamma'}$ is the ratio of statistical weights for the internal states of the molecules; e.g., $g_J = 2J + 1$ for rotational state J.

Substituting the microscopic reversibility relationship 6–8 into equation 6–6 gives the rigorous form of the detailed balancing principle, viz.,

$$\frac{k[AB(\Gamma) \to CD(\Gamma'); T]}{k[CD(\Gamma) \to AB(\Gamma'); T]} = \left(\frac{\mu'}{\mu}\right)^{3/2}\left(\frac{g_{\Gamma'}}{g_\Gamma}\right)e^{-\Delta E_{int}/k_B T} \qquad (6\text{-}10)$$

Since the entropy change ΔS is just the logarithm of the ratio of the statistical weights, the right-hand side of equation 6–10 can be written as $\exp(\Delta S/k_B - \Delta E_{int}/k_B T) = \exp(-\Delta G°/k_B T) = K_{eq}(T)$, which recovers the form of detailed balancing given earlier in chapter 2. Thus, detailed balancing is contingent upon the existence of Boltzmann velocity distributions, which define a kinetic temperature for the system; if in addition, averaging over internal states Γ is carried out, then a Boltzmann distribution among these variables is required as well. The principle of microscopic reversibility always holds, however, no matter what the degree of disequilibrium in the system, since it is based on the invariance of time reversal alone.

6.3 THE MICROSCOPIC-MACROSCOPIC CONNECTION

The hierarchy of relationships between microscopic quantities such as cross sections and state-to-state probabilities and macroscopic quantities such as rate coefficients is summarized in Figure 6–2.[5] Several comments on the figure, as well as an indication of its optimum pathways, are in order. First, the multiple levels of averaging make clear the absurdity of attempting to go "up the ladder", i.e., from rate coefficients to details of the potential energy surface U. Even reversing a single stage of averaging is fraught with difficulties. For example, the deconvolution of $k(T)$ to yield $\sigma_R(v)$ would be determinate only if rate data were available with essentially infinitely good precision over an infinite temperature range. Indeed, attempting to sample the reaction cross section with a thermal equilibrium distribution of reactants has been likened to measuring an absorption spectrum without using a monochromator:[1] it would be as if one varied the temperature of a black-body light source and attempted to extract the wavelength dependence of the extinction coefficient from the change in total transmittance at different color temperatures! Since rate data do not possess the requisite range or precision, it is generally not possible to extract microscopic details such as energy-dependent cross sections from conventional kinetics experiments. The preferable approach is to calculate microscopic quantities from some model and then perform the downward averaging for comparison with measured quantities.

The current state of the art in experimental techniques, which are described in

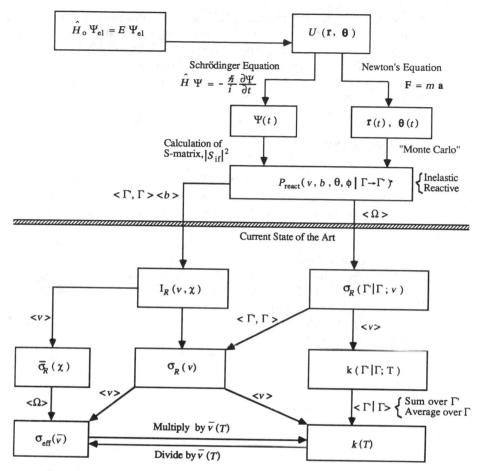

Figure 6–2. "Flowchart" from the intermolecular potential energy surface to thermally averaged rate coefficients. The bracket notation is used to indicate averaging over a variable; i.e., $< O >$ denotes $\int \ldots f(O)dO$. [Adapted from I. W. M. Smith, *Kinetics and Dynamics of Elementary Gas Reactions* (London: Butterworths, 1980).]

greater detail in chapter 9, permits measurements of state-to-state rate coefficients $k(\Gamma' \mid \Gamma; T)$ and of angle- and velocity-dependent cross sections $I_R(v, \theta)$. Measurements in which relative velocity, scattering angle, and product and reactant states may all be simultaneously determined are only now becoming possible using combinations of laser and molecular-beam techniques. Such measurements are essential for stringent tests of computed potential energy surfaces.

A final point worth considering is whether all of the intermediate details are really necessary if one is interested only in, say, the temperature dependence of the rate coefficient for a particular chemical process. Such considerations have led to the development of *information theory,* which is considered in detail in chapter 13. A second simplifying approach is *statistical rate theory* (transition state theory), which seeks to go from the potential U to the rate coefficient $k(T)$ in a single jump. This approach is described in chapter 10. First, however, we shall examine how microscopic quantities such as scattering cross sections and transition probabilities may be calculated from intermolecular potential energy surfaces. Although simple model potentials are frequently used in such calculations, modern computational techniques permit the use of accurate, *ab initio* potential functions. In the next chapter, we shall learn how these functions may be obtained.

REFERENCES

1. J. I. Steinfeld and J. L. Kinsey, *Progr. Reaction Kinetics 5*, 1 (1970).
2. K. Shuler, J. C. Light, and J. Ross, *Equilibrium, Transport, and Collision Processes in Gases and Plasmas*, IDA-ARPA Re-entry Physics Series (New York: Academic Press, 1967).
3. A. Ben-Shaul, Y. Haas, K. L. Kompa, and R. D. Levine, *Lasers and Chemical Change* (Berlin: Springer-Verlag, 1981).
4. J. C. Polanyi and J. L. Schreiber, "The Dynamics of Bimolecular Reactions" in *Physical Chemistry, an Advanced Treatise: Vol. 6A, Kinetics of Gas Reactions*, edited by H. Eyring, D. Henderson, and W. Jost (New York: Academic Press, 1974).
5. I. W. M. Smith, *Kinetics and Dynamics of Elementary Gas Reactions* (London: Butterworths, 1980).

CHAPTER 7
Potential Energy Surfaces

The potential energy surfaces considered in this chapter are based on the Born-Oppenheimer separation of nuclear and electronic motion.[1] A justification for this separation is the disparity in the electron and nuclear masses, which results in very slow nuclear motion compared to electronic motion. With the Born-Oppenheimer separation, each electronic state of the chemical reactive system has a potential energy surface. The primary focus of this chapter will be the *ground electronic state* potential energy surface. In the last section potential energy surfaces will be considered for excited electronic states.

For a nonlinear molecule, consisting of N atoms, the potential energy surface depends on 3N-6 independent coordinates, and depicts how the potential energy changes as relative coordinates of the atomic nuclei involved in the chemical reaction are varied. An analytic function which represents a potential energy surface is called a *potential energy function*. Understanding the relationship between properties of the potential energy surface and the behavior of the chemical reaction is a central issue in chemical kinetics. Both the macroscopic thermal rate constant for a chemical reaction and its microscopic counterpart such as a quantum mechanical state-to-state reaction probability may be interpreted in terms of a potential energy surface.

One of the simplest possible chemical reactions is rupture of a diatomic bond as in $HF \rightarrow H \cdot + F \cdot$. In this case, the dependence of the potential energy on the internuclear bond length r is a one-dimensional potential energy curve $V(r)$ illustrated in Figure 7–1. If three or more atoms participate in the chemical reaction, the potential energy will depend on more than one coordinate and a multidimensional surface results instead of a curve.

The forces between atoms which participate in a chemical reaction may be determined from the potential energy function. An interatomic force is defined by the negative of the derivative of the potential with respect to the internuclear coordinate. For the diatomic potential the force is given by

$$F_r = -\frac{dV(r)}{dr}.$$ (7–1)

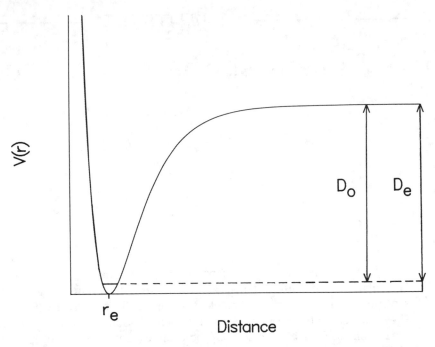

Figure 7–1. Potential energy curve for a diatomic molecule. r_e is the equilibrium bond length, D_e the classical dissociation energy.

From the plot in Figure 7-1, it is clear that the force is dependent on r and goes through zero at the potential minimum at r_e.

7.1 LONG-RANGE POTENTIALS

Long-range classical electrostatic interactions between two particles depend on the inverse of the interparticle distance. A familiar example is the Coulombic potential between two charges $Z_1 e$ and $Z_2 e$ separated by a distance r; i.e.,

$$V(r) = \frac{Z_1 Z_2 e^2}{r} \qquad (7-2)$$

For the interaction of a charge $Z_1 e$ and a dipole μ_2, the potential is

$$V(r) = -\frac{Z_1 e \mu_2 \cos \theta}{r^2} \qquad (7-3)$$

where $\cos \theta$ is the cosine of the angle between μ_2 and $\mathbf{r}$. Similar to the above charge-charge and charge-dipole interactions, dipole-dipole interactions vary as r^{-3}, dipole-quadrupole as r^{-4}, and quadrupole-quadrupole as r^{-5}. These latter three interactions depend on the orientation of the multipoles as well as the interparticle distances.

Even if a molecule does not have a permanent dipole, a dipole may be induced by the electric field of an ion. The induced dipole and the electric field are related by

$$\mu = \alpha_2 \mathbf{E} \tag{7-4}$$

where α_2 is the polarizability of the nonpolar molecule and has units of volume. The magnitude of the ion's electric field is given by

$$E = -d[Z_1 e/r]/dr = Z_1 e/r^2, \tag{7-5}$$

so that the induced dipole is $\mu = \alpha_2 Z_1 e/r^2$. As a result of the ion's electric field the Coulomb force on the induced dipole is

$$F = \mu \cdot \frac{d\mathbf{E}}{dr} \tag{7-6}$$

To find the ion-induced dipole interaction energy at a given r, the differential $dV(r) = -\mathbf{F} \cdot \mathbf{dr}$ must be integrated from a separation of infinity, where the potential is zero to r;

$$V(r) = -\int_{\infty}^{r} \mathbf{F} \cdot \mathbf{dr} = -\int_{\infty}^{r} F \, dr \tag{7-7}$$

Combining equations 7–4 through 7–6 and inserting them into equation 7–7 gives

$$V(r) = -\alpha_2 (Z_1 e)^2 / 2r^4 \tag{7-8}$$

A dipole as well as an ion can induce a dipole in a nonpolar molecule. For this case, the attractive potential varies by $1/r^6$, and a detailed calculation gives

$$V(r, \theta) = \frac{-\alpha_2 \mu_1^2 (3 \cos^2 \theta + 1)}{2r^6} \tag{7-9}$$

where θ is the angle between the permanent dipole and the inter-particle axis, α_2 is the polarizability of the nonpolar molecule, and μ_1 is the permanent dipole.

Dipole moments and polarizabilities for various molecules are listed in Tables 7-1 and 7-2. It is useful to estimate the magnitudes of the ion-ion, ion-dipole, ion-induced dipole, and dipole-dipole interactions discussed above. For illustration we will use the representative values for μ and α of 1.5D and 3 Å^3, respectively, and $Z = 1$ and θ equal to zero so that the dipoles are aligned along the interparticle axis. With these parameters and r of 5 and 10 Å, we obtain the interaction energies in Table 7-3.[2] Only the ion-ion interaction energy is comparable to a typical covalent single bond energy of 380-420 kJ/mol. The dipole-dipole interaction, which can be present in the interaction of two neutral molecules, is comparable to the thermal energy $\frac{3}{2} k_B T$ at 300 K.

Equations 7–3 and 7–9 give the instantaneous interaction energy at a particular orientation of two particles. For a large number of particles there is an average interaction energy $\langle V(r) \rangle$, which is obtained by averaging the interaction potential at a fixed r, $V(\theta_1, \theta_2; r)$, over all possible orientations θ_1 and θ_2 of the two interacting particles. At each orientation the potential must be weighted by the Boltzmann fac-

TABLE 7-1 EXPERIMENTAL DIPOLE MOMENTS.[a]

Molecule	Dipole moment (debyes)[b]
C_3H_8	0.08
H_2O_2	2.2
H_2O	1.85
H_2S	0.97
NaCl	9.0
KCl	10.3
HCl	1.12
CH_3D	0.0006
CH_2Cl_2	1.60
SO_2	1.63
HCN	2.98

[a] From Natl. Bur. Stand. U.S. Publ. NSRDS-NBS 10, 1967; A. L. McClellan, *Tables of Experimental Dipole Moments*, Vol. 1 (Freeman, New York, 1963) and Vol. 2(Rahara Enterprises, 1974).

[b] 1 debye = 10^{-18} esu-cm.

TABLE 7-2 EXPERIMENTAL POLARIZABILITIES.[a]

Molecule	Polarizability ($Å^3$)
Ar	1.63
N_2	1.76
F_2	1.3
I_2	10.2
CH_4	2.6
C_3H_8	6.29
CO	1.98
O_2	1.60
HCl	2.60
CO_2	2.63
NH_3	2.22
C_6H_6	10.4
CH_3Cl	4.53
$CHCl_3$	8.50

[a] From N. J. Bridge and A. D. Buckingham, *Proc. Roy. Soc. (London)* A295, 334 (1966).

TABLE 7-3 LONG-RANGE INTERACTION ENERGIES.[a]

	$r = 5$ Å	$r = 10$ Å
Ion-ion. Eq. (7-2)	278	139
Ion-dipole, Eq. (7-3)	17.4	4.35
Ion-induced dipole, Eq. (7-8)	3.34	0.21
Dipole-induced dipole, Eq. (7-9)	0.052	0.00081

[a] Interaction energies are in units of kJ/mol, and are taken from reference 2.

tor $\exp[-V(\theta_1, \theta_2; r)/k_B T)]$. This averaging may alter n in the r^{-n} term of the potential. As discussed above, the instantaneous dipole-dipole interaction is proportional to r^{-3}. However, the average dipole-dipole interaction energy with the proper Boltzmann weighting depends on r^{-6} and is

$$\langle V(r) \rangle = -\frac{2}{3} \frac{\mu_1^2 \mu_2^2}{k_B T r^6} \tag{7-10}$$

Equation 7–9 can be generalized to represent the total dipole-induced dipole interaction potential between two polar molecules. Averaging this potential over all orientations of the two dipoles μ_1 and μ_2 and Boltzmann weighting gives

$$\langle V(r) \rangle = -\frac{(\mu_1^2 \alpha_2 + \mu_2^2 \alpha_1)}{r^6} \tag{7-11}$$

For this interaction averaging over all orientations does not affect the r^{-n} term in the potential. The average potential in equation 7–11 is half the value in equation 7–9 for $\theta = 0$.

Even when a net charge or permanent dipole moment is absent from the molecule, there remain London or dispersion forces. In contrast to those above, these forces cannot be understood in terms of classical electrostatics. They are quantum mechanical in origin and arise from the correlated motion of the electrons in the two molecules. Although the time-averaged electron distribution of a nonpolar molecule is symmetric, fluctuations in the distribution can lead to a transient dipole moment, which can induce a dipole in a neighboring molecule. These fluctuations become correlated and an attractive interaction results. Quantum mechanical calculations show that the interaction energy for a dispersion force is proportional to $1/r^6$.

If the particles are neutral and only dipoles are considered, the long-range potential is a sum of the average dipole-dipole interaction in equation 7–10, the average dipole-induced dipole interaction in equation 7–11, and the dispersion forces, each of which depends on r^{-6}. For most cases the dispersion contribution is the largest and is the only interaction when the particles are nonpolar. Including average dipole-quadrupole and quadrupole-quadrupole interactions gives rise to terms in the long-range potential which depend on r^{-8} and r^{-10}, respectively.

Long-range potentials are important in kinetics. For example, the Coulombic potential, equation 7–2, is appropriate for a molecule dissociating to ions (e.g., $KCl \rightarrow K^+ + Cl^-$) and the charge-induced dipole potential, equation 7–8, is the basis of the Langevin theory for ion-molecule reactions (chapter 8, section 8.4).

7.2 EMPIRICAL INTERMOLECULAR POTENTIALS

To describe an intermolecular interaction which does not involve the formation of a covalent bond, as for Ar_2, $(H_2O)_2$, or $Li^+(H_2O)$, the r^{-n} long-range attractive potentials described above have been combined with empirical models for the intermediate and short-range regions of the potential. The degree of precision required of the model is determined by the quantity to be represented. For example, a model with

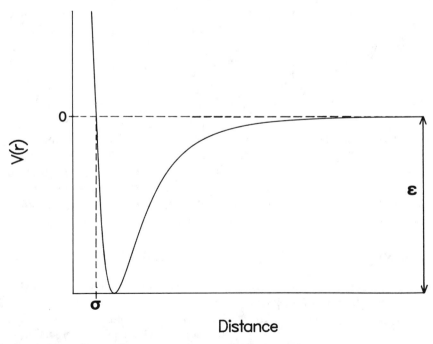

Figure 7–2. (a) Hard-sphere potential energy curve. (b) Lennard-Jones (6–12) potential energy curve.

more detail is required to fit molecular beam scattering experiments (chapter 9, section 9.1) than is required for transport properties such as diffusion, heat conductivity, thermal diffusion, or viscosity.

The simplest interparticle potential is that for the hard-sphere model, which attempts to account for the short-range repulsive forces but not for the long-range attraction. This potential energy function is given by

$$V(r) = 0 \qquad r > \sigma$$
$$V(r) = \infty \qquad r < \sigma \tag{7-12}$$

and is depicted in Figure 7-2(a). The two hard-sphere particles come into contact at the internuclear distance σ and the potential energy rises to ∞ at this point.

Interparticle potentials with more detail are constructed by combining either of the repulsive terms $V(r) = ae^{-br}$ or $V(r) = a/r^m$ with an attractive r^{-n} term to form

$$V(r) = ae^{-br} - c/r^n \tag{7-13}$$

or

$$V(r) = a/r^m - c/r^n \tag{7-14}$$

One of the most celebrated forms of equation (7-14) is the Lennard-Jones "6-12" potential

$$V(r) = 4\epsilon\left[\left(\frac{\sigma}{r}\right)^{12} - \left(\frac{\sigma}{r}\right)^{6}\right] \tag{7-15}$$

where the r^{-6} attractive term represents the dispersion forces, as well as dipole-dipole and dipole-induced dipole interactions. The repulsive part, though purely empirical and chosen for mathematical convenience, does fit many potentials. The potential has a minimum at $r_m = 2^{1/6}\sigma$, with $V(r_m) = -\epsilon$. A depiction of the Lennard-Jones potential is given in Figure 7-2(b). Values of r_m and ϵ for noble-gas atoms are given in Table 7-4.

TABLE 7-4 POTENTIAL PARAMETERS FOR NOBLE GAS MIXTURES.[a]

	He	Ne	Ar	Kr	Xe
He	$\left(\begin{array}{c}10.8\\2.97\end{array}\right)$	22	29.4	30.2	28
Ne	3.00	$\left(\begin{array}{c}42.3\\3.08\end{array}\right)$	70	73	75
Ar	3.47	3.43	$\left(\begin{array}{c}143.2\\3.76\end{array}\right)$	167.5	187.4
Kr	3.67	3.58	3.88	$\left(\begin{array}{c}199.9\\4.01\end{array}\right)$	231.1
Xe	3.95	3.75	4.06	4.18	$\left(\begin{array}{c}282.3\\4.36\end{array}\right)$

[a] From G. Scoles, *Ann. Rev. Phys. Chem. 31*, 81 (1980). Upper triangle: well depth(K). Lower triangle: position of potential minimum r_m (Å).

7.3 MOLECULAR BONDING POTENTIALS

The above potential functions describe nonbonding interparticle and intermolecular interactions between atoms, ions, and molecules. Different potential functions are required for bonding interactions between atoms in molecules. For a diatomic molecule the only bonding interaction is a bond stretch between the two atoms. The simplest representation of the stretching potential is that of the harmonic oscillator model, i.e.

$$V(r) = V(r_e) + \frac{1}{2}f_r(r - r_e)^2 \qquad (7-16)$$

The vibrational frequency for the harmonic stretch is $\nu_e = (f_r/\mu)^{1/2}/2\pi$, where μ is the diatomic reduced mass. The vibrational energy levels are $E(n) = (n + 1/2)h\nu_e$.

The harmonic oscillator model is an accurate representation of the stretching potential only for small displacements of r from the equilibrium bond length r_e. Differences between the true potential and the harmonic are called "anharmonic effects." These differences are magnified at large r, where, as shown in Figure 7–1, the anharmonic potential approaches D_e while the harmonic potential increases to large positive values.

A way to include anharmonicity is to represent the potential in a Taylor series expansion,

$$V(r) = V(r_e) + \left(\frac{\partial V}{\partial r}\right)_{r=r_e}(r - r_e) + \frac{1}{2}\left(\frac{\partial^2 V}{\partial r^2}\right)_{r=r_e}(r - r_e)^2 + \frac{1}{6}\left(\frac{\partial^3 V}{\partial r^3}\right)_{r=r_e}(r - r_e)^3$$

$$+ \frac{1}{24}\left(\frac{\partial^4 V}{\partial r^4}\right)_{r=r_e}(r - r_e)^4 + \ldots \qquad (7-17)$$

The first derivative term equals zero at the minimum $r = r_e$. The second, third, and fourth derivative terms are called quadratic (harmonic), cubic, and quartic force constants, respectively. Many terms must be included in equation 7–17 to accurately represent the potential energy for large displacements in r from r_e.

A useful approximation to a bond stretch potential for all values of r is the Morse function:

$$V(r) = D_e[1 - \exp\{-\beta_e(r - r_e)\}]^2. \qquad (7-18)$$

The three parameters in the Morse function D_e, β_e, and r_e are usually chosen to fit the bond dissociation energy, the harmonic vibrational frequency ν_e, and the equilibrium bond length. The parameters D_e and β_e are related to the harmonic force constant f_r through the relation $f_r = 2D_e\beta_e^2$. Vibrational energy levels for the Morse oscillator are

$$E(n) = (n + 1/2)h\nu_e - (n + 1/2)^2 h\nu_e x_e, \qquad (7-19)$$

where the anharmonicity term x_e is given by

$$x_e = h\nu_e/4D_e \qquad (7-20)$$

The difference between the classical and quantum dissociation energies D_e and D_0 (See Figure 7-1) is the zero point energy for the diatomic molecule. The harmonic oscillator zero point energy is $h\nu_e/2$, while that for the anharmonic Morse oscillator is $h\nu_e(1 - x_e/2)/2$.

Two shortcomings of the Morse function are that in representing actual diatomic potentials it is not sufficiently repulsive at small r and too attractive at large r. Several different functions have been suggested as improvements to the Morse function.[3-5] A widely used procedure is to represent the β_e term in the exponential as a polynomial in r;[6] e.g.,

$$\beta_e = \beta_0 + a(r - r_e) + b(r - r_e)^2 + c(r - r_e)^3. \qquad (7\text{--}21)$$

7.4 INTERNAL COORDINATES AND NORMAL MODES OF VIBRATION

For molecules with more than two atoms, accurate representation of the potential energy becomes more difficult, since more coordinates and potential parameters are required. To depict the potential either conceptually or mathematically, it is natural and convenient to use internal coordinates such as bond stretches and angle bends. A nonlinear molecule consisting of N atoms has 3N-6 independent internal coordinates. Additional internal coordinates are called "redundant coordinates." The analytic representation of a potential energy surface is often facilitated by including redundant internal coordinates. In general, no difficulty arises in including redundant internal coordinates in a potential energy function as long as the function correctly represents the molecular symmetry.

A nonlinear triatomic molecule such as water, shown in Figure 7–3(a), has three independent internal coordinates which may be chosen as two bond stretches and a valence angle bend. A comparison with the diagrams for formaldehyde in Figures 7–3(b) and 7–3(c) shows that increasing the size of a molecule by only one atom can lead to a significant increase in the complexity of the potential energy. A potential energy function for formaldehyde could include three bond stretching coordinates, three valence angle bends, and an out of plane bend. Figure 7–3(b) illustrates the three valence angle bends and Figure 7–3(c) shows the out of plane bend. Note that a redundant valence angle bend angle is included in this representation of

Figure 7–3. Different types of internal coordinates.

the formaldehyde potential energy function. An important internal coordinate for many molecules is the torsion (dihedral) angle, which is depicted for hydrogen peroxide in Figure 7–3(d).

In chapter 11, section 11.7, internal coordinates are used to define analytic potential energy functions for chemical reactions. They can also be used to represent the potential energy for a polyatomic molecule near its potential energy minimum by expanding the potential in a Taylor series as described by equation 7–17 for a diatomic. Expressing the potential energy for the water molecule, Figure 7–3(a), in this manner gives

$$V = V_0 + \frac{1}{2} f_{rr}(r - r_e)^2 + \frac{1}{2} f_{\bar{r}\bar{r}}(\bar{r} - \bar{r}_e)^2 + \frac{1}{2} f_{r\bar{r}}(r - r_e)(\bar{r} - \bar{r}_e) + \frac{1}{2} f_{\theta\theta}(\theta - \theta_e)^2$$

$$+ \frac{1}{2} f_{r\theta}(r - r_e)(\theta - \theta_e) + \frac{1}{2} f_{\bar{r}\theta}(\bar{r} - \bar{r}_e)(\theta - \theta_e) + \dots \qquad (7-22)$$

where the anharmonic cubic, quartic, and higher order terms have been neglected. The $f_{\bar{r}\bar{r}}$, $f_{r\theta}$ and $f_{\bar{r}\theta}$ nondiagonal terms in equation 7–22 are often called "interaction terms," since they describe how the displacement of a particular internal coordinate affects the potential of the remaining coordinates. Force constants may be determined from *ab initio* calculations[7] and by fitting experimental vibrational frequencies.[8] Representative internal coordinate quadratic force constants are listed in Table 7-5.

Normal mode coordinates of vibration, which assume a harmonic (i.e., quadratic) potential energy function as in equation 7–22, are often used to characterize the potential energy minimum for a polyatomic molecule. For a molecule with N atoms the most straightforward way to determine the normal mode coordinates and their frequencies is to diagonalize the $3N \times 3N$ mass-weighted Cartesian force constant matrix [$\mathbf{F}$],[9] i.e.

TABLE 7–5 DIAGONAL INTERNAL COORDINATE FORCE CONSTANTS.[a]

Diagonal force constant	Value[b]
H-C stretch	4.55, 4.64, 4.70
C-C stretch	4.39
C=O stretch	12.99
C-O stretch	5.70
HCH bend	0.540, 0.550
HCC bend	0.645, 0.656
CCC bend	1.130, 1.084, 1.086
HCO or CCO bend	1.166
COC or COH bend	0.651

[a] The force constants were derived from pyruvic acid, acetone, acetic acid, methyl acetate, formic acid, methyl formate, and straight chain aliphatic hydrocarbons: J. H. Schachtschneider and R. G. Snyder, *Spectrochim. Acta 19*, 117 (1963); H. Hollenstein and Hs. H. Günthard, *J. Mol. Spectrosc. 84*, 457 (1980).

[b] The stretching force constants are in units of mdyn/Å, and the bending force constants in units of mdyn-Å/rad².

$$\tilde{\mathbf{L}}\,\mathbf{F}\,\mathbf{L} = \Lambda \qquad\qquad (7\text{--}23)$$

where $\tilde{\mathbf{L}}$ is the transpose of the matrix $\mathbf{L}$ with $\tilde{\mathbf{L}}\,\mathbf{L} = 1$. Solving this equation is a standard problem in linear algebra. The solution gives Λ, which is a diagonal matrix of the 3N eigenvalues λ_i, and $\mathbf{L}$ the eigenvector matrix which defines the transformation between normal mode coordinates Q and displacements of the mass-weighted Cartesian coordinates q from the equilibrium geometry q_0; i.e., $[q - q_0] = \mathbf{L}\,\mathbf{Q}$. Included in Λ are the eigenvalues for translation and external rotation, which are equal to zero. The remaining nonzero eigenvalues are for vibration and equal $4\pi^2 \nu_i^2$, when ν_i is the normal mode frequency. The normal modes of vibration for water are illustrated in Figure 7–4.

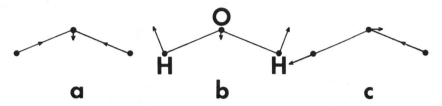

Figure 7–4. H_2O normal modes of vibration: (a) symmetric stretch; (b) symmetric bend; (c) asymmetric stretch.

The mass-weighted Cartesian coordinates $q_{\alpha i}$ are related to Cartesian coordinates $s_{\alpha i}$ according to

$$q_{\alpha i} = \sqrt{m_i}\, s_{\alpha i} \qquad i = 1, \dots, N$$
$$\alpha = x, y, z \qquad\qquad (7\text{--}24)$$

The elements of the force constant matrix $\mathbf{F}$ are denoted by $f_{\alpha i, \alpha' i'}$ and are equal to the second derivative of the potential energy with respect to $q_{\alpha i}$ and $q_{\alpha' i'}$ i.e. $\partial^2 V / \partial q_{\alpha i} \partial q_{\alpha' i'}$. The second derivatives are determined from the potential energy function, e.g., equation 7–22, by writing the internal coordinates as functions of the Cartesian coordinates. The force constants $f_{\alpha i, \alpha' i'}$ are symmetric with respect to interchanging the indices αi and $\alpha' i'$.

To illustrate a normal mode analysis consider the stretching vibration of a diatomic molecule with the harmonic potential $V = f(r - r_e)^2 / 2$. (A normal mode analysis of the stretching vibrations for a linear triatomic molecule is included in the problem set). By inserting $x_2 - x_1$ into the potential energy function for the internuclear distance r, the second derivatives of the potential with respect to the Cartesian coordinates x_1 and x_2 are found to be $\partial^2 V / \partial x_1^2 = f$, $\partial^2 V / \partial x_2^2 = f$, and $\partial^2 V / \partial x_1 \partial x_2 = \partial^2 V / \partial x_1 \partial x_2 = -f$. Thus, the mass-weighted Cartesian force constant matrix is

$$\mathbf{F} = \begin{vmatrix} f/m_1 & -f/(m_1 m_2)^{1/2} \\ -f/(m_1 m_2)^{1/2} & f/m_2 \end{vmatrix} \qquad\qquad (7\text{--}25)$$

If equation 7–23 is written in the form

$$(\mathbf{F} - \Lambda)\,\mathbf{L} = 0. \qquad\qquad (7\text{--}26)$$

it is apparent that the two eigenvalues λ_1 and λ_2, are found by solving the secular determinant

$$\begin{vmatrix} f/m_1 - \lambda & -f/(m_1 m_2)^{1/2} \\ -f/(m_1 m_2)^{1/2} & f/m_2 - \lambda \end{vmatrix} = 0 \qquad (7-27)$$

The resulting two eigenvalues are $\lambda_1 = 0$ for translation and $\lambda_2 = f/\mu$ for vibration, where the reduced mass μ equals $m_1 m_2/(m_1 + m_2)$. Since $\lambda = 4\pi^2 \nu^2$, the harmonic vibrational frequency is $\nu = (f/\mu)^{1/2}/2\pi$.

Each column of $\mathbf{L}$ is the eigenvector for an eigenvalue λ_i. To determine an eigenvector insert the value for λ_i into equation 7–26 to obtain 3N simultaneous homogeneous linear equations, which for the diatomic molecule considered here are

$$\begin{aligned} (f/m_1 - \lambda)\ell_1 - [f/(m_1 m_2)^{1/2}]\ell_2 &= 0 \\ -[f/(m_1 m_2)^{1/2}]\ell_1 + (f/m_2 - \lambda)\ell_2 &= 0 \end{aligned} \qquad (7-28)$$

Using the normalization condition $(\ell_1^2 + \ell_2^2) = 1$, equation 7–28 can be solved to give $\ell_{11} = [m_1/(m_1+m_2)]^{1/2}$ and $\ell_{12} = [m_2/(m_1 + m_2)]^{1/2}$ for $\lambda_1 = 0$, and $\ell_{21} = [m_2/(m_1 + m_2)]^{1/2}$ and $\ell_{22} = -[m_1/(m_1 + m_2)]^{1/2}$ for $\lambda_2 = f/\mu$. Since the normal coordinates are given by $\mathbf{Q} = \tilde{\mathbf{L}}[q - q_0]$, ℓ_{11} and ℓ_{12} give the mass-weighted Cartesian coordinate displacements for normal mode 1 (a translation), while ℓ_{21} and ℓ_{22} give the mass-weighted Cartesian coordinate displacements for normal mode 2 (a vibration).

It becomes exceedingly difficult to determine the matrices Λ and $\mathbf{L}$ analytically as the dimensionality of the problem is increased. For polyatomic molecules, these matrices may be obtained numerically with standard computer programs which are readily available.[10] Values for quadratic force constants (Table 7-5) can be found by varying them in a normal mode analysis to fit experimental harmonic frequencies.

The energy levels for a normal mode are given by $E_i = (n_i + 1/2)h\nu_i$. A summation of the individual normal mode energies gives the normal mode approximation to the total vibrational energy of the molecule, i.e.

$$E = \sum_i E_i = \sum_i (n_i + 1/2)h\nu_i \qquad (7-29)$$

7.5 POTENTIAL ENERGY SURFACES

As shown in Figure 7–1, the potential energy versus a diatomic's internuclear distance defines a curve for a diatomic molecule. In a chemical reaction, three or more atoms must interact; thus the relevant coordinates define a multidimensional surface. Since it is difficult, if not impossible, for most people to visualize surfaces with more than three dimensions, methods must be used to reduce the dimensionality of the problem. For even the simplest system, the three-atom reaction A + BC, the potential surface is a four-dimensional hypersurface (three independent coordinates plus the energy). This can be reduced to a three-dimensional surface in real space by constraining one of the coordinates, e.g., fixing the angle between the two bonds to a specified value. This value is often taken to be 180°, so that the surface represents collinear A—B—C. This gives the potential surface as a function of the two independently variable bond lengths. Even such a three-dimensional surface is inconvenient to represent on a planar blackboard or the pages of a book; thus the surface is generally presented in the form of a contour diagram. This approach of reducing the potential surface for a triatomic (or any polyatomic) to a three-dimensional surface in real space does not place any restrictions on the surface itself but is merely used for graphical representation.

There are two general types of potential energy surfaces for a collinear A + BC → AB + C reaction. One has a potential energy barrier at intermediate A-B and B-C distances, while the other has a potential energy well (minimum). Contour maps for these two types of reactions are depicted in Figure 7–5. Some special A + BC reactions may have both minima and potential energy barriers. The dashed line in Figure 7–5(a) and 7–5(b) represents the reaction path $r^{\dagger}$ (the term *reaction coordinate* is a number representing a particular position along the reaction path). In the reactant configuration, A + BC, the reaction path is primarily the A-B distance, while it is the B-C distance for the products, AB + C. In the intermediate region of the surface, the reaction path contains components of both of these bond distances. At the maximum in the potential energy along the reaction path in Figure 7–5(a) there is a pass or saddlepoint. There is a well at the potential energy minimum in Figure 7–5(b). A potential energy profile results from plotting the potential versus reaction path. The reaction path potential energy profiles for the contour maps in Figure 7–5 are plotted in Figure 7–6.

The saddlepoint in Figure 7–5(a) is a stationary point on the surface. Thus, at the saddlepoint, as for a potential energy minimum, the derivative of the potential energy is zero with respect to each Cartesian coordinate; i.e., $\partial V / \partial q_i = 0$. The harmonic vibrational frequencies and eigenvectors at the saddlepoint are found in the same manner as for a potential energy minimum, viz., by use of equations 7–23 through 7–28. At the saddlepoint, the potential energy surface curves downward along the reaction path, so the corresponding force constant is negative. Since the frequency is proportional to the square root of the force constant, this implies an imaginary frequency. Thus, in contrast to the minimum where all 3N-6 frequencies are real, one of the 3N-6 frequencies is imaginary at the saddlepoint. This frequency corresponds to infinitesimal motion along the reaction path. The remaining 3N-7 frequencies and their eigenvectors define harmonic vibrational motion orthogonal to the reaction path. From the 3N-7 real frequencies the zero-point energy at the saddlepoint can be calculated.

A potential energy surface may also be represented by a perspective. Such a perspective and its relationship to a contour map are illustrated in Figure 7–7. Potential energy perspectives and contour maps are described by topographical terms such as pass, saddlepoint, valley, well, depression, ridge, etc., which are evident in Figure 7–7. Hikers and mountain climbers should be particularly adept at analyzing potential energy surfaces. Many different perspectives and contour maps may be constructed for a reaction with more than three atoms, since different sets of coordinates may be constrained. Several types of perspectives and maps for polyatomic systems are depicted in the next section.

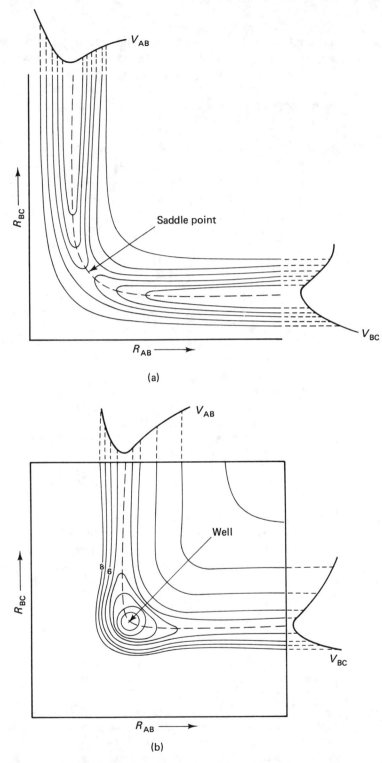

Figure 7–5. Depiction of collinear A + BC → AB + C contour maps. The dashed lines are reaction paths. (a) Contour map with a potential energy barrier; (b) contour map with a potential energy minimum.

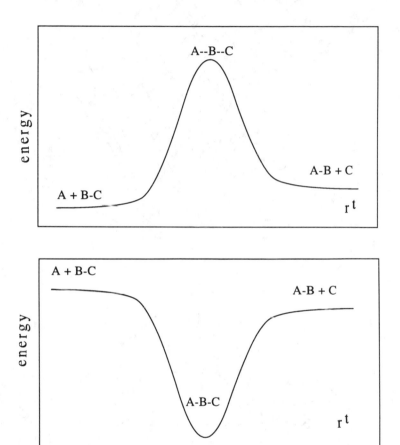

Figure 7–6. Potential energy profile (i.e., potential along the reaction path) for the contour maps in Figure 7–5(a) and 7–5(b).

Figure 7–7. A three-dimensional perspective (a) and contour map (b) for a model chemical reaction. The solid line is the reaction path. [Adapted from G. M. Maggiora and R. E. Christofferson, in *Transition States of Biochemical Processes*, ed. R. D. Gandour and R. L. Schowen (New York: Plenum, 1978).]

7.6 AB INITIO CALCULATION OF POTENTIAL ENERGY SURFACES

Potential energy surfaces may be determined by *ab initio* electronic structure calculations. If one makes the Born-Oppenheimer approximation, discussed in the introduction to this chapter, the molecular wave function is written as

$$\Psi = \psi_e(r, R,)\psi_n(R) \tag{7-30}$$

where ψ_e is the electronic wave function, which depends on the electron coordinates **r** and nuclear coordinates **R**, and ψ_n is the nuclear wavefunction. Solving the Schrodinger Equation for the molecule, i.e.

$$\hat{H}\Psi = E\Psi$$

$$(\hat{H}_e + \hat{H}_n)\psi_e\psi_n = (E_e + E_n)\psi_e\psi_n$$

at a particular nuclear configuration **R** gives the electronic energy E_e and the nuclear-nuclear repulsion energy V_{nn} for the configuration. The sum of these two energies is the total potential energy. By performing the calculation at many different nuclear configurations a complete potential energy surface may be determined.[1]

Sophisticated *ab initio* methods, which are based on the variational theorem,[1] have been developed for calculating potential energy surfaces. The Self-Consistent-Field (SCF) method often gives good structural and force constant information for a molecule near its equilibrium geometry, and is widely used for evaluating these properties.[11] However, the SCF method usually gives an inaccurate potential energy surface for a chemical reaction in which bonds are broken and formed so that there are extensive geometry changes. This is because the SCF approximation neglects instantaneous correlations between electronic motions, and, thus, it often can not describe changes in the wavefunction ψ_e as bonds are broken and formed.[12] This difficulty is well known for reactions as simple as $H_2 \rightarrow 2H$ and $H + H_2 \rightarrow H_2 + H$.[12] To treat electron correlation correctly and obtain an accurate potential energy surface requires a Generalized-Valence-Bond (GVB),[13] Multiconfiguration-SCF(MCSCF),[14] Configuration-Interaction (CI)[15] or perturbation[16] treatment. Computer programs are available for calculating *ab initio* potential energy surfaces with these treatments.[17]

Ab initio calculations have provided much of the information known about potential energy surfaces. The accuracy of the calculated result is determined by the number of basis functions and the treatment of electron correlation, both of which depend on the number of electrons in the molecular system. For many reactions the calculated structures for potential energy minima are as accurate as those found experimentally.[18] Also, the *ab initio* and experimental harmonic vibrational frequencies often agree to within 5%.[19] Of course, the agreement usually improves as the total number of electrons decreases. For reactions which are comprised of three or fewer second row atoms, activation energies and heats of reaction are usually accurate to within several kJ/mole.[20] A very accurate potential energy surface has been calculated for the $H + H_2 \rightarrow H_2 + H$ reaction.[21] Extensive calculations have also been performed for the $F + H_2 \rightarrow FH + H$ surface,[22] illustrated in Figures 7–8 and

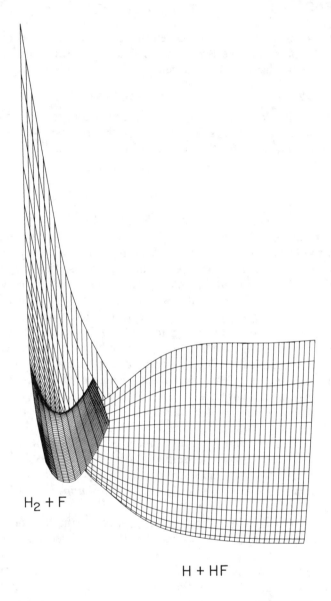

$H_2 + F$

$H + HF$

XBL723-2655

Figure 7–8. Plot of the electronic energy of linear FH_2 as a function of F-H and H-H distances. Each small section bounded by four sides represents a square region in space 0.075 bohrs on a side. Thus the F-H distance ranges from 1.4 to 8.0 bohrs and the H-H distance from 1.0 to 7.6 bohrs. [Adapted from C. F. Bender, et al., *Science 176*, 1412 (1972). Copyright 1972 by the AAAS.]

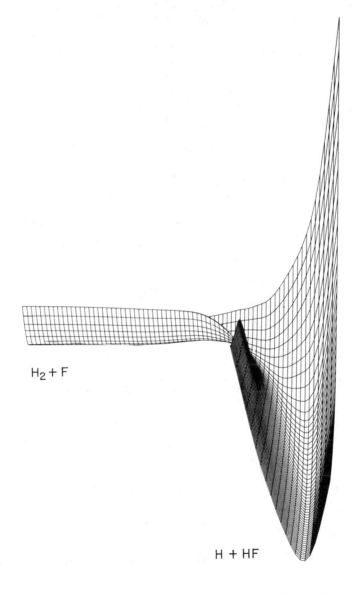

H₂ + F

H + HF

XBL723-2654

Figure 7-8. (*continued*)

7-9. Other reactions which have received considerable study are listed in Table 7-6 and two additional ones are discussed below.

Ab initio calculations have been widely used to calculate interaction energies between alkali ions and ligands.[23,24] The potential surface for this interaction is often depicted by rotating the ion in a plane about the ligand which is fixed in its equilibrium geometry. Such a surface for the $Li^+ + H_2O$ interaction is given in Figure 7-10.

Distance along minimum-energy path

Figure 7–9. (a) Contour map of the linear FH_2 surface. The F-H distance varies from 1.4 to 8.0 bohrs and the H-H distance from 1.0 to 7.6 bohrs. [Adapted from C. F. Bender, S. V. O'Neil, P. K. Pearson, and H. F. Schaefer, *Science 176,* 1412 (1972)]; (b) Potential energy profile for the collinear FH_2 system. [Adapted from I. N. Levine, *Physical Chemistry* (McGraw-Hill, New York, 1988). Reproduced with permission of McGraw-Hill.]

TABLE 7–6 POTENTIAL ENERGY SURFACES FROM *AB INITIO* CALCULATIONS.

Reaction	Reference
$H + H_2 \rightarrow H_2 + H$	Reference 21 in text
$F + H_2 \rightarrow FH + H$	Reference 22 in text
$H + CO \rightarrow HCO$	J. M. Bowman. J. S. Bittman, and L. B. Harding, *J. Chem. Phys. 85,* 911 (1986).
$O(^3P) + H_2 \rightarrow OH + H$	J. M. Bowman, A. F. Wagner, Walch, and T. H. Dunning, Jr., *Chem. Phys 81,* 1739 (1984); B. C. Garrett and D. G. Truhlar, *Int. J. Quantum Chem. 29,* 1463 (1986).
$H_2CO \rightarrow H + HCO$ $H_2 + CO$	J. D. Goddard, Y. Yamaguchi, and H. F. Schaefer III, *J. Chem. Phys, 75,* 3459 (1981); M. J. Frisch, R. Krishnan, and J. A. Pople, *J. Phys. Chem. 85,* 1467 (1981); M. Dupuis, W. A. Lester, Jr., B. H. Lengsfield, and B. Liu, *J. Chem. Phys. 79,* 6167 (1983).
$Be + HF \rightarrow BeF + H$	S. Chapman and S. Green, *Chem. Phys. 78,* 93 (1983).

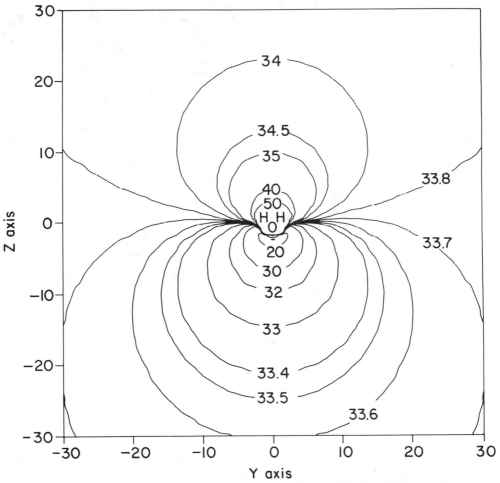

Figure 7–10. $Li^+ + H_2O$ potential energy contour diagram. The X and Y axes define the distance of Li^+ from O. The Li^+ ion is in the plane of H_2O molecule. Energy is in units of kcal/mol. [Adapted from W. L. Hase and D.-F. Feng, *J. Chem. Phys. 75,* 738(1981)].

A somewhat different approach is used to represent *ab initio* calculations of the H + CH_3 potential energy surface.[25] As shown in Figure 7–11, the two internal coordinates which are used in representing the 3-dimensional surface are the bond length r and the angle χ. This angle is a displacement from the C_{3v} symmetry axis in a plane that includes C and two H atoms, and bisects the HCH angle of the remaining two H-atoms; i.e.

The coordinates of the CH_3 group are optimized versus r, but not versus χ.

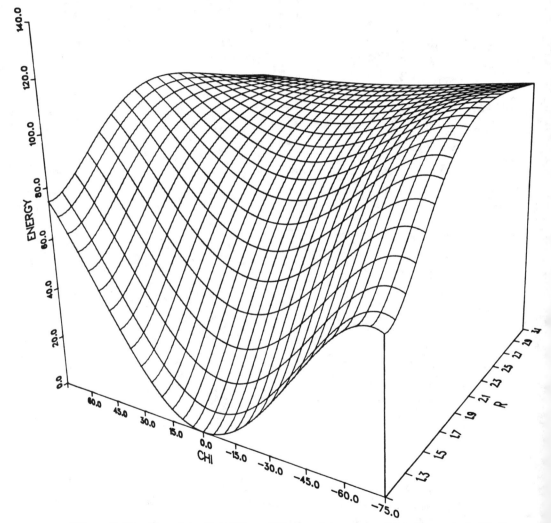

Figure 7–11. Plot of the H + CH_3 → CH_4 potential energy surface, where r is the H–C distance (Å) and χ is the H–CH_3 angle (degrees). The geometry of the CH_3 group is optimized at each value of r. Energy is in kcal/mol. [Adapted from W. L. Hase, S. L. Mondro, R. J. Duchovic, and D. M. Hirst, *J. Am. Chem. Soc. 109*, 2916(1987)].

7.7 ANALYTIC POTENTIAL ENERGY FUNCTIONS

Dynamical calculations, whether classical or quantum mechanical, are facilitated by analytic potential energy functions. A variety of analytic functions have been used in such calculations. A function may be based on an empirical model for the chemical reaction. Parameters for the empirical functions may be chosen to fit experimental equilibrium geometries, vibrational frequencies, bond energies, and activation energies, and the results of *ab initio* calculations. *Ab initio* results are often used to connect smoothly regions of the surface which are known experimentally such as the reactant and product asymptotic limits, and potential energy minima. The parameters for empirical surfaces are often of physical significance.

A dynamical calculation can also be performed by using a table of potential energy points. Spline functions [26] may be used to connect analytically the points in the table. One of the drawbacks of using such a table is that it can not be easily cast into an empirical potential energy function, which is often useful in interpreting the calculated reaction dynamics. Also, in contrast to a table of points, an empirical function can be easily modified by adjusting several parameters.

Rather simple analytic potential energy functions are those based on the r^{-n} intermolecular potential terms discussed in section 7.1. Such potentials have been used to describe ion-molecule (e.g., $Li^+ + H_2O$) and molecule-molecule intermolecular interactions (e.g., $NH_3 + NH_3$), for cases where the intermolecular interaction causes only minor changes in the molecular equilibrium geometry and vibrational frequencies. The complete intermolecular potential is represented by a sum of atom-atom two-body potentials similar to those in equations 7–13 and 7–14. Such potential functions have been used in studies of the structure and thermodynamics of solutions,[27] and of ion-molecule reaction kinetics in the gas phase [28] and in solution.[29]

As an example, consider the intermolecular potential between a metal ion M and a ligand L. The potential is given by

$$V_{M,L} = \sum_i V_{im}, \tag{7-32}$$

where i is a sum over the atoms on the ligand and V_{im} is a two-body potential between M and one of the ligand atoms. A widely used form for V_{im} is

$$V_{im} = \frac{a_{im}}{r^{12}} - \frac{b_{im}}{r^6} + \frac{c_{im}}{r} \tag{7-33}$$

The first two terms comprise the Lennard-Jones "6–12" potential. The remaining term represents an electrostatic interaction, either repulsive or attractive, between the metal atom and ligand atom. *Ab initio* quantum mechanical calculations are used to derive the parameters a_{im}, b_{im}, and c_{im}.[23,24] The *ab initio* potential energy surface for $Li^+ + H_2O$ in Figure 7–10 is accurately described by equations 7–32 and 7–33.

A potential energy function, similar to that given above, may be used to describe the interaction between two molecules A and B. Here a sum of two-body interactions between the *a* atoms contained in A and the *b* atoms contained in B is required: i.e.

$$V_{A,B} = \sum_a \sum_b V_{a,b} \qquad (7\text{--}34)$$

An expression similar to equation 7–33 may be used for $V_{a,b}$ and *ab initio* calculations may be used to derive the potential energy parameters.[30,31]

The most widely used empirical analytic potential function for A + BC atom-diatom reactive scattering is the London-Eyring-Polanyi-Sato (LEPS) surface.[32] This function has its genesis in the semiempirical valence bond approach of London.[33] However, over the years the original function has been modified by including empirical functions and parameters, and has acquired an empirical character.

One widely used form of the LEPS function is given by

$$V(r_{AB}, r_{AC}, r_{BC}) = \frac{Q_{AB}}{1 + S_{AB}} + \frac{Q_{BC}}{1 + S_{BC}} + \frac{Q_{AC}}{1 + S_{AC}} - \left\{ \frac{1}{2} \left[\left(\frac{J_{AB}}{1 + S_{AB}} - \frac{J_{BC}}{1 + S_{BC}} \right)^2 \right. \right.$$
$$\left. \left. + \left(\frac{J_{BC}}{1 + S_{BC}} - \frac{J_{AC}}{1 + S_{AC}} \right)^2 + \left(\frac{J_{AC}}{1 + S_{AC}} - \frac{J_{AB}}{1 + S_{AB}} \right)^2 \right] \right\}^{1/2}$$

$$(7\text{--}35)$$

Q and J are the Coulomb and exchange integrals, respectively, which occur in the Heitler-London valence bond theory. They are determined from the Morse and anti-Morse functions via

$$Q_{AB} + J_{AB} = D_{AB}\{\exp[-2\beta_{AB}(r_{AB} - r_{AB}^0)] - 2\exp[-\beta_{AB}(r_{AB} - r_{AB}^0)]\} \qquad (7\text{--}36)$$

$$Q_{AB} - J_{AB} = \frac{1}{2}D_{AB}\{\exp[-2\beta_{AB}(r_{AB} - r_{AB}^0)] + 2\exp[-\beta_{AB}(r_{AB} - r_{AB}^0)]\} \qquad (7\text{--}37)$$

The three parameters S_{AB}, S_{BC}, and S_{AC} in equation 7–35 give the LEPS function flexibility. They may be chosen to fit the location of the potential energy barrier with respect to the B-C and A-B distances, and the height of the barrier. The original LEPS form of Sato[34] has $S_{AB} = S_{BC} = S_{AC}$. In the extended LEPS form[35] these S parameters are allowed to differ.

Though the LEPS function has received wide use, it has severe limitations. Since it has only three adjustable parameters, it has limited flexibility and does not correctly represent surfaces for a number of AB + C reactions. Often, when trying to fit a barrier height and saddlepoint location, spurious wells appear in the surface.[32] This is particularly evident for the D + HCl system.[36] Clearly, the LEPS function can not be used to represent highly accurate *ab initio* calculations.

Another widely used empirical potential energy function for A + BC → AB + C reactions is the BEBO ("bond energy − bond order") model of Johnston and Parr.[37] However, like the LEPS model it also lacks flexibility. The BEBO model is developed from a number of empirical relationships. One is the equation proposed by Pauling[38] to relate bond length r and bond order n:

$$r = r_s - 0.26 \ln n \qquad (7\text{--}38)$$

where r_s is the single-bond length and 0.26 has units of Å. A log-log correlation between bond energy and bond order is assumed, so that the bond energy D is given by

$$D = D_s n^p \tag{7-39}$$

where the single-bond energy is D_s and p is a constant. This relationship works quite well for CC, CO, and OO bonds. Furthermore, it is postulated that the bond order is conserved so that

$$n_{AB} + n_{BC} = 1. \tag{7-40}$$

The potential energy is written as the dissociation energy of the BC bond minus the sum of the energies of the partially formed AB bond and partially broken BC bond plus the A-C repulsion energy:

$$V = D_{BC} - D_{BC}(n_{BC})^{p_{BC}} - D_{AB}(n_{AB})^{p_{AB}} + V(r_{AC}) \tag{7-41}$$

An anti-Morse function, equation 7–37, may be used for $V(r_{AC})$. Equations 7–38 through 7–41 specify the potential energy along the minimum energy path. Additional terms must be added to give the variation in the potential energy for displacements from the reaction path.

Another procedure for writing analytic potential energy functions uses the concept of a switching function. An example is the $A + BC \rightarrow AB + C$ analytic function developed by Blais and Bunker,[39] and based on the Morse function and a switching function. The analytic function is

$$V = D_{AB}\{1 - \exp\left[-\beta_{AB}(r_{AB} - r^0_{AB})\right]\}^2 + D_{BC}\{1 - \exp\left[\beta_{BC}(r_{BC} - r^0_{BC})\right]\}^2$$

$$+ D_{BC}[1 - \tanh(ar_{AB} + c)] \exp\left[-\beta_{BC}(r_{BC} - r^0_{BC})\right] \tag{7-42}$$

$$+ D_{AC} \exp\left[-\beta_{AC}(r_{AC} - r^0_{AC})\right].$$

The first two terms are Morse potentials for the A-B and B-C bonds. The third term attenuates the attraction between B and C as A approaches by means of the switching function $[1 - \tanh(ar_{AB} + c)]$, which ranges from zero for large r_{AB} to unity for smaller r_{AB}. The final term supplies a repulsive interaction between A and C.

Murrell and co-workers[40] have developed a function for triatomic systems based on a many-body expansion of the potential energy with parameters for the function chosen to match experimental energies and vibrational frequencies, and *ab initio* data. The potential is a sum of three two-body terms and a three-body term, and is written as

$$V = V_{AB}(r_{AB}) + V_{BC}(r_{BC}) + V_{AC}(r_{AC}) + V_{ABC}(r_{AB}, r_{BC}, r_{AC}), \tag{7-43}$$

The role of the three-body term is to fill the difference between the exact potential and the sum of the diatomic parts. It goes to zero if any atom of the triatomic is removed to infinity. The three-body term is expressed as a polynomial (quartic if necessary) multiplied by a product of switching functions:

$$V_{ABC}(r_{AB}, r_{BC}, r_{AC}) = AP(s_1, s_2, s_3) \prod_{i=1}^{3} \left(1 - \tanh \frac{1}{2} \gamma_i s_i\right) \tag{7-44}$$

The s_i are displacements in the bond lengths from a reference configuration; e.g., $s_1 = r_{AB} - r^0_{AB}$. The switching function $(1 - \tanh \gamma_i s_i / 2)$ goes to zero for large s_i. Parameters in the function are A, the γ_i, and the coefficients in the polynomial P. In principle, this method can be generalized to tetra-atomic molecules by writing the potential as a sum of two-body and three-body terms, plus a four-body term.

A number of analytic potential functions have been derived for polyatomic reactive systems and some are listed in Table 7–7. In one approach, (e.g., H + C_2H_4 and H + CH_3) polyatomic analytic potential functions have been constructed from internal coordinates, switching functions, and extensions of BEBO concepts. To illustrate, consider the H + $CH_3 \rightleftarrows CH_4$ system depicted in Figure 7–11. The H–r–C—H diagonal quadratic bending force constant f_χ and equilibrium angle χ_0 along the minimum energy path can be written as

$$f_\chi = S_f(r) f_\chi^{CH_4} \tag{7–45}$$

and

$$\chi_0 = \chi_0^{CH_4} + [\chi_0^{CH_4} - 90.00][S_\chi(r) - 1.0]. \tag{7–46}$$

In these equations, $S_f(r)$ and $S_\chi(r)$ are switching functions which depend on the rupturing CH bond length r. They are equal to unity at the CH equilibrium distance r_e and go to zero for large r. Parameters for the switching functions may be fit to *ab initio* data. By including other functions similar to those in Equations 7–45 and 7–46, a complete analytic function results for the H + $CH_3 \rightleftarrows CH_4$ system.

TABLE 7–7 POLYATOMIC ANALYTIC POTENTIAL ENERGY SURFACES.

Reaction	Reference
$NO + O_3 \rightarrow O_2 + NO_2$	R. Viswanathan and L. M. Raff, *J. Phys.* 87, 3251 (1983); S. Chapman. *J. Chem. Phys.* 74, 1001 (1980).
$SiH_4 \rightarrow SiH_2 + H_2$ $\rightarrow SiH_3 + H$	R. Viswanathan, D. L. Thompson, and L. M. Raff, *J. Chem Phys.* 80, 4230 (1984).
$Rb + CH_3I \rightarrow RbI + CH_3$	D. L. Bunker and E. A. Goring-Simpson, *Chem. Soc. Faraday Discuss.* 55, 93 (1973).
$H + CH_4 \rightarrow H_2 + CH_3$	T. Joseph, R. Steckler, and D. G. Truhlar, *J. Chem. Phys.* 87, 7036 (1987).
$C_2H_6 \rightarrow 2 CH_3$ $\rightarrow H + C_2H_5$	E. R. Grant and D. L. Bunker, *J. Chem. Phys.* 68, 628 (1978).
$H + C_2H_4 \rightarrow C_2H_5$	W. L. Hase, G. Mrowka, R. J. Brudzynski, and C. S. Sloane, *J. Chem. Phys.* 69, 3548 (1978).
$H + CH_3 \rightarrow CH_4$	R. J. Duchovic, W. L. Hase, and H. B. Schlegel, *J. Phys. Chem.* 88, 1339 (1984).
$CH_3CF_3 \rightarrow HF + CH_2CF_2$	R. M. Benito and J. Santamaria, *Chem. Phys. Lett.* 73, 21 (1980).
$OH + H_2 \rightarrow H_2O + H$	G. C. Schatz and H. Elgersma, *Chem. Phys. Lett.* 73, 21 (1980).
$H + CO_2 \rightarrow OH + CO$	G. C. Schatz, M. S. Fitzcharles, and L. B. Harding, *Far. Disc. Chem. Soc.* 84, 255 (1987).

7.8 EXPERIMENTAL DETERMINATION OF POTENTIAL ENERGY SURFACE PROPERTIES

The determination of interatomic potentials for weakly bound systems (e.g., Ar + Ar) from molecular beam elastic scattering experiments (see chapter 9, section 9.2) is an example of how experiment may be used to derive information about potentials. Here the potential is a curve. For a polyatomic chemical reactive system, where the potential is not a curve but a multidimensional surface, the derivation of potential information from experiment is often less direct. Measurements of kinetic rate parameters, vibrational/rotational transition frequencies, and chemical reaction dynamics are used to deduce potential energy surface properties. In chapter 10, it will be shown how transition state theory may be used to estimate the tightness or looseness of the saddlepoint molecular configuration as compared to that of the reactants. Also, a value for the zero-point energy difference between the saddlepoint and reactants may be determined from the thermal activation energy using transition state theory.

Potential energy surface properties may be determined from experimental measurements of vibrational and rotational energy levels. For small polyatomic molecules it is possible to extract harmonic normal mode vibrational frequencies from the experimental anharmonic $n = 1 \leftarrow 0$ normal mode transition energies (the harmonic frequencies are usually approximately 5 per cent larger than the anharmonic $1 \leftarrow 0$ transition frequencies). Using a normal mode analysis as described by equations 7–23 through 7–28, internal coordinate quadratic force constants (e.g., Table 7–5) may be determined for the molecule by fitting the harmonic frequencies.

For a diatomic molecule the Rydberg-Klein-Rees (RKR) method may be used to determine the potential energy curve $V(r)$ from the experimental vibrational energy levels. This method is based on the Einstein-Brillouin-Keller (EBK) semiclassical quantization of the action integral of the old quantum theory.[41] It may also be derived from the Wentzel-Kramer-Brillouin (WKB) semiclassical limit of quantum mechanics.[42] The classical vibrational energy of a diatomic molecule is $E = p_r^2/2\mu + V(r)$, for which the EBK quantization condition is

$$\left(n + 1/2\right)h = 2 \int_{r_{min}}^{r_{max}} p(r) \, dr = 2(2\mu)^{1/2} \int_{r_{min}}^{r_{max}} [E - V(r)]^{1/2} \, dr \qquad (7\text{--}47)$$

Values of the inner and outer potential energy turning points, r_{min} and r_{max}, are evaluated for each level n, from which the complete $V(r)$ curve is determined.

In contrast to the above RKR method for diatomics, a general practical method has not been developed for determining potential energy surfaces from experimental anharmonic vibrational energy levels of polyatomic molecules. Methods which have been used are based on an analytic representation of the potential energy surface. At low levels of excitation the surface may be represented as a sum of quadratic, cubic, and quartic terms; viz., Eqs. (7–17) and (7–22). A vibrational basis set is chosen for the molecule, and the quantum mechanical variational method is used to calculate vibrational energy levels from the analytic potential.[43] Parameters in the potential are varied until the experimental vibrational energy levels are fit. This variational method is only practical at low levels of excitation. To fit potential surface parameters to highly excited levels, a vibrational self-consistent-field (SCF) method has been proposed.[44] In one version of this approach, the SCF theory and EBK semiclassical quantization theory have been combined.[45]

Dynamical properties of chemical reactions are also used to determine potential energy surfaces. The relationship between reaction dynamics and potential energy surfaces is discussed in considerable detail in chapters 8, 9, and 11. Here this relationship is briefly described. For a bimolecular reaction the dependence of the observed cross section on translational, vibrational and/or rotational energy, velocity and angular momentum, scattering angles, and product energy partitioning characterize the reaction's potential energy surface.[46] Dynamical calculations, such as classical trajectories, performed on a potential energy surface can be used to reproduce the experimental dynamical results. By varying adjustable parameters in an analytic function depicting the surface, it is often possible to determine what surface properties (shapes) are required to fit the experimental results. In unimolecular dissociation reactions the partitioning of available energy to vibration, rotation, and translation in the products may be used to deduce qualitative shapes of potential surfaces.[47]

7.9 DETAILS OF THE REACTION PATH

Reaction paths for chemical reactions have been illustrated in Figure 7–5. A reaction path may be found by starting at the saddlepoint and moving towards reactants and products in infinitesimal steps, with the kinetic energy removed after each step. In mass-weighted Cartesian coordinates, (see chapter 8, section 8.3.1), this motion traces out the steepest descent path and can be found by following the negative gradient vector from the saddlepoint.[48] If the surface does not have a saddlepoint, as for many association reactions, the reaction path can be found by choosing a sufficiently large separation between the two reactants and initializing the motion at that point. By following the negative gradient vector one finds the variation in the Cartesian coordinates as the chemical system transverses the reaction path. Internal coordinates such as bond lengths, valence angle bends, and dihedral angles can be determined at any point on the reaction path from the Cartesian coordinates.

There are 3N degrees of freedom for the N-atom chemical system moving along the reaction path. Six of the degrees of freedom are for overall translation and rotation, and one is for the reaction coordinate. The remaining 3N-7 degrees of freedom are vibrations orthogonal to the reaction coordinate. Harmonic normal mode frequencies for those vibrations may be determined at any point along the reaction path.[49] However, only at saddlepoints and potential minima are they determined by diagonalizing the mass-weighted Cartesian force constant matrix $\mathbf{F}$, equation 7–23. At all other points along the reaction path the derivatives of the potential with respect to the Cartesian coordinates ($\partial V / \partial q_{\alpha i}$) are not zero, and instead a projected force constant matrix $\mathbf{F^P}$ must be diagonalized.[49] $\mathbf{F^P}$ is formed from $\mathbf{F}$ by projecting out the direction along the reaction path and the directions corresponding to infinitesimal rotations and translations.

The eigenvalues and eigenvectors of $\mathbf{F^P}$ are determined in the same manner as those for $\mathbf{F}$; i.e., equations 7–23 through 7–28, except $\mathbf{F^P}$ replaces $\mathbf{F}$. $\mathbf{F^P}$ has seven zero eigenvalues which correspond to infinitesimal translations and rotations, and motion along the reaction path. The remaining 3N-7 nonzero eigenvalues give the harmonic frequencies $\nu_i(r^\dagger)$ for the vibrations orthogonal to the reaction path. By evaluating $\mathbf{F^P}$ as a function of the reaction path, the $\nu_i(r^\dagger)$ may be determined along the reaction path. The harmonic frequencies orthogonal to the $H_2CO \rightarrow H_2 + CO$ reaction path are plotted in Figure 7–12. The potential energy along the reaction path is called the vibrationally adiabatic ground state potential, which at the harmonic level is given by

$$V_a^G(r^\dagger) = V(r^\dagger) + \sum_{i=1}^{3N-7} h\nu_i(r)/2 \qquad (7\text{--}48)$$

In this expression, $V(r^\dagger)$ is the classical potential energy.

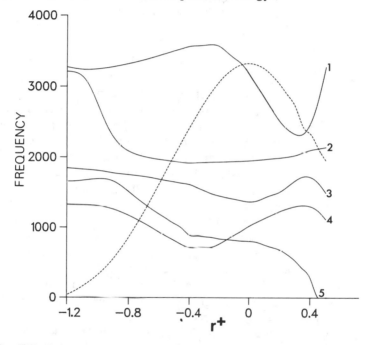

Figure 7–12. Plots of vibrational frequencies (cm^{-1}) vs. reaction coordinate ($amu^{1/2} \cdot$ Å) for the reaction $H_2CO \rightarrow H_2 + CO$. Saddle point is at $r^\dagger = 0$, and positive and negative $r^\dagger$ lead to products and reactants, respectively. At the H_2CO minimum, $\nu_1 - \nu_5$ are the asymmetric CH stretch, symmetric CH stretch, CO stretch, out-of-plane bend, and CH_2 symmetric deformation. The dashed line depicts the vibrationally adiabatic ground-state potential. [Adapted from W. L. Hase, *Acc. Chem. Res. 16*, 258(1983).]

7.10 POTENTIAL ENERGY SURFACES OF ELECTRONICALLY EXCITED MOLECULES

Up to this point, we have considered potential energy surfaces for only the ground electronic state of molecular systems. However, within the framework of the Born-Oppenheimer approximation, every molecular system has a multitude of electronically excited states, each with its own potential energy surface. Each state is identified by its spin angular momentum and orbital symmetry. The electronic states for diatomic systems are further classified by the component of electronic orbital angular momentum along the internuclear axis. Since the electronic density between nuclei is different for each electronic state, each state has its own equilibrium geometry, sets of vibrational frequencies, and bond dissociation energies.

One of the ramifications of the Born-Oppenheimer approximation is that potential energy curves for states with the same orbital symmetry and spin do not cross; i.e., the noncrossing rule.[50,51] Energy curves determined from simple (approximate) molecular orbital or valence bond wave functions χ_j do cross. However, to obtain accurate Born-Oppenheimer wave functions ϕ_i and potential energy curves one has to account for electron correlation. This is done by writing the ϕ_i as a mixture of the χ_j which are of the same spin and symmetry, a procedure called "configuration interaction";[15] i.e.,

$$\phi_i = \sum_j c_{ij}\chi_j \qquad (7\text{--}49)$$

Perturbation theory tells us that in general this mixing is most pronounced at nuclear configurations where the simple curves have similar energies. Thus, the mixing between the approximate wave functions is a function of the nuclear geometry.

An illustration of the noncrossing rule is shown in Figure 7–13 where the four lowest $^2A'$ states* of LiH_2 are treated. Potential energy curves for the accurate ϕ_i wave functions are shown in Figure 7–13(a), while those for the approximate χ_j wave functions are shown in Figure 7–13(b). In the region of crossing between the approximate energy curves, the approximate electronic wave functions strongly mix. The result of such an interaction is always to push the approximate potential energies apart, and this interaction is strongest at the *avoided crossing* point at $r = r_0$.

The accurate potential energy curves with mixing are called "adiabatic potentials" $V(r)$. The term *diabatic* is used to describe the approximate potential curves $U(r)$. At an avoided crossing the electronic character of a wave function for an adiabatic potential can change significantly. Consider the highest energy adiabatic potential in Figure 7–13(a). At large Li-H_2 separations the wave function has ionic character. At shorter separations it becomes covalent. The wave functions which are ionic at short Li-H_2 separations are those for low energy adiabatic potentials.

Spin-orbit couplings and a breakdown in the Born-Oppenheimer approximation can lead to a transition between adiabatic potentials. A treatment of such a transition is given in chapter 8, section 8.3.5. These two perturbations can also cause transitions between potential curves of different electronic symmetries and/or spin which do cross. In the *adiabatic approximation* in chemical kinetics, it is assumed that transitions between adiabatic curves are negligible, so that nuclear motion evolves on a single adiabatic potential energy surface.

In the context of the Born-Oppenheimer approximation, adiabatic correlation rules may be used to predict reaction pathways.[51] A correlation diagram for the states of $O + H_2$ and $OH + H$ is shown in Figure 7–14. The ground state reactants

*The group-theoretical notation used to denote the symmetry of electronic states is discussed in J. I. Steinfeld, *Molecules and Radiation* (Cambridge, MA: MIT Press, 1985), chapter 6.

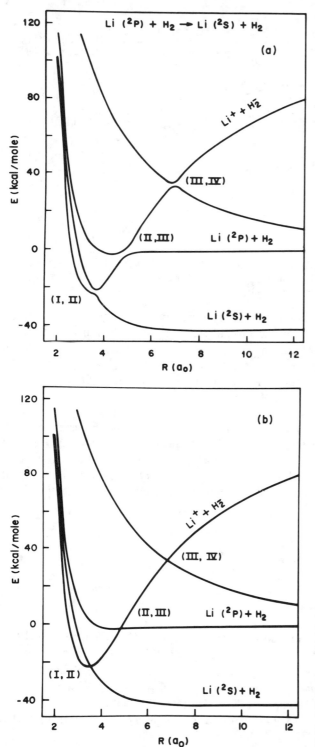

Figure 7–13. Four lowest $^2A'$ potential energy curves for LiH_2 as a function of Li–H_2 separation. The orientation of the H_2 internuclear axis is 60° relative to the Li–H_2 direction, and the H_2 internuclear distance is $1.4a_o$ (a_o = bohr radius). (a) Adiabatic representation. (b) Diabatic representation. [Adapted from J. C. Tully, *Dynamics of Molecular Collisions,* Part B, ed. W. H. Miller (New York: Plenum, 1976).]

Figure 7–14. Correlation diagram connecting the states of O + H$_2$ and OH + H. [Adapted from I. W. M. Smith, *Kinetics and Dynamics of Elementary Gas Reactions* (London: Butterworths, 1980)].

[O(^{3}P) + H$_2$($^1\Sigma_g^+$)] correlate adiabatically with both the ground state products OH + H and excited OH plus the ground state H-atom. Within the adiabatic approximation, H$_2$O can not be formed by an O(^{3}P) + H$_2$($^1\Sigma_g^+$) collision. Water can be formed by a collision with electronically excited O(^{1}D$_2$) and ground state H$_2$. However, the resulting vibrationally and rotationally excited H$_2$O has sufficient energy to dissociate to ground state OH + H.

One of the limitations of such adiabatic correlations is that they give no indication of the potential energy barriers for chemical reactivity. Experiments indicate there are large barriers on the ^{3}A′ and ^{3}A″ surfaces connecting O(^{3}P) + H$_2$($^1\Sigma_g^+$) and the ground state products. Formation of the ground state products from O(^{1}D$_2$) + H$_2$($^1\Sigma_g^+$) occurs without a barrier. The reaction pathway also appears to involve formation of vibrationally/rotationally excited H$_2$O as a reaction intermediate.

REFERENCES

1. D. M. Hirst, *Potential Energy Surfaces* (London: Taylor and Francis, 1985), p. 1.

2. R. S. Berry, S. A. Rice, and J. Ross, *Physical Chemistry* (New York: Wiley, 1980), p. 413.

3. D. Steele, E. R. Lippincott, and J. T. Vanderslice, *Rev. Mod. Phys. 34*, 239 (1962).

4. P. Huxley and J. N. Murrell, *J. Chem. Soc. Farad. 279*, 323 (1983).

5. J. N. Murrell, S. Carter, P. Huxley, S. C. Farantos, and A. J. C. Varandas, *Molecular Potential Energy Functions* (Chichester: Wiley, 1984).

6. P. J. Kuntz and W. N. Whitton, *J. Chem. Phys. 64*, 3624 (1976).

7. Peter Pulay, in *Applications of Electronic Structure Theory*, edited by H. F. Schaefer III (New York: Plenum, 1977), p. 153.

8. J. L. Duncan, D. C. McKean, and P. D. Mallinson, *J. Mol. Spectrosc.* 45, 221 (1973).

9. S. Califano, *Vibrational States* (New York: Wiley, 1976), p. 21.

10. H. L. Sellers, L. B. Sims, L. Schafer, and D. E. Lewis, *QCPE 13*, 339 (1980).

11. J. A. Pople, in *Applications of Electronic Structure Theory*, edited by H. F. Schaefer III (New York: Plenum Press, 1977), p. 1.

12. Hirst, p. 94.

13. W. A. Goddard III, T. H. Dunning, Jr., W. J. Hunt, and P. J. Hay, *Acc. Chem. Res.* 6, 368 (1973).

14. P. E. M. Siegbahn, J. Almlöf, A. Heiberg, and B. O. Roos, *J. Chem. Phys.* 74, 2384 (1981).

15. S. R. Langhoff and E. R. Davidson, *Int. J. Quantum Chem.* 8, 61 (1974); I. Shavitt, in *Methods of Electronic Structure Theory*, edited by H. F. Schaefer III (New York: Plenum, 1977), p. 189.

16. J. A. Pople, J. S. Binkley, and R. Seeger, *Int. J. Quantum Chem., Symp.* 10, 1 (1976); H. B. Schlegel, *J. Chem. Phys.* 84, 5340 (1986).

17. J. S. Binkley, R. A. Whiteside, R. Krishnan, R. Seeger, D. J. DeFrees, H. B. Schlegel, S. Topiol, L. R. Kahn, and J. A. Pople, *QCPE 13*, 406 (1980).

18. W. J. Hehre, L. Radom, P. v. R. Schleyer, and J. A. Pople, *Ab Initio Molecular Orbital Theory* (New York: Wiley, 1986), pp. 135–226.

19. Langhoff and Davidson, pp. 226–260.

20. Langhoff and Davidson, pp. 261–323.

21. B. Liu, *J. Chem. Phys.* 80, 581 (1984); F. Mentch and J. B. Anderson, *J. Chem. Phys.* 80, 2675 (1984); and D. M. Ceperly and B. J. Alder, *J. Chem. Phys.* 81, 5833 (1984).

22. H. F. Schaefer III, *J. Phys. Chem.* 89, 5336 (1985).

23. E. Clementi and H. Popkie, *J. Chem. Phys.* 57, 1077 (1972).

24. G. Corongiu, E. Clementi, E. Pretsch, and W. Simon, *J. Chem. Phys.* 72, 3096 (1980).

25. W. L. Hase, S. L. Mondro, R. J. Duchovic, and D. M. Hirst, *J. Am. Chem. Soc.* 109, 2916 (1987).

26. L. F. Shampine and R. C. Allen, Jr., *Numerical Computing* (Toronto: Saunders, 1973), p. 54.

27. W. L. Jorgensen, J. Chandrasekhar, J. D. Madura, R. W. Impey, and M. L. Klein, *J. Chem. Phys.* 79, 926 (1983).

28. K. N. Swamy and W. L. Hase, *J. Am. Chem. Soc.* 106, 4071 (1984).

29. J. P. Bergsma, B. J. Gertner, K. R. Wilson, and J. T. Hynes, *J. Chem. Phys.* 86, 1356 (1987).

30. O. Matsuoka, E. Clementi, and M. Yoshimine, *J. Chem. Phys.* 64, 1351 (1976).

31. W. L. Jorgensen and M. Ibrahim, *J. Am. Chem. Soc.* 102, 3309 (1980).

32. C. A. Parr and D. G. Truhlar, *J. Phys. Chem.* 75, 1844 (1971).

33. F. London, "Probleme der modernen Physik" *Sommerfeld Festschrift* (1928), p. 104.

34. S. Sato, *J. Chem. Phys.* 23, 592, 2465 (1955).

35. P. J. Kuntz, E. M. Nemeth, J. C. Polanyi, S. D. Rosner, and C. E. Young, *J. Chem. Phys.* 44, 1168 (1966).

36. A. Persky and M. Baer, *J. Chem. Phys.* 60, 133 (1974).

37. H. S. Johnston and C. Parr, *J. Am. Chem. Soc.* 85, 2544 (1963); H. S. Johnston, *Gas Phase Reaction Rate Theory* (New York: Ronald, 1966), pp. 177–183.

38. L. Pauling, *J. Am. Chem. Soc. 69,* 542 (1947).

39. N. C. Blais and D. L. Bunker, *J. Chem. Phys. 37,* 2713 (1962).

40. K. S. Sorbie and J. N. Murrell, *Molec. Phys. 29,* 1387 (1975); J. N. Murrell, K. S. Sorbie, and A. J. C. Varandas, *Molec. Phys. 32,* 1359 (1976).

41. I. C. Percival, *Adv. Chem. Phys. 36,* 1 (1977); D. W. Noid, M. L. Koszykowski, and R. A. Marcus, *Ann. Rev. Phys. Chem. 32,* 267 (1981).

42. L. I. Schiff, *Quantum Mechanics* (New York: McGraw-Hill, 1968), p. 268.

43. N. C. Handy and S. Carter, *Chem. Phys. Lett. 79,* 118 (1981); B. Maessen and M. Wolfsberg, *J. Chem. Phys. 80,* 4651 (1984); H. Romanowski, J. M. Bowman, and L. B. Harding, *J. Chem. Phys. 82,* 4155 (1985); S. Carter and N. C. Handy, *Mol. Phys. 57,* 175 (1986).

44. J. M. Bowman, *J. Chem. Phys. 68,* 608 (1978); K. M. Christoffel and J. M. Bowman, *Chem. Phys. Letters 85,* 220 (1982).

45. M. A. Ratner and R. B. Gerber, *J. Phys. Chem. 90,* 20 (1986).

46. R. D. Levine and R. B. Bernstein, *Molecular Reaction Dynamics* (New York: Oxford University Press, 1974), chapters 4 and 6.

47. W. L. Hase and R. J. Wolf, in *Potential Energy Surfaces and Dynamics Calculations,* edited by D. G. Truhlar (New York: Plenum, 1981), p. 37.

48. K. Fukui, *J. Phys. Chem. 74,* 4161 (1970); K. Fukui, S. Kato, and H. Fujimoto, *J. Am. Chem. Soc. 97,* 1 (1975); N. Kato, H. Kato, and K. Fukui, *J. Am. Chem. Soc. 99,* 684 (1977); H. F. Schaefer, *Chem. Brit. 11,* 227 (1975); K. Ishida, K. Morokuma, and A. Komornicki, *J. Chem. Phys. 66,* 2153 (1977).

49. W. H. Miller, N. C. Handy, and J. E. Adams, *J. Chem. Phys. 72,* 99 (1980); S. Kato and K. Morokumo, *J. Chem. Phys. 73,* 3900 (1980).

50. W. Kauzmann, *Quantum Chemistry* (New York: Academic Press, 1957), pp. 142, 401.

51. C. A. Mead, *J. Chem. Phys. 70,* 2276 (1979).

BIBLIOGRAPHY

Intermolecular Potentials

HIRSCHFELDER, J. O., CURTISS, C. F., and BIRD, R. R. *Molecular Theory of Gases and Liquids.* New York: Wiley, 1954. Chapters 12–14.

KAUZMANN, W. *Quantum Chemistry.* New York: Academic Press, 1957. Chapter 13.

MARGENAU, H., and KESTNER, N. R. *Theory of Intermolecular Forces.* 2d ed. Oxford: Pergamon, 1969.

SMITH, I. W. M. *Kinetics and Dynamics of Elementary Gas Reactions.* Boston: Butterworths, 1980. Chapter 2.

WESTON, R. E., JR., and SCHWARZ, H. A. *Chemical Kinetics.* Englewood Cliffs, NJ: Prentice-Hall, 1972. Chapter 2.

Vibrational Force Constants and Frequencies

CALIFANO, S. *Vibrational States.* New York: Wiley, 1976.

STEINFELD, J. I. *Molecules and Radiation.* Cambridge, MA: MIT Press, 1985. Chapters 4 and 8.

WILSON, E. B., JR., DECIUS, J. C., and CROSS, P. C. *Molecular Vibrations*. New York: McGraw-Hill, 1955.

Properties of Potential Energy Surfaces

LAIDLER, K. J. *Theories of Chemical Reaction Rates*. New York: McGraw-Hill, 1969. Chapter 2.

MAGGIORA, G. M., and CHRISTOFFERSEN, R. E. in *Transition States of Biochemical Processes*, edited by R. G. Gandour and R. L. Schowen, p.119. New York: Plenum. 1978.

SALEM, L. *Electrons in Chemical Reactions: First Principles* (New York: Wiley, 1982).

Ab Initio and Analytical Potential Energy Surfaces

HEHRE, W. J., RADOM, L., SCHLEYER, P. v. R., and POPLE, J. A. *Ab Initio Molecular Orbital Theory*. New York: Wiley, 1986.

HIRST, D. M. *Potential Energy Surfaces*. London: Taylor and Francis, 1985.

KUNTZ, P. J. in *Dynamics of Molecular Collisions, Part B.*, edited by W. H. Miller, p. 53. New York: Plenum, 1976.

MURRELL, J. N. *Gas Kinetics and Energy Transfer, Vol. 3*. London: The Chemical Society, 1978. Chapter 5.

SCHAEFER, H. F. *Electronic Structure of Atoms and Molecules*. New York: Addison-Wesley, 1972.

TRUHLAR, D. G., STECKLER, R., and GORDON, M. S. *Chem. Rev. 87*, 217 (1987).

Potential Energy Surface Properties from Experiment

BUNKER, D. L. *Acc. Chem. Res. 7*, 195 (1974).

CARNEY, G. D., SPRANDEL, L. L., and KERN, C. W. *Adv. Chem. Phys. 37*, 305 (1978).

HERZBERG, G. *Molecular Spectra and Molecular Structure II* New York: Van Nostrand and Reinhold, 1945.

POLANYI, J. C. *Acc. Chem. Res. 5*, 16 (1972).

Reaction Path

FUKUI, K. *Acc. Chem. Res. 14*, 363 (1981).

MILLER, W. H. *J. Phys. Chem. 87*, 3811 (1983).

Potential Energy Surfaces for Excited Electronic States

DONOVAN, R. J., and HUSAIN, D. *Chem. Rev. 70*, 489 (1970).

MAHAN, B. H. *Acc. Chem. Research 8*, 55 (1975).

SHULER, K. E. *J. Chem. Phys. 21*, 624 (1952).

TULLY, J. C. in *Dynamics of Molecular Collisions, Part B*, edited by W. H. Miller, p. 217. New York: Plenum, 1976.

Bibliography

243

PROBLEMS

7.1 Compute the factor for converting a quadratic force constant in units of millidyne/angstrom (mdyn/Å) to kJ/mol/Bohr^2.

7.2 Using equations 7–4 through 7–7 derive equation 7–8.

7.3 Derive an expression for the quadratic force constant of a Morse oscillator in terms of D_e and β.

7.4 For a Lennard-Jones (6–12) potential determine the internuclear distance at the potential energy minimum. Using this value, derive an expression for the quadratic force constant.

7.5 From the force constants in Table 7-5, calculate harmonic frequencies for the diatoms C—H, C—C, and C═O.

7.6 Show that for a harmonic oscillator, the EBK quantization condition equation 7–47 leads to the expression $E = (n + 1/2)h\nu$.

7.7 What are the relationships between internal bond stretch and angle bend coordinates and Cartesian coordinates of the atomic nuclei?

7.8 A linear triatomic molecule such as HCN has two bond-stretching internal coordinates, r_{HC} and r_{CN}. Assume the potential energy for the linear internal motion is given by

$$V = f_{HC}(r_{HC} - r_{HC}^0)^2/2 + f_{CN}(r_{CN} - r_{CN}^0)^2/2$$

with $f_{HC} = 5.0$ mdyn/Å, $f_{CN} = 14.0$ mdyn/Å, $r_{HC}^0 = 1.08$ Å, and $r_{CN}^0 = 1.15$ Å. Using equations 7–23 through 7–28 calculate the normal mode frequencies and eigenvectors for linear motion. One of the eigenvalues, that for translation, will be zero.

7.9 **(a)** Sketch potential-energy contour maps for the following:
   (i) Isomerization of H—N≡C → H—C≡N. Your sketch should be in the r_{HC}, θ plane where θ is given by

   (ii) Dissociation of H—C≡N → H+C≡N. Your sketch should be in the r_{HC}, r_{CN} plane for a linear configuration.
   (b) Discuss significant differences between the two potential energy surfaces.

7.10 What are the major differences between the F + H$_2$ → FH + H and F + C$_2$H$_4$ → C$_2$H$_3$F + H potential energy surfaces?

7.11 Describe the difference between electronically adiabatic and electronically nonadiabatic processes. What is the noncrossing rule?

7.12 What is the mathematical property common to both minima and saddlepoints on potential energy surfaces?

7.13 **(a)** How is the harmonic zero-point energy of a molecule or saddlepoint (transition state) determined? Give the mathematical expression for calculating the zero-point energy.
   (b) Describe with both an equation and diagram the relationship between the harmonic quantum mechanical threshold energy and the classical mechanical threshold energy.

7.14 A widely used LEPS function is that for the "Muckermann V" F + H$_2$ → FH + H potential energy surface; P. A. Whitlock and J. T. Muckermann, *J. Chem. Phys. 61*, 4624 (1974). The LEPS parameters are

$D_{FH} = 591.1$ kJ/mol $\qquad D_{HH} = 458.2$ kJ/mol

$\beta_{FH} = 2.2189$ Å^{-1} $\qquad \beta_{HH} = 1.9420$ Å^{-1}

$r°_{FH} = 0.917$Å $\qquad r°_{HH} = 0.7419$Å

$S_{FH} = 0.167$ $\qquad S_{HH} = 0.106$

Using the above parameters and equations 7–35 through 7–37 make a contour diagram for the F + H$_2$ → FH + H reaction like that shown in Figure 7–8(a).

Dynamics of Bimolecular Collisions

Once an intermolecular potential energy function for a reactive system is available, either from experiment or theory, scattering amplitudes and cross sections for that system may be found by carrying out appropriate calculations. Classical scattering theory requires the solution of Hamilton's equations of motion on the potential surface (see section 8.3.1); alternatively, a quantum mechanical solution for the wave functions may be effected by solving Schrödinger's equation using the said potential function (see section 8.3.2). Before considering these methods, however, let us first examine a few very simple collision models. Even though these models are only crude approximations to actual reactive scattering, they have proven to be very useful in the development of chemical-kinetic theories because they lead to analytic expressions which provide an intuitive grasp of the more complex real situations.

8.1 SIMPLE COLLISION MODELS

8.1.1 Constant Cross Section

The simplest possible picture of an encounter between two molecules is that shown in Figure 8–1, in which it is assumed that the molecules are structureless hard spheres and that the collision may be described by a cross section with an energy-independent magnitude

$$\sigma_R(E) = \text{constant} = \pi d^2 \qquad (8\text{–}1)$$

where $d = \frac{1}{2}(d_A + d_B)$ is the arithmetic average of the "hard-sphere" diameters of molecules A and B. The collision model employed is that the probability of a "reaction" between A and B occurring is $P_{\text{react}} = 1$ if the intermolecular distance $r_{AB} \leq d$ and $P_{\text{react}} = 0$ otherwise. The effective activation energy for such a reaction is $E_{\text{act}} = 0$. The rate coefficient may be found by substituting equation 8–1 for the cross section into equation 6–7a. Since σ is independent of v, it may be taken outside the integral to give

$$k(T) = \left(\frac{\mu}{2\pi k_B T}\right)^{3/2} \sigma \int_0^\infty v e^{-\mu v^2 / 2k_B T} 4\pi v^2 \, dv$$

$$= \bar{v}\sigma$$

or

$$k(T) = \left(\frac{8 k_B T}{\pi \mu}\right)^{1/2} \cdot \pi d^2 = \left(\frac{8\pi k_B T}{\mu}\right)^{1/2} d^2 \qquad (8\text{-}2)$$

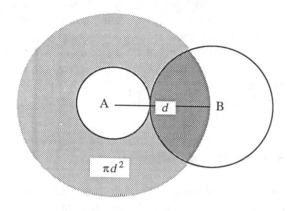

Figure 8–1. Collision of two hard-sphere molecules.

For a generic molecular cross section ($d \approx 4 \times 10^{-8}$ cm) and a typical molecular velocity ($\bar{v} \approx 5 \times 10^4$ cm sec^{-1}), equation 8–2 gives a "hard-sphere" or "gas-kinetic" collision rate coefficient $k_{coll} \approx 2.5 \times 10^{-10}$ cm^3 molecule^{-1} sec^{-1}. In other commonly used units,

$$k_{coll} = \left(\frac{6.022 \times 10^{23} \text{ molecule mole}^{-1}}{10^3 \text{ cm}^3 \text{ liter}^{-1}}\right)(2.5 \times 10^{-10} \text{ cm}^3 \text{ molecule}^{-1} \text{ sec}^{-1})$$

$$= 1.5 \times 10^{11} \text{ liter mole}^{-1} \text{ sec}^{-1}$$

If we multiply instead by the ratio $(6.022 \times 10^{23}$ molecule mole$^{-1})/(22,421$ cm^3 mole^{-1} @STP), we obtain a relaxation rate $(p\tau)^{-1} = 6.7 \times 10^9$ atm^{-1} sec^{-1} at 273 K, or approximately 1×10^7 torr^{-1} sec^{-1} at that temperature. This means that τ_{coll}, the mean free time between gas-kinetic collisions, is about 0.1 μsec at a pressure of 1 torr. The corresponding collision frequency would be $\omega = \tau_{coll}^{-1} = 10^7$ sec^{-1}.

One frequently encounters in the literature values of the "effective cross section" for a reaction defined as $\sigma_{eff} = k_{meas}/\bar{v}$. This relationship between rate coefficient and cross section is of course valid only for the hard-sphere model. An even cruder approximation is to define the "reaction probability" as

$$P_{react} = \frac{k_{meas}}{k_{coll}} = \frac{\sigma_{eff}}{\sigma_{gas-kinetic}} \approx \frac{\sigma_{eff}}{50\text{Å}^2} \qquad (8\text{-}3)$$

A glance back at Figure 6–2, however, shows that equation 8–3 glosses over three stages of averaging between the reaction probability and the rate coefficient!

One should also be careful about other concepts employed in simple collision models, e.g., "mean free time between collisions." Recall that molecules interact with an ensemble of velocities and instantaneous intermolecular distances and therefore possess a *distribution* of time intervals between collisions. This distribution can be expressed as a Poisson function of n, the number of collisions per time interval T, given by

$$P(n) = e^{-\gamma}\gamma^n/n! \qquad (8\text{--}4)$$

with mean $\langle n \rangle = \gamma = T/\tau_{\text{coll}}$. Thus, in any gas phase system, some molecules may survive many multiples of the mean free time before undergoing a collision. And this is often important—for example, in measurements of quenching of luminescence or photochemical reaction following excitation by a short laser pulse. In a similar vein, the notion of molecules as structureless hard spheres is an oversimplification: molecules are, after all, *not* simply hard spheres, and they possess long-range attractive forces. This will be discussed in greater detail later on in the chapter.

8.1.2 Old Collision Theory (Reactive Hard Spheres)

The hard-sphere collision model just presented is clearly inadequate for describing chemical reactions. For one thing, it predicts a dependence of $T^{1/2}$ for the rate coefficient, while experimentally measured rate coefficients typically follow an Arrhenius behavior, i.e., exp $(-E_{\text{act}}/RT)$. Furthermore, the magnitudes of hard-sphere collision rates are the gas-kinetic limit, while most rate coefficients are only a small fraction of that. Clearly, a more elaborate model must be introduced to account for these observations.

Such a model is that of reactive hard spheres. Consider a collision between hard-sphere molecules A and B, with relative velocity $\mathbf{v}$ and kinetic energy (in the center-of-mass coordinate system) $E = \frac{1}{2}\mu v^2$. The hard-sphere minimum approach distance is $d = \frac{1}{2}(d_A + d_B)$, as before. We define an *impact parameter b*, given by the perpendicular distance between parallel lines drawn in the direction of $\mathbf{v}$ and through the centers of molecules A and B, as shown in Figure 8–2. Since $\mathbf{v}$ is a vector quantity, we can resolve it into components $v_\perp$ along the line of centers and $v_\parallel$ perpendicular to $v_\perp$. Even though the kinetic energy E is a scalar quantity, it can also be expressed in terms of these velocity components, viz., $E = \frac{1}{2}\mu(v_\parallel^2 + v_\perp^2) = \frac{1}{2}\mu v_\parallel^2 + \frac{1}{2}\mu v_\perp^2$. The first term is that part of the energy which is tangential to the surfaces of the interacting spheres; the second term, denoted by E_c, represents energy "directed" along the line of centers which is presumed to be effective in bringing about reaction. This "energy along the line of centers" is given from a simple geometrical construction as

$$\frac{E_c}{E} = 1 - \frac{b^2}{d^2} \qquad (b \le d)$$

$$(8\text{--}5)$$

$$\frac{E_c}{E} = 0 \qquad (b > d)$$

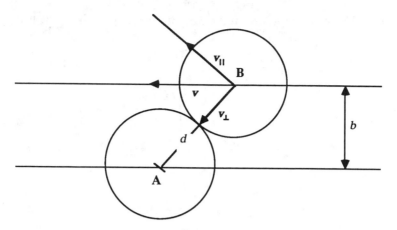

Figure 8–2. The reactive hard-sphere model.

Assuming an energy-dependent reaction probability

$$P_R(E_c) = 0 \qquad (E_c < E^*)$$
$$P_R(E_c) = 1 \qquad (E_c \geq E^*)$$

(8–6)

where E^* is the minimum-energy threshold for the reaction to occur, and using a relationship between reaction probability and cross section (equation 8–10), to be proven in section 8.2, we find the following expression for the cross section:

$$\sigma_R(E) = \int_0^\infty P_R(E(b)) \cdot 2\pi b\, db = \begin{cases} 0 & \text{if } E < E^* \\ \pi d^2\left(1 - \dfrac{E^*}{E}\right) & \text{if } E \geq E^* \end{cases}$$

(8–7)

The energy dependence of $\sigma_R(E)$ is shown in Figure 8–3.

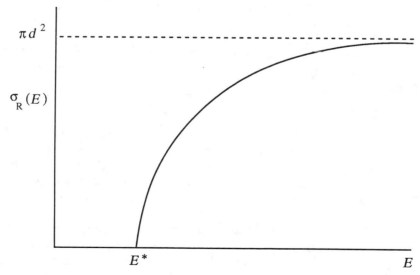

Figure 8–3. Energy dependence of the reactive cross section in the reactive hard-sphere model.

Using equation 8–7 for $\sigma_R(E)$ in equation 6–7b to calculate the thermal rate coefficients yields

$$k_{\text{coll}} = \frac{1}{k_B T}\left(\frac{8}{\pi \mu k_B T}\right)^{1/2} \int_0^{\infty} E\sigma_R(E)e^{-E/K_B T}\, dE$$

$$k_{\text{coll}} = \frac{1}{k_B T}\left(\frac{8}{\pi \mu k_B T}\right)^{1/2} \int_0^{\infty} \pi d^2 (E - E^*)e^{-E/k_B T}\, dE \qquad (8\text{–}8)$$

$$k_{\text{coll}} = \pi d^2 \left(\frac{8k_B T}{\pi \mu}\right)^{1/2} e^{-E^*/k_B T}$$

Note that k_{coll} consists of three factors, viz.,

$$(\text{hard-sphere cross section}) \times (\text{mean } \bar{v}(T)) \times (\text{Arrhenius factor})$$

Differentiating the last of equations 8–8 with respect to $1/T$ affords a useful comparison with the phenomenological rate coefficient. We have

$$\frac{d(\ln k_{\text{coll}})}{d(1/T)} = cT^{3/2} - \frac{E^*}{k_B}$$

while

$$\frac{d(\ln k_{\text{obs}})}{d(1/T)} = -\frac{\tilde{E}_{\text{act}}}{R}$$

Thus, the activation energy is seen to be equal to the threshold energy of the reactive hard-sphere model, give or take a few $k_B T$ per molecule. (The latter will be stated more precisely in chapter 10.) Equation 8–8 also predicts an explicit form for the pre-exponential factor equal to $\pi d^2 \bar{v} \approx 2.5 \times 10^{-10}$ cm^3 molecule^{-1} sec^{-1}, i.e., the gas-kinetic collision rate.

The major problem with the reactive hard-sphere model is that the observed pre-exponential factor is sometimes much less than the gas-kinetic collision rate! This can be taken care of in an ad hoc way by using $P_R(E_c) = p$ in equation 8–6, rather than $P_R(E_c) = 1$, for $E_c = E^*$. The quantity p, known as the *steric factor*, carries right through the integration in equations 8–8 to give a modified rate coefficient

$$k_{\text{coll}} = \pi d^2 \left(\frac{8k_B T}{\pi \mu}\right)^{1/2} p e^{-E^*/k_B T}$$

For the moment we can consider p a purely empirical correction which in some way is meant to take into account the idea that molecules must usually be in a fairly specific relative orientation in order to react. This orientation requirement is related to the concept of an *entropy* of activation, which is discussed in chapter 10, section 10.5.

While the simple models introduced in this section are only qualitative, they do display the essential elements of a bimolecular rate constant: a collision must occur, some critical energy must be exceeded, and the relative orientations of the colliding particles must be appropriate. Additional useful comments on these models are given by Nordman and Blinder.[1]

8.2 TWO-BODY CLASSICAL SCATTERING

To begin a more rigorous discussion of reactive scattering, let us consider collisions of two structureless particles interacting through a central potential $U(r)$, where r is the distance between the centers of the two particles. Although such an interaction describes only collisions between atoms, such as helium or argon, it is a useful starting point for introducing a number of important concepts and quantities which are important for understanding the description of more complex scattering processes. Furthermore, only this "two-body" problem can be solved analytically: collisions involving three or more particles, which are in fact required for a chemical reaction, can be treated only numerically or through the use of approximations. Additional discussion of two-body scattering may be found in a number of introductory texts on chemical kinetics listed in the bibliography, as well as in textbooks on classical mechanics.[2]

The scattering process in laboratory fixed coordinates is shown in Figure 8–4. As previously noted, the interparticle potential energy function $U(r)$ depends only

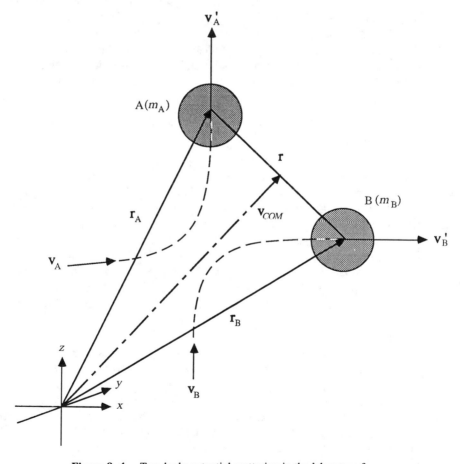

Figure 8–4. Two-body potential scattering in the laboratory frame.

on the distance $r = |\mathbf{r}_A - \mathbf{r}_B|$. The total energy (kinetic plus potential) in the system is

$$E = \frac{1}{2}m_A|\mathbf{v}_A|^2 + \frac{1}{2}m_B|\mathbf{v}_B|^2 + U(r) + E_{A,\text{internal}} + E_{B,\text{internal}}$$

Several generic types of collisions can occur. In an *elastic* collision, the A and B particles retain their identities and no energy is exchanged with internal degrees of freedom of the particles. Since the total energy E is conserved, we must have

$$\frac{1}{2}m_A|\mathbf{v}_A|^2 + \frac{1}{2}m_B|\mathbf{v}_B|^2 + U(r) = \frac{1}{2}m_A|\mathbf{v}_A'|^2 + \frac{1}{2}m_B|\mathbf{v}_B'|^2 + U(r)$$

where the primes refer to quantities (velocities, in this case) that exist after the collision.

Inelastic collisions are those in which some of the internal energy in the A and B particles is exchanged between the particles or is coupled into translation. In this case, all the terms in the expression for the total energy must be retained. In monatomic atoms, this energy can only take the form of electronic excitation; molecular species possess vibrational and rotational degrees of freedom as well. We have already noted that *reactive* collisions, i.e., those in which the colliding particles change their identities, cannot strictly take place in a two-body collision. Nevertheless, we shall use this simple dynamical model to derive relationships between some important scattering parameters, such as reaction probability and reactive cross section. In this regard, a *charge-exchange* collision

$$A^+ + B \longrightarrow A + B^+$$

is a reactive process which can be well described in terms of two-body scattering.

A simpler analytic representation of the two-body collision is obtained in a transformed coordinate system. The simplest such transformation is to the *center-of-mass* coordinate system, in which the origin is located at $(m_A\mathbf{r}_A + m_B\mathbf{r}_B)/(m_A + m_B)$ and is moving at constant velocity $\mathbf{v}_{\text{COM}}$. In the center-of-mass system, the collision appears to follow a simple linear trajectory as the particles approach each other, contact, and then separate. For a central potential the particle motion is confined to a plane, and a more suitable coordinate representation for potential scattering is the *fixed-center-of-force* coordinate representation. In this representation, a composite particle AB is presumed to have a mass equal to the collision-reduced mass $\mu = m_A m_B/(m_A + m_B)$ and to scatter off a point center of infinite mass through the interaction potential $U = U(r = |\mathbf{r}_A - \mathbf{r}_B|)$. The fixed-center-of-force coordinate system is shown in Figure 8–5.

The relevant variables in this representation, in addition to the interparticle distance r (shown here as the distance from the composite particle AB to the center of force) and the relative velocity $v = dr/dt$, are the impact parameter b (equivalent to the impact parameter b in the reactive hard-sphere model shown in Figure 8–2), the scattering angle θ, and the deflection angle χ. The azimuthal angle ϕ is also shown, but since the coordinate system is cylindrically symmetric, ϕ will be eliminated by a simple integration.

$$d\Omega = \sin \chi d\chi d\phi$$

$$\mu = \frac{m_A m_B}{m_A + m_B}$$

$$v = |\mathbf{v}_A - \mathbf{v}_B| = \frac{dr}{dt}$$

$$r = |\mathbf{r}_A - \mathbf{r}_B|$$

Figure 8–5. Two-body potential scattering in the fixed-center-of-force coordinate representation.

We now proceed to develop the relationships among these variables. The principal relationship derives from the conservation of particle flux. This requires that every system entering from the left in Figure 8–5, with velocity v_0 and impact parameter between b and $b + db$, must leave the scattering center to the right with velocity v' and a deflection angle between χ and $\chi + d\chi$. The explicit relationship between χ and b is determined by the potential function $U(r)$, to be shown shortly. For a hypothetical reactive process (e.g., A + B → C + D), the differential reactive scattering cross section is

$$I_R(\chi, \phi; v\Gamma) = P_R(v, b; \Gamma)\frac{d(\pi b^2)}{d\Omega} \qquad (8\text{–}9)$$

where the *reaction probability* P_R depends on velocity v, internal state Γ, and impact parameter b, and the differential ratio $d(\pi b^2)/d\Omega$ maps the area of the annular region on the left-hand side of Figure 8–5 onto that on the right-hand side. Using equation 6–2, we find the corresponding expression for the total reactive cross section σ_R by integrating over the solid angle $d\Omega$:

$$\sigma_R(v, \Gamma) = \iint I_R(\chi, \phi; v\Gamma)\, d\Omega = \int_0^\infty P_R(v, b; \Gamma)2\pi b\, db \qquad (8\text{–}10)$$

We next proceed to find the *deflection function* $\chi(b)$. To do so, we first write the expression for the total energy in center-of-force coordinates, i.e.,

$$E = \frac{1}{2}\mu v^2 + U(r) + \frac{L^2}{2\mu r^2} \qquad (8\text{–}11)$$

where $L = \mu v_0 b = \mu v' b'$ is the classical *angular momentum* associated with the collision trajectory. Both E and L are conserved quantities. The value of E can be readily determined from the boundary conditions of the problem: as the distance r → ∞ (corresponding to the time $t \to -\infty$), any well-behaved potential function $U(r)$ must asymptotically approach zero, because at infinite separation the interaction between two particles must cease to exist. Also, the term $L/2\mu r^2$ obviously goes to zero. Therefore, $E = \frac{1}{2}\mu v_0^2$. L is also equal to the product of the moment of inertia $I = \mu r^2$ and the angular velocity $\omega = d\theta/dt$; this fact will be used in the next step of the calculation.

For a simple central force field, the trajectory $\theta(t)$ can be found by solving the set of equations

$$L = \mu r^2 \frac{d\theta}{dt} \tag{8-12}$$

and

$$E = \frac{1}{2}\mu\left(\frac{dr}{dt}\right)^2 + \frac{L^2}{2\mu r^2} + U(r) \tag{8-13}$$

for the conserved quantities E and L. Doing so, we have

$$d\theta = \frac{L^2}{\mu r^2}\, dt$$

and

$$dr = -\left[\frac{2}{\mu}\left(E - U(r) - \frac{L^2}{2\mu r^2}\right)\right]^{1/2} dt$$

so that

$$dt = -\left[\frac{2}{\mu}\left(E - U(r) - \frac{L^2}{2\mu r^2}\right)\right]^{-1/2} dr$$

We find $\theta(t)$ by integrating:

$$\theta(t) = \int_0^\theta d\theta$$

$$= \int_{-\infty}^t \left(\frac{L}{\mu r^2}\right) dt$$

$$= -\int_\infty^r \frac{L/\mu r^2}{\left[\frac{2}{\mu}\left(E - U(r) - \frac{L^2}{2\mu r^2}\right)\right]^{1/2}}\, dr$$

Using $L = \mu v_0 b = b(2\mu E)^{1/2}$, we have

$$\theta(t) = -b \int_\infty^r \frac{dr}{r^2\left[1 - \frac{b^2}{r^2} - \frac{U(r)}{E}\right]^{1/2}} \tag{8-14}$$

For any trajectory, there will be a *closest approach distance* r_c at which $r(t)$ has its minimum value. Typically, this occurs when the two particles are interacting via the repulsive part of the potential and all of the kinetic energy has been converted into potential energy. Since $r(t)$ has a minimum at r_c, the velocity must be zero at that point in the trajectory. It is convenient to define $t = 0$ for $r(t) = r_c$ and $\theta(0) = \theta_c$. From Figure 8–5, the *deflection angle* χ is just $\pi - 2\theta_c$. We therefore have an expression for the deflection angle as a function of the impact parameter b and collision energy E, viz.,

$$\chi(E, b) = \pi - 2b \int_{r_c}^{\infty} \frac{dr}{r^2 \left[1 - \dfrac{b^2}{r^2} - \dfrac{U(r)}{E} \right]^{1/2}} \tag{8–15}$$

Let us now proceed to find $\chi(b)$ for some simple model potentials.

(a) *Hard-sphere potential*. This is just the potential considered earlier in Chapter 7, section 7.2. It can be written explicitly as

$$U(r) = 0 \qquad (r > d)$$

$$U(r) = \infty \qquad (r \leq d)$$

Clearly, $r_c = d$ for collisions at any energy. Using equation 8–15, we have

$$\chi(E, b) = \pi - 2b \int_{d}^{\infty} \frac{dr}{r^2 [1 - (b^2/r^2) - 0]^{1/2}}$$

$$= \pi - 2b \int_{d}^{\infty} \frac{dr}{r(r^2 - b^2)^{1/2}}$$

$$= \pi - 2b \left\{ \frac{1}{b} \cos^{-1}(b/r) \right\}_{r=d}^{r=\infty} \qquad \text{(see reference 3)}$$

$$= \pi - 2[\cos^{-1}(b/\infty) - \cos^{-1}(b/d)]$$

$$= \pi - \pi + 2\cos^{-1}(b/d)$$

so that

$$\chi(E, b) = 2\cos^{-1}(b/d) \tag{8–16}$$

For impact parameters greater than the hard-sphere diameter d, equation 8–16 requires taking the inverse cosine of a quantity greater than one, which does not exist. In this situation, $\chi = 0$. For impact parameters less than d, $\chi = 2\cos^{-1}(b/d)$ and varies over the range 0 to π. Note that χ is independent of E for any b and d. The scattering in center-of-force coordinates is shown in Figure 8–6(a), and the corresponding deflection function in Figure 8–6(b). The cross section for this potential can be calculated most conveniently by referring back to Figure 8–5. Recall that all particles scattered from the range between b and $b + db$ must end up with scattering angles between χ and $\chi + d\chi$. This maps the annular area $2\pi b$ into $I(E, \chi)2\pi \sin \chi d\chi$. (The factor 2π in the range comes from integrating the azimuthal angle ϕ over 2π radians.) Equating the two areas gives

$$I(E, \chi)2\pi \sin \chi d\chi = 2\pi b db$$

or

$$I(E, \chi) = \frac{b}{|\sin \chi (d\chi/db)|} \tag{8-17}$$

For the hard-sphere potential with $\chi(b) = 2 \cos^{-1}(b/d)$, we obtain the simple results $I(\chi) = \frac{1}{4}d^2$, which is independent of E and χ (in the center of mass), and

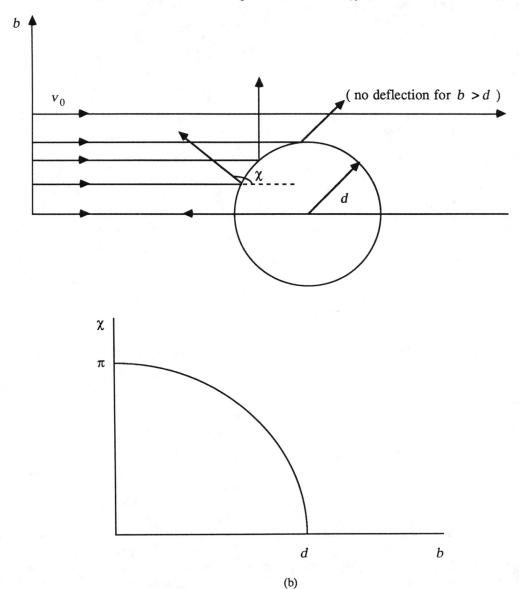

(b)

Figure 8-6. (a) Scattering from a hard-sphere potential in center-of-force coordinates. (b) Deflection function $\chi(b)$ for a hard-sphere potential.

$\sigma(E) = \pi d^2$, just the geometrical cross section, which is also independent of E. The latter is just the result (equation 8–1) we found earlier on a simple intuitive basis. Deriving it in this manner confirms our intuition and, more importantly, provides an example of the use of scattering theory in which a simple analytic form can be obtained.

(b) *Lennard-Jones potential*. A more realistic intermolecular potential energy function is the Lennard-Jones potential, discussed in chapter 7 and given by

$$U(r) = 4\epsilon \left[\left(\frac{\sigma}{r} \right)^{12} - \left(\frac{\sigma}{r} \right)^{6} \right]$$

The deflection function $\chi(b)$ for this potential can be calculated with the use of equation 8–15, but the arithmetic becomes too complicated to reproduce here in detail.[4,5] The results are shown in Figure 8–7 in terms of the reduced impact parameter

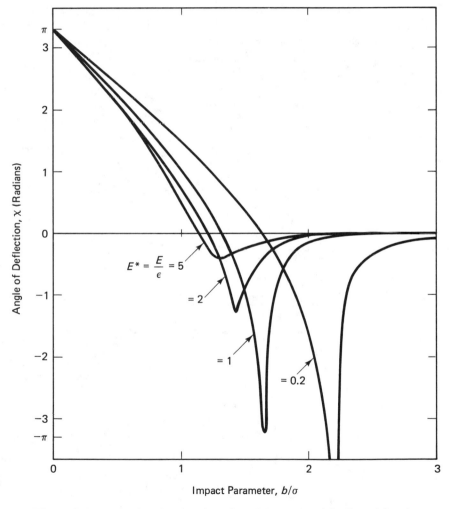

Figure 8–7. Deflection function for Lennard-Jones potential. The minima in $\chi(b/\sigma)$ correspond to the "rainbow" angle discussed in the text.

$b^* = b/\sigma$ and for a range of reduced energies $E^* = E/\epsilon$ between 5 and 0.2. There are three distinct regions of scattering behavior:

1. For nearly head-on collisions ($b \to 0$), the deflection function is very similar to that for a hard-sphere collision shown in Figure 8–6(a), because the impinging particle passes through the attractive region of the potential and interacts with the repulsive r^{-12} wall. Almost all potential scattering has such a region of *specular* reflection.

2. For *large* impact parameters, the scattering is also similar to that for a hard sphere, namely, the impinging particle is hardly scattered at all, because the attractive $(-r^{-6})$ part of the potential is rapidly falling to zero in this region. This limit is reached more rapidly at higher values of E^*.

3. In the intermediate region of b^* and E^*, we encounter behavior which is not found for hard-sphere scattering. The deflection angle χ may take on large *negative* values, even exceeding $-\pi$. This results from the impinging particle interacting strongly with the *attractive* part of the Lennard-Jones potential, resulting in trajectories such as those shown in Figure 8–8. Figure 8–8(a) illus-

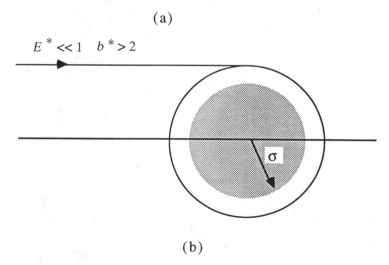

(a)

(b)

Figure 8–8. Trajectories in Lennard-Jones scattering. (a) Rainbow scattering, $\chi = -\pi$. (b) Example of an orbiting collision.

Dynamics of Bimolecular Collisions   Chap. 8

trates a deflection $\chi = -\pi$, which occurs at a value of $b^* \approx 1.7$ for $E^* = 1$. This behavior is an example of *rainbow* scattering, which will be discussed shortly. Figure 8–8(b) shows the very large negative deflection angles which can occur for small E^* and large b^*. In this case, the impinging particle appears to go into orbit around the scattering center, so such trajectories are called *orbiting* collisions. In most instances, the orbits are unstable and the particles will eventually separate.

The cross section may be found from the deflection function with the use of equation 8–17. The angular dependence of the cross section is shown in Figure 8–9 for a value of E^* on the order of unity. Each of the previously described three regions of the deflection function maps onto a portion of the cross section. Region 1 corresponds to $b \to 0$ and $\chi \to \pi$; since both b and $\sin \chi$ approach zero in a uniform manner, the cross section remains well defined and approaches the limiting hard-sphere value $d^2/4$. In region 2, b increases without limit while χ (and therefore $\sin \chi$) goes to zero; also, $d\chi/db \to 0$. This leads to a strong divergence of I at small χ. In region 3, there is a minimum in the deflection function at some value of χ (not necessarily $-\pi$), at which the derivative $d\chi/db \to 0$. The cross section also diverges at this value of χ, which is called the *rainbow angle* χ_r. The term arises from the mathematical similarity between this phenomenon and the scattering of light from raindrops in the atmosphere which produces the rainbow. Both this divergence and that in region 2 are an artifact of classical mechanics; for example, the classical attractive potential always has some finite (but very small) value at arbitrarily large r,

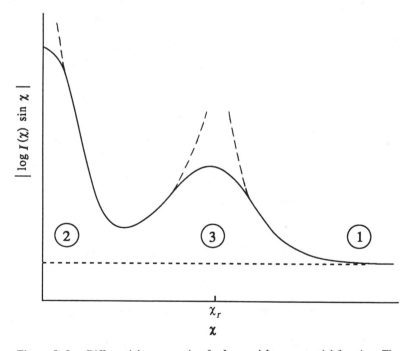

Figure 8–9. Differential cross section for Lennard-Jones potential function. The dashed parts of the curve show the classical divergences.

so there is always some nonzero scattering angle. Quantum mechanics damps out these divergences, however; for example, the uncertainty principle places a lower limit on the deflection angle that can be measured with a given apparatus, and the cross section rounds off at that value of χ.

Before leaving the topic of the Lennard-Jones potential, let us momentarily return to the correction to the gas-kinetic collision rate due to attractive intermolecular forces, mentioned earlier in section 8.1.1. The importance of this correction was noted by Kohlmaier and Rabinovitch[6] and has been evaluated rigorously by Hirschfelder et al.[7] We outline their results here.

The collision rate coefficient between a and b molecules in a gas is

$$Z_{\text{coll}} = \Omega_{ab}^{(2,2)*} \overline{v}_{ab} \pi \sigma_{ab}^2 \tag{8-18}$$

where

$$\sigma_{ab} = \frac{1}{2}(\sigma_a^{\text{LJ}} + \sigma_b^{\text{LJ}})$$

is the arithmetic average of the Lennard-Jones σ parameters, and the "omega integral" is defined as

$$\Omega_{ab}^{(2,2)*} = \frac{2\pi k_B T}{\mu_{ab}} \int \int_0^\infty e^{-\gamma_{ab}^2} \gamma_{ab}^7 (1 - \cos^2 \chi) b \, db \, d\gamma_{ab} \tag{8-19}$$

The variables in equation 8–19 are

$$\gamma_{ab} = \left(\frac{\mu_{ab}}{2k_B T}\right)^{1/2} v_{ab}$$

and $\chi(b)$ is the deflection function, defined by equation 8–15. Tables of $\Omega_{ab}^{(2,2)*}$ for the Lennard-Jones potential function are given in Hirschfelder et al.,[7] and the de-

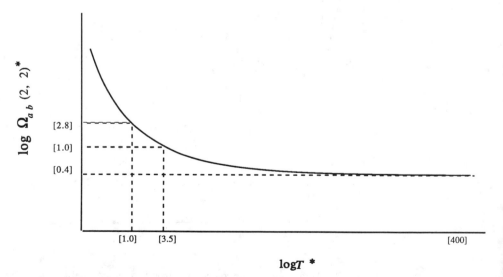

Figure 8–10. Log-log plot of $\Omega_{ab}^{(2,2)*}$ vs. reduced temperature $T^* = k_B T/\epsilon$.

pendence of the integral on a reduced temperature $T^* = k_B T/\epsilon$ is shown in Figure 8–10. At small values of T^* (low temperatures, or sticky molecules with large attractive terms), Z_{coll} is *larger* than the calculated hard-sphere collision rate. At large values of T^*, on the other hand, corresponding to high temperatures or for collisions with He atoms, Z_{coll} may actually be less than the hard-sphere estimate. A series of measurements on vibrational relaxation of polyatomic molecules in supersonically cooled molecular beams indicates that the Lennard-Jones collision rate is a limiting rate for inelastic collisions[8].

8.3 COMPLEX SCATTERING PROCESSES

While the two-body scattering dynamics described in the preceding section has been useful in introducing a number of important concepts and methods of scattering theory, there is obviously no way to describe a chemical reaction in terms of only two interacting particles; a minimum of three is required. Unfortunately, the three-body scattering problem is one for which no analytic solution is known. Accordingly, we turn to either approximations or numerical analysis to solve this problem. In this section, we shall consider two methods which are of particular importance in theoretical chemical kinetics: classical trajectory calculations with Monte Carlo sampling, and solution of the time-dependent Schrödinger equation on the potential energy surface.

8.3.1 Classical Trajectory Calculations

A potential surface for a chemical reaction is expressed as a function of the relative coordinates of the constituent atoms (see chapter 7). With such a function available, it is possible to solve the classical equations of motion on the surface for the collision trajectory, which is just the set of time-dependent coordinates of the individual atoms. In general, the solution requires numerical integration of the Hamilton equations of motion for each coordinate Q_j and its conjugate momentum $P_j = \mu_j \, dQ_j/dt$:

$$\frac{dQ_j}{dt} = \frac{\partial \mathcal{H}}{\partial P_j}, \frac{dP_j}{dt} = -\frac{\partial \mathcal{H}}{\partial Q_j} = -\frac{\partial U}{\partial Q_j} \qquad (8\text{–}20)$$

These equations are equivalent to Newton's second law of motion, $\mathbf{F} = m\mathbf{a}$, where the force $\mathbf{F}$ is a gradient of the potential energy and the acceleration $\mathbf{a}$ is the second time derivative of the corresponding coordinate.

The essential content of the problem is contained in the Hamiltonian function $\mathcal{H}$, which is the sum of kinetic and potential energies:

$$\mathcal{H} = T + V$$

The kinetic energy T is a function of relative velocities, while the potential energy V is just the potential $U(\{\mathbf{r}\})$ expressed in terms of relative coordinates. This separation into position- and velocity-dependent terms is what allows us to simplify the derivative $\partial \mathcal{H}/\partial Q_j$ in equation 8–20 to $\partial U/\partial Q_j$.

The simplest example of a classical trajectory is that for a collinear AB + C collision complex, for which there are only two coordinates, the A–B distance r_1 and the B–C distance r_2. The trajectories for such a system can be represented on a contour diagram such as that shown in Figure 7–5, or even simulated by rolling a steel ball on a machined facsimile of the surface. Even for such a simple system, however, a complication is introduced by transforming from internal molecular coordinates to Cartesian space. The origin of this problem may be seen by writing out the Hamiltonian in detail:

$$\mathcal{H} = T + V$$

$$\mathcal{H} = \frac{1}{2}\frac{m_A m_B}{m_A + m_B}\dot{r}_1^2 + \frac{1}{2}\frac{(m_A m_B)m_C}{m_A + m_B + m_C}\left(\dot{r}_2 + \frac{m_A}{m_A + m_B}\dot{r}_1\right)^2$$

$$+ U(r_1, r_2) \qquad (8\text{–}21)$$

$$\mathcal{H} = \frac{1}{2(m_A + m_B + m_C)}[m_A(m_B + m_C)\dot{r}_1^2 + 2m_A m_B \dot{r}_1 \dot{r}_2 + m_C(m_A + m_B)\dot{r}_2^2]$$

$$+ U(r_1, r_2) \qquad (8\text{–}22)$$

The first kinetic energy term in equation 8–21 represents the relative motion of A with respect to B (with reduced mass μ_{AB}), while the second represents motion of AB with respect to C (with reduced mass μ_{AB+C}). We see that when equation 8–21 is expanded to give equation 8–22, a cross term in $\dot{r}_1\dot{r}_2$ remains. This cross term must be eliminated if the motion of a representative point across the surface is to be that of the "sliding mass" alluded to previously. This can be done by a further coordinate transformation to a *mass-weighted coordinate system*, known as the *affine transformation*,[9,10] which diagonalizes the kinetic energy of the system. Such a transformation is shown in Figure 8–11. We let $r_1 = x - y \tan\theta$ and $r_2 = \alpha y \sec\theta$. Then if

$$\alpha = \left[\frac{m_A(m_B + m_C)}{m_C(m_A + m_B)}\right]^{1/2}$$

and

$$\sin\theta = \frac{\alpha m_C}{m_B + m_C} = \left[\frac{m_A m_C}{(m_A + m_B)(m_B + m_C)}\right]^{1/2}$$

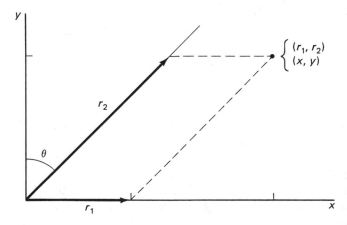

Figure 8–11. Affine transformation to mass-weighted coordinate system.

$T(\dot{r}_1, \dot{r}_2)$ becomes $\frac{1}{2}m\dot{x}^2 + \frac{1}{2}m\dot{y}^2$, with $m = m_A(m_B + m_C)/(m_A + m_B + m_C)$. (The reader is encouraged to verify this by direct substitution.) This transformation gives rise to the *skewed coordinate system* often seen in representations of potential surface contour diagrams. For F + H–H, $\theta = 43.6°$ and $\alpha = 0.7255$. The skewed surface, derived from that depicted in Figure 7–8(a), is shown in Figure 8–12.

The power of the classical trajectory method lies in its ability to predict the details of reactive scattering for complex systems on almost any potential surface for which analytic solutions cannot be obtained. Its major drawback is that it is very computation-intensive, even for simple systems. For example, for a three-body system, such as H + H_2 or F + H_2, there are $3 \times 3 = 9$ degrees of freedom in the problem. Three of these can be immediately eliminated by going to center-of-mass coordinates. But this still leaves six independent coordinates which, with their six conjugate momenta, give rise to 12 coupled linear first-order differential equations having the form given by equation 8–20. The full Hamiltonian for the three-body problem is

$$\mathcal{H}(\mathbf{P}, \mathbf{Q}) = \frac{1}{2\mu_{BC}}(P_1^2 + P_2^2 + P_3^2) + \frac{1}{2\mu_{A,BC}}(P_4^2 + P_5^2 + P_6^2)$$

$$+ U(Q_1, \ldots, Q_6) \qquad (8\text{--}23)$$

These equations can be readily integrated by one of the numerical techniques discussed in chapter 2. Because each trajectory may involve a large number of successive integration steps, cumulative errors ("global error") may become a

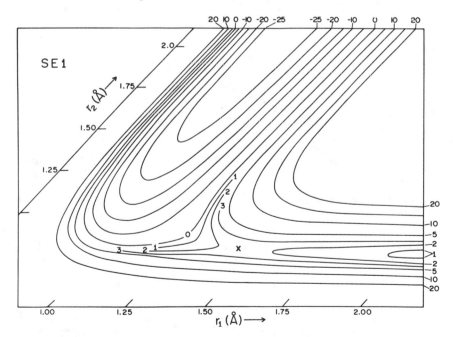

Figure 8–12. F + H–H potential surface in mass-weighted coordinate system. [From J. C. Polanyi and J. L. Schreiber, *Faraday Disc. Chem. Soc.* 62, 269 (1977)].

problem. A good way of verifying the accuracy of the integration is to run the trajectory *backwards* from the final computed state: by virtue of microscopic reversibility, the initial conditions should be accurately reproduced. The major problem in implementing classical trajectories is that the typical experiment samples a wide distribution of initial conditions. Since each trajectory is run for a specified set of initial coordinates and momenta, a large number of trajectories must be run over a range of initial conditions in order to simulate a particular experimental situation. An efficient choice of initial conditions is essential to getting useful information from trajectory calculations.

The conditions which must be specified can be identified by reference to Figure 8–13. Three of the coordinates refer to the diatomic species BC. These can be written, along with their conjugate momenta, in the form

$$Q_1^0 = R \sin \theta \cos \phi \qquad P_1^0 = -P(\sin \phi \cos \eta - \cos \phi \cos \theta \sin \eta)$$

$$Q_2^0 = R \sin \theta \sin \phi \qquad P_2^0 = P(\cos \phi \cos \eta - \sin \phi \cos \theta \sin \eta)$$

$$Q_3^0 = R \cos \theta \qquad P_3^0 = P(\sin \theta \sin \eta)$$

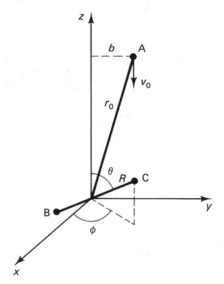

Figure 8–13. Classical coordinates for an A + BC collision.

In these equations, R is the initial B–C distance, P is the rotational angular momentum $J(J + 1)h^2/R^2$, and the angles θ, ϕ, and η specify the initial orientation of the BC molecule in space. For specifying the coordinates and momenta of the impinging atom A, it is convenient to take the initial relative velocity v_r^0 in the yz-plane. This gives the following simple forms:

$$Q_4^0 = 0 \qquad\qquad P_4^0 = 0$$

$$Q_5^0 = b \qquad\qquad P_5^0 = 0$$

$$Q_6^0 = -(r^2 - b^2)^{1/2} \qquad\qquad P_6^0 = \mu_{A,BC} v_r^0$$

In most trajectory calculations, the reactant molecule *BC* is specified to be in an initial vibrational and rotational state, which determines P and allows R to be set to the maximum bond extension compatible with the total vibrational energy. The initial relative velocity v_r^0 may be varied systematically, or it may be chosen at random from a Boltzmann distribution $f(v) = 4\pi v^2 e^{-\mu v^2/2k_BT}$. The orientation angles, which specify rotational phase, are selected at random, as is the impact parameter b. Finally, the vibrational phase is accounted for by selecting r_0 at random from the interval between r_0 and $r_0 + v_r^0 \tau$, where τ is the vibrational period of BC. In any case, r_0 must be chosen so that the intermolecular interaction is initially zero, and the trajectory is integrated forward until $r(t)$ is sufficiently large that the interaction is once again zero. At that point the final state of the system is determined by inspecting which atom is bound to which other atom, and final energies and angular momenta are determined by calculating classical expectation values over a few cycles of the motion. The use of random sampling for many of the initial conditions has given rise to the term "Monte Carlo method" in connection with these calculations.

The basic quantity derived from such calculations is the reaction probability for a specified initial relative velocity (i.e., initial relative kinetic energy), rotational state, and impact parameter. This is operationally defined as the ratio of the number of reactive trajectories to the total number of trajectories after a sufficiently large number of trajectories has been run:

$$P_R(v_r^0, J, b) = \lim_{\substack{N\to\infty \\ (\$\to 0)}} \frac{N_R(v_r^0, J, b)}{N_{\text{total}}(v_r^0, J, b)} \tag{8–24}$$

Although a large number of trajectories is required to obtain good statistics, the practical limitation is usually the availability and cost of computer time. The reactive cross section is found by averaging the reaction probability over the impact parameter and rotational state:

$$\sigma_R(v_r^0, J) = 2\pi \int_0^{b_{\max}} P_R(v_r^0, J, b) b \, db$$

$$\sigma_R^0(v_r^0) = \sum_{J=0}^{J_{\max}} \sigma_R(v_r^0, J) f_B(J; T)$$

Finally, the thermally averaged rate coefficient is the velocity-averaged, velocity-weighted cross section, as discussed in chapter 6:

$$k(T) = \int_0^\infty dv_A \int_0^\infty dv_{BC} f_A(v_A) f_{BC}(v_{BC}) v_r^0 \sigma_R(v_r^0) \tag{8–25}$$

In the averaging equations shown above, we have assumed that trajectories have been run for a random sampling of initial relative velocities, rotational states, and impact parameters, and have included the appropriate distribution functions explicitly. The reason for doing this is practical: it turns out that most reactive collisions tend to occur at small b and large v_r^0, so that sampling at random from the distributions $f(b) = 2\pi b$ and $f(v) = 4\pi v^2 e^{-\mu v^2/2k_B T}$ would tend to give a preponderance of nonreactive trajectories. It is more efficient to sample from a random distribution of b's and v_r^0's and then account for ensemble averages by weighting with the proper distribution function. This type of biasing, known as *importance sampling*,[12–14] is an important aid in performing efficient trajectory calculations.

The first fully realized classical trajectory calculation of a rate coefficient was by Karplus and co-workers[15] on $H + H_2$. Representative reactive and nonreactive trajectories from that work are shown in Figure 8–14. One microscopic quantity which can be determined by inspection of such plots is the "collision duration," which can be estimated as that period of time during which the approaching or departing atom deviates significantly from straight-line trajectories. From the plots in the figure, this appears to be on the order of 10^{-14} sec. Most "impulsive" collisions have durations on this order; longer lived collision complexes are also possible and are discussed briefly in section 8.3.3.

(a)

(b)

Figure 8–14. $H + H_2$ scattering trajectories, represented as plots of interatomic distance vs. time. (a) Nonreactive trajectory with $J_i = 0$, $v_i = 0$, $v_r^0 = 1.32 \times 10^6$ cm sec^{-1}. (b) Reactive trajectory with same initial conditions. The difference between trajectories (a) and (b) appears to be a minuscule difference in the vibrational phase of the original (BC) hydrogen molecule. The "crossing" of the plots in (b) indicates rotational excitation in the products, with $J' \approx 5$. [From Karplus et al., *J. Chem. Phys.* **43**, 3259 (1965).]

The trajectory results for the H + H$_2$ reaction probability and reactive cross section are shown in Figure 8–15(a) and (b), respectively. Note that the reaction probability does not at all resemble the old-collision-theory, reactive-hard-sphere model; the b-dependence is much more like $P_R^0 \cos (\pi b/2b_{max})$, with $P_R^0 \approx 0.4$ and $b_{max} \approx 0.98$ Å. Nevertheless, the reactive cross section shown in Figure 8–15(b) approximates very well to the reactive-hard-sphere form. The computed thermally

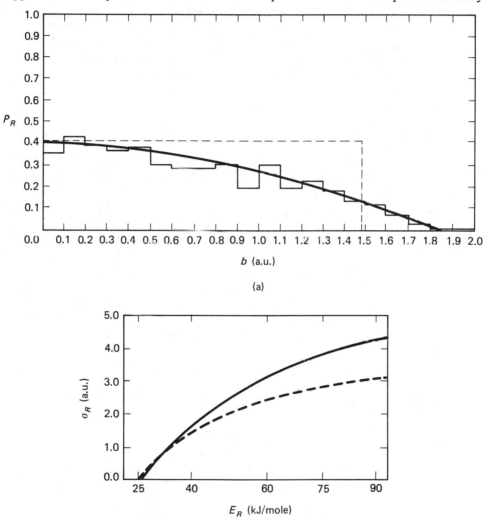

(a)

(b)

Figure 8–15. Results of classical trajectory calculations for H + H$_2$ reaction. (a) Reactive probability P_R vs. impact parameter. The solid curve is a fit to the empirical function $P_R^0/\cos (\pi b/2b_{max})$, while the dashed curve is the reactive-hard-sphere form (see equation 8–6) with $b^* = 1.5$ a.u. and steric factor $p = 0.4$. (b) Reaction cross section vs. relative kinetic energy. The solid line is the (smoothed) result of the classical trajectory calculations, while the dashed curve is the reactive-hard-sphere (old collision theory) result given in equation 8–7. [From Karplus et al., *J. Chem. Phys. 43*, 3259 (1965).]

averaged rate coefficient for the H + H$_2$ reaction, over the range 300 K to 1,000 K, is within a factor of two or three of the experimentally determined values.* On the one hand, such an *ab initio* calculation of a rate coefficient is a remarkable *tour de force;* on the other hand, it illustrates vividly how rate constants, like all thermally averaged properties of a reaction, are very insensitive to the details of interaction potentials and molecular dynamics.

In fact, the real usefulness of trajectory calculations is their ability to provide, in great detail, information on product final states, scattering angles, and other dynamical quantities, which may then be compared with experimental results obtained by using the techniques to be discussed in the next chapter. Trajectory calculations at this level have been carried out for a number of systems, including F + H$_2$.[11] Tested and documented trajectory programs are available from the Quantum Chemistry Program Exchange at the University of Indiana.

8.3.2 Quantum Scattering Calculations

In the preceding section, we have described how solution of the classical equations of motion may be used to find reactive probabilities and cross sections. Such calculations are more properly termed *quasiclassical* because the quantization of vibrational and rotational energy levels is accounted for after the calculation by "binning" product states possessing suitable ranges of internal energies into groups according to corresponding v and J quantum numbers. We know, however, that atoms and molecules obey the laws of quantum mechanics—specifically, that their motion is governed by a wave equation (the Schrödinger equation) rather than by the deterministic laws of classical motion. The questions we now address are, first, how do we set up the scattering problem quantum mechanically and, second, does the quantum solution display significant differences from the classical treatment?

We begin with the time-dependent Schrödinger equation,

$$\mathcal{H}\psi = i\hbar \frac{\partial \psi}{\partial t} \tag{8-26}$$

The wave function ψ depends on both the coordinates $\mathbf{r}$ and the time t. The Hamiltonian function is

$$\mathcal{H} = \frac{p^2}{2\mu} + U(\mathbf{r}) \tag{8-27}$$

The potential function $U(\mathbf{r})$ is the same as in the classical treatment, but since the momentum is a differential operator, viz.,

$$\hat{p}_j = \frac{\hbar}{i} \frac{\partial}{\partial q_j} \tag{8-28}$$

the entire Hamiltonian is an operator rather than an algebraic function such as equation 8–21. That is to say, $\mathcal{H}$ is written as

$$\mathcal{H} = -\frac{\hbar^2}{2\mu} \nabla^2 + U(\mathbf{r}) \tag{8-29}$$

*For a discussion of experimental techniques for determining H + H$_2$ reaction rates, the reader is referred to Johnston.[16]

The second major difference from classical mechanics is that not all coordinates and momenta can be specified simultaneously, as was the case in setting up initial conditions for the classical trajectory calculations. This is a consequence of the Heisenberg uncertainty relationship

$$\Delta p_j \Delta q_j > \hbar \tag{8-30}$$

which follows from the operator nature of the momentum given by equation 8-28. Thus, in Figure 8-5, for example, we cannot specify both the impact parameter b and the relative velocity $\mathbf{v}$ for an individual collision.

If we wish, we can retain the center-of-force coordinate system used in Figure 8-5, but the interacting-particle representation must then be replaced by a wave picture, as shown in Figure 8-16. We can specify only the wave vector $\mathbf{k} = \mathbf{p}/\hbar$; the interaction potential $U(r)$ remains the same as in the classical calculation. In the case of elastic scattering, such as the interaction of two rare-gas atoms, the time-dependent wave function can be written as a simple product, viz.,

$$\psi(\mathbf{r}, t) = \psi(\mathbf{r})e^{-iEt/\hbar}$$

where $E = (\hbar k)^2/2\mu$. The time-independent part of the wave function may be represented as a superposition of an incoming plane wave and an outgoing spherically scattered wave, i.e.,

$$\psi(r) = e^{i\mathbf{k}\cdot\mathbf{r}} + f(\Omega)\frac{e^{ikr}}{r} \tag{8-31}$$

where Ω denotes the angular variables θ and ϕ.

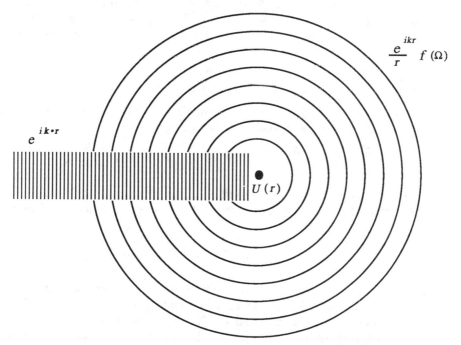

Figure 8-16. Quantum-mechanical representation of elastic scattering in a center-of-force coordinate system.

Experimental observables in quantum mechanics are expressed as expectation values of the wave function. The pertinent expectation value for the scattering problem is the *current* corresponding to the flux of molecules, dN_c/dt, in equation 6–1. This is defined by

$$\hat{j} = \frac{\hbar}{2i\mu}(\psi^*\nabla\psi - \psi\nabla\psi^*) \qquad (8-32)$$

For the incident plane wave $\psi = \psi_0 = e^{i\mathbf{k}\cdot\mathbf{r}}$,

$$\hat{j}_{\text{incident}} = \hbar\mathbf{k}/\mu$$

For the outgoing scattered wave given by equation 8–31,

$$\hat{j}_{\text{scattered}} = \frac{\hbar\mathbf{k}}{\mu}\frac{|f(\Omega)|^2}{r^2}$$

Therefore, the differential cross section for scattering is

$$I(\theta, \phi) = \frac{d\sigma}{d\Omega} \propto |f(\Omega)|^2 \qquad (8-33)$$

To find $f(\Omega)$, we must solve the time-independent Schrödinger equation

$$\mathcal{H}\psi(r, \Omega) = -\frac{\hbar^2}{2\mu}\nabla^2\psi + U\psi = E\psi \qquad (8-34)$$

There are many techniques for effecting a solution, but the one most appropriate for this problem is a *partial wave analysis*. We expand the wave function ψ in partial waves corresponding to the orbital angular momentum ℓ. (For large $\ell, \hbar\ell = \mu v b$, so $\ell = |\mathbf{k}|b$.) This expansion has the form

$$\psi = \sum_\ell P_\ell(\cos\theta)\frac{Y_\ell(r)}{r} \qquad (8-35)$$

where the P_ℓ are Legendre polynomials which solve the angular part of the Schrödinger equation. Equation 8–34 then becomes

$$-\frac{\hbar^2}{2\mu}\frac{d^2}{dr^2}Y_\ell + \left[\frac{\hbar^2\ell(\ell+1)}{2\mu r^2} + U(r)\right]Y_\ell = EY_\ell \qquad (8-36)$$

Note that the term in $\ell(\ell+1)$ just takes the place of the centrifugal potential term in equation 8–13. Solution of equation 8–36 gives the asymptotic form of Y_ℓ at large r as

$$Y_\ell = a_\ell \sin\left(kr - \frac{\pi\ell}{2} + \eta_\ell\right) \qquad (8-37)$$

where η_ℓ is the collision-induced *phase shift* of the wave function.

At this point in the treatment of quantum scattering, it is useful to introduce the *scattering operator* or *S*-matrix. This is just an operator that takes the incoming plane wave ϕ_0 into the asymptotic, outgoing scattered wave $\psi(r \to \infty)$. Since for elastic scattering, the two waves differ merely by a phase shift η_ℓ, in this case the *S*-matrix has the simple form

$$S^{\ell}(k) = \exp(2i\eta_\ell) \qquad (8\text{--}38)$$

In terms of the S-matrix, the scattering amplitude $f(\Omega)$ can be expressed as

$$f(\Omega) = \frac{i}{2k} \sum_{\ell=0}^{\infty} (2\ell + 1)(1 - S^{\ell}(k))P_\ell(\cos\theta) \qquad (8\text{--}39)$$

Using equation 8–33 and the orthonormality property of the Legendre polynomials leads to an expression for the cross section:

$$\sigma = \int d\Omega \, |f(\Omega)|^2 = \frac{\pi}{k^2} \sum_{\ell=0}^{\infty} (2\ell + 1)|1 - S_\ell(k)|^2$$

$$\sigma = \frac{4\pi}{k^2} \sum_{\ell=0}^{\infty} (2\ell + 1) \sin^2 \eta_\ell \qquad (8\text{--}40)$$

While the preceding treatment can be generalized to more complex situations, it is most suitable for elastic scattering, in which neither energy nor constituent particles are exchanged between the collision partners. The results of this treatment are useful chiefly for inverting molecular beam-scattering data to obtain intermolecular potential functions, e.g., in rare gas–rare gas collisions. The energy dependence of such scattering cross sections is implicit, appearing in the phase shifts η_ℓ. The principal limitation of this calculation is that one may need to include 100 or more partial waves (values of ℓ) for molecule-molecule collisions. Additional details of the theory may be found in texts on quantum mechanics.

In the case of inelastic or reactive scattering, the situation is even more complex, because the molecular basis states change between the initial and final states of the collision. The potential $U(r, R)$ includes terms coupling internal coordinates R with relative translational coordinates r. The total energy

$$E = \frac{\mathcal{H}_0 \Phi}{\Phi} + \frac{\hbar^2 k^2}{2\mu}$$

is, of course, conserved in the collision. ($\mathcal{H}_0$ is the internal Hamiltonian of the molecule, e.g., BC in an A + BC reaction, and Φ is the asymptotic form of the wave function as $t \to \infty$ and $r \to \infty$.) The cross section for such an inelastic or reactive collision can be expressed in terms of matrix elements of the scattering operator in the (i, j) basis

$$\frac{d\sigma(i \to j)}{d\Omega} = \left| \sum_\ell \frac{(\ell + \frac{1}{2})}{k_i} [S_{ij}^\ell - \delta_{ij}]P_\ell(\cos\theta) \right|^2 \qquad (8\text{--}41)$$

where the Kronecker δ function in equation 8–41 is equal to zero for $i \neq j$, i.e., if the collision is reactive or inelastic. For $i = j$, the collision is elastic and we regain the factor $1 - S_{ij}^\ell$ which appears in equation 8–39.

"Coupled-channel" calculations of the type just described have been carried out for only a few systems, because the large number of basis states which must be included makes the calculations rather formidable. The calculations on the H + H$_2$ reaction by Schatz and Kupperman[17] are a good example. Results of quantum scattering calculations carried out to date have indicated several important differences from purely classical calculations. The possibility of tunneling through a potential barrier (see chapter 10) is one such effect. The consequence of tunneling is the appearance of a threshold cross section slightly below the barrier height energy, which in turn affects the activation energy for the reaction and can lead to non-Arrhenius behavior of the rate coefficient. Interference between scattered waves can lead to amplitude fluctuations in the differential cross section $I(\Omega)$ and also rounds off the classical divergences indicated in Figure 8–9. Classical trajectories also experience difficulties at threshold: when the total energy is just slightly above the minimum energy required to cross the potential barrier, the vibrational zero-point energy tends to be converted into translational kinetic energy, which is not allowed quantum mechanically. With these exceptions, however, it is found that classical trajectories usually provide a good approximation to the motion of the corresponding quantum mechanical wave packet across the potential energy surface.

8.3.3 Long-lived Collision Complexes

From Figure 8–14, we estimated the "collision duration" (a concept which is not well defined quantum mechanically) to be about 10^{-14} sec. If the potential surface possesses a strong attractive well, a long-lived collision complex may be formed. Classically, this would result from "orbiting" collisions such as the trajectory shown in Figure 8–8(b). The quantum analogue is a "resonance" which can occur when the collision energy matches that of a bound state in the potential well. When such a complex collision takes place, products tend to be isotropically scattered in the center-of-mass system, and there may also be near-statistical energy release in the reaction products. An example is discussed in chapter 11.

8.3.4 Ion-Molecule Scattering

Thus far, the discussion has focused almost exclusively on reactions between neutral, i.e., uncharged, atoms and molecules. An important class of reactions involves charged species (ions) interacting with neutral species. Such reactions occur in ionized gas plasmas, radiation chemistry, electric discharges (as in electron-beam pumped lasers), flames, and the upper atmosphere. They can also occur within a mass spectrometer, giving rise to additional peaks which complicate the interpretation of mass spectra. For dynamical studies (see chapter 9), the charge provides a convenient "handle" on the ion by means of which the kinetic energy in the collision can be varied over a wide range. The theoretical treatment of such collisions, first developed by Langevin[18] in 1905, is quite straightforward,[19] and we proceed to develop it here.

An ion (M^+ or M^-) and a neutral molecule interact through the ion-induced-dipole potential

$$U(r) = -\frac{1}{2}\frac{\alpha e^2}{r^4} \tag{8–42}$$

where α is the polarizability of the neutral molecule and e is the elementary electronic charge (4.8×10^{-10} esu). The collision is best analyzed in the center-of-force coordinate system (see section 8.2, and cf. especially Figure 8–5). The effective potential, including the centrifugal term, is

$$V_{\text{eff}}(r) = U(r) + \frac{L^2}{2\mu r^2} = -\frac{1}{2}\frac{\alpha e^2}{r^4} + \frac{L^2}{2\mu r^2} \tag{8-43}$$

where $E = \frac{1}{2}\mu v_0^2$ and $L = \mu v_0 b$. We can thus write

$$V_{\text{eff}}(r) = E\left(\frac{b}{r}\right)^2 - \frac{1}{2}\frac{\alpha e^2}{r^4}$$

A plot of $V_{\text{eff}}(r)$ given by equation 8–43 is shown in Figure 8–17.

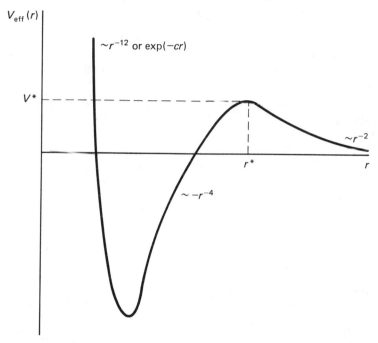

Figure 8–17. Effective potential for the interaction between an ion and a neutral, nonpolar molecule.

Note that the potential has a maximum at $r^* = (\alpha e^2/E)^{1/2}/b$, with $V^*(r^*) = \frac{1}{2}E^2 b^4/\alpha e^2$. Whenever $E < V^*$, the centrifugal barrier cannot be penetrated (if tunneling is neglected), and so r_c always remains greater than r^*. The collision is presumed to be elastic, in that the internal states of the ion and the target molecule remain unchanged. For $E \geq V^*$, the particle can move inside the centrifugal barrier and indeed can penetrate to the inner repulsive wall of the potential. In this case, we assume that reaction occurs with a probability of unity; if E is exactly equal to V^*, then "orbital capture" can occur as shown in Figure 8–8(b).

The critical impact parameter at which reaction can occur is found from the expression for V^* to be

$$b^* = (2\alpha e^2/E)^{1/4}$$

Since the reaction probability is stipulated to be unity for $b < b*$, the cross section for reaction is just

$$\sigma_L(E) = \pi (b*)^2 = \pi \left(\frac{2\alpha e^2}{E}\right)^{1/2} \tag{8-44}$$

The quantity σ_L is often referred to as the *Langevin cross section*. We can estimate its magnitude by using a typical polarizability for small polyatomic molecules. (From Table 7–2, $\alpha = 3 \times 10^{-24}$ cm^3 and $E = k_B T = 4 \times 10^{-14}$ erg/molecule.) With these values, equation 8–44 gives

$$\sigma_L = (3.5 \times 10^{-29})^{1/2} = 2 \times 10^{-14} \text{ cm}^2$$

The Langevin cross section is thus on the order of a few hundred square angstroms. It depends only on the collision energy E and the polarizability of the collision partner; no other chemical or structural information is incorporated into the expression. As in the reactive-hard-sphere model (see section 8.1.2), this represents an upper limit for the ion-molecule reaction cross section; for a given system, the accurate cross section will depend on the permanent dipole moment, activation barriers along the reaction coordinate, and other chemical properties of the ion and the molecule.

The rate constant for the Langevin model may be found by inserting the equation 8–44 for the cross section into equation 6–6 and substituting $v = (2E/\mu)^{1/2}$. This gives

$$k = \int v\sigma (v) f (v) \, dv$$

$$= \int \left(\frac{2E_c}{\mu}\right)^{1/2} \left(\frac{2\alpha e^2}{E}\right)^{1/2} \pi f (v) \, dv$$

Noting that the $E^{1/2}$ factors cancel and that the velocity distribution function $f(v)$ is normalized gives the simple result

$$k = \left(\frac{4\pi^2 \alpha e^2}{\mu}\right)^{1/2} \tag{8-45}$$

which is velocity (relative kinetic energy)-independent.

Detailed discussions of ion-molecule reactions may be found in several review articles.[20-23]

8.3.5 Electronically Nonadiabatic Processes: Surface Crossing

The scattering processes which we have been considering thus far have all been *adiabatic*, in the sense that they take place on a single electronic potential energy surface. Another important class of reactions is *nonadiabatic* reactions, which involve two interacting potential surfaces and the possibility of moving between these surfaces during a collision. Examples of such processes include reactions involving electronically excited species, such as

$$N*(^2P) + O_2(^3\Sigma) \longrightarrow NO(^2\Pi, v) + O(^3P, {}^1D)$$

and near-resonant electronic-energy-transfer reactions, such as

$$I(^2P_{3/2}) + O_2^*(a^1\Delta) \longrightarrow I*(^2P_{1/2}) + O_2(^3\Sigma)$$

The latter reaction is important in the chemically pumped oxygen-iodine laser system.[24] Potential energy surfaces for electronically excited molecules have been described previously in chapter 7.

The simplest description of electronically nonadiabatic processes was developed by Landau[25] and Zener[26] in the early 1930s. The Landau-Zener model estimates the probability of surface crossing or surface hopping at the avoided crossing between two interacting one-dimensional potential energy curves (see Figure 8–18). This probability has the form[27]

$$P_{LZ} = \exp\left(-\frac{4\pi V_{12}^2}{hv|s_1 - s_2|}\right) \tag{8–46}$$

where $v = dr/dt$, the velocity at the crossing point r_0; s_1 and s_2 are the slopes dU/dr

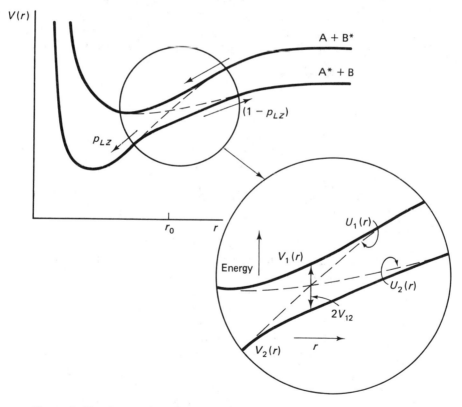

Figure 8–18. Interacting potential energy curves, showing detail of avoided-crossing region. An electronically nonadiabatic trajectory is shown which crosses with probability P_{LZ} from the upper to the lower surface on the way in and remains on the lower surface with probability $1 - P_{LZ}$ on the way out.

of the *unperturbed* potential curves at r_0, which must be calculated from the potential model; and V_{12} is one-half the splitting between the actual potential curves at r_0.

The net nonadiabatic process involves two traversals of the avoided-crossing region, one as the particles approach each other and a second as they separate. In order for the collision products to end up on a potential surface different from the one on which they started, one of these traversals must be nonadiabatic with probability P_{LZ}, and the other must be adiabatic with probability $1 - P_{LZ}$. The net transition probability in the Landau-Zener model is thus

$$p_{12}(v) = 2P_{LZ}(1 - P_{LZ}) \tag{8–47}$$

and the thermally averaged nonadiabatic rate coefficient is

$$k(T) = \frac{\pi r_0^2 \int p_{12}(v) f(v) v \, dv}{\int f(v) \, dv} \tag{8–48}$$

The velocity dependence in equation 8–46 gives rise to a very non-Arrhenius form for $k(T)$. This model has been applied to a number of electronically nonadiabatic systems, including the $O_2^* - I$ transfer process mentioned earlier.[28]

More exact treatments of these processes are based on the classical trajectory method described in section 8.3.1. The interacting potential energy curves are approximated as sections of the actual surfaces normal to the collision trajectory, and the Landau-Zener model is used to estimate the surface-hopping probability at the crossing seams. Further details may be found in Tully.[29]

REFERENCES

1. C. E. Nordman and S. M. Blinder, *J. Chem. Ed. 51*, 790 (1974).

2. H. Goldstein, *Classical Mechanics* (Reading, Mass.: Addison-Wesley, 1950).

3. C. D. Hodgman, ed., *CRC Standard Mathematical Tables,* 12th Ed., p. 284, Integral No. 21 (Cleveland, Ohio: Chemical Rubber Publishing Co., 1960).

4. H. Pauly, "Collision Processes, Theory of Elastic Scattering", in *Physical Chemistry: An Advanced Treatise,* Vol. VI-B, edited by H. Eyring, D. Henderson, and W. Jost (New York: Academic Press, 1975), pp. 553–628.

5. H. Pauly, "Elastic Scattering Cross Sections I: Spherical Potentials", in *Atom-Molecule Collision Theory,* edited by R. B. Bernstein (New York: Plenum Press, 1979), pp. 111–199.

6. G. H. Kohlmaier and B. S. Rabinovitch, *J. Chem. Phys. 38*, 1692 (1963).

7. J. O. Hirschfelder, C. F. Curtiss, and R. B. Bird, *Molecular Theory of Gases and Liquids* (New York: Wiley, 1954), pp. 1126–1127.

8. S. H. Kable, J. W. Thoman, Jr., and A. E. W. Knight, *J. Chem. Phys. 88*, 4748 (1988).

9. R. D. Levine and R. B. Bernstein, *Molecular Reaction Dynamics* (Oxford: University Press, 1974), pp. 98–100.

10. B. H. Mahan, *J. Chem. Ed. 51*, 308 (1974).

11. J. C. Polanyi and J. L. Schreiber, *Faraday Disc. Chem. Soc. 62*, 26 (1977).

12. I. W. M. Smith, "Quasiclassical Trajectory Methods", in *Modern Gas Kinetics: Theory,*

Experiment, and Application, edited by M. J. Pilling and I. W. M. Smith (Oxford: Blackwell, 1987), pp. 55–72.

13. M. B. Faist, J. T. Muckerman, and F. E. Schubert, *J. Chem. Phys. 69,* 4087 (1978).
14. J. T. Muckerman and M. B. Faist, *J. Phys. Chem. 83,* 79 (1979).
15. M. Karplus, R. N. Porter, and R. D. Sharma, *J. Chem. Phys. 43,* 3259 (1965).
16. H. S. Johnston, *Gas Phase Reaction Rate Theory* (New York: Ronald Press, 1966), pp. 158–169.
17. G. C. Schatz and A. Kupperman, *J. Chem. Phys. 65,* 4668 (1976).
18. P. Langevin, *Ann. Chem. Phys. 5,* 245 (1905).
19. A. Henglein, "Elastic and Reactive Scattering of Ions on Molecules", in *Physical Chemistry: An Advanced Treatise,* Vol. VI-B, edited by H. Eyring, D. Nenderson, and W. Jost (New York: Academic Press, 1975), pp. 509–551.
20. P. Ausloos, "Ion-molecule reactions in radiolysis and photoionization of hydrocarbons", in *Progr. Reaction Kinetics,* edited by G. Porter (Oxford: Pergamon Press, 1970), pp. 113–179.
21. M. T. Bowers, ed., *Gas Phase Ion Chemistry* (New York: Academic Press, 1979).
22. A. G. Brenton, R. P. Morgan, and J. H. Beynon, *Ann. Rev. Phys. Chem. 30,* 51 (1979).
23. D. C. Clary, *Mol. Phys. 53,* 3 (1984).
24. W. E. McDermott, N. R. Pchelkin, D. J. Benard, and R. R. Bousek, *Appl. Phys. Letts. 32,* 469 (1978).
25. L. Landau, *Physik Z. Sowjetunion 2,* 46 (1932).
26. C. Zener, *Proc. Roy. Soc. London A137,* 696 (1933); ibid. *A140,* 660 (1933).
27. H. Eyring, J. Walter, and E. W. Kimball, *Quantum Chemistry* (New York: Wiley, 1944), pp. 326–330.
28. J. I. Steinfeld and D. G. Sutton, *Chem. Phys. Letts. 64,* 550 (1979).
29. J. C. Tully, in *Dynamics of Molecular Collisions,* Part B, edited by W. H. Miller (New York: Plenum Press, 1976), pp. 217–267.

PROBLEMS

*8.1 Consider the following "single beam" experiment, which enables the total cross section for elastic scattering to be measured directly. A beam of molecules of species B at a high translational energy E is passed through a gas of species G which is at very low pressures; E is sufficiently large that the atoms of G are essentially stationary with respect to B. Let the velocity of B be v.

    (a) From the definition of the cross section, show that the total number of deflections (collisions) of B per unit time per G atom is $\sigma n_B v$, where n_B is the density of B and σ is the total elastic cross section for B–G collisions. (Assume that only elastic collisions are possible.)

    (b) Show that the total number of B–G collisions per unit time in a volume V is $\sigma V v n_B n_G$, where n_G is the density of G.

    (c) Suppose that the volume in question has (constant) area A and length dx. Suppose also that the intensity of B is $J = $ (density of B) $\cdot v$ and is initially $J = J_0 = n_B v$. Show that the change in intensity over dx is $dJ = -J\sigma n_G \, dx$.

    (d) Show that the overall change in intensity over a finite distance x is given by

$$\ln (J_0/J) = n_G \sigma X$$

It is plain that the measurement of this intensity change enables the total cross section to be determined from one experiment (see also chapter 9, problem 1.)

8.2 Gas viscosity is given by

$$\eta = \frac{5\pi}{32\sqrt{2}} \frac{\mu \bar{v}}{Q}$$

[R. D. Present, *Kinetic Theory of Gases,* New York: McGraw-Hill, 1958], where the average velocity $\bar{v} = (8k_B T/\pi \mu)^{1/2}$, μ is the collision-reduced mass, and

$$Q = \frac{1}{32}\left(\frac{\mu}{2k_B T}\right)^4 \int_0^\infty dv \int_0^{2\pi} d\phi \int_0^\pi d\theta \, \sin^3\theta \, I(\theta) \, v^7 \exp\left(-\frac{\mu v^2}{4k_B T}\right) dv$$

Given that the differential cross section of a hard sphere of diameter d is $I(\theta) = d^2/4$, find the viscosity of a gas of hard spheres.

*8.3 Show that Hamilton's equations of motion

$$-\frac{\partial H}{\partial q_i} = \frac{dp_i}{dt} \qquad \frac{\partial H}{\partial p_i} = \frac{dq_i}{dt}$$

with $H(q_i, p_i) = \Sigma_i \, (p_i^2/2\mu_i) + V(q_1, \ldots, q_N)$ are equivalent to Newton's law

$$\frac{d^2 q_i}{dt^2} = -\frac{\partial V}{\partial q_i}$$

*8.4 Consider motion in two dimensions, with positions x, y and momenta p_x, p_y. Suppose that the potential V is a function only of $r = (x^2 + y^2)^{1/2}$. Find the expression for the Hamiltonian in terms of the new positions and momenta

$$r = (x^2 + y^2)^{1/2}$$

$$\theta = \tan^{-1}(y/x)$$

$$p_r = p_x \cos\theta + p_y \sin\theta$$

and

$$p_\theta = -p_x r \sin\theta + p_y r \cos\theta$$

(i.e., the radial and angular momenta). Hence, show from Hamilton's equations that angular momentum is conserved.

8.5 Suppose molecules A and B are characterized by Boltzmann velocity distributions with temperatures T_A and T_B, respectively ($T_A \neq T_B$). If d is the A–B collision diameter, show that the average rate of bimolecular collisions per unit volume is

$$\pi d^2 \sqrt{\frac{8k_B T_A}{\pi m_A} + \frac{8k_B T_B}{\pi m_B}} \cdot C_A \cdot C_B$$

8.6 Let d and D denote the collision diameters of A–B and AB–C, respectively. The function

$$h(d - r)h(D - R)$$

where

$$r = |\mathbf{R}_A - \mathbf{r}_B|$$

and

$$R = \left| \mathbf{r}_C - \frac{m_A \mathbf{r}_A + m_B \mathbf{r}_B}{m_A + m_B} \right|$$

is 1 if A, B, and C are experiencing a three-body collision, and is zero otherwise. Average the time derivative of the function over Boltzmann distributions of molecules A, B, and C, and show therefrom that the average three-body collision rate per unit volume is

$$\bar{v}_{AB}\, \pi d^2 \frac{4}{3} \pi D^3 + v_{AB-C}\, \pi D^2 \frac{4}{3} \pi d^3$$

where

$$\bar{v}_{AB} = \sqrt{\frac{8k_B T}{\pi} \frac{(m_A + m_B)}{m_A m_B}}$$

and

$$\bar{v}_{AB}-C = \sqrt{\frac{8kT}{\pi} \frac{(m_A + m_B) + m_C)}{(m_A + m_B)m_C}}$$

*8.7 For the case of classical scattering of two particles with a repulsive coulomb potential given by $V = +B/r$,

$$\chi(E, b) = 2\,\mathrm{cosec}^{-1} \left[1 + \left(\frac{2bE}{B} \right)^2 \right]^{1/2}$$

Show, then, that

$$I(E, \chi) = \left(\frac{B}{4E} \right)^2 \mathrm{cosec}^4 \left(\frac{\chi}{2} \right)$$

This result is that used (and derived) by Rutherford in his original α-scattering experiments. Note that the classical total cross section is infinite and that the same result is obtained in an exact quantum treatment.

*8.8 Consider a reaction between two dilute species in an inert gas, and assume that only one initial and one final state are involved. The rate coefficient is then related to the total cross section by

$$k = 8\pi\mu^{-1/2}\,(2\pi k_B T)^{-3/2} \int_0^\infty E\sigma(E)\,\exp\left(-E/k_B T\right)\,dE$$

The "old collision theory" cross section is

$$\sigma(E) = 0 \qquad \text{if } E < E_0$$

$$\sigma(E) = \pi d^2 \left(1 - \frac{E_0}{E} \right) \qquad \text{if } E \geq E_0$$

(a) Evaluate k for T = 1000, 1,500 and 2,000 K for a system where $\mu = 10$ daltons, $d = 0.1$ nm, and $E_0 = 100$ kJ mol^{-1}. N.B. units!

(b) Plot your three values from (a) on an Arrhenius plot and comment.

*8.9 The classical deflection function is given by

$$\chi(b) = \pi - 2b \int_{r_0}^{\infty} dr\, r^{-2} \left[1 - \frac{V(r)}{E} - \frac{b^2}{r^2} \right]^{-1/2}$$

where E is the kinetic energy and r_0 is the classical distance of closest approach. Derive $\chi(b)$ for a hard-sphere potential given by $V(r) = 0$ if $r \geq d$ and $V(r) = \infty$ if $r < d$.

*8.10 Ion-molecule reactions such as $N^+ + O_2 \rightarrow NO^+ + O$ are important in the upper atmosphere and in interstellar space. In this problem, we use a simple model, called the *Langevin model*, for *reactive* scattering in such reactions. The following diagram shows results for the reaction $D_2^+ + N \rightarrow ND^+ + D$; the points are from experiment, while the line is the Langevin model prediction [from McClure et al., *J. Chem. Phys.* **66**, 2079 (1977)].

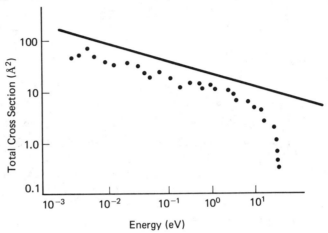

(a) Plainly, the Langevin model breaks down at high energies. Suggest a reason for this.

(b) Evaluate b^* for the $N^+ + D_2$ reaction, assuming that $\alpha = 10^{-24}$ cm^3 and the translational energy $E = k_B T$ for $T = 300$ K. Comment on your value of b^* compared to the value of a typical molecular van der Waals radius, i.e., the value of d in the Lennard-Jones potential $V(r) = 4\epsilon[(d/r)^{12} - (d/r)^6]$.

*8.11 For a particular collinear reaction $A + BC \rightarrow AB + C$, it is found that the classical Hamiltonian can be represented by

$$\mathcal{H} = (p_s^2 + p_u^2)/2m + \tfrac{1}{2}ku^2 + A \exp(-Bs^2)$$

where (s, u) are the curvilinear reaction coordinates.

(a) Sketch a contour diagram of the potential in (i) (s, u) space and (ii) (R_{AB}, R_{BC}) space.

(b) Consider the classical probability of reaction $\chi(p_s, p_u, u, s = \infty)$. This probability is always 0 or 1; find the conditions for it to be 0 and those when it is 1.

(c) Would the rate coefficient show an Arrhenius form? Justify your answer.

(d) Give two reasons why the preceding Hamiltonian cannot be accurate for a real system.

*8.12 Consider two particles colliding under influence of a *purely repulsive* potential (e.g., $V = C/r^{12}$). Sketch the deflection function $\chi(b)$ and classical cross section $I(\chi)$, explaining your answers briefly. No mathematical derivation is required. Note that $\chi(b)$ and $I(\chi)$ are significantly different from their counterparts for, say, a Lennard-Jones potential.

*8.13 (a) For the attractive part of a Lennard-Jones potential, one finds that $\chi(b) \approx -C/b^6$, where $\chi(b)$ is the deflection function and C is a constant. Given this result, derive an expression for the classical differential cross section $I(\chi)$.

(b) Will the expression derived in (a) be valid for large or small angles and for large or small energies? Explain your answers briefly.

*8.14 Consider the following cross sections, the first two for the reactive system K + HI and the third for the isoelectronic K + Xe:

$$K + HI \rightarrow K + HI \qquad K + HI \rightarrow KI + H \qquad K - Xe$$

(a) Why are no "wiggles" present for the K–HI elastic cross section shown in the first diagram?

(b) Why does the elastic cross section for K–HI drop off much more rapidly at large χ than it does for K–Xe?

(c) Does reaction occur mainly at small or large impact parameters?

*These problems have been provided by Dr. R. G. Gilbert, Department of Theoretical Chemistry, University of Sydney, Australia.

Experimental Determination of New Kinetic Parameters

In the preceding chapters, we developed the relationships between macroscopic chemical-kinetic observables (specifically, rate constants) and microscopic quantities of interest in chemical dynamics (namely, reactive scattering cross sections and reaction probabilities). We then saw how the latter quantities may be derived from a potential energy surface for the reacting species. In this chapter, we shall consider some of the important experimental techniques for measuring such quantities.[1] These include molecular beam scattering and state-resolved spectroscopic techniques. We then examine the reaction $F + H_2$, which has been characterized in extensive detail by use of these techniques. Afterwards, some general principles relating potential energy surfaces to reaction properties are discussed, and we consider how application of the principle of detailed balancing can be used to relate specific energy release in exothermic reactions to selective energy consumption in endothermic reactions. Finally, we look at how detailed knowledge of state-specific rate constants stimulated the development of chemically pumped laser systems.

9.1 MOLECULAR BEAM SCATTERING

One of the first approaches to moving the experimental study of chemical reactions from the level of macroscopic kinetics to that of detailed dynamical studies was the crossed molecular beam method. This is essentially a realization of the scattering experiment discussed in section 6.1. The principles of the method can be discussed with the aid of Figure 9–1. The experiment consists of intersecting two beams of molecules (A and B) and catching the product molecules C in a suitable detector. While this may seem simple, there are a number of constraints which must be met in order for the experiment to be successful.

The first of these constraints reflects the fact that the critical pieces of information obtained from the experiment are the angular distribution of product molecules and the velocity with which these product molecules leave the scattering center. In order for this information to be preserved, the product molecules must not undergo

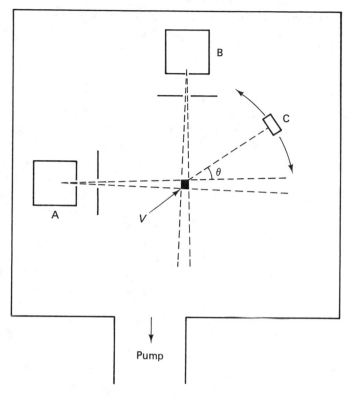

Figure 9–1. Schematic of a molecular beam-scattering apparatus. The reactant molecules, A and B, intersect in a small scattering volume V. Product molecules C are collected in the detector, which can be rotated around the scattering center. Various devices may be inserted in the beam paths A $\rightarrow$ V, B$_i$ $\rightarrow$ V, and V $\rightarrow$ C to select reactant and/or product species according to velocity or other properties. [From Steinfeld and Kinsey, *Progr. Reaction Kinetics 5,* 1 (1970).]

collisions with background gas molecules before they reach the detector. This means that the mean free path in the gas must be greater than the distance from the scattering center to the detector, which may be on the order of 10 to 30 cm. That is to say,

$$\text{Mean free path} = (2^{1/2} \pi n d^2)^{-1} > 30 \text{ cm}$$

Therefore, the gas density must be

$$n < 1/(1.4 \times 3 \times 30 \times 30 \times 10^{-16})$$
$$< 3 \times 10^{12} \text{ cm}^{-3} \approx 10^{-4} \text{ torr}$$

This is not a severe constraint; molecular beam-scattering chambers ("cans", in the trade) are typically run at pressures below 10^{-6} torr. In fact, *differential pumping* is frequently employed to reduce pressures in parts of the chamber to 10^{-10} torr or less, for reasons that will be described shortly.

The second constraint is more serious and has to do with the net flux of product molecules. This can be estimated[1] from equation 6–1, which is reproduced as follows for convenience:

$$\frac{dN_C}{dt} = n_A f_A(v_A) n_B f_B(v_B) I_R(v, \theta, \phi) v V \qquad (9\text{–}1)$$

Let us take typical beam densities of 10^{11} molecules cm^{-3} and make use of the entire velocity distribution $f(v)$. Also, let the reactive cross section $I_R = 10^{-15}$ cm^{-2}, the average relative velocity $v = 10^5$ cm sec^{-1}, and the beam intersection volume $V = 10^{-2}$ cm^3. This gives a total production rate of 10^{10} C molecules per second. Only a small fraction of these are intercepted by the detector, however, which may subtend a solid angle element $d\Omega = 10^{-4}$, so that a flux of 10^6 molecules sec^{-1} *at the detector* may be anticipated. This may still seem like a lot, but it must be considered in light of currently available detection techniques.

The simplest collection technique that can be imagined is actual collection of the product. However, 10^6 molecules sec^{-1} amounts to only a few nanograms of material per year, or perhaps a few monolayers per year coated onto a plate. Nevertheless, this method has been used on occasion. A far more efficient procedure is to convert the molecular current into an electrical current which can then be measured. There are two principal ways of doing this. Historically, the first method to be employed was a thermal ionization technique: when alkali metal atoms impinge on a hot metal filament, they lose an electron which can be collected as a negative current. Furthermore, by judiciously damaging the filament with hydrocarbon vapors, it can be rendered selective, so that only atomic alkalis, and not their salts, will be detected. This surface ionization technique is very efficient; unfortunately, it is applicable only to the alkali metals (K, Rb, Cs), which explains why virtually all the early molecular beam studies were carried out using these species. A more general method, the so-called universal detector, uses mass spectrometry: electron-impact ionization followed by a quadrupole field to select individual ions by their masses. The problem with this technique, however, is that the ionization efficiency is only about 10^{-4}, so that the original 10^6 product molecules per second have now become a few hundred ion counts per second. In order to measure these accurately, the detection region of the beam chamber must be pumped down to 10^{-10} torr or less to eliminate background ions, as mentioned earlier.

An alternative detection technique makes use of laser-excited fluorescence of the product molecules[2] and will be discussed further in section 9.2. While this may be a much more efficient detection method than electron impact ionization, it, too, is restricted to a small set of product molecules.

The first beam experiments were done using *effusive* beams, that is, molecular flow from low-pressure gas reservoirs or ovens. It was found that much higher beam fluxes could be obtained using *supersonic* sources,[3] which result from releasing a high-pressure gas through a small nozzle. Under such conditions, the flow is hydrodynamic rather than molecular. One consequence of this kind of flow is that the molecules are accelerated to many times their thermal velocity in the forward direction; hence the name supersonic. Another consequence is that the energy required for this acceleration is drawn from the transverse velocity components and the internal (vibrational, rotational) energy of the molecules. The net effect is a fast but very cold molecular beam. This cooling has been put to advantage in the spectroscopy of complicated molecules;[4] one disadvantage is that, at these low internal temperatures, the molecules tend to stick together to form clusters. This can be inconvenient if one is trying to perform scattering experiments on individual, uncomplexed molecules. Molecular beamists are nothing if not resourceful, however, and an industry has sprung up which studies the spectroscopy and scattering of molecular clusters.

Other than supersonic acceleration, there are few methods available for controlling the relative velocities of the beam particles. Amdur and co-workers[5] used an ionization–acceleration–neutralization scheme to produce fast beams of inert gas atoms, and thus to measure repulsive interaction potentials. This method works well only for high particle velocities, however. Portions of the velocity distributions can be selected by inserting rotating slotted disks in the beam paths which pass only molecules moving at certain speeds; doing so, however, throws away a substantial fraction of the total beam flux and reduces even further the signal at the detector.

To see what sort of information can be obtained from a molecular beam-scattering experiment, let us consider one of the very first such reactions studied,[6] viz.,

$$K + CH_3I \longrightarrow KI + CH_3$$

First, we need to account for the total energy available in the reaction system. The major contribution is made by the reaction exothermicity ΔH^0, which can be estimated as the difference in energy between the newly formed K–I bond and the broken H_3C–I bond, approximately 88 kJ/mol. To this is added the relative translational energy of the particles in the two beams and the mean thermal excitation of the reactants, in this case vibration and rotation of the CH_3I. This total available energy Q must be partitioned between the translational energy of the products and their internal, i.e., vibrational and rotational, energy. If thermal effusive beams are employed, the reactant translational energy is about 10 kJ/mol, and CH_3I vibration and rotation add another 2 kJ/mol, to give

$$Q \approx 88 + 10 + 2 = 100 \text{ kJ/mol}$$

Interpretation of molecular beam-scattering experiments is most easily done with the aid of a Newton diagram, shown for the reaction $K + CH_3I$ in Figure 9–2. This is essentially a velocity space construction in the laboratory rest frame. The two initial velocity vectors are at right angles to each other, since this is the orientation of the beam sources in the experimental apparatus; the K atom vector is much longer than that of CH_3I, because it is produced in a heated oven and in fact contributes most of the relative translational kinetic energy in the reaction. The velocity of the center of mass, a constant of the motion, is given by

$$v_{COM} = \frac{m_1 v_1 + m_2 v_2}{m_1 + m_2} = \frac{m_3 v_3' + m_4 v_4'}{m_3 + m_4} \tag{9–2}$$

where the subscripts 1 and 2 refer to the reactants (K and CH_3I) and the subscripts 3 and 4 and the primed velocities refer to the products (KI and CH_3). The relative velocities in the center-of-mass coordinate system are

$$w_R = v_1 - v_2$$

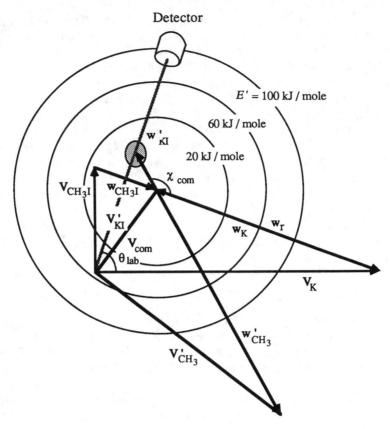

Figure 9–2. Newton diagram for the reaction K + CH$_3$I. The concentric circles are labeled by the amount of translational energy in the products.

and

$$\mathbf{w}'_R = \mathbf{v}'_3 - \mathbf{v}'_4 \qquad (9\text{–}3)$$

The relative velocity components are found by weighting these quantities by the inverse mass ratio, i.e.,

$$\mathbf{w}_1 = \mathbf{w}_R \frac{m_2}{M} \qquad \mathbf{w}'_3 = \mathbf{w}'_R \frac{m_4}{M}$$

$$\mathbf{w}_2 = -\mathbf{w}_R \frac{m_1}{M} \qquad \mathbf{w}'_4 = -\mathbf{w}'_R \frac{m_3}{M} \qquad (9\text{–}4)$$

where the total mass $M = m_1 + m_2 = m_3 + m_4$.

We now make use of the two basic conservation laws of classical mechanics to find the range of final velocity vectors and scattering angles (θ_{lab}, χ_{COM}) compatible with the available reaction energy. The first law, conservation of linear momentum, is automatically satisfied by the definitions of the relative velocity components given in equation 9–4:

$$m_1 w_1 = -m_2 w_2$$
$$m_3 w'_3 = -m_4 w'_4 \tag{9-5}$$

The second law, conservation of total energy, provides an important additional constraint:

$$m_1 w_1^2 + m_2 w_2^2 = 2E_{\text{transl}}$$
$$m_3 w_3'^2 + m_4 w_4'^2 = 2E'_{\text{transl}} = Q - E'_{\text{vib,rot}} \tag{9-6}$$

The concentric circles in Figure 9–2 show the allowable ranges of w'_{KI} for various specified values of E'_{transl}.

What is observed experimentally is that the KI product peaks at angles θ of 70°–90° in the laboratory frame. This is consistent with at least two possible scenarios:

1. a very low value of E'_{transl}, say, 10–20 kJ/mol, with final velocity vectors peaked sharply *backwards* in the center-of-mass frame.
2. a value of E'_{transl} near the maximum of 100 kJ/mol, with χ_{COM} near 90°.

The molecular beam method, in its simplest form, cannot distinguish between these possibilities, or, for that matter, any intermediate possibilities. On the other hand, the kinematic constraints expressed by equations 9–5 and 9–6 preclude the possibility of the KI product being forward-scattered in the center-of-mass frame, and, indeed, no such forward scattering is observed.

In order to pin down the reaction dynamics, additional measurements must be performed. These include:

1. Measurement of the product CH_3, but this is very difficult.
2. Velocity analysis of the product KI, by inserting rotating slotted disks between the beam intersection volume and the detector.
3. Measurement of out-of-plane scattering. The experiment described uses coplanar beam sources and detectors. Because of the cylindrical symmetry of the center of mass (see Figure 8–5), there will be significant out-of-plane scattering. In the first scenario presented previously, the KI product will be confined to small out-of-plane angles, while in the second scenario, the out-of-plane scattering will be a much larger fraction of the total.
4. As an alternative to these kinematic techniques, the product *internal* energy distributions can be measured by a technique such as laser-induced fluorescence.

Results of the investigations of a number of alkali halide-forming reactions, primarily using methods (2) and (4), reveal that most of the product is in fact backward-scattered, with a small value of E'_{transl}. Most of the reaction exothermicity appears in vibration of the newly formed metal-halogen bond. One way of presenting the data is by transforming to center-of-mass coordinates using the Jacobian transformation[7]

$$I_R(\chi) = I_R(\theta_{\text{lab}}) \left| \frac{d\Omega_{\text{lab}}}{d\Omega_{\text{COM}}} \right| \tag{9-7}$$

A presentation of this type is shown in Figure 9–3.

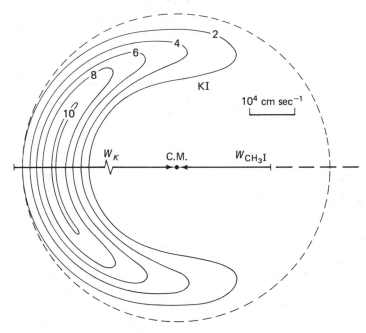

Figure 9–3. Contour representation of reactive scattering in center-of-mass coordinates for K + CH$_3$I system at a translational energy $E_T = 11.6$ kJ/mol. [From A. M. Rulis and R. B. Bernstein, *J. Chem. Phys. 57*, 5497 (1972)].

Newton Diagram analysis provides information on the kinematics of the reaction. We also want to find the magnitude of the reactive cross section, and there are several methods available for doing this.[8] The most direct method would appear to be simply measurement of the total product flux; combined with a knowledge of particle densities in the beams, use of equation 9–1 should then provide values for the differential cross section I_R. The equation for I_R can then be integrated with respect to the scattering angle to provide the total reactive cross section σ_R. There are a number of problems with this procedure, however. The most important is the difficulty of attaining absolute calibration of the detectors, so that absolute particle fluxes cannot be established with precision. Also, it is not generally possible to measure I_R over a complete range of scattering angles, and the beam intersection volume V is difficult to define precisely.

Perhaps the most accurate way of determining total reactive scattering cross sections is by carrying out *elastic* scattering measurements on the same system. The principle is as follows. Most molecular interactions can be approximated by a Lennard-Jones or similar potential function. The deflection function, and therefore the elastic scattering cross section, from a Lennard-Jones potential is well known (see Figure 8–9). Thus, since scattering at small deflection angles, corresponding to

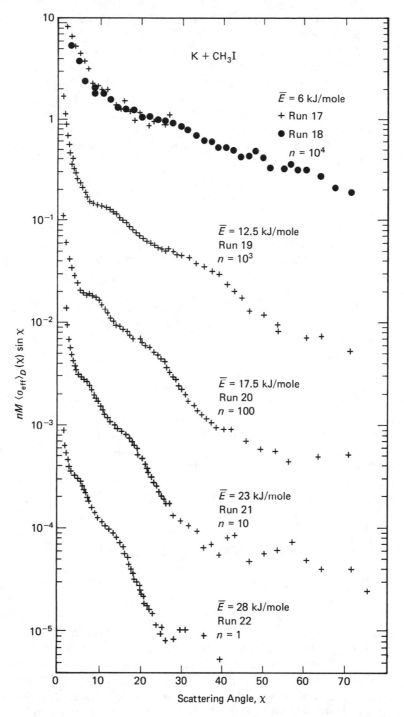

Figure 9–4. Experimental differential elastic scattering cross sections for the system K + CH₃I. The factor of n is introduced to displace successive plots vertically. [From M. Ackerman, et al., Ninth Symp. (Int'l.) on Combustion (New York: Academic Press, 1963).]

large impact parameters, is essentially completely elastic, a potential function can be fit to small-angle scattering in a reactive system. Alternatively, a simulant can be used to determine the elastic scattering, such as Xe atoms for CH_3I. When either of these methods is employed, it is observed that elastic scattering at large angles falls below the predicted curve in reactive systems, the difference being due to reactive scattering at the smaller impact parameters. Figure 9–4 shows typical data[9] for K + CH_3I. The procedure is then to calculate the reaction probability at a specified relative kinetic energy $\bar{E}$ over a range of scattering angles χ from the calculated elastic scattering cross section $I_{el}(\bar{E}, \chi)$ and the observed cross section $I_{obs}(\bar{E}, \chi)$ using the formula

$$P_R(\bar{E}, \chi) = \frac{I_{el}(\bar{E}, \chi) - I_{obs}(\bar{E}, \chi)}{I_{el}(\bar{E}, \chi)} \qquad (9\text{–}8)$$

Subsequent application of equations 8–9 and 8–10 allows determination of the total reactive cross section σ_R.

For the reaction K + CH_3I, the reaction probability P_R is found to be nearly equal to unity at small impact parameters and drops off a little ways beyond the Lennard-Jones interaction parameter σ_{LJ}. The drop-off is not as sharp as in the reactive-hard-sphere collision model: the reaction threshold energy is 5–10 kJ/mol, and the cross section attains a maximum value of approximately 40 A^2 at relative kinetic energies above 50 kJ/mol.

Molecular beam-scattering measurements have also been applied to the reaction D + $H_2 \rightarrow$ HD + H, discussed in chapter 8, using a "universal" mass spectrometer detector. The experiments, which were carried out over a period of 25 years,[10] provided complete angular and velocity resolution of the product. The results were in very good agreement with both classical trajectory and quantum mechanical calculations for this system. The total reactive cross section at a relative collision energy of 1.5 eV (145 kJ/mol) was found to be 1.7 $\pm$ 0.8 $Å^2$, using the absolute calibration method. This figure agrees quite well with an independent determination using the hot-atom technique;[11] the result of these experiments[12] gave $\sigma_R = 1.1 \pm 0.2$ $Å^2$. The experimental results are also in remarkably good agreement with the classical trajectory calculations described in chapter 8.

9.2 STATE-RESOLVED SPECTROSCOPIC TECHNIQUES

Another important class of techniques for determining microscopic properties of chemical reactions is spectroscopic measurement. Spectroscopic measurement techniques complement the molecular beam method discussed in the preceding section in that they generally provide thermally averaged rate coefficients rather than scattering cross sections. Their advantage over molecular beam scattering is that they are capable of extremely high energy resolution, so that rotational and vibrational state distributions of the product can be determined. In principle, the combination of spectroscopic detection with molecular beam techniques can furnish both energy and angular resolution. Although experiments using these techniques are difficult, they are now being carried out in well-equipped laboratories. The principles of the methods discussed in this section are essentially the same as those described in chapter 3 on the techniques of kinetic measurements, which should be reviewed at this point.

9.2.1 Infrared Chemiluminescence

Observation of radiation from excited product molecules is a clear indication that these molecules are produced in excited states. If the excitation is primarily vibrational, then the radiation will appear in the infrared region of the spectrum, i.e., between 3 μm and 15 μm (see chapter 3). This technique has been applied extensively to hydrogen-halogen reactions, producing HX molecules in excited (v, J) states. In order to be able to identify populations in individual states, two experimental criteria must be met. First, the spectral bandwidth at the detector must be sufficiently narrow to resolve the emission bands; to accomplish this, the infrared filter indicated in Figure 3–8 must be replaced by a monochromator or a tunable filter, such as a circular variable filter. The second requirement is dictated by the rapid deactivation of excited vibration-rotation states due to collisions. In order to observe the *nascent* distribution of reaction products, that is, the distribution unrelaxed by collisions, the molecules must be able to radiate prior to suffering a deactivating collision. Since radiative lifetimes in the infrared are on the order of milliseconds, this means in turn that the experiments must be carried out at extremely low pressures, which of course reduces the signal strength. One way of ensuring that both of these conditions are met is through the use of the *arrested relaxation* method devised by Polanyi.[13] In this technique, the reaction is carried out in a chamber whose walls are maintained at a very low temperature, e.g., 20 K (by means of liquid H_2). This has the effect of *cryopumping* the chamber so that all products and unreacted starting materials are condensed on the walls; only those product molecules that radiate before being pumped away are observed. The low temperature also has the effect of reducing thermal noise and black-body background in the detector and filters, making it less difficult to observe weak infrared chemiluminescence signals. Results on product distributions in the reaction $F + H_2$, obtained using this technique, will be discussed in section 9.3.

9.2.2 Laser-induced Fluorescence

The LIF technique described in chapter 3 is very well suited to the determination of product state distributions, since the spectroscopic resolution is afforded by the narrow-band dye laser. The principal limitation, of course, is that the technique is restricted to those product species which display visible- or ultraviolet-excited fluorescence. Also, in order to observe the nascent distribution, the molecular density must be kept sufficiently low so that collisional relaxation does not occur. Improvements in laser technology are steadily expanding the range of species that can be studied by LIF. For example, the $H + D_2$ cross section measurements reported by Gerlach-Meyer[12] were done by detecting the H and D atoms with vacuum-ultraviolet radiation at the Lyman-α wavelength, generated by frequency-tripling the output of an excimer-pumped dye laser.

9.2.3 Raman Techniques

An alternative to LIF which has proven useful in a number of cases is the use of Raman spectroscopy for product detection. Of course, spontaneous Raman scattering is much too weak to be useful for observing the low densities of product molecules required to suppress collisional relaxation, but laser-based Raman techniques, such as stimulated Raman gain and coherent anti-Stokes Raman spectroscopy (CARS),[14] can provide the necessary sensitivity. As with LIF, the state resolution is provided by the narrow-bandwidth lasers used to excite the Raman spectrum. The measurements of product distributions in the reaction $H + D_2$ reported by Gerrity and Valentini[15] were carried out using the CARS technique to detect the HD product molecules.

9.2.4 Resonant Multiphoton Ionization

The resonant multiphoton ionization technique described in Chapter 3 is also a very good probe for product state distributions, since ionization currents can be detected with high sensitivity. While single-photon ionization spectra are typically broad and featureless, offering little or no state specificity, the presence of a resonant intermediate state in the MPI process permits isolation of individual features with a resolu-

Figure 9–5. Resonant multiphoton ionization spectrum of NO in the $v = 1$ state produced in the flowing afterglow reaction of nitrogen atoms with O_2. The upper trace is the experimental spectrum, and the lower trace is the synthesized spectrum calculated for saturated two-photon resonant ionization [From I. C. Winkler, et al., *J. Chem. Phys. 85,* 890 (1986).]

tion comparable to that of LIF. An example is shown in Figure 9–5, which illustrates detection of nitric oxide in excited vibrational states produced by the reaction[16]

$$N(^4S) + O_2 \longrightarrow NO(v) + O$$

In addition to the vibrational states, rotational and spin-orbit states are resolved in the MPI spectrum.

In a further development of product detection by MPI, a microchannel plate is used to image the spatial distribution of the photoions.[17] This imaging technique provides direct information on angular and energy distribution of the products.

9.2.5 Fourier Transform Doppler Spectroscopy

One limitation of the spectroscopic techniques discussed in this section is that, while they provide internal-state resolution, they give no information on product velocity distributions. A way to obtain that information has been developed[18] using laser-induced fluorescence detection of scattered reaction products with a narrow band laser. A Fourier transform of the Doppler profile of the fluorescence excitation features gives the three-dimensional velocity distribution.

9.3 AN EXAMPLE OF STATE-TO-STATE KINETICS: THE F + H₂ REACTION

One reaction for which the microscopic rates have been characterized in considerable detail[19] is

$$F + H_2 \longrightarrow HF + H$$

The reaction proceeds along the potential energy surface depicted in Figures 7–8 and 7–9. Not only is the surface exothermic, but it is also highly asymmetric, in that the local maximum along the reaction coordinate (the transition state, as defined in chapter 10) occurs while the reaction complex is very close to the $F + H_2$ configuration. The consequence of this is that most of the reaction exothermicity is channeled into vibration of the product HF molecule. This may be understood by examining the trajectory shown in Figure 9–6(a); energy is released while the H–F distance is decreasing, and this energy appears in vibration of the H–F bond.

The reaction exothermicity may be calculated from the difference between the H–F and H–H bond dissociation energies, viz., 5.869 eV − 4.478 eV = 1.391 eV = 130 kJ/mol. In addition, there is a small activation barrier along the reaction coordinate, amounting to an additional 5 kJ/mol. If this entire amount of energy were channeled into product vibration, it would be sufficient to produce HF in the $v = 3$ state, as indicated in Figure 9–6(b). In addition to product vibration, some of the energy can appear as rotation of the HF molecule.

The HF product energy distributions have been measured[20] using the infrared chemiluminescence technique described in the preceding section. The partition of energy among the various degrees of freedom of the reaction product can be conveniently expressed by the fractions f_v, f_R, and f_T, where, for $F + H_2$, the average values are

$$\langle f_v \rangle = \frac{\text{product vibrational energy}}{Q} = 0.66$$

$$\langle f_R \rangle = \frac{\text{product rotational energy}}{Q} = 0.08$$

and

$$\langle f_T \rangle = \frac{\text{product translational energy}}{Q} = 0.26$$

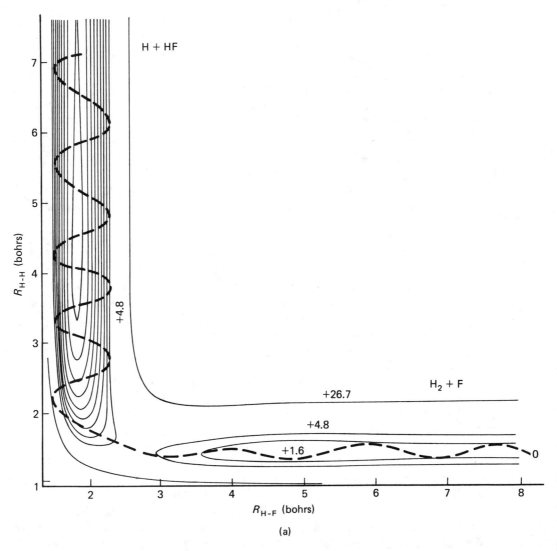

Figure 9–6. (a) F + H₂ reactive potential energy surface with a typical trajectory leading to vibrational excitation of HF.

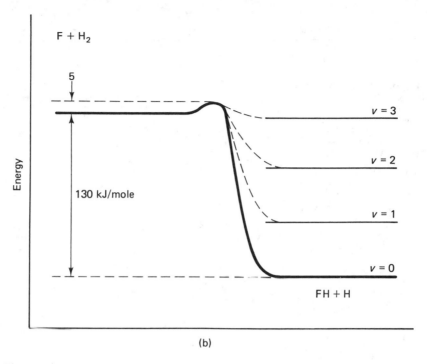

Figure 9–6. (b) Profile along $F + H_2$ reaction coordinate, showing accessible HF vibrational energy levels.

Q is the total amount of energy available in the reaction and is equal to the sum of the reaction exothermicity, the activation energy, and any initial reactant excitation; clearly,

$$f_v + f_R + f_T = 1 \qquad (9-9)$$

A convenient way of displaying the preceding quantities is the *triangle plot*, shown in Figure 9–7 for the $F + H_2$ reaction. In such a plot, f_v and f_R are plotted on orthogonal axes; then, by virtue of equation 9–9, the f_T scale must lie along the bisector of the triangle. The contours on the plot indicate the region of f-space in which most of the product is found. Of course, it must be remembered that the accessible energy states are not continuously distributed on the f_v–f_R plane, because vibrational and rotational energies are quantized. The value of the contour line at any discrete v, J-state gives the probability of a HF molecule being produced in that state. Plainly, for the $F + H_2$ reaction, the HF molecules are produced with appreciable vibrational excitation, but relatively little rotation. By contrast, a statistical distribution of reaction products would favor rotational and translational excitation of the products because of the higher statistical weights associated with those degrees of freedom. $F + H_2$ is thus an excellent example of *specific energy release* in a chemical reaction, in which energy is channeled into degrees of freedom having relatively low statistical weights. The departure from the statistical, or "prior," distribution is considered quantitatively in chapter 13.

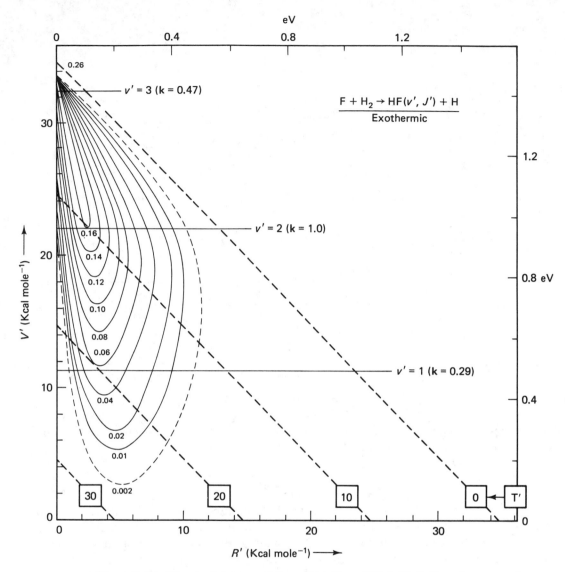

Figure 9–7. Triangle plot representation of F + H$_2$ → HF(v', J') + H product energy distribution. The v' = 1, 2, and 3 levels are drawn in at their quantized energies. [Adapted from Polanyi and Schreiber, *Faraday Disc. Chem. Soc. 62*, 267 (1977).]

Other measurements of the rates and product distributions of the F + H$_2$ reaction have been made using flow-tube (see chapter 3) and molecular beam techniques. The results of the latter deserve comment. The reaction

$$F + D_2 \longrightarrow DF(v) + D \qquad (9\text{--}10)$$

was studied in a crossed molecular beam,[21] and the DF product translation energy was deduced from the scattering angle by means of a Newton diagram. In this reaction, vibrational levels up to $v = 4$ can be populated because the DF vibration frequency is 27% smaller than that of HF. Assuming that there is little rotational excitation in the product, the vibrational energy distribution can be obtained using equation 9–9. The results are as follows:

	v	0	1	2	3	4	Reference
		Branching ratios, $k(v)/k_{total}$, for F + $D_2 \rightarrow$ DF(v) + D					
Infrared chemiluminescence		—	0.10	0.24	0.38	0.27	20
Molecular beam		0.01	0.02	0.04	0.21	0.73	21

The molecular beam results clearly differ from those obtained by infrared chemiluminescence. The latter, of course, cannot detect molecules in $v = 0$, but all measurements and calculations indicate that few molecules are produced in that state. The difference is that there is peak at $k(3)$, while the molecular beam results show a monotonic increase with most molecules produced in $v = 4$. The reason for the difference is that the beam measurements were carried out using fast F and D_2 beams, so that the center-of-mass collision energy was about 11 kJ/mol. By contrast, the infrared chemiluminescence measurements were carried out at or below room temperature, so that the average relative kinetic energy was only about 2.3 kJ/mol. This illustrates the importance of taking account of experimental conditions when attempting to compare measurements using different techniques.

9.4 SOME GENERAL PRINCIPLES CONCERNING ENERGY DISPOSITION IN CHEMICAL REACTIONS

In order to attain a greater understanding of the dynamics of simple reactions of the kind A + BC, such as the F + H_2 reaction discussed in the preceding section, Polanyi has carried out an extensive series of calculations on model potential surfaces.[22-24] In particular, classical trajectory calculations are carried out on a simple LEPS potential surface. The height and location of the energy barrier in the reaction coordinate are systematically varied, and the effects on the results of the trajectory calculations are noted. The purpose of such calculations is not to simulate an actual chemical system, but rather to identify those features of the potential energy surface that affect energy consumption and disposal in bimolecular reactions.

The feature of the energy surface that appears to be most important in determining energy distribution in reaction products is the location of the barrier along the reaction coordinate. An "early" or "attractive" barrier, which occurs in the entrance channel while the reactants are approaching each other, generally tends to favor vibrational excitation of the product. Conversely, a "late" barrier, which occurs in the exit channel where the products are separating, tends to lead to low product vibrational excitation. The basis for this is relatively easy to see. In the first case, the energy of the exothermic reaction is released while the A–B distance, which is the new chemical bond, is changing. In the second case, the energy is released only after the A–B bond has formed, and the B–C distance is changing; this motion corresponds to separation of products, so the energy is channeled into relative translation.

In order to make such visualizations work reliably, however, we must use properly mass-weighted coordinates (see chapter 8). The affine transformation which accounts for the various mass combinations tends to compress the angle between the entrance and exit channels, coupling the vibrational and translational degrees of freedom. Thus, the product energy distribution in a reaction in which a light atom attacks a heavy molecule (such as $H + F_2$) differs from that in which the masses are reversed ($F + H_2$).

The location of the energy barrier also plays an important role in selecting the distribution of reactant energy most likely to lead to reaction. Translational energy is most effective for passage across an "early" barrier in the entrance channel, whereas

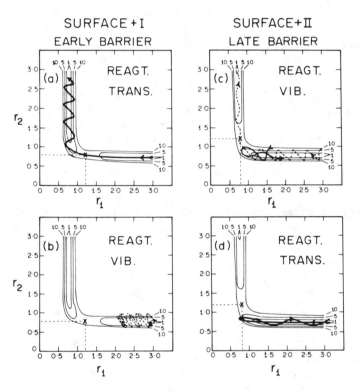

Figure 9–8. Specimen trajectories on a model potential surface, illustrating selective energy consumption (barrier height = 29.3 kJ mol^{-1} in all cases). In (a) and (b), the potential energy barrier is located in the reactant (entrance) channel side of the reaction coordinate. In (a), molecules with 37 kJ mol^{-1} of translational energy easily surmount the barrier, and 79% of that energy is converted into product vibration. In (b), molecules with 60 kJ mol^{-1} of vibrational energy are unable to cross the barrier. In (c) and (d), the barrier is located in the product (exit) channel. Now molecules with 67 kJ mol^{-1} of translational energy (d) cannot cross the barrier, while molecules with 31 kJ mol^{-1} of vibrational energy (c) can do so, if the vibrational phase is favorable. Note that reactant vibrational energy is converted into product translational energy in this case. [From J. C. Polanyi and W. H. Wong, *J. Chem. Phys. 51,* 1439 (1969).]

reactant vibrational energy far in excess of the barrier height may be ineffective for reaction. Conversely, a "late" barrier is best surmounted by vibrational rather than translational energy in the reactants. The dynamical basis for these tendencies can be seen with the aid of Figure 9–8. In the case of the early barrier, a vibrationally excited molecule (b) is just too busy rattling from side to side and does not have enough energy left to reach the top of the barrier; the molecule in (a), however, has all its energy in motion along the reaction coordinate and easily surmounts the barrier.* In the late barrier case, vibrationally excited molecules (c) with the correct phase can find their way to the top of the barrier, while molecules with rapid r_{AB} motion (d) simply slam into the repulsive inner wall of the potential surface and bounce back into the entrance channel.

Polanyi[24] has pointed out that these general principles were anticipated in 1955 by Hammond,[25] who postulated that for endothermic reactions the transition state** resembles the reaction products, while for exothermic reactions the transition state resembles the reactants. In the endothermic case this implies that the maximum of the potential barrier occurs in the product channel (late barrier), so that vibrational energy in the reactants will be more effective than translational energy in crossing the barrier. In exothermic reactions, where the barrier occurs in the reactant channel (early barrier), translational energy is sufficient to reach the product side. Bauer[26] has reviewed the effects of energy accumulation and disposal on chemical reaction, and how laser excitation of reactants may influence the rates.

9.5 DETAILED BALANCE REVISITED

We have encountered the principle of detailed balance several times in this text. In chapter 2, it was used to relate the forward and reverse rate constants of a reversible or opposing set of reactions (equation 2–17). In chapter 6, we saw how detailed balancing could be derived from the more general principle of microscopic reversibility, which embodies the invariance of classical physics under time reversal (equation 6–10). Here, we shall use an explicit form of the latter equation to find the relationship between detailed state-to-state rate constants for an exothermic reaction and those for the reverse endothermic reaction at the same total energy. This relationship was derived by Polanyi and co-workers[27] in 1969. It may be written as

$$k_r(v', J') = \frac{2\bar{J} + 1}{2J' + 1} \left(\frac{\mu}{\mu'}\right)^{3/2} \left(\frac{\bar{E}_T}{E'_T}\right)^{1/2} k_f(v', J') \qquad (9-11)$$

where relative translational energy $\bar{E}_T = E_{act} + \frac{3}{2}RT$, $E'_T = \bar{E}_T - \Delta H°$, $\bar{J}$ is the most populated rotational level at the initial temperature T, and μ' and J' refer to the products of the exothermic, or forward, reaction. Essentially, equation 9–11 says that, if the exothermic reaction favors products with high internal (v' or J') and low

*There is probably a moral here, if one looks hard enough.
**The concept of a transition state is thoroughly discussed in chapter 10; for the present purpose, we can simply regard it as the configuration of the reacting species at the top of the potential energy barrier in the reaction coordinate.

translational excitation (as in the F + H$_2$ reaction), then the reverse or endothermic reaction at the same total energy will be favored by the same distribution of energy in the reactants (HF + H in this case). That is, the *specific energy disposal* of the exothermic process is mapped into *selective energy consumption* in the reverse, endothermic process.

Figure 9–9 shows the energy consumption map for the reaction HF + H → F + H$_2$ at a total energy of 140 kJ/mol. Clearly, reaction is most probable from the $v = 3$ state of HF, with little rotational or translational excitation. Such states would be very inefficiently populated by simple thermal heating of a mixture of HF and H atoms; there would instead tend to be equipartition of energy, with approximately equal amounts of excitation in translational, rotational, and vibrational degrees of freedom. Observations such as these have led to suggestions for "laser-selective chemistry,"[28] in which reactions could be promoted by excitation of specific molecular states with a narrow-band laser. Attempts to find such reactions have not been very successful, however, primarily because of the rapid inter- and intramolecular relaxation processes which redistribute energy within the molecule.

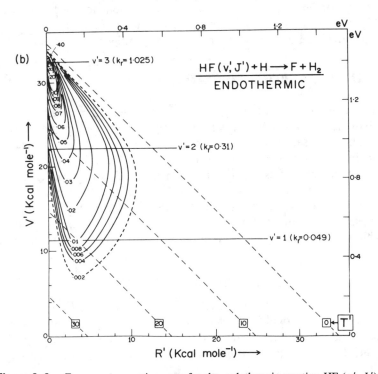

Figure 9–9. Energy consumption map for the endothermic reaction HF (v', J') + H → F + H$_2$ at a total energy of 140 kJ/mol, obtained by applying equation 9–11 to the data in Figure 9–7. [From Polanyi and Tardy, *J. Chem. Phys. 51,* 5717 (1969).]

9.6 CHEMICAL LASERS

The product HF vibrational population distribution described in section 9.3 is remarkable in that more molecules are produced in excited than in ground vibrational states. That is, $n(v = 2) > n(v = 1) > n(v = 0)$ for the nascent distribution. Such a distribution is said to be *inverted* with respect to the Boltzmann equilibrium distribution, in which the low-lying energy states are preferentially populated. A population inversion can be used to create a laser[29] if an allowed radiative transition connects the inverted levels, as is certainly the case for HF. Polanyi[30] first suggested the use of hydrogen-halogen reactions to create a laser pumped by an exothermic chemical reaction, which he termed a "chemical laser." Following this suggestion, a number of such systems were developed, with the hydrogen-fluorine system being one of the most efficient. The output of this laser occurs on HF vibration-rotation bands in the $2.8–3.0$-μm region of the infrared.

The behavior of the HF chemical laser depends on a complex interplay of chemical kinetics, gas flow, and optics. Pimentel and co-workers have made use of this interaction to learn about kinetic processes from the behavior of the HF laser under various operating conditions. The HF laser is also notable in that it can be scaled to very high output powers by simply increasing the flow of gases through the laser cavity. Some of these characteristics will be discussed further in chapter 14.

Chemical lasers operating on inverted electronic transitions have also been an object of study. These lasers would radiate in the visible rather than the infrared region of the spectrum. Although a number of chemiluminescent reactions are known, for most of them the inversion density (i.e., the amount of energy channeled into electronic excitation of the products) is too small to be of practical use for a laser.

A good summary of the practical aspects of chemical lasers may be found in a handbook edited by Gross and Bott.[31]

9.7 STATE-TO-STATE CHEMICAL KINETICS CAN BE HAZARDOUS TO YOUR HEALTH!

In this chapter we have seen that the techniques of modern experimental and computational physical chemistry can produce an enormous amount of detail on energy-, angle-, and state-resolved chemical reaction rates and cross sections. Indeed, it may now be possible to generate too much detail: the quantitative information produced by such measurements may be greater than the capacity of human minds to assimilate, or of traditional journals to publish. Attempts have thus been made to systematize the data by finding some general rules and principles that govern the behavior of chemical reactions. In chapter 13 we shall consider the information-theoretic approach to compacting such information by reducing large arrays of state-to-state kinetic quantities to one or a few parameters. Before doing so, however, we shall take up two more traditional approaches to chemical kinetics. These approaches treat reacting systems as being close to thermal equilibrium, rather than far removed from it. For bimolecular reactions, this treatment results in transition state theory, discussed in chapter 10; for unimolecular reactions, the corresponding treatment is the Rice-Ramsperger-Kassel-Marcus, or RRKM, theory, developed in chapter 11.

REFERENCES

1. Y. T. Lee, *Science 236,* 793 (1987).
2. J. L. Kinsey, *Ann. Rev. Phys. Chem. 28,* 349 (1977).
3. J. B. Anderson, R. P. Andres, and J. B. Fenn, in *Molecular Beams,* edited by J. Ross *Adv. Chem. Phys. 10,* 275–317 (1966).
4. J. I. Steinfeld, *Molecules and Radiation* (Cambridge, MA: MIT Press, 1985) pp. 307–309.
5. I. Amdur and J. E. Jordan in *Molecular Beams,* edited by J. Ross, *Adv. Chem. Phys. 10,* 29–73 (1966).
6. D. R. Herschbach, *Disc. Faraday Soc. 33,* 149 (1962).
7. R. D. Levine and R. B. Bernstein, *Molecular Reaction Dynamics* (Oxford: University Press, 1974), pp. 60–63.
8. J. I. Steinfeld and J. L. Kinsey, *Progr. Reaction Kinetics 5,* 1 (1970).
9. M. Ackerman, E. F. Greene, A. L. Moursund, and J. Ross, *Ninth Symp. (Int'l.) on Combustion,* (New York: Academic Press, 1963), p. 669.
10. R. Gotting, H. R. Mayne, and J. P. Toennies, *J. Chem. Phys. 85,* 6396 (1986).
11. R. Wolfgang, *Progr. Reaction Kinetics 3,* 97 (1965).
12. U. Gerlach-Meyer, K. Kleinermanns, E. Linnebach, and J. Wolfrum, *J. Chem. Phys. 86,* 3047 (1987).
13. K. G. Anlauf, P. J. Kuntz, D. H. Maylotte, P. D. Pacey, and J. C. Polanyi, *Disc. Faraday Soc. 44,* 183 (1967); J. C. Polanyi et al., *J. Chem. Phys. 57,* 1547, 1561, 1574, 4988 (1972).
14. Steinfeld, *op. cit.,* pp. 429–438.
15. D. P. Gerrity and J. J. Valentini, *J. Chem. Phys. 81,* 1298 (1984).
16. I. C. Winkler, R. A. Stachnik, J. I. Steinfeld, and S. M. Miller, *J. Chem. Phys. 85,* 890 (1986).
17. D. W. Chandler and P. L. Houston, *J. Chem. Phys. 87,* 1445 (1987).
18. J. L. Kinsey, *J. Chem. Phys. 66,* 2560 (1977).
19. J. C. Polanyi and J. L. Schreiber, *Faraday Disc. Chem. Soc. 62,* 267 (1977).
20. J. C. Polanyi and K. Woodall, *J. Chem. Phys. 57,* 1574 (1972).
21. T. P. Schafer, P. E. Siska, J. M. Parson, F. P. Tully, Y. C. Wang, and Y. T. Lee, *J. Chem. Phys. 53,* 3385 (1970).
22. J. C. Polanyi, *Accts. Chem. Research 5,* 161 (1972).
23. J. C. Polanyi and J. L. Schreiber, "The Dynamics of Bimolecular Reactions" in *Kinetics of Gas Reactions: Physical Chemistry—An Advanced Treatise,* Vol. 6A, edited by H. Eyring, W. Jost, and D. Henderson (New York: Academic Press, 1974) p. 383.
24. J. C. Polanyi, *Science 236,* 680 (1987).
25. G. S. Hammond, *J. Am. Chem. Soc. 77,* 334 (1955).
26. S. H. Bauer, *Chem. Reviews 78,* 147 (1978).

27. K. G. Anlauf, D. H. Maylotte, J. C. Polanyi, and R. B. Bernstein, *J. Chem. Phys. 51*, 5716 (1969); J. C. Polanyi and D. C. Tardy, *ibid.*, p. 5717.

28. A. M. Ronn, *Scientific American* v. 290, No. 5, 114 (May, 1979).

29. Steinfeld, *op. cit.*, pp. 308–320.

30. J. C. Polanyi, *Appl. Opt. Suppl. 2*, 109 (1965).

31. R. W. F. Gross and J. Bott, eds., *Handbook of Chemical Lasers* (New York: Wiley-Interscience, 1976).

PROBLEMS

*9.1 The total cross section for elastic scattering can be obtained from a single-beam experiment, in which a beam of molecules of species A at high velocity is passed through a rarefied gas of species B and the attenuation of the beam is measured. Although the molecules of the rarefied gas have random thermal velocities, if the velocity of the beam is sufficiently high, then the *relative* velocity, v_{AB}, is well approximated by that of the beam A. Let the densities of the beams be n_A and n_B, respectively, and let the beam of A have area α. Consider the attenuation in the beam intensity over a small increment of length dx.

B. Brunetti, G. Liuti, E. Luzzatti, and F. Vecchiocattivi [*J. Chem. Phys. 74*, 6734 (1981)] carried out single-beam studies for a beam of O_2 passed through an O_2 bath gas. Their data show that at 0.01 Pa pressure and a temperature of 85 K, with a path length of 10 cm, for a beam velocity of 1.0×10^3 m s^{-1}, the beam has an attenuation of 92% (i.e., $I/I_0 = 0.08$). For a velocity of 1.5×10^3 m s^{-1}, they found an attenuation of 86%.

(a) Evaluate σ for these two velocities.

(b) Do you think that the values obtained in (a) correspond to an ordinary molecular "size"?

(c) You will notice that σ decreases with increasing velocity in these experiments. Explain this in terms of collision dynamics.

(d) Brunetti et al. used their data on $\sigma(E)$ to deduce the O_2/O_2 intermolecular potential function. List two physical properties and one chemical property of oxygen that could be predicted from this potential function.

9.2 When a beam of Rb atoms from a source at 600 K intersects a beam of CH_3I at 90° from a source at 300 K (with no velocity selection), reaction occurs according to the formula

$$Rb + CH_3I \longrightarrow CH_3 + RbI$$

and the final kinetic energy E' of relative motion of the products peaks at about 8.3 kJ/mole. Assuming that the angle between the final relative velocity v'_r and the velocity v_c of the center of mass is 90°,

(a) Construct a vector diagram (Newton diagram) and indicate the in-plane laboratory angle where the peak intensity of RbI would be expected to appear.

(b) Account for any major differences between the figure obtained in (a) and that for $K + CH_3I$ (Figure 9–2 and 9–3).

(c) Assume that the in-plane peak intensity predicted in (a) has been experimentally observed. What additional experimental observations would be expected if a substantial fraction of the products had values of E' equal to about 21 kJ/mole?

9.3 A direct measurement of the threshold energy for the $D + H_2$ reaction may be obtained by photolyzing DI in the presence of H_2 and measuring the appearance of DH [Kupperman and White, *J. Chem. Phys. 44,* 4353 (1966)].

(a) If DI is photolyzed by ultraviolet light having a wavelength of 3,000 Å, what is the velocity of the D atom in the laboratory coordinate system? (The bond dissociation energy of DI is 3.098 eV.)

(b) What would be the relative kinetic energy in the collision center-of-mass system if the D atom produced in (a) struck an H_2 molecule? Assume the H_2 molecule is at rest in the laboratory frame.

(c) If the *longest* photolysis wavelength at which HD is observed is 3,340 Å, what is the energy threshold for the $D + H_2$ reaction? How does this value compare with the experimental activation energy and the theoretical potential barrier height?

*9.4 An experimental colleague in the Department of Quintessential Chemistry makes the following proposal to you for a research project. Correct his errors and misconceptions, suggest an explanation for the observed effect, and suggest how your explanation could be checked by a proper calculation. "We pass protons from a radioactive source into ethylene at low pressure and monitor the formation of C_2H_5. From the observed rate coefficient and from the hard-sphere formula $k = \pi d^2 (8 k_B T / m \pi)^{1/2}$, we deduce the value of the total cross section $\sigma = \pi d^2$. This assumes that the critical energy for the reaction is zero. The resulting cross section is for the reaction $H^+ + C_2H_4 \longrightarrow C_2H_5^+$; it is found to be considerably greater than the hard-sphere cross section for typical systems. We suspect that this is due to tunneling. We would like to carry out classical trajectory calculations to confirm this. For these trajectory calculations, we can use a potential function obtained from infrared spectral studies of $C_2H_5^+$ in polar solvents."

9.5 Reactions involving N_2, O_2, N, and O are important processes in the upper atmosphere. Answer each of the following questions concerning these reactions, using the potential-energy diagrams at the end of this problem where appropriate. Show briefly how you arrived at each numerical result requested in parts (a) through (i); the expected precision is that to which you can read off the diagrams. Note that 1 eV = 8,065.73 cm^{-1} = 23,061 cal/mole = 11, 605 degrees K$\cdot k_B$ = 1.6×10^{-12} ergs/molecule.

(a) Estimate the energy released in the reaction

$$O_2(X^3\Sigma_g^-) + N(^4S°) \longrightarrow NO(X^2\Pi) + O(^3P) \tag{1}$$

(b) Estimate the energy released in the reaction

$$O_2(X^3\Sigma_g^-) + N(^2D°) \longrightarrow NO(X^2\Pi) + O(^3P) \tag{2}$$

(c) What is the highest vibrational level, v_{max}'', of $NO(X^2\Pi)$ which could be populated in reaction (1)? In reaction (2)?

(d) Is the reaction

$$N_2(X^1\Sigma_g^+) + O(^3P) \longrightarrow NO(X^2\Pi) + N(^4S°) \tag{3}$$

likely, on energetic grounds?

(e) Is the reaction

$$N_2^*(A^3\Sigma_u^+) + O(^3P) \longrightarrow NO(X^2\Pi) + N(^4S°) \tag{4}$$

likely, on energetic grounds?

(f) Is NO stable with respect to disproportionation to N_2 and O_2?

(g) Estimate the longest wavelength of light that can be used to excite fluorescence from an electronically excited state of NO. Assume that the NO is originally in $v'' = 0$ of the X state.

(h) Estimate the longest wavelength of light that would be effective for ionization of NO in a one-photon transition.

(i) Estimate the longest wavelength of light that would be effective for resonant multi-photon ionization of NO.

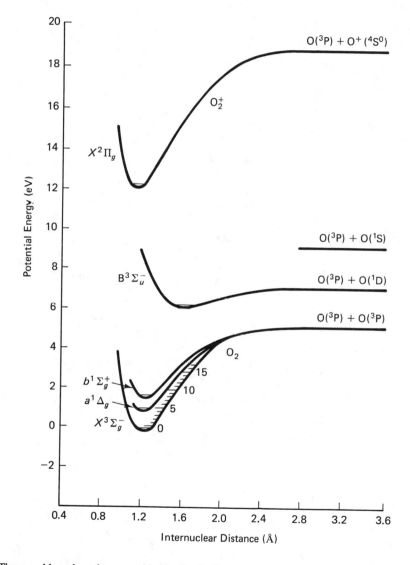

*These problems have been provided by Dr. R. G. Gilbert, Department of Theoretical Chemistry, University of Sydney, Australia.

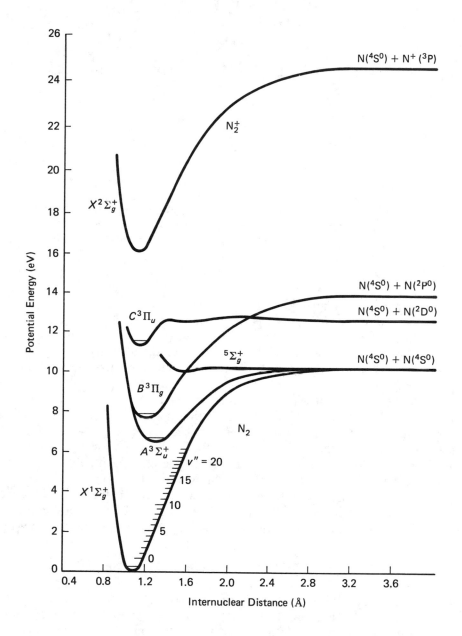

Statistical Approach to Reaction Dynamics: Transition State Theory

In the preceding chapters, we have seen how reaction rates can be calculated from molecular dynamics, given a potential energy surface for the reaction. These calculations, besides being complex and frequently difficult, are often inexact because (1) classical dynamics is only an approximation to the quantum behavior of atomic systems, and (2) sampling of initial conditions is limited. Exact quantum calculations are possible, but they are limited to reactive systems and energies for which the total number of available quantum states is small. Also, the results of such classical and quantum calculations are frequently far more detailed than the available experimental data, which may consist only of a thermal rate constant. Information on state-to-state reactive cross sections, scattering angles, and product energy partitioning may be generated in the calculations, but those properties are measured in only the most painstaking experiments.

The level of detail in a theoretical model should be consistent with that of the experimental observations to which the model will be applied. One approach to matching the information content of theory and experiment will be discussed in chapter 13; in this chapter, an approach is described which is based on the concept of a *transition state* intermediate between reactants and products. The reactants are assumed to be in thermal equilibrium with the transition state. The resulting theory, called *transition-state theory*, takes into account only the statistical properties of reactive systems, not the microscopic details of molecular collisions. Transition-state theory rate constants are determined using the principles of statistical mechanics.

10.1 MOTION ON THE POTENTIAL SURFACE

For a chemical reaction involving N atoms there are $3N - 6$ vibrational degrees of freedom at the saddle point. As discussed in section 7.5, one of these degrees of freedom corresponds to infinitesimal motion along the reaction path. The remaining $3N - 7$ degrees of freedom define vibrational motion orthogonal to the reaction path.

To study properties of the saddle point in more detail, consider a potential energy contour diagram for a collinear chemical reaction such as the one shown in Figure 7–5(a). For simplicity, consider three identical atoms (such as H + H-H) in a linear configuration (Figure 10–1). This collinear system has two orthogonal internal degrees of freedom, for which the coordinates are

$$\xi = r_2 + r_1$$

$$s = r_2 - r_1$$

The saddle point has a symmetric configuration at which $r_2 = r_1 = r_0$, and $\xi = 2r_0$ and $s = 0$. As shown in Figure 10–1, at the saddle point s defines the reaction path and ξ is perpendicular to the reaction path. Curves of the potential energy versus s, $U(s)$, and versus ξ, $U(\xi)$, are shown in Figures 10–2(a) and 10–2(b), respectively, for these s and ξ. For the reaction path potential $U(s)$, there is a maximum at the

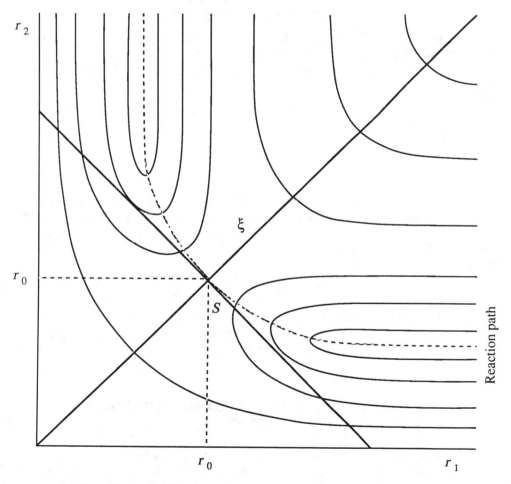

Figure 10–1. Saddle-point region of a symmetrical potential energy surface.

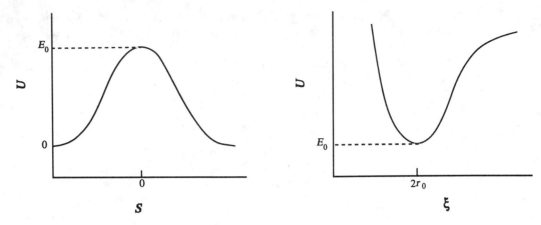

Figure 10–2. Sections of the potential energy surface showing (a) the maximum along the reaction path and (b) the minimum perpendicular to the reaction path.

saddle point. The potential for the coordinate orthogonal to the reaction path, $U(\xi)$, has a minimum at the saddle point.

The first and second derivatives of the potential at the saddle point are

$$\frac{\partial U}{\partial s} = 0, \frac{\partial^2 U}{\partial s^2} < 0$$

and

$$\frac{\partial U}{\partial \xi} = 0, \frac{\partial^2 U}{\partial \xi^2} > 0$$

since $\partial^2 U / \partial s^2 < 0$, the vibrational frequency for motion along the reaction path will be imaginary (see chapter 7, section 7.5). The vibrational frequency for the ξ coordinate will be positive. In the vicinity of the saddle point, a Taylor's series expansion of the potential is

$$U(\xi, s) \approx \frac{1}{2}C_\xi(\xi - 2r_0)^2 - \frac{1}{2}C_s s^2 + E_0. \tag{10–1}$$

The transition state theory (TST) is based on the postulate that the rate of transformation of systems from reactants to products is given solely by passage in the forward direction from coordinates $r_1 > r_2(s < 0)$ to $r_1 < r_2$ $(s > 0)$. We now proceed to develop that theory.

10.2 BASIC POSTULATES AND STANDARD DERIVATION OF TRANSITION STATE THEORY

Transition state theory, introduced by Eyring[1] and by Evans and Polyani[2] in 1935, provided the first theoretical attempt to determine absolute reaction rates. In this theory, a transition state separating reactants and products is used to formulate an expression for the thermal rate constant. The relationship between transition state theory and dynamical theories was first discussed by Wigner[3] who emphasized that the theory was a model essentially based on classical mechanics. A number of assumptions are made in deriving the TST rate expression. The two most basic are the separation of electronic and nuclear motions, equivalent to the Born-Oppenheimer approximation in quantum mechanics, and the assumption that the reactant molecules are distributed among their states in accordance with the Maxwell-Boltzmann distribution. However, the following additional assumptions, which are unique to the theory, are also required:

1. Molecular systems that have crossed the transition state in the direction of products cannot turn around and reform reactants.
2. In the transition state, motion along the reaction coordinate may be separated from the other motions and treated classically as a translation.
3. Even in the absence of an equilibrium between reactant and product molecules, the transition states that are becoming products are distributed among their states according to the Maxwell-Boltzman laws.

In a rigorous formulation of transition state theory, the third assumption is not necessary, since it follows from the first.

It is customary to derive transition state theory by postulating a quasi-equilibrium between transition state species and reactants. This approach focuses on the third assumption listed. Another approach follows the dynamical formulation of Wigner and uses the first assumption, which is considered the "fundamental assumption" of transition state theory. Both derivations are given here, and they use the conventional transition state theory model in which the transition state is located at the saddle point on the potential energy surface. Later in the chapter, variational transition state theory, a more accurate method for choosing the transition state, is discussed.

In many presentations the term *activated complex* is used synonymously with "transition state." It is avoided here, however, because first, the word "complex" implies an entity which has a chemically significant lifetime, which the transition state does not have, and second, the collision complex which may be formed when two molecules collide (as in a molecular beam experiment) is often called an activated complex.

The third assumption, the "quasi-equilibrium hypothesis," is based on a simple physical interpretation. Suppose we have an elementary reaction

$$A + B \longrightarrow X^{\ddagger} \longrightarrow C \qquad (10\text{--}2)$$

where $X^{\ddagger}$ is the transition state, A and B are reactants, and C is the product. Suppose also that it is possible to define a small region at the top of the potential energy barrier such that all systems entering from the left pass through to products without turning back, and similarly all products entering from the right must pass through to reactants. This small region is often referred to as the *dividing surface* and is orthogonal to the reaction pathway. We consider two parallel dividing surfaces separated by a very small distance δ (see Figure 10–3). Systems within this small distance are by definition transition states; to the right are products, to the left reactants. Conse-

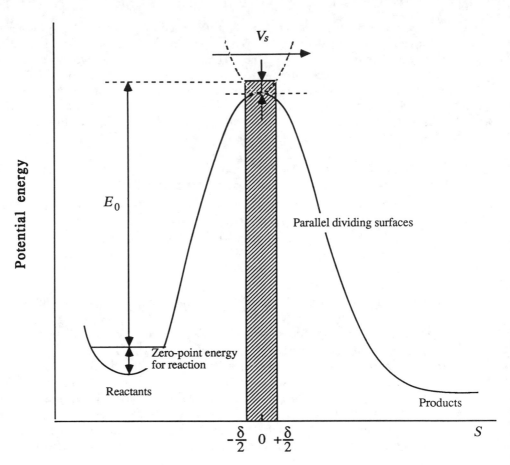

Reaction coordinate

Figure 10–3. Reaction coordinate profile.

quently, for a system in which reactants are in equilibrium with the products, there are two types of transition states: those moving from reactant to products, and those moving in the opposite direction from products to reactant. We designate the concentration of these by $N_f^{\ddagger}$ and $N_b^{\ddagger}$, where f and b stand for forward and backward, respectively. At equilibrium the rate of the forward reaction must be the same as the backward reaction; consequently, the concentrations of the transition states moving toward reactants and toward products must be equal. Hence,

$$N_f^{\ddagger} + N_b^{\ddagger} = N^{\ddagger} = K^{\ddagger}[\text{A}][\text{B}] \qquad (10\text{–}3)$$

Now, if all the products are suddenly removed from equilibrium, that is, if

$$N_b^{\ddagger} = 0 \qquad (10\text{–}4)$$

then, since the forward rate of reaction must be the same regardless of the back reaction rate, it follows that the concentration of the forward-moving transition states is the same as it would be for full equilibrium; hence,

$$N_f^{\ddagger} = K^{\ddagger}[A][B]/2 \qquad (10\text{--}5)$$

This is the "quasi-equilibrium hypothesis" of transition state theory.

To calculate the net rate of reaction, the rate (average velocity) at which transition states pass over the barrier to products is needed; i.e., we derive

$$\frac{dN}{dt}(\text{reactants} \longrightarrow \text{products}) = \frac{\delta N^{\ddagger}}{\delta t} \qquad (10\text{--}6)$$

The number of transition states per unit volume having velocity between v and $v + dv$ in *one* direction is represented by $\delta N^{\ddagger}$. The average time δt for the $\delta N^{\ddagger}$ transition states to cross the barrier is equal to the thickness δ of the dividing surface divided by the average velocity $\bar{v}_s$ at which the transition states traverse the dividing surface; that is,

$$\delta t = \frac{\delta}{\bar{v}_s} \qquad (10\text{--}7)$$

Hence,

$$\frac{dN}{dt} = \delta N^{\ddagger}\frac{\bar{v}_s}{\delta} \qquad (10\text{--}8)$$

From equation 10–5, the number of transition states crossing the dividing surface in the direction of products is one-half the total number $N^{\ddagger}$ of transition states, i.e.,

$$\delta N^{\ddagger} = \frac{N^{\ddagger}}{2} \qquad (10\text{--}9)$$

Therefore, the number of transition states crossing the barrier per unit volume in unit time is

$$\frac{dN}{dt} = \frac{N^{\ddagger}}{2}\frac{\bar{v}_s}{\delta} \qquad (10\text{--}10)$$

On the assumption that there is an equilibrium distribution of velocities, the average velocity of the transition state moving in one direction, e.g., the forward direction, is

$$\bar{v}_s = \frac{\int_0^{\infty} v_s e^{(-\mu_s v_s^2/k_B T)}\, dv_s}{\int_0^{\infty} e^{(-\mu_s v_s^2/2k_B T)}\, dv_s} = \left(\frac{2k_B T}{\pi \mu_s}\right)^{1/2} \qquad (10\text{--}11)$$

where μ_s is the reduced mass for motion through the dividing surface. Inserting this value for $\bar{v}_s$ into equation 10–10 gives

$$\frac{dN}{dt} = \frac{N^{\ddagger}}{2}\left(\frac{2k_B T}{\pi \mu_s}\right)^{1/2}\frac{1}{\delta} \qquad (10\text{--}12)$$

Since the transition state is in equilibrium with the reactants, it is possible to obtain $N^{\ddagger}$ from equilibrium statistical mechanics, namely, through

$$K_c^{\ddagger} = \frac{N^{\ddagger}}{[A][B]} \qquad (10\text{--}13)$$

This statement of equilibrium does *not* imply that the transition state is long-lived. This equilibrium constant can be expressed in terms of partition functions as

$$K_c^{\ddagger} = \frac{N^{\ddagger}}{[\text{A}][\text{B}]} = \frac{Q_{\text{tot}}^{\ddagger}}{Q_\text{A} Q_\text{B}} e^{-E_0/k_B T} \tag{10-14}$$

where $Q_{\text{tot}}^{\ddagger}$ is the equilibrium partition function per unit volume for the transition state, Q_A and Q_B are the same functions for the reactants, and E_0 is the energy of the lowest level of the transition state relative to that of the lowest level of the reactants (see Figure 10–3). Combining equations 10–12 and 10–14, we have, for the reaction rate,

$$\frac{dN}{dt} = \left(\frac{2k_B T}{\pi \mu_s}\right)^{1/2} \frac{1}{2\delta} \frac{Q_{\text{tot}}^{\ddagger}}{Q_\text{A} Q_\text{B}} e^{-E_0/k_B T}[\text{A}][\text{B}] \tag{10-15}$$

In the approaches of Eyring[1] and Evans and Polanyi,[2] the partition function for the reaction coordinate in the transition state is considered to be a translational function. Since this motion is assumed to be separable (as a consequence of the second assumption), the translational partition function is therefore separable as

$$Q_{\text{tot}}^{\ddagger} = Q_s Q^{\ddagger} \tag{10-16}$$

where Q_s is the partition function for the reaction coordinate motion and $Q^{\ddagger}$ is the partition function for all other $3N - 1$ degrees of freedom in the transition state. The translational partition function for motion in *one* dimension in a system of length δ is

$$Q_s = (2\pi \mu_s k_B T)^{1/2} \, \delta/h \tag{10-17}$$

This Q_s is not divided by 2, since it is the partition function for all the transition states (see equation 10–14). Substituting this expression into equation 10–16 and then the result into equation 10–15 gives

$$\frac{dN}{dt} = \frac{k_B T}{h} \frac{Q^{\ddagger}}{Q_\text{A} Q_\text{B}} e^{-E_0/k_B T}[\text{A}][\text{B}] \tag{10-18}$$

From chapter 1, we know that the experimental rate equation for a bimolecular reaction is

$$\frac{dN}{dt} = k[\text{A}][\text{B}] \tag{10-19}$$

Comparison of equation 10–18 and equation 10–19 gives the absolute rate coefficient for A + B → products,

$$k = k_{\text{abs}} = \frac{k_B T}{h} \frac{Q^{\ddagger}}{Q_\text{A} Q_\text{B}} e^{-E_0/k_B T} \tag{10-20}$$

Note that the artificial constructs δ and μ_s, as well as the factor $1/2$, have cancelled out of the absolute rate theory expression. The ratio $k_B T/h$, having the units of frequency and the magnitude 6.25×10^{12} sec^{-1} (for $T = 300$ K), is frequently termed the *frequency factor*. It is of the magnitude of the encounter frequency between the molecules in liquids, or of slow molecular vibrations (300 K is equivalent to 200 cm^{-1}), but the period $(k_B T/h)^{-1}$ is a little longer than gas-kinetic collision durations.

In order to be able to evaluate equation 10–20, we must be able to calculate the transition state partition function $Q^{\ddagger}$, using the statistical mechanical techniques discussed subsequently in section 10.6. To do that, we need to know the structural parameters of the transition state (specifically, the moment of inertia $I^{\ddagger}$), as well as its $3N - 7$ vibrational frequencies $\{\nu^{\ddagger}\}$. In practice, such parameters can only be estimated approximately. With detailed information about potential surfaces now becoming available from *ab initio* theory, however, we may expect to see more precise calculations of rate coefficients from transition state theory.

The energy E_0 is the difference in zero point energy between the transition state and the reactants. It is useful to point out the relationship between E_0 and the experimental activation energy E_a. As shown by Tolman, Fowler, and Guggenheim,[4]

$$E_a = \langle E_r(T) \rangle - \langle E(T) \rangle \qquad (10-21)$$

where $\langle E_r(T) \rangle$ is the average energy of molecules undergoing reaction and $\langle E(T) \rangle$ is the average energy of all reactant molecules. In transition state theory, $\langle E_r(T) \rangle$ is given by the average energy of the transition states plus E_0.

10.3 DYNAMICAL DERIVATION OF TRANSITION STATE THEORY

In the preceding derivation of transition state theory, we have postulated a dividing surface perpendicular to the reaction coordinate at the saddle point on the potential surface. The reactants exist on one side of the dividing surface, and products exist on the other. An important step in the derivation of the rate is the calculation of the number of transition states that lie at the critical surface, as well as the calculation of their velocity as they move across the saddle point. The dynamical derivation of transition state theory[5,6] follows the same line of argument. However, in this derivation the equilibrium flux through the transition state is evaluated. By using the "fundamental assumption" of transition state theory, this flux is equated to the reactive flux. Since the dynamical derivation uses classical mechanical language, the classical origin of transition state theory becomes evident. Appendix 2 gives a general discussion of classical statistical mechanics.

With the assumption that the reactants are in a state of thermodynamic equilibrium, one can determine the number of activated species that lie in a volume element of phase space on the dividing surface. For a reactant system defined by n atoms, there are six dimensions for each atom in the system (three coordinates and three conjugate momenta), and the volume is specified as $d\tau = dq_1^{\ddagger} \cdots dq_{3n}^{\ddagger} dp_1^{\ddagger} \cdots dp_{3n}^{\ddagger}$, where $q_1^{\ddagger}$ is the reaction coordinate and $p_1^{\ddagger}$ is its conjugate momentum. As shown by equations 2–1 through 2–4 in Appendix 2, the number of states in the volume element $d\tau$ is $d\tau/h^{3n}$. The fraction of reactant species lying in this volume element on the dividing surface is given by the following standard expression from classical statistical mechanics (equation A2–26 in Appendix 2):

$$\frac{dN*}{N} = \frac{e^{-H/k_BT}dp_1^{\ddagger} \cdots dp_{3n}dq_1^{\ddagger} \cdots dq_{3n}/h^{3n}}{\int^R e^{-H/k_BT}dp_1 \cdots dp_{3n}dq_1 \cdots dq_{3n}/h^{3n}} \qquad (10\text{--}22)$$

In this expression, H is the classical Hamiltonian of the system which defines the total energy expressed in coordinates and conjugate momenta. The asterisk on $dN*$ denotes that only one volume element on the dividing surface is under consideration. The differential for all volume elements on the dividing surface will be denoted by dN.

The integral in the denominator of equation 10–22 is over all values of coordinates and momenta which are associated with the reactant (R) molecules. As shown by equation A2–25 in Appendix 2, this integral is the partition function of the reactants. For reactants A and B, it can be written as $Q_A V Q_B V$, where the Q's are partition functions per unit volume and V is the volume of the container. The quantity N is the total number of reactant pairs and equals $N_A N_B$, the product of the numbers of individual A and B molecules. Thus, for an A + B bimolecular reaction, equation 10–22 becomes

$$dN* = \frac{[A][B]e^{-H/k_BT}dp_1^{\ddagger} \cdots dp_{3n}^{\ddagger}dq_1^{\ddagger} \cdots dq_{3n}^{\ddagger}/h^{3n}}{Q_A Q_B} \qquad (10\text{--}23)$$

where [A] and [B] have replaced the two N/V terms.

If the fundamental assumption is made that molecular systems do not recross the transition state, the reactant-to-product flux across the transition state becomes the reactive flux. The rate of passage of a representative point through the phase-space volume on the dividing surface associated with $dN*$ is then

$$\frac{dN*}{dt} = [A][B]\frac{dq_1^{\ddagger}}{dt}\frac{e^{-H/k_BT}dq_2^{\ddagger} \cdots dq_{3n}^{\ddagger}dp_1^{\ddagger} \cdots dp_{3n}^{\ddagger}/h^{3n}}{Q_A Q_B} \qquad (10\text{--}24)$$

To obtain the total rate of reaction, integration must be carried out over all the phase space on the dividing surface. Also, this integration must be over all *positive* values of $p_1^{\ddagger}$, since it is only this motion which takes the system from the reactant to the product side of the dividing surface. To carry out the integration, the total Hamiltonian at the dividing surface is separated into two parts and is given by

$$H = H^{\ddagger} + H' \qquad (10\text{--}25)$$

where $H^{\ddagger}$ defines the energy of the reactant system along the reaction coordinate $q_1^{\ddagger}$ and H' gives the energy along all coordinates orthogonal to the reaction coordinate.

The Hamiltonian for translational motion along the reaction coordinate is

$$H^{\ddagger} = \frac{(p_1^{\ddagger})^2}{2\mu_1} \qquad (10\text{--}26)$$

where μ_1 is the reduced mass and

$$p_1^{\ddagger} = \mu_1 \frac{dq_1^{\ddagger}}{dt} \qquad (10\text{--}27)$$

Inserting equations 10–25 through 10–27 into equation 10–24 and integrating over all the phase-space volume on the dividing surface gives

$$\frac{dN}{dt} = \frac{[A][B]\dfrac{1}{\mu_1 h}\displaystyle\int_0^\infty p_1^\ddagger e^{-(p_1^{\ddagger 2}/2\mu_1 k_B T)}\,dp_1^\ddagger \int e^{-H'/k_B T}\,dq_2^\ddagger \cdots dq_{3n}^\ddagger dp_2^\ddagger \cdots dp_{3n}^\ddagger/h^{3n-1}}{Q_A Q_B}$$

(10–28)

which is the total reactant-to-product flux through the dividing surface. The result of integrating over $p_1^\ddagger$ gives

$$\frac{dN}{dt}\frac{1}{[A][B]} = \frac{k_B T}{h}\frac{\int e^{-H'/k_B T}\,dq_2^\ddagger \cdots dq_s^\ddagger dp_2^\ddagger \cdots dp_s^\ddagger/h^{3n-1}}{Q_A Q_B}$$

(10–29)

If the potential energy at the transition state is E_0, one can write

$$H' = E_0 + H''$$

(10–30)

and make this substitution in the integral in equation 10–29. By comparing the resulting integral with equation A2–25 in Appendix 2, the integral is clearly recognized as the partition function for the transition state's $3n - 1$ degrees of freedom multiplied by $\exp(-E_0/k_B T)$. Thus, one gets

$$\frac{dN}{dt}\frac{1}{[A][B]} = \frac{k_B T}{h}\frac{Q^\ddagger V}{Q_A Q_B}\exp(-E_0/k_B T)$$

(10–31)

where $Q^\ddagger$ is the transition state's partition function per unit volume. Bringing V to the left side, we obtain the rate of disappearance of reactant pairs expressed in concentration units. The resulting expression for the rate constant is

$$\frac{\text{Rate}}{[A][B]} = \frac{1}{V}\frac{dN}{dt}\frac{1}{[A][B]} = \frac{k_B T}{h}\frac{Q^\ddagger}{Q_A Q_B}\exp(-E_0/k_B T)$$

(10–32)

which is the same as equation 10–20 found before, but derived with fewer assumptions.

In the dynamical derivation of transition state theory, the quasi-equilibrium assumption (the third assumption) is not required. In fact, this assumption should be viewed as a corollary of the fundamental assumption that trajectories do not recross the transition state. To illustrate this, consider the situation where there is thermal equilibrium between the reactants and the products. Obviously, there will be a thermal equilibrium at the transition state for trajectories crossing toward products as well as for those crossing toward reactants. Now consider the situation at the transition state when all trajectories emanating from the reactants are removed. If trajectories do not recross the transition state, none of the trajectories crossing the transition state toward reactants will have been removed. As a result, thermal equilibrium will be maintained between the reactants and trajectories proceeding through the transition state toward products in the absence of thermal equilibrium between the reactants and the products. Thus, the quasi-equilibrium assumption follows from the fundamental assumption of TST that trajectories do not recross the dividing surface.

10.4 QUANTUM MECHANICAL EFFECTS IN TRANSITION STATE THEORY

The foregoing dynamical derivation illustrates that transition state theory is a classical mechanical model. The theory is exact if classical trajectories passing through the transition state do not turn back and form reactant molecules again. Is there an equivalent model for a quantum mechanical transition state theory? Regretfully, the answer is no. Advances have been made toward a truly quantum transition state theory,[6-7] but much additional work needs to be done.

A basic difficulty in the generalization of classical transition state theory to quantum mechanics is that the reaction criterion cannot be formulated as the condition that a trajectory pass through a critical surface. This is connected with the fact that, in a quantum mechanical observation, the coordinates and momenta of a system cannot be assigned simultaneously. Consider the Heisenberg uncertainty relations

$$\Delta p \Delta q \gtrsim \hbar$$

and

$$\Delta E \Delta t \gtrsim \hbar$$

for the reaction coordinate. If Δp is replaced by $\hbar/\lambda$, where λ is the de Broglie wavelength, we have the relation $\Delta q > \lambda$. Thus, the uncertainty in the value of the reaction coordinate at the transition state must be larger than the wavelength associated with the motion along the reaction coordinate. In other words, quantum mechanically, the transition state is not localized. The uncertainty relation $\Delta E \Delta t \gtrsim \hbar$ may be analyzed in a similar way. For the thermal averaging in transition state theory to have a meaning, it is necessary that the translational energy E in the reaction coordinate be much less than $k_B T$. As a result, the lifetime Δt of the transition state must be larger than $\hbar/k_B T$. Therefore, in the quantum case, the transition state cannot be considered a definite configuration of nuclei during an infinitesimal interval of time. Thus, for the calculation of the reactive flux in the quantum case, it is necessary to consider explicitly the dynamics of the trajectories' motion in the region Δq or to follow the evolution with time of the system for time Δt.

In classical transition state theory, the potential energy is constant and the reaction coordinate motion is separated from the remaining internal motions at the localized position along the reaction coordinate which defines the transition state. Quantum mechanical delocalization of the transition state along the reaction coordinate can lead to two problems. First, if the potential is not flat in the region Δq, so that the system is not freely moving, it is incorrect to treat the reaction coordinate as being a classical translational motion. Rather, the potential will usually have a concave-down shape, and quantum mechanical tunneling (described in more detail shortly) will occur through the potential.

The second problem is more critical. If there is curvature along the reaction coordinate in the region Δq, the reaction coordinate is not separable from the remaining internal degrees of freedom.[6,8-10] Thus, the rate constant expression cannot be factored into a frequency $k_B T/h$ for the reaction coordinate and a partition function for the remaining degrees of freedom, as is done in equation 10–16. Also, since the curvature couples the reaction coordinate with the remaining modes, tunneling cannot be treated as a one-dimensional reaction coordinate barrier-penetration problem. Instead, there will be a multitude of tunneling paths which involve all the coordinates.[11-13]

Clearly, then, there are many difficulties in deriving a quantum mechanical transition state theory expression which does not include the separable reaction coordinate approximation. To obtain a working transition state theory expression, the separable approximation is assumed and corrections are made to the classical rate constant of equation 10–20 to account for quantum effects. It is well known that quantum partition functions are required for most vibrational motions. By contrast, for translational and most rotational motions, classical mechanical partition functions are accurate. The exception for rotation is at low temperatures and for molecules like H_2 and D_2 with small moments of inertia. With the vibrations treated quantum mechanically, the barrier for reaction is not the classical barrier, but the difference in zero-point energies between the transition state and the reactants. Finally, a correction for reaction coordinate motion must be included to account for tunneling effects.

To correct for tunneling, we have to consider the problem of crossing the barrier quantum mechanically rather than classically as done in previous sections. According to quantum mechanics, there is a probability that a particle finds its way from reactants to products, and that probability varies continuously with energy; for $E < E_0$ there is a finite nonzero probability, and for $E > E_0$ it approaches unity. The probabilities can be calculated for any barrier from the Schrödinger equation

$$\frac{\partial^2 \Psi}{\partial x^2} + \frac{8\pi^2 m}{h^2}[E - V(s)]\Psi = 0 \tag{10-33}$$

We have assumed a one-dimensional barrier, which is allowable since, with the separable approximation, there is reaction only along one coordinate, that is, the reaction pathway. A solution of equation 10–33 is obtained for a suitable choice of potential $V(s)$ to describe the barrier. The resulting expression for the probability of tunneling through the barrier is given by[14]

$$G(E) = \exp\left(-\frac{4\pi \sqrt{\mu}}{h} \int_{s_1}^{s_2} (V(s) - E)^{1/2} \, ds\right) \tag{10-34}$$

where $G(E)$, called the *permeability function*, depends on the exact shape of the barrier along the reaction coordinate, and s_1 and s_2 are turning points, i.e., the coordinates of the reaction path for which $V(s) = E$. The tunneling correction is obtained by integrating over a Maxwell-Boltzmann distribution,[15,16] i.e.,

$$Q_{\text{tunnel}} = e^{E_0/k_B T} \int_0^\infty \frac{1}{k_B T} e^{-E/k_B T} G(E) \, dE \tag{10-35}$$

Even for simple barrier shapes, a closed solution for the permeability function is difficult to obtain; and, in any case, the integration over all energies must be carried out numerically. However, one barrier shape for which an exact expression for $G(E)$

is known and is physically meaningful is the symmetrical Eckart barrier.[17] The potential for this barrier has the form

$$V(s) = E_0 \operatorname{sech}^2 (\pi s / \ell) \tag{10–36}$$

The permeability function for this barrier is

$$G(E) = \frac{\cosh (4\pi\alpha) - 1}{\cosh (4\pi\alpha) + \cosh (2\pi\delta)} \tag{10–37}$$

where

$$\alpha = \frac{1}{h} (2mE)^{1/2}$$

and

$$\delta = \frac{1}{h} \left(8mE_0 - \frac{h^2}{4\ell^2} \right)^{1/2}$$

and where ℓ is the width of the barrier. For small barrier heights and widths (1.25–1.77Å), the tunneling correction approaches an asymptotic form[18] given by

$$Q_{\text{tunnel}} = 1 - \frac{1}{24} \left(\frac{h\nu_s}{k_B T} \right)^2 \left(1 + \frac{k_B T}{E_0} \right) \tag{10–38}$$

where ν_s is the imaginary frequency of the transition state at the top of the barrier (see Section 10.1). For reactions involving $H + H_2$ and $D + D_2$, some tunneling corrections are given in Table 10–1. As is evident from the trends in the tunneling correction factors, quantum mechanical tunneling is important, and the lighter isotope reacts more rapidly.

The experimental evidence for tunneling comes from Arrhenius plots of $\ln k$ versus $1/T$ (see Figure 10–4). If the Arrhenius plots show significant nonlinearity, then tunneling may be the cause. The curvature in the Arrhenius plots of $\ln k$ vs. $1/T$ tends to be concave upward when tunneling is important, because the calculated reaction rate deviation is positive and increases for large values of $1/T$.

TABLE 10–1 TUNNEL EFFECT CORRECTION FACTORS FOR H AND D ATOM REACTIONS.[18]

	Q_{tunnel}	
$T(°K)$	$H + H_2$	$D + D_2$
200	5.180	3.090
250	3.706	2.353
300	2.900	1.950
400	2.093	1.546
600	1.507	1.253
800	1.297	1.148
1000	1.198	1.094
1200	1.142	1.071

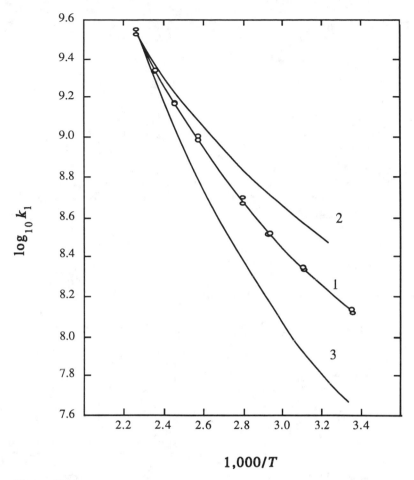

Figure 10–4. Arrhenius plots of ln k vs. $1/T$. Curve 1: experimental results for the reaction $H + H_2 \rightarrow H_2 + H$ exhibiting non-Arrhenius behavior. Curve 2: theoretical curve for an Eckart barrier with $E_o = 15.05$, $\nu_s = 2,573i$. Curve 3: theoretical curve for an Eckart barrier with $E_o = 14.35$, $\nu_s = 1,964i$. [Adapted from W. R. Schulz and D. J. LeRoy, *J. Chem. Phys. 42*, 3869 (1965)].

For some reactions, additional tunneling paths besides the one along the reaction path must be included. Several approximate ways to include these have been considered.[11–13]

10.5 THERMODYNAMIC FORMULATION OF TRANSITION STATE THEORY

The transition state derived rate constant can be reformulated in thermodynamic terms, since it is sometimes more useful to work with the rate constant in this form than with partition functions. The rate constant expression given by equation 10–20, viz.,

$$k = \frac{k_B T}{h} \frac{Q^\ddagger}{Q_A Q_B} e^{-E_0/k_B T}$$

can be written as

$$k = \frac{k_B T}{h} K_c^\ddagger \qquad (10\text{--}39)$$

where

$$K_c^\ddagger = \frac{Q^\ddagger}{Q_A Q_B} e^{-E_0/k_B T} \qquad (10\text{--}40)$$

which is the equilibrium constant for formation of the transition state. If the equilibrium constant is expressed in terms of the molar Gibbs standard free energy using the van t'Hoff relation

$$\Delta G_0^\ddagger = -RT \ln K_c^\ddagger \qquad (10\text{--}41)$$

then equation 10–20 can be written as

$$k = \frac{k_B T}{h} e^{-\Delta G_0^\ddagger/RT} \qquad (10\text{--}42)$$

$\Delta G_0^\ddagger$ may be expressed in terms of standard enthalpy and entropy changes by

$$\Delta G_0^\ddagger = \Delta H_0^\ddagger - T\Delta S_0^\ddagger \qquad (10\text{--}43)$$

allowing equation 10–42 to be divided into two terms:

$$k = \frac{k_B T}{h} e^{\Delta S_0^\ddagger/R} e^{-\Delta H_0^\ddagger/RT} \qquad (10\text{--}44)$$

For rate constants expressed in units of mol/dm^3 (concentration) and seconds (time), the appropriate standard state for $K_c^\ddagger$, $\Delta G_0^\ddagger$, $\Delta H_0^\ddagger$, and $\Delta S_0^\ddagger$ is 1 mol/dm^3. [1 m^3 = 1,000 dm^3.]

Equation 10–44 is similar to the Arrhenius equation

$$k = A e^{-E_a/RT} \qquad (10\text{--}45)$$

and the thermodynamic parameters can be related to the Arrhenius parameters. The Arrhenius activation energy is defined by

$$\frac{d(\ln k)}{dT} \equiv \frac{E_a}{RT^2} \qquad (10\text{--}46)$$

Taking the logarithm of equation 10–39 and differentiating with respect to T gives

$$\frac{d(\ln k)}{dT} = \frac{1}{T} + \frac{d(\ln K_c^\ddagger)}{dT} \qquad (10\text{--}47)$$

Since $K_c^\ddagger$ is a concentration equilibrium constant, its variation with temperature is given by the Gibbs-Helmholtz equation

$$\frac{d(\ln K_c^\ddagger)}{dT} = \frac{\Delta E_0^\ddagger}{RT^2} \qquad (10\text{--}48)$$

Inserting this equation into equation 10–47 and comparing the result with equation 10–46 gives

$$E_a = RT + \Delta E_0^\ddagger \qquad (10\text{–}49)$$

The relationship between the standard thermodynamic energy and enthalpy for a constant-pressure process is

$$\Delta H_0^\ddagger = \Delta E_0^\ddagger + P(\Delta V_0^\ddagger) \qquad (10\text{–}50)$$

since $H = E + PV$. The quantity $\Delta V_0^\ddagger$ is known as the *standard volume of activation*. With equation 10–50, equation 10–49 becomes

$$E_a = \Delta H_0^\ddagger + RT - P(\Delta V_0^\ddagger) \qquad (10\text{–}51)$$

For a unimolecular reaction, there is no change in the number of molecules in going from the reactants to the transition state, so $\Delta V_0^\ddagger$ is zero. Also, to a very good approximation, $\Delta V_0^\ddagger$ is zero for reactions in solution or in a condensed phase. Therefore, in these cases,

$$E_a = \Delta H_0^\ddagger + RT \qquad (10\text{–}52)$$

Insertion of this equation into equation 10–44 leads to

$$k = \frac{k_B T}{h} e^{(1 + \Delta S_0^\ddagger/R)} e^{-E_a/RT} \qquad (10\text{–}53)$$

Hence,

$$A = \frac{k_B T}{h} e^{(1 + \Delta S_0^\ddagger/R)} \qquad (10\text{–}54)$$

For gas phase reactions other than unimolecular, the relationships between the Arrhenius parameters and thermodynamic terms are different from the relationships just given. If the ideal gas relation is assumed, i.e.,

$$P\Delta V_0^\ddagger = \Delta n^\ddagger RT \qquad (10\text{–}55)$$

then, from equation 10–51, one obtains

$$E_a = \Delta H_0^\ddagger + RT - \Delta n^\ddagger RT = \Delta H_0^\ddagger + (1 - \Delta n^\ddagger)RT \qquad (10\text{–}56)$$

and the Arrhenius pre-exponential factor is

$$A = e^{-(\Delta n^\ddagger - 1)} \frac{k_B T}{h} e^{\Delta S_0^\ddagger/R} \qquad (10\text{–}57)$$

10.6 APPLICATIONS OF TRANSITION STATE THEORY

10.6.1 Evaluating Partition Functions (Statistical Mechanics)

Determining rate constants from the canonical level of transition state theory is an invaluable aid in both elucidating mechanisms and evaluating rate constants. Calculating the rate constant for a reaction requires that we know how to evaluate the partition functions that arise out of the transition state theory derivation. Real molecules may have translational, rotational, vibrational, and electronic energies. Therefore, the total partition function associated with the internal motion for each molecule is given by the product of partition functions associated with independent rotational, vibrational, electronic, and translational motions:

$$Q = Q_{rot} Q_{vib} Q_{elec} Q_{trans}$$

To calculate the individual partition functions, one needs to know moments of inertia, vibrational frequencies, and electronic states. Such information for reactants A and B can be obtained from spectroscopic data, but, for the transition state, we must make use of semiempirical or *ab initio* calculations.

10.6.1.1 Electronic partition function.

The electronic partition function is given by

$$Q_{elec} = \sum_i g_i e^{-E_i/k_B T} \tag{10–58}$$

where g_i is the degeneracy and E_i is the energy above the lowest state of the system. In most reactions, few electronic energy levels other than the ground state must be considered. For reactions involving doublet and triplet systems, the degeneracy factor is the corresponding spin degeneracy which would contribute a factor of two or more to the calculated rate constant.

10.6.1.2 Translational partition function.

For the translational motion for a particle of mass m moving in a one-dimensional box of length l, the energy derived from the Schrödinger equation is

$$E_n = \frac{n^2 h^2}{8ml^2} \tag{10–59}$$

where n is the quantum number for nondegenerate energy levels and h is Planck's constant. The partition function for one dimension is

$$Q_{trans} = \sum_{n=1}^{\infty} \exp\left(\frac{-n^2 h^2}{8ml^2 k_B T}\right) \tag{10–60}$$

When the energy levels are closely spaced, that is, when their separation $E_{n+1} - E_n = (h^2/8ml^2)(2n + 1)$ is $\ll k_B T$, the sum can be replaced by an integral:

$$Q_{trans} = \int_0^{\infty} \exp\left(\frac{-n^2 h^2}{8ml^2 k_B T}\right) dn$$

$$Q_{trans} = \frac{1}{2}\left(\frac{8m\pi l^2 k_B T}{h^2}\right)^{1/2} = \frac{(2\pi m k_B T)^{1/2} l}{h} \tag{10–61}$$

The total translation energy in three dimensions is given by

$$E_{trans} = E_{trans,x} + E_{trans,y} + E_{trans,z} \tag{10–62}$$

so that the total translational partition function is a product of one-dimensional partition functions for each component, i.e.,

$$Q_{trans} = Q_{trans,x} Q_{trans,y} Q_{trans,z} \qquad (10\text{--}63)$$

or

$$Q_{trans} = \frac{(2\pi m k_B T)^{3/2} l^3}{h^3} \qquad (10\text{--}64)$$

Since l^3 is the volume V,

$$\frac{Q_{trans}}{V} = (2\pi m k_B T/h^2)^{3/2} \qquad (10\text{--}65)$$

10.6.1.3 Vibrational partition function. To find partition functions for the vibrational contribution, we evaluate

$$Q_{vib} = \sum_i g_i \exp\left(-E_i/k_B T\right) \qquad (10\text{--}66)$$

For a simple harmonic oscillator, the vibrational energy has nondegenerate levels

$$E_v = \left(v + \frac{1}{2}\right) hc\tilde{\nu}_s \qquad (10\text{--}67)$$

where $\tilde{\nu}_s$ (in cm^{-1}) is the harmonic frequency of a single vibrational level and v is the quantum number. Each nondegenerate level has multiplicity 1, so to evaluate the partition function for the vibrational contribution, we must sum over all levels. Taking the zero-point level as the zero of energy, this leads to

$$Q_{vib} = \frac{1}{1 - \exp\left(-hc\tilde{\nu}/k_B T\right)} \qquad (10\text{--}68)$$

A polyatomic molecule has $s = 3N - 6$ vibrational modes if it is nonlinear and $3N - 5$ modes if it is linear; therefore, the vibrational partition function for a polyatomic molecule is

$$Q_{vib} = \prod_{i=1}^{s} \frac{1}{1 - \exp\left(-hc\tilde{\nu}_i/k_B T\right)} \qquad (10\text{--}69)$$

10.6.1.4 Rotational partition function. For a linear rigid rotor, the energy levels are given by

$$E_J = \frac{J(J + 1)\hbar^2}{2I} \qquad (10\text{--}70)$$

where $\hbar^2/2I$ is the rotational constant and $J = 0, 1, 2, \ldots$ is a rotational quantum number. The multiplicity of each level is $2J + 1$, so

$$Q_{rot} = \sum_{J=0}^{\infty} (2J + 1) \exp\left[-J(J + 1)\hbar^2/2Ik_B T\right] \qquad (10\text{--}71)$$

If the level intervals are small compared to $k_B T$, the sum can replaced by an integral:

$$Q_{rot} = \int_0^\infty (2J + 1) \exp\left[-J(J + 1)\hbar^2/2Ik_B T\right] dJ$$

so

(10–72)

$$Q_{rot} = \frac{8\pi^2 Ik_B T}{h^2}$$

Equation 10–72 is also a reasonably good approximation for real nonrigid molecules. For a polyatomic molecule with moments of inertia I_a, I_b, and I_c about its principal axes, the rotational partition function is

$$Q_{rot} = \pi^{1/2} \left(\frac{8\pi^2 I_a k_B T}{h^2}\right)^{1/2} \left(\frac{8\pi^2 I_b k_B T}{h^2}\right)^{1/2} \left(\frac{8\pi^2 I_c k_B T}{h^2}\right)^{1/2}$$

(10–73)

Usually, in a sample at room temperature the lowest vibrational level is populated, so the rotational partition function should be evaluated using parameters for the structure of this state. For reactions involving molecules in excited vibrational states, it is appropriate to use rotational constants for the particular vibrational state.

10.6.2 Symmetry and Statistical Factors

If molecules involved in a reaction have elements of symmetry, then in calculating the rate expression, the symmetry must be accounted for in the partition functions. For molecules which have rotational symmetry, it is customary to divide the rotational partition function for each molecule by its appropriate symmetry number σ, defined as the number of equivalent arrangements that can be obtained by rotating the molecule. For example, consider Cl_2. The number of identical atoms in the molecule is 2, and by rotating we get two equivalent arrangements: $Cl^1 - Cl^2$ and $Cl^2 - Cl^1$. Therefore, the symmetry number is 2. For planar NO_3 radical, the symmetry number is 6. In taking into account the symmetry numbers in the rotational partition function, the standard procedure is to divide by the symmetry number. This procedure is correct when calculating rate constants, but it has some limitations. A good example which illustrates this is the reaction

$$H + H_2 \longrightarrow [H \cdots H\text{–}H]^\ddagger \longrightarrow H_2 + H$$

The symmetry number for H_2 is 2, and so is the symmetry number of the complex; consequently, the rate constant is

$$k_1 = \frac{k_B T}{h} \frac{Q^\ddagger/2}{(Q_H/1)(Q_{H_2}/2)} e^{-E_0/k_B T}$$

(10–74)

The factors cancel out to give

$$k_1 = \frac{k_B T}{h} \frac{Q^\ddagger}{Q_H Q_{H_2}} e^{-E_0/k_B T}$$

(10–75)

If we compare the rate obtained for the reaction

$$D + H_2 \longrightarrow [D \cdots H\text{–}H]^\ddagger \longrightarrow DH + H$$

we see that the symmetry number for the complex is 1; therefore,

$$k_2 = \frac{k_B T}{h} \frac{Q^{\ddagger}/1}{(Q_D/1)(Q_{H_2}/2)} e^{-E_0/k_B T} \tag{10-76}$$

and we have

$$k_2 = \frac{k_B T}{h} \frac{2Q^{\ddagger}}{Q_D Q_{H_2}} e^{-E_0/k_B T} \tag{10-77}$$

The conclusion is that the second reaction is favored over the first by a factor of 2. However, since both reactions are extracting hydrogen atoms from H_2, the rates cannot differ by a factor of 2; this factor should appear in both equations!

The best way to resolve this dilemma is by using a statistical factor, defined as the number of different transition states that can be formed if all identical atoms are labeled to distinguish them from one another. Thus, for the reaction

$$H^1 + H^2 - H^3 \begin{cases} \nearrow [H^1 \cdots H^2 \cdots H^3]^{\ddagger} \longrightarrow H^1 - H^2 + H^3 \\ \searrow [H^1 \cdots H^3 \cdots H^2]^{\ddagger} \longrightarrow H^1 - H^3 + H^2 \end{cases} \tag{10-78}$$

the statistical factor for forming the transition state is 2. Similarly, the reaction

$$H^1 + D^2 - D^3 \begin{cases} \nearrow [H^1 \cdots D^2 \cdots D^3]^{\ddagger} \longrightarrow H^1 - D^2 + D^3 \\ \searrow [H^1 \cdots D^3 \cdots D^2]^{\ddagger} \longrightarrow H^1 - D^3 + D^2 \end{cases} \tag{10-79}$$

also has a statistical factor of 2. Therefore, the correct rate constant for *both* reactions is given by equation 10–75. In general, with this definition, we can omit symmetry numbers from the partition functions and multiply the rate expression by the statistical factor $L^{\ddagger}$, i.e.,

$$k = L^{\ddagger} \frac{k_B T}{h} \frac{Q^{\ddagger}}{Q_A Q_B} e^{-E_0/k_B T} \tag{10-80}$$

Great care must be used in the choice of a transition state with the correct degree of symmetry. For further reading, see Bishop and Laidler,[19] Pollak and Pechukas,[20] and Coulson.[21]

10.6.3 Collisions between Atoms

The simplest application of transition state theory is to collisions between two atoms A and B, where a rate constant k_c may be calculated for attaining a critical A–B separation r_c. In this simple case, the energy profile along the reaction coordinate is identical to the one-dimensional interatomic potential curve; we shall see that this leads to a cancellation of partition functions and a particularly simple result. The problem is somewhat artificial, because if the intermediate AB retains the energy of bond formation, it will dissociate after one vibrational period. If there is a mechanism for releasing this energy, however, a stable molecule may be formed and k_c can be related to the rate of formation of AB. An example of this is mentioned at the end of the section.

The transition state $AB^{\ddagger}$ for a collision between two atoms has three translational and two rotational degrees of freedom. The A–B relative center-of-mass motion (vibration) is the reaction coordinate. Thus, the transition state partition function includes contributions from translation and rotation and is

$$Q^{\ddagger} = \frac{[2\pi(m_A + m_B)k_B T]^{3/2}}{h^3} \frac{8\pi^2 I^{\ddagger} k_B T}{h^2} \tag{10-81}$$

where

$$I^{\ddagger} = \frac{m_A m_B}{m_A + m_B} r_c^2 \tag{10-82}$$

The atoms A and B have only translational contributions to their partition functions. After inserting these partition functions into equation 10–20, we find the transition state theory value for k_c to be

$$k_c = \pi r_c^2 \left(\frac{8k_B T}{\pi\mu}\right)^{1/2} e^{-E_0^{\ddagger}/k_B T} \tag{10-83}$$

This equation is identical with equation 8–8, derived for a line-of-centers collision model.

A case in which an atom-atom collision can lead to a real chemical event is radiative recombination. Consider the collision between an electronically excited atom A* and a ground-state atom B. Once inside the critical separation r_c, the electronically excited diatom (AB)* may emit a photon of radiation to form the stable ground-state molecule AB, i.e.,

$$AB^* \longrightarrow AB + h\nu \tag{10-84}$$

The rate constant for formation of AB by this process can be written as the rate constant k_c multiplied by a probability of radiative emission from AB*, which in turn depends on the Franck-Condon overlap density and the oscillator strength of the transition. Some examples of this process are discussed by Smith.[22]

10.6.4 Application: The F + H₂ Reaction

The reaction $F + H_2 \rightarrow HF + H$ has been of particular interest in chemical kinetics because it is the rate-limiting step in the chain reaction $H_2 + F_2 \rightarrow 2HF$, which plays an important role in the underlying kinetics of the HF chemical laser (see chapters 9 and 14). It is also of special theoretical interest, because it is one of the simplest examples of an exothermic chemical reaction.

We can calculate the rate of reaction of the F atom with an H_2 molecule by transition state theory using equation 10–20. Consider a linear activated complex

Properties of this complex are calculated using a LEPS surface[23,24,25] and are summarized in Table 10-2. There should be three real vibrational frequencies and one imaginary frequency. The frequency at 4007.6 cm^{-1} corresponds to the H_2 stretch against the F atom in the complex, and the degenerate bending frequency is 397.9 cm^{-1}. The imaginary frequency at 310.8i corresponds to passage over the barrier. Using this information, together with properties of the reactants, we can calculate a rate constant for the F + H_2 reaction. First, we can rewrite the expression for partition functions in equation 10-20 as a product of partition function ratios for the translational, vibrational, rotational and electronic motion:

$$k = L^{\ddagger} \frac{k_B T}{h} \left(\frac{Q^{\ddagger}}{Q_F Q_{H_2}} \right)_{\text{vib}} \left(\frac{Q^{\ddagger}}{Q_F Q_{H_2}} \right)_{\text{rot}} \left(\frac{(Q^{\ddagger}/V)}{(Q_F/V)(Q_{H_2}/V)} \right)_{\text{trans}} \left(\frac{Q^{\ddagger}}{Q_F Q_{H_2}} \right)_{\text{elec}} e^{-E_0/RT} \qquad (10-85)$$

These ratios can be evaluated separately.

TABLE 10-2 PROPERTIES OF THE REACTANTS AND TRANSITION STATE FOR THE F + H_2 REACTION.

Parameters	F $\cdots$ H$_a$ − H$_b$	F	H$_2$
r_2(F−H),Å	1.602		r(H−H) = 0.7417Å
r_1(H−H),Å	0.756		
$\bar{\nu}_1$, cm^{-1}	4007.6		4395.2
$\bar{\nu}_2$, cm^{-1}	397.9		
$\bar{\nu}_3$, cm^{-1}	397.9		
$\bar{\nu}_4$, cm^{-1}	310.8i		
E_0(kJ/mole)	6.57		
m(amu)	21.014	18.9984	2.016
I(amu Å^2)	7.433		0.277
g_{elec}	4	4	1

Next, we note that the electronic degeneracy for the fluorine atom in its ground electronic state ($^2P_{3/2}$) is 4, and that for the linear FH$_2$ complex ($^2\Pi$) is the same. We neglect contributions from the ($^2P_{1/2}$) spin-orbit state of F at 404 cm^{-1}. The degeneracy for the H$_2$ molecule in its ground state ($^1\Sigma_g^+$) is 1. Consequently, the electronic partition function contribution is unity. The translational partition function ratio, using equation 10-65, is

$$\left(\frac{(Q^{\ddagger}/V)}{(Q_F/V)(Q_{H_2}/V)} \right)_{\text{trans}} = \left(\frac{m_F + m_{H_2}}{m_F m_{H_2}} \right)^{3/2} \left(\frac{h^2}{2\pi k_B T} \right)^{3/2}$$

and inserting the values given in Table 10-2 gives

$$\left(\frac{(Q^{\ddagger}/V)}{(Q_F/V)(Q_{H_2}/V)} \right)_{\text{trans}} = (6.01 \times 10^{39} \text{ kg}^{-3/2})(6.926 \times 10^{-71} \text{ kg}^{3/2} \text{ m}^3)$$

so that

$$\left(\frac{(Q^{\ddagger}/V)}{(Q_F/V)(Q_{H_2}/V)} \right)_{\text{trans}} = 4.16 \times 10^{-31} \text{ m}^3 \qquad (10-86)$$

The rotational and vibrational partition functions are dimensionless, and contain no contribution from the mono-atomic fluorine species. The rotational partition function ratio is

$$\left(\frac{Q^{\ddagger}}{Q_{H_2}}\right)_{rot} = \left(\frac{8\pi^2 k_B T I^{\ddagger}/h^2}{8\pi^2 k_B T I_{H_2}/h^2}\right) = \frac{I^{\ddagger}}{I_{H_2}} \tag{10-87}$$

Substituting numerical values, we obtain

$$\left(\frac{Q^{\ddagger}}{Q_{H_2}}\right)_{rot} = 26.8 \tag{10-88}$$

In evaluating the vibrational partition function ratio we obtain the expression

$$\left(\frac{Q^{\ddagger}}{Q_{H_2}}\right)_{vib} = \frac{1 - \exp(-h\nu_{H_2}/k_B T)}{\prod\limits_{i=1}^{3}[1 - \exp(-h\nu_i^{\ddagger}/k_B T)]} \tag{10-89}$$

so that

$$\frac{h\nu_{H_2}}{k_B T} = \frac{(6.626 \times 10^{-34}\,\text{J}\cdot\text{s})(4395.2\ \text{cm}^{-1})(2.99 \times 10^{10}\ \text{cm/s})}{(1.281 \times 10^{-23})\,\text{J}\cdot\text{K}^{-1} \times 300\text{K}}$$

$$= 21.0$$

Consequently, the term $1 - \exp(-h\nu_{H_2}/k_B T)$ is close to unity. For the complex, we have the real vibrational frequencies

i	$\bar{\nu}_i$, cm^{-1}	$h\nu_i/k_B T$	$[1 - exp(h\nu_i/k_B T)]$
1	4007.6	19.2	≈ 1.0
2, 3	397.9	1.9	0.850

Thus, the vibrational partition ratio term is

$$\left(\frac{Q^{\ddagger}}{Q_{H_2}}\right)_{vib} = \frac{1.0}{[1 - \exp(-h\nu_1/k_B T)][1 - \exp(-h\nu_2/k_B T)]^2} \tag{10-90}$$

$$= 1.39$$

The statistical factor $L^{\ddagger}$ contributes a factor of 2. Thus, inserting the corresponding values for the partition function ratios, we obtain for the rate constant

$$k = (1.95 \times 10^{-16}\ \text{m}^3\ \text{s}^{-1}) \exp(-E_0/k_B T)$$

Since these are molecular units, we multiply by Avogadro's number, $6.022 \times 10^{23}\ \text{mol}^{-1}$, to obtain

$$k = 1.17 \times 10^8\ \text{mol}^{-1}\ \text{m}^3\ \text{s}^{-1}\ e^{-6570/8.31T}$$

$$= 1.17 \times 10^{11} \exp(-790/T)\ \text{liter mole}^{-1}\ \text{sec}^{-1}$$

The experimental data[26,27] are best represented by $k(T) = 2 \times 10^{11} \exp(-800/T)$. This is very reasonable agreement, for such a simple model.

10.7 MICROCANONICAL TRANSITION STATE THEORY

In the previous sections we were concerned with the development of a statistical theory of reaction rates in which the population of the reactant molecular states is defined by the Boltzmann distribution and a single temperature. A collection of molecules forming such a distribution is referred to as a *canonical* ensemble. The underlying basis of the Boltzmann distribution is that all molecular states at the same energy are equally probable. An assembly of molecules, for which there is an equal population of all states at a particular energy, is called a *microcanonical* ensemble. In the microcanonical approach to transition state theory, a rate constant is determined for the microcanonical ensemble of reactant molecules.[9,28,29]

The microcanonical rate constant $k(E)$ is for reactants whose total energy is E. The canonical transition state theory rate constant $k(T)$ is a Boltzmann average of the $k(E)$ and is given by

$$k(T) = \int_0^\infty k(E)P(E)\, dE, \tag{10-91}$$

where the reactants have energy, E, $P(E)$ is the normalized Boltzmann probability, and $k(E)$ is the specific rate constant for reactants with total energy between E and $E + dE$.

To derive the microcanonical transition state theory rate constant $k(E)$, it is assumed, as in canonical transition state theory, that the reaction coordinate is separable from the remaining degrees of freedom. Thus, the Hamiltonian is written as in equation 10–25. Microcanonical transition states are reactive systems which have a total energy $H = E$ and a value for the reaction coordinate q_1 which lies between $q_1^\ddagger$ and $q_1^\ddagger + dq_1^\ddagger$. The reaction coordinate potential at the transition state is E_0.

For a bimolecular reaction, the combined states of the reactant molecules at total energy E define a supermolecule. In a unimolecular reaction the supermolecule is simply the single reactant. The fraction of supermolecules that lie at the transition state (i.e., at q_1 ranging from $q_1^\ddagger$ to $q_1^\ddagger + dq_1^\ddagger$) with reaction coordinate momenta in the range from $p_1^\ddagger$ to $p_1^\ddagger + dp_1^\ddagger$ may be found from statistical mechanics. From classical statistical mechanics (equation A2–7 in Appendix 2), this fraction is

$$\frac{dN(q_1^\ddagger, p_1^\ddagger)}{N} = \frac{dq_1^\ddagger\, dp_1^\ddagger \displaystyle\int_{H=E-E_1^\ddagger-E_0} \cdots \int dq_2^\ddagger \cdots dq_{3n}^\ddagger dp_2^\ddagger \cdots dp_{3n}^\ddagger}{\displaystyle\int_{H=E} \cdots \int dq_1 \cdots dq_{3n} dp_1 \cdots dp_{3n}} \tag{10-92}$$

where $E_1^\ddagger$ is the translational energy in the reaction coordinate. In this equation, $dq_1^\ddagger dp_1^\ddagger$ divided by Planck's constant is the number of translational states in the reaction coordinate and the surface integral in the numerator divided by h^{3n-1} is the density of states for the $3n - 1$ degrees of freedom orthogonal to the reaction coordinate. Similarly, the surface integral in the denominator is the density of reactant states multiplied by h^{3n}.

All supermolecules that lie within $q_1^\ddagger$ and $q_1^\ddagger + dq_1^\ddagger$ and with positive $p_1^\ddagger$ will cross the transition state toward products in the time interval $dt = \mu_1 dq_1^\ddagger/p_1^\ddagger$. Thus, from equation 10–92, the reactant-to-product flux through the transition state for momentum $p_1^\ddagger$ is

$$\frac{dN(q^\ddagger_1, p^\ddagger_1)}{dt} = \frac{N \frac{p^\ddagger_1}{\mu_1} dp^\ddagger_1 \underset{H=E-E^\ddagger_1-E_0}{\int} \cdots \int dq^\ddagger_2 \cdots dq^\ddagger_{3n} dp^\ddagger_2 \cdots dp^\ddagger_{3n}}{\underset{H=E}{\int} \cdots \int dq_1 \cdots dq_{3n} dp_1 \cdots dp_{3n}}$$ (10–93)

The translational energy along the reaction coordinate is given by equation 10–26, so that $dE^\ddagger_1 = p^\ddagger_1 dp^\ddagger_1 / \mu_1$. Therefore, equation 10–93 may be written

$$\frac{dN(q^\ddagger_1, p^\ddagger_1)}{dt} = \frac{N \, dE^\ddagger_1 \underset{H=E-E^\ddagger_1-E_0}{\int} \cdots \int dq^\ddagger_2 \cdots dq^\ddagger_{3n} dp^\ddagger_2 \cdots dp^\ddagger_{3n}}{\underset{H=E}{\int} \cdots \int dq_1 \cdots dq_{3n} dp_1 \cdots dp_{3n}}$$ (10–94)

To find the total reaction flux, equation 10–94 must be integrated between the limits $E^\ddagger_1$ equal to 0 and $E - E_0$, so that

$$\frac{dN}{dt} = \frac{N \int_0^{E-E_0} \left[\underset{H=E-E^\ddagger_1-E_0}{\int} \cdots \int dq^\ddagger_2 \cdots dq^\ddagger_{3n} dp^\ddagger_2 \cdots dp^\ddagger_{3n} \right] dE^\ddagger_1}{\underset{H=E}{\int} \cdots \int dq_1 \cdots dq_{3n} dp_1 \cdots dp_{3n}}$$ (10–95)

which gives

$$\frac{dN}{dt} = \frac{N \int_{H=0}^{H=E-E_0} \cdots \int dq^\ddagger_2 \cdots dq^\ddagger_{3n} dp^\ddagger_2 \cdots dp^\ddagger_{3n}}{\underset{H=E}{\int} \cdots \int dq_1 \cdots dq_{3n} dp_1 \cdots dp_{3n}}$$ (10–96)

Since $dN/dt = k(E)N$, the ratio of integrals in equation 10–96 is the microcanonical transition state theory rate constant $k(E)$. This becomes readily apparent (see Appendix 2) when the numerator is identified as the sum of states at the transition state $G^\ddagger(E - E_0)$ multiplied by h^{3n-1}. As discussed, the denominator is the reactant density of states times h^{3n}. Thus, $k(E)$ becomes

$$k(E) = \frac{G^\ddagger(E - E_0)}{hN(E)}$$ (10–97)

The reactant density $N(E)$ has units of inverse energy for a unimolecular reaction and (energy $\times$ volume)$^{-1}$ for a bimolecular reaction.

The canonical transition state theory rate constant results from inserting equation 10–97 into equation 10–91. Since, from equation A1–3 in Appendix 1, the Boltzmann probability $P(E)$ equals $N(E) \exp(-E/k_B T)/Q$, we have

$$k(T) = \frac{1}{hQ} \int_{E_0}^{\infty} G^\ddagger(E - E_0) \exp(-E/k_B T) \, dE$$ (10–98)

where Q is the total partition function for the reactants. Writing $E = E^\ddagger + E_0$, we obtain, for equation 10–98,

$$k(T) = \frac{e^{-E_0/k_B T}}{hQ} \int_0^{\infty} G^\ddagger(E^\ddagger) \exp(-E^\ddagger/k_B T) \, dE^\ddagger$$ (10–99)

The integration with respect to $E^{\ddagger}$ may be done by parts or, more simply, by noting that the integral is the Laplace transform* of $G^{\ddagger}(E^{\ddagger})$.[30] The resulting value for the integral is $k_B T Q^{\ddagger}$, where $Q^{\ddagger}$ is the partition function for the $3n - 1$ degrees of freedom orthogonal to the reaction coordinate. A comparison of the resulting $k(T)$ with k_{abs} in equation 10–20 shows that the two are the same.

10.8 VARIATIONAL TRANSITION STATE THEORY

As discussed in section 10.3, classical transition state theory gives the exact rate constant (in a classical universe) if the net rate of reaction equals the rate at which trajectories pass through the transition state. These two rates are equal, however, only if trajectories do not recross the transition state: any recrossing makes the reactive flux smaller than the flux through the transition state. Thus, the classical transition state theory rate constant may be viewed as an upper bound to the correct classical rate constant.

The effect of recrossings on the reaction rate is illustrated by the six schematic trajectories in Figure 10–5. The left side and right side represent the reactants and products, respectively. The solid line shows the transition state dividing surface (TSDS) at the saddle point, while the dashed line is an alternative transition state dividing surface. There are six crossings of the saddle point in the direction from reactant to products, and transition state theory counts all of them as contributing to the reactive flux. However, only two trajectories in fact contribute to the reactive flux. Thus, in this example the transition state theory rate constant is too large by a factor of three. This is reflected in the fact that, of the six trajectories in the figure, only numbers 1 and 4 cross the saddle point once.

The alternative transition state denoted by the dashed line suggests how transition state theory may be improved. There are only three crossings for the alternative transition state in the direction from reactants to products, as compared with the two reactive trajectories. Thus, this transition state gives a rate constant that is only 50 percent too large. Clearly, then, the alternative transition state is a much better bottleneck for the reaction. Accordingly, one can consider different positions for the transition state along the reaction path and calculate the rate constant corresponding to each. The minimum rate so obtained is the closest to the truth, assuming that quantum effects related to tunneling and nonseparability are negligible. This procedure is called *variational transition state theory*.[4,10,19,31-38] There are both 'canonical and microcanonical versions of variational transition state theory. In the canonical approach, the minimum in the canonical transition state theory constant given by equation 10–20 is found along the reaction path

$$\frac{dk(T)}{dq^{\ddagger}} = 0 \qquad\qquad (10-100)$$

Since the canonical rate constant is related to the free energy according to equation 10–42, the canonical variational transition state will be located at the *maximum* in the free energy along the reaction path.

*See chapter 2 for a discussion of Laplace transforms.

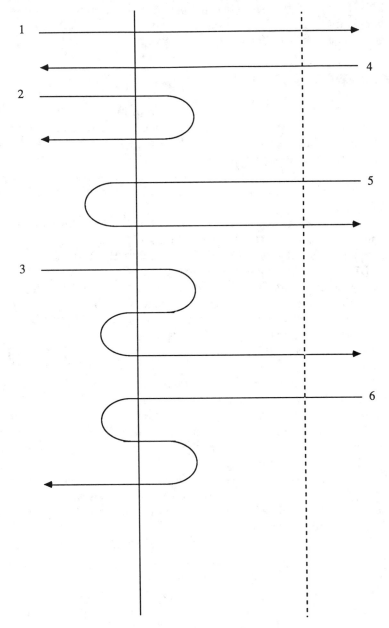

Figure 10–5. Schematic trajectories showing different ways of crossing the transition state indicated by the solid vertical line. An alternative transition state is identified by the dashed vertical line.

Canonical variational transition state theory gives an approximate upper bound to the rate constant. To obtain the correct upper bound, microcanonical variational transition state theory must be used. Here, the minimum in the microcanonical rate constant given by equation 10–97 is found along the reaction path

$$\frac{dk(E)}{dq^{\ddagger}} = 0 \qquad (10\text{--}101)$$

so that a different transition state is found for each energy. The thermal rate constant results from a Boltzmann average of the microcanonical rate constants as given by equation 10–97. The individual microcanonical rate constants are minimized at the point along the reaction path where the sum of the states $G^{\ddagger}(E - E_0)$ is a minimum.

Intuitively, the canonical and microcanonical results are what one expects. The best bottleneck along the reaction path in the canonical sense should be where the free energy is a maximum. In the microcanonical approach, the reaction bottleneck is located where there is the smallest sum of states.

To apply canonical variational transition state theory, the transition state's partition function (or free energy) must be calculated as a function of the reaction path. In the microcanonical approach, the transition state's sum of states must be calculated along the reaction path. Each of these calculations requires values for the classical potential energy, zero-point energy, vibrational frequencies, and moments of inertia as a function of the reaction path. The extraction of this information from a potential energy surface is described in chapter 7, section 7.9.

Variational transition state theory is particularly important for reactions without saddle points, e.g., $H + CH_3 \rightarrow CH_4$. Results of a canonical variational transition state theory calculation[39] for this reaction are listed in Table 10–3. A particularly interesting result is the shorter H–CH_3 bond length and thus "tighter" transition state as the temperature is increased. The vibrational frequency for the degenerate H–CH_3 rocking motions increases as the transition state tightens. A tightening of the transition state with increase in temperature is a common result for reactions without potential energy barriers. The calculated rate constant for $H + CH_3 \rightarrow CH_4$ is in good agreement with the experimental value of $2.8 \times 10^8 \ m^3 mol^{-1} s^{-1}$ for $T = 300\text{--}600 \ K$.

TABLE 10–3 CANONICAL VARIATIONAL TRANSITION STATE THEORY RESULTS FOR THE REACTION $H + CH_3 \rightarrow CH_4$.

Temp (K)	$R^{\ddagger a}$	$\nu_{rock}^{\ddagger b}$	E_0^{c}	k^{d}
200	3.54	95	−0.51	1.24
400	3.33	138	−1.05	1.56
600	3.19	173	−1.62	1.70
800	3.09	204	−2.22	1.78
1000	3.00	233	−2.84	1.83

[a] H–CH_3 bond length (Å) at the transition state.
[b] Harmonic vibrational frequency for the degenerate H–CH_3 rocking modes (cm^{-1}).
[c] Difference in classical potential energy between the transition state and reactants.
[d] Rate constant in units of $10^8 m^3 mol^{-1} s^{-1}$.

10.9 CRITIQUE OF TRANSITION STATE THEORY

In this chapter, we have considered an approach to the theory of rate processes which is both microscopic and statistically based. While this theory has been remarkably successful in systematizing a large amount of data, it is appropriate to look back and ask ourselves how well justified are the basic premises of the theory. The following points need to be considered:

1. Transition state theory makes use of classical mechanics in evaluating partition functions and canonical averages. We have considered, in section 10.4, the difficulties of arriving at a quantum mechanical formulation of transition state theory and the effect of tunneling as a specifically quantum phenomenon. Tunneling appears to be important for reactions involving very light species, such as H or D atoms.

2. Transition state theory simplifies the motion on the potential surface to one-dimensional motion along the reaction coordinate. However, the actual motion may include components normal to the reaction coordinate. To the extent that this is so, the assumption of separability of coordinates (see section 10.2) is invalidated. The reaction path may lie above the minimum energy configuration across the potential surface, with the result that the effective activation energy is somewhat higher than the barrier height E_0.

3. The use of harmonic potentials at the saddle point is reasonably accurate at low temperatures, since large-amplitude motions are not frequently encountered in the transition state configuration. At higher temperatures, however, when the average energies of the molecules and the transition state are higher, anharmonicity corrections could be important.

4. By the first assumption in section 10.2, systems pass through the transition state only once. In that case, exactly one-half of the systems yield products (see equation 10–9). By contrast, it is often found experimentally that the effective yield of products is much smaller than this, due to multiple crossings of the transition state region (see section 10.8). This is commonly represented by an empirical "transmission coefficient" $\kappa < 1$, analogous to the "steric factor" $p < 1$ in the old collision theory (see chapter 8). To the extent that recrossing the transition state region (or, equivalently, failure to react in a bimolecular encounter) can be attributed to the need for the molecules to approach each other in a particular orientation or relative configuration, this can also be represented by a negative entropy of activation $\Delta S^{\ddagger}$, which "soaks up all the p".[40]

5. Again, the local equilibrium postulate (the first assumption of section 10.2) is a consequence of the absence of any recrossing of the transition state. But if recrossing occurs, as evidenced by a transmission coefficient less than 1, then local equilibrium is not maintained and the calculation of equilibrium partition functions as in section 10.6.1 may not be valid.

6. A fundamental assumption in transition state theory is that both the reactants and the products are maintained in Maxwell-Boltzmann equilibrium. The examples presented in chapter 9, however, amply demonstrate that in bimolecu-

lar gas reactions one can have both selective energy consumption (from reactants) and specific energy release (in products), that is, highly non-Boltzmann distributions over energy states. The equilibrium condition can be maintained if relaxation of both reactants and products is rapid compared with the reaction rate.[41] It might appear that transition state theory would be more applicable to reactions in liquid solutions than in low-pressure gases, because relaxation times in solution may be on the order of picoseconds. However, while there is a region of low solvent density and viscosity in which rapid relaxation may be significant, in general, solvent effects lead to deviations from transition state theory, as is discussed in chapter 12. These effects are manifested as the *diffusion limit* (see chapter 4) of reactants toward each other, which is on the order of 10^{-10} sec.

Notwithstanding all of these limitations, transition state theory does a remarkably good job describing a wide variety of chemical reactions.

REFERENCES

1. H. Eyring, *J. Chem. Phys. 3,* 107 (1935).
2. M. G. Evans and M. Polyani, *Trans. Faraday Soc. 31,* 875 (1935).
3. E. Wigner, *Trans. Faraday Soc. 34,* 29 (1938).
4. R. H. Fowler and E. A. Guggenheim, *Statistical Thermodynamics* (Cambridge: University Press, 1939).
5. J. Horiuti, *Bull. Chem. Soc. Jpn. 13,* 210 (1938).
6. B. H. Mahan, *J. Chem. Ed. 51,* 709 (1974); W. H. Miller, *Acc. Chem. Res. 9,* 306 (1976).
7. W. H. Miller, *J. Chem. Phys. 61,* 1823 (1974).
8. H. S. Johnston and D. Rapp, *J. Am. Chem. Soc. 83,* 1 (1961).
9. P. Pechukas and F. J. McLafferty, *J. Chem. Phys. 58,* 1622 (1973).
10. D. G. Truhlar and B. C. Garrett, *Acc. Chem. Res. 13,* 440 (1980).
11. R. A. Marcus and M. E. Coltrin, *J. Chem. Phys. 67,* 2609 (1977).
12. W. H. Miller, N. C. Handy and J. E. Adams, *J. Chem. Phys. 72,* 99 (1980).
13. R. T. Skodje, D. G. Truhlar and B. C. Garrett, *J. Chem. Phys. 85,* 3019 (1981); B. C. Garrett and D. G. Truhlar, *J. Chem. Phys. 81,* 309 (1984).
14. E. E. Nikitin, *Theory of Elementary Atomic and Molecular Processes in Gases* (Oxford: Clarendon, 1974), pp. 20–23.
15. H. S. Johnston, *Gas Phase Reaction Rate Theory* (New York: Ronald, 1966) pp. 38–47; B. C. Garrett and D. G. Truhlar, *J. Phys. Chem. 83,* 2921 (1979).
16. D. L. Bunker, *Theory of Elementary Gas Reaction Rates* (New York: Pergamon, 1966) pp. 29–32.
17. H. S. Johnston and J. Heicklen, *J. Phys. Chem. 66,* 532 (1966).
18. I. Shavitt, *J. Chem. Phys. 31,* 1359 (1959).
19. D. M. Bishop and K. J. Laidler, *Trans. Faraday Soc. 66,* 1685 (1970).

20. E. Pollak and P. Pechukas, *J. Am. Chem. Soc. 100*, 2984 (1978).

21. D. R. Coulson, *J. Am. Chem. Soc. 100*, 2992 (1978).

22. I. W. M. Smith, *Kinetics and Dynamics of Elementary Gas Reactions* (Boston: Butterworths, 1980) pp. 224–227.

23. R. L. Jaffe and J. B. Anderson, *J. Chem. Phys. 54*, 2224 (1971).

24. H. F. Schaefer, *J. Phys. Chem. 89*, 5336 (1985).

25. M. J. Frisch, B. Liu, J. S. Binkley, H. F. Schaefer and W. H. Miller, *Chem. Phys. Lett. 114*, 1 (1985).

26. K. H. Homann, W. C. Solomon, J. Warnatz, H. Gg. Wagner, and C. Zetzsch, *Ber. Bunsenges. Phys. Chem. 74*, 585 (1970).

27. N. Cohen and K. Westberg, *J. Phys. Chem. Ref. Data 12*, 531 (1983).

28. B. C. Garrett and D. G. Truhlar, *J. Chem. Phys. 70*, 1593 (1979).

29. R. A. Marcus, *J. Chem. Phys. 45*, 2138 (1966); ibid. 2630 (1966).

30. W. Forst, *Theory of Unimolecular Reactions* (New York: Academic Press, 1973) p. 154.

31. J. C. Keck, *J. Chem. Phys. 32*, 1035 (1960).

32. J. C. Keck, *Adv. Chem. Phys. 13*, 85 (1967).

33. D. L. Bunker and M. Pattengill, *J. Chem. Phys. 48*, 772 (1968).

34. J. B. Anderson, *J. Chem. Phys. 58*, 4684 (1973).

35. J. B. Anderson, *J. Chem. Phys. 62*, 2446 (1975).

36. M. Quack and J. Troe, *Ber. Bunsenges. Phys. Chem. 78*, 240 (1974); ibid. *81*, 329 (1977).

37. W. J. Chesnavich, T. Su and M. T. Bowers, *J. Chem. Phys. 72*, 2641 (1980).

38. W. L. Hase, *Acc. Chem. Res. 16*, 258 (1983).

39. W. L. Hase, S. L. Mondro, R. J. Duchovic and D. M. Hirst, *J. Am. Chem. Soc. 109*, 2916 (1987).

40. I. Amdur, private communication.

41. B. Widom, *Adv. Chem. Phys. 5*, 353 (1963).

BIBLIOGRAPHY

Basic Postulates and Derivation of Transition State Theory

BAUER, S, H. *J. Chem. Ed. 63*, 377 (1986).

GLASSTONE, S., LAIDLER, K. J., and EYRING, H. *The Theory of Rate Processes*. New York: McGraw-Hill, 1941.

JOHNSTON, H. S. *Gas Phase Reaction Rate Theory*. New York: Ronald, 1966.

LAIDLER, K. J. *Theories of Chemical Reaction Rates*. New York: McGraw-Hill, 1969.

LEVINE, R. D., and BERNSTEIN, R. B. *Molecular Reaction Dynamics*. Oxford: Oxford University Press, 1974.

NIKITIN, E. E. *Theory of Thermally Induced Gas Phase Reactions*. Bloomington, IN: Indiana University Press, 1966.

Statistical Thermodynamics

DAVIDSON, N. *Statistical Mechanics.* New York: Harper and Row, 1973.

HILL, T. L. *Introduction to Statistical Mechanics.* Reading, MA: Addison-Wesley, 1972.

RICE, O. K. *Elements of Statistical Mechanics.* San Francisco, CA: Freeman, 1967.

Quantum Mechanical Tunneling

BELL, R. P. *The Tunnel Effect in Chemistry.* London: Chapman and Hall, 1980.

Microcanonical and Variational Transition State Theory

HORIUTI, J., and KECK, J. C. *Adv. Chem. Phys. 13*, 85 (1967).

PECHUKAS, P. In *Dynamics of Molecular Collision (Part B),* edited by W. H. Miller. New York: Plenum, 1976, p. 299.

PECHUKAS, P. *Ann. Rev. Phys. Chem. 32*, 159 (1981).

SMITH, I. W. M. *Kinetics and Dynamics of Elementary Gas Reactions.* London: Butterworths, 1980, p. 129.

TRUHLAR, D. G., and GARRETT, B. C. *Ann. Rev. Phys. Chem. 35*, 159 (1984).

TRUHLAR, D. G., HASE, W. L., and HYNES, J. T. *J. Phys. Chem. 87*, 2664 (1983).

WARDLAW, D. W., and MARCUS, R. A., *Adv. Chem. Phys. 70*, 231 (1987).

PROBLEMS

10.1 Find expressions for s and ξ in a collinear A + B−C system for a point (r_1, r_2) where the reaction path on the potential surface is taken to be a circle of radius R centered at $r_1 = a_0$, $r_2 = b_0$. Over what range of r_1 and r_2 would this be a satisfactory approximation to the true reaction path?

*10.2 Consider two collinear reactions of the form A + BC → AB + C which have similar potential surfaces except that the transition state frequencies are 3000 cm^{-1} for the first reaction and 200 cm^{-1} for the second.
  (a) Qualitatively sketch the potential contours for each.
  (b) Which has the larger frequency factor? (A qualitative argument only is required here.)

*10.3 Consider the collinear reaction

$$M + F_2 \rightarrow MF + F$$

where M = H or D. Using transition state theory, find an expression for the *ratio* of H:D rate coefficients at a given temperature. Note that

$$k^{\text{TST}} = \frac{k_B T}{h} \frac{Q^{\ddagger}}{Q} e^{-E_0/k_B T}$$

and assume that the reactant and transition states are harmonic oscillators with force constants k_R and k_T, respectively. Also, $\nu = (k/m)^{1/2}/(2\pi c)$, and the transformation from Cartesian to (q_1, q_2) coordinates in a collinear ABC system yields $H = (p_1^2 + p_2^2)/2m + V$, where $m = [m_A m_B m_C/(m_A + m_B + m_C)]^{1/2}$.

*10.4 Consider the reaction A + BC → AB + C *in three dimensions*. Write down all the degrees of freedom (translational, rotational, and vibrational) of (i) the reactants and (ii) the transition state. Take the transition state to be linear (i.e., it will have two rotational degrees of freedom, not three as would a nonlinear molecule). Use the formulae for partition functions given in section 10.6.1.

(a) What are the units of each of the partition functions?

(b) Show that equation 10–20 gives the correct units for a second-order rate coefficient.

(c) Consider the specific reaction H + Br_2 → HBr + Br. Specify precisely what data are required to carry out an evaluation of the rate coefficient using equation 10–20.

10.5 On the basis of transition state theory, what temperature dependence of the pre-exponential factor would be expected for each of the following reactions?

(a) $H_2 + F_2 \rightarrow 2HF$

(b) $NO_2 + F_2 \rightarrow NO_2F + F$

(c) $2NOCl \rightarrow 2NO + Cl_2$

(d) $2NO_2 + O_2 \rightarrow 2NO_3$

(e) $C_2H_4 + HCl \rightarrow C_2H_5Cl$

10.6 Use transition state theory to determine an expression for the pre-exponential term and the rate constant for the following hypothetical reactions:

RXN 1: atom + linear → linear activated complex

RXN 2: atom + linear → nonlinear activated complex

RXN 3: atom + nonlinear → nonlinear activated complex

Express the pre-exponential factor in terms of appropriate partition functions. Also, cite one real chemical reaction of the kind of each of the three hypothetical reactions.

10.7 The following reaction was done experimentally at various temperatures in order to obtain rate constants:

$$O + HO_2 \longrightarrow OH + O_2$$

$$T_1 = 298 \text{ K} \qquad k_1 = 6.1 \times 10^{-11} \text{ cm}^3/\text{molecule} \cdot \text{s}$$

$$T_2 = 229 \text{ K} \qquad k_2 = 7.57 \times 10^{-11} \text{ cm}^3/\text{molecule} \cdot \text{s}$$

(a) Using the experimental data, find the activation energy E_{act}.

(b) Calculate the following thermodynamic properties at $T = 298$ K using transition state theory: $\Delta E^{\ddagger}, \Delta H^{\ddagger}, \Delta S^{\ddagger}, \Delta G^{\ddagger}, A$.

10.8 (a) Using transition state theory, formulate the rate constant for the exchange reaction

$$H + H_2 \; \rightleftharpoons \; H_3^{\ddagger} \; \longrightarrow \; H_2 + H \qquad (1)$$

in terms of the appropriate functions. Assume that $H_3^{\ddagger}$ has a linear symmetric configuration with a ground state electronic degeneracy of 2 and a symmetry number of 2. The electronic degeneracies for ground state H_2 and H are 1 and 2, respectively.

For the $H_3^{\ddagger}$ transition state, theoretical calculations suggest that the H–H distance is 0.93 Å and the fundamental frequencies are 2193, 978, and 978 cm^{-1}.

The barrier height may be taken to be 37.6 kJ/mole. The H–H distance in H_2 is 0.74 Å, and its fundamental vibrational frequency is 4395 cm^{-1}.

Calculate numerical values for the transition state theory rate constants for reaction 1 at 300 and 1,000 K. The experimentally measured rate coefficient is $(2.1 \pm 0.6) \times 10^{-12}$ cm^3 molecule^{-1} sec^{-1} at 1,000 K.

(b) Calculate pre-exponential factors for an "old collision theory" rate constant for the reaction $H + H_2 \rightarrow H_2 + H$, and compare the rates calculated in this way at 300 K and 1,000 K with the values found in part (a). Assume a hard-sphere diameter $d = 2.5$ Å.

(c) Estimate the Wigner tunneling correction for the $H + H_2$ reaction.

(d) Compare your results with any other calculated rate coefficients (classical or quantum) you can readily find in the literature.

10.9 Repeat problem 7 for the reaction $D + D_2 \rightarrow D_2 + D$. Assume that the geometry and potential surface for $D + D_2$ are the same as those for $H + H_2$, and use isotope-effect relationships to find the vibrational frequencies and moments of inertia.

10.10 Consider the reaction

$$R-H + Cl \longrightarrow R + HCl$$

where R is an alkyl group. Using the following reactant data and assumptions as well as the BEBO scheme discussed in chapter 7 to estimate the necessary molecular parameters for the transition state, calculate the Arrhenius A-factor for this reaction at 400 K. *Data:* The H–Cl bond length and vibrational frequency are 1.27 Å and 3000 cm^{-1}, respectively. The R–H bond length is 1.09 Å and the stretching and bending frequencies are 3000 cm^{-1} and 1500 cm^{-1}, respectively. The mass and moment of inertia of the R group (assumed to be spherical) are 30 amu and 20 amu Å^2, respectively. Assume that the R–H and H–Cl bond orders are both $\frac{1}{2}$ at the transition state, and make any other reasonable assumptions that you find necessary.

10.11 The reaction of CH with N_2 is of importance in the chemistry of planetary atmospheres and hydrocarbon flames. By understanding this basic reaction, other complex systems such as combustion in the environment can be studied. The reaction is

$$CH + N_2 \longrightarrow (CHN_2)^{\ddagger} \longrightarrow HCN + N$$

Using transition state theory, find the partition functions for each molecule and then the rate constant at 298 K. Also, specify all the degrees of freedom and the temperature dependence of the A factor. From experiment, the second-order rate constant at 298 K has been found to be $k_{exp} = 7.1 \times 10^{-14}$ cm^3/molecule·sec.

PROPERTIES OF THE TRANSITION STATE + MOLECULES.

	CH	N_2	HCN$_2$ (assume linear adduct)
M (amu = g/mol)	13.019	28.013	41.032
I (10^{-40} g cm^2)	1.935	13.998	73.2
E_o			≈ 221 kJmol
ν (cm^{-1})	2733	2330	3130; 2102; 1252; 1170; 564; 401
g_{elec}	2	1	2

*These problems have been provided by Dr. R. G. Gilbert, Department of Theoretical Chemistry, University of Sydney, Australia.

CHAPTER 11

Unimolecular Reaction Dynamics

Developments in the field of chemical kinetics often arise from the interplay and exchange of ideas between theoretical and experimental research. This relationship is perhaps more vibrant and alive in unimolecular reaction dynamics than in any other area of kinetics. The development of unimolecular theoretical models has been fostered by new experimental studies which have provided more detail about chemical reactivity and molecular motion than was previously available. Likewise, formerly unknown microscopic properties of unimolecular reactions have been discovered by theoretical computer simulation studies and later observed in experiments designed to probe for such properties.

A unimolecular reaction is denoted by

$$A^* \longrightarrow \text{Products} \tag{11-1}$$

where the asterisk indicates that A contains sufficient internal vibrational energy to decompose. This energy is denoted by the symbol E and must be greater than the unimolecular decomposition threshold energy E_0. There are three general types of potential energy profiles for unimolecular reactions (see Figure 11-1). One type is for an isomerization reaction, such as

$$CH_3NC \longrightarrow CH_3CN \tag{11-2}$$

where there is a substantial potential energy barrier separating the two isomers. The other two types are for a unimolecular dissociation reaction. In one case, as for example,

$$C_2H_5Cl \longrightarrow HCl + C_2H_4 \tag{11-3}$$

or

$$C_2H_5 \longrightarrow H + C_2H_4 \tag{11-4}$$

there is an activation energy for the reverse recombination reaction. In the other, for example,

$$C_2H_6 \longrightarrow 2CH_3 \tag{11-5}$$

there is no activation energy for recombination.

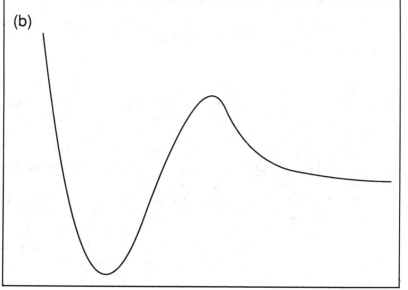

Figure 11–1. Schematic potential energy profiles for three types of unimolecular reaction. (a) Isomerization. (b) Dissociation where there is a high energy barrier and a large activation energy for reaction in both the forward and reverse directions.

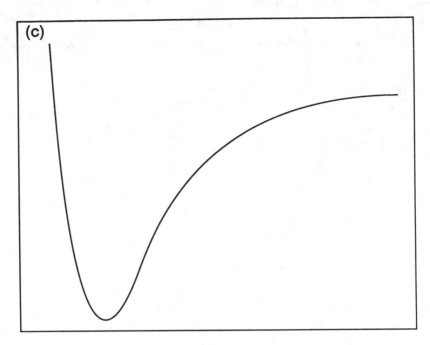

(c)

Reaction Coordinate

Figure 11–1. (c) Dissociation where the potential energy rises monotonically as for rotational ground state species, so that there is no barrier to the reverse association reaction.

To understand the dynamics and kinetics of unimolecular reactions a number of different topics must be treated and mastered. Some of the most important are (1) the formation of an energized molecule with energy E, (2) the intramolecular vibrational/rotational motion of highy energized molecules; (3) unimolecular decomposition lifetime distributions and rate constants; (4) intermolecular energy transfer to and from highly energized molecules; and (5) energy partitioning between unimolecular dissociation fragments.

There are a number of different ways to present these topics. The one partially adopted here is based on a historical perspective of the development of unimolecular reaction dynamics. Its strength is that the special relationship between theory and experiment which has existed in this field is emphasized. In the presentation given, the primary focus will be on unimolecular reactions which occur on the ground-state potential energy surface.

11.1 FORMATION OF ENERGIZED MOLECULES

Before a unimolecular reaction can occur, a molecule must be energized above the reaction threshold energy E_0. A number of different experimental methods have been used to accomplish this energization. For example, energization can take place by transfer of energy in a bimolecular collision, as in the reaction

$$CH_3NC + Ar \longrightarrow CH_3NC^* + Ar \qquad (11-6)$$

Another method which involves molecular collisions is chemical activation. Here the collision is reactive, and the molecule is excited by potential energy release as in the reaction

$$CH_3 + CF_3 \longrightarrow CH_3CF_3^* \qquad (11-7)$$

The reaction coordinate potential energy profile for the $CH_3 + CF_3$ system is similar to the one shown in Figure 11–1(c). The excited $CH_3 + CF_3$ molecule can redissociate to radicals, or go on to form new products such as $CH_2{=}CF_2 + HF$.

Absorption of electromagnetic radiation is another means of preparing vibrationally excited molecules. A widely used method involves initial electronic excitation by absorption of one photon of visible or ultraviolet radiation. After this excitation, many molecules undergo a rapid internal conversion to the ground electronic state S_0, which converts the energy of the absorbed photon into vibrational energy. Such an energization scheme is depicted in Figure 11–2 for formaldehyde, where the complete excitation/decomposition mechanism is

$$H_2CO(S_0) + h\nu \longrightarrow H_2CO(S_1) \longrightarrow H_2CO^*(S_0) \longrightarrow H_2 + CO \quad (11-8)$$

Molecules also acquire vibrational energy by the absorption of infrared photons. In contrast to visible and ultraviolet radiation, the energy provided by a single infrared photon is usually much less than the unimolecular threshold energy E_0.

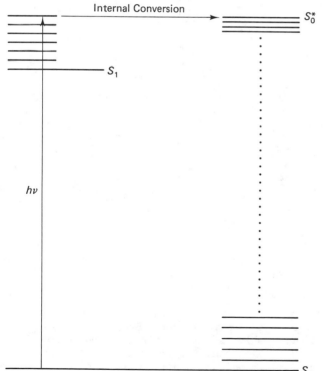

Figure 11–2. Preparation of a vibrationally excited ground-state molecule S_0^* by internal conversion from a vibrational level of S_1.

However, by using a sufficiently intense monochromatic radiation source such as an infrared laser, a molecule may be compelled to absorb multiple infrared photons so that its total vibrational energy is in excess of E_0.[1] Chapter 14, section 14.4.1, discusses this further.

Two additional excitation methods, both of which employ lasers, have also been developed: overtone excitation[2] and stimulated emission pumping (SEP)[3]. In overtone excitation, states are prepared by direct single-photon absorption in which an MH or MD bond contains n quanta of vibrational energy. M is a massive atom in contrast to H or D. For states with large n, the energy in the bond may exceed the molecule's unimolecular threshold. Overtone excitation of CH and OH bonds has been used to decompose molecules. The limitation to the technique is the very weak oscillator strengths of overtone absorptions.

Stimulated emission pumping is an even more selective excitation process than overtone excitation. SEP involves the use of two lasers; the excitation mechanism is

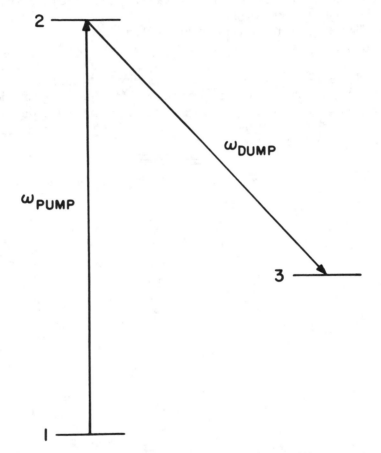

Figure 11–3. Three-level stimulated emission pumping (SEP) scheme, showing PUMP and DUMP transitions. The PUMP laser excites a vibrational/rotational level of an excited electronic state. The DUMP laser then stimulates emission from this state to an excited vibrational/rotational level of the ground electronic state. [Adapted from C. Kittrell, et al. *J. Chem. Phys. 75*, 2056 (1981).]

depicted in Figure 11–3. The first laser prepares a vibrational level of an excited electronic state. Before this state undergoes a spontaneous radiative or nonradiative transition, a second laser with a lower energy photon stimulates emission of light. The stimulated emission forms a highly excited vibrational level S_0. Because of the high resolution of the lasers, individual rotation-vibration states of S_0 may often be prepared.

The foregoing excitation methods differ in the extent of the vibrational state selectivity inherent in their energization processes. Later in this chapter it will be seen that, as the vibrational selectivity of the experiment is increased, an accurate unimolecular rate theory often requires a more detailed description of molecular properties.

11.2 SUM AND DENSITY OF STATES

At the high energies to which unimolecular reactants are energized, there is often a large number of vibrational states in an energy interval dE. The number of these vibrational states is denoted by $W(E)$, which is related to the sum of the states, $G(E)$, and the density of the states, $N(E)$. $G(E)$ is defined as the number of vibrational states in the energy range 0 to E. The number of states in the energy interval from E to $E + dE$ is $W(E) = N(E)dE$. Thus, the density of the states, $N(E)$, is given by $[G(E + dE) - G(E)]/dE$.

The calculation of $G(E)$ and $N(E)$ may be illustrated by using the low-energy harmonic vibrational energy levels of H_2O depicted in Figure 11–4. For $E = 5000$ cm^{-1} $G(E)$ is 6, and it rises to 19 for $E = 10000$ cm^{-1}. Clearly $N(E)$ is not a continuous function. For example, in the constant-width energy intervals of 2000–3000, 3000–4000, and 7000–8000 cm^{-1}, there are 0, 3, and 4 states, respectively. For higher energies and for molecules with more vibrational degrees of freedom, $G(E)$ becomes very large and approximates a continuous function. This is illustrated in Figure 11–5, where values of $G(E)$ for CH_4 are plotted against E. Large values of harmonic $G(E)$ and $N(E)$ are easily enumerated on a computer with the very efficient Beyer-Swinehart algorithm.[4] Continuous densities of states are illustrated in Figure 11–6.

If all the frequencies in the molecule are identical, evaluating the harmonic $W(E)$ and $G(E)$ terms becomes an elementary combinatorial problem. Suppose the molecule has s oscillators each with frequency ν and the total molecular energy above the zero-point level is $E = jh\nu$. Then $W(E)$ and $G(E)$ are found by considering the number of ways of distributing the j quanta among the s oscillators; specifically,

$$W(E) = \frac{(j + s - 1)!}{j!(s - 1)!} \tag{11–9}$$

and

$$G(E) = \frac{(j + s)!}{j!s!} \tag{11–10}$$

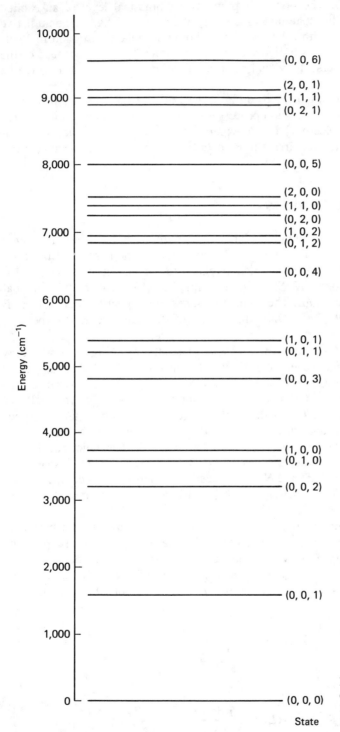

Figure 11–4. Harmonic vibrational levels for H_2O. ν_1 (asymmetric stretch) = 3750 cm^{-1}, ν_2 (symmetric stretch) = 3650 cm^{-1}, and ν_3 (bend) = 1600 cm^{-1}.

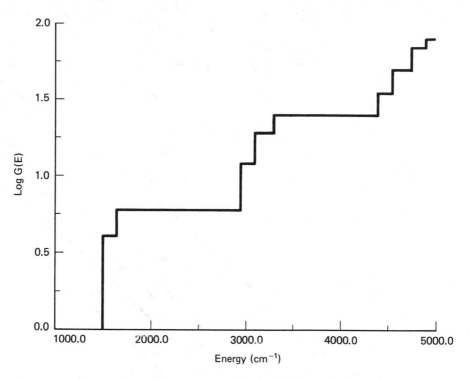

Figure 11–5. Direct count of $G(E)$ for CH_4. Energy is in cm^{-1}.

It is worthwhile to consider the classical equivalents to $G(E)$ and $N(E)$, since they are used in unimolecular theories and provide a basis for analytic approximations to the quantum mechanical sum and density of states. As shown in Appendix 2, the classical sum of states is

$$G(E) = \frac{E^s}{s! \prod_{i=1}^{s} h\nu_i} \qquad (11\text{--}11)$$

The classical density of states $dG(E)/dE$ is then

$$N(E) = \frac{E^{s-1}}{(s-1)! \prod_{i=1}^{s} h\nu_i} \qquad (11\text{--}12)$$

As illustrated in Table 11–1, the classical expression for the sum severely underestimates the quantum mechanical harmonic value. Similar errors are observed for the density of states. The error in the classical approximation decreases with energy, and, according to the principle of correspondence, classical and quantum theories should become equivalent in the high-energy limit. However, for energies of interest in unimolecular reactions, the classical sum and density of states are woefully inadequate.

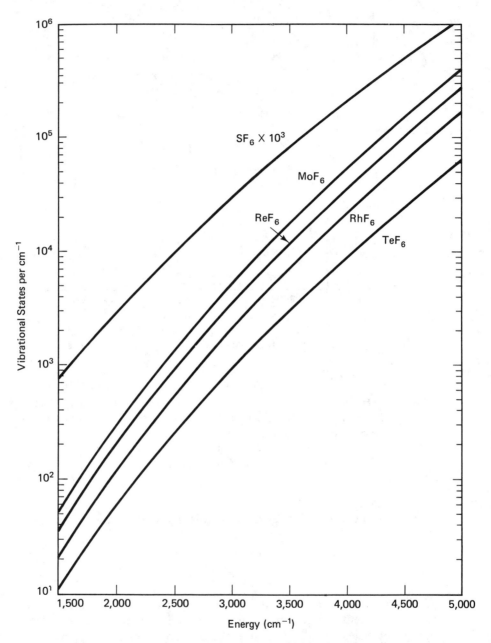

Figure 11–6. Density of vibrational states for MoF_6, ReF_6, RhF_6, TeF_6, and SF_6. [Adapted from D. Jackson, *Statistical Thermodynamic Properties of Hexafluoride Molecules*, Report LA-6025-MS, Los Alamos Scientific Laboratory (1975).]

TABLE 11–1 VALUES OF HARMONIC $G(E)$ FOR CYCLOPROPANE BY VARIOUS METHODS[a]

E (kJ/mol)	Classical (11–11)	Marcus-Rice (11–13)	Whitten-Rabinovitch (11–14)	Exact count
41.8	0.00	545×10^2	7.17×10^2	8.02×10^2
125	0.02	20.9×10^6	2.65×10^6	2.69×10^6
209	8.20×10^2	21.5×10^8	6.15×10^8	6.12×10^8
418	1.72×10^9	9.94×10^{12}	5.90×10^{12}	5.84×10^{12}
627	8.58×10^{12}	4.03×10^{15}	3.02×10^{15}	3.00×10^{15}
837	3.61×10^{15}	4.27×10^{17}	3.56×10^{17}	3.54×10^{17}

[a]Taken from Table 5.11 of reference 25. The cyclopropane harmonic vibrational frequencies are 3221(6), 1478(3), 1118(7), 878(3), 749(2) cm^{-1}. The values in parentheses are the degeneracies associated with each frequency.

To improve on the classical approximation, Marcus and Rice[5] suggested that the classical energy should be measured from the minimum of the vibrational potential. Thus, the quantum sum or density at energy E should be compared with the classical sum or density at energy $E + E_z$, where E_z is the molecular zero-point energy of $\Sigma \, h\nu_i/2$. The resulting "semiclassical" expression for the sum of states is

$$G(E) = \frac{(E + E_z)^s}{s! \displaystyle\prod_{i=1}^{s} h\nu_i} \qquad (11\text{–}13)$$

The semiclassical $N(E)$ is obtained by differentiating equation 11–13 with respect to E. Semiclassical values for $G(E)$ are given in Table 11–1. Plainly, this approach gives a more accurate approximation than the classical expression given by equation 11–11; however, the number of states is overestimated.

Although the Marcus-Rice expression is not sufficiently accurate, it leads to a greatly improved semiclassical type of approach. Rabinovitch and co-workers[6] suggested that only an appropriate fraction of the zero-point energy should be included in the classical energy. Specifically, they proposed that

$$G(E) = \frac{(E + aE_z)^s}{s! \displaystyle\prod_{i=1}^{s} h\nu_i} \qquad (11\text{–}14)$$

The empirical factor a, which has the limits of 0 and 1 for $E \to 0$ and $E \to \infty$, respectively, is written as a function of $E' = E/E_z$ and the dispersion in the molecular vibrational frequencies. Explicitly,

$$a = 1 - \beta w(E') \qquad (11\text{–}15)$$

where

$$\beta = \frac{s - 1}{s} \frac{\langle \nu^2 \rangle}{\langle \nu \rangle^2}$$

and

$$w = [5.00E' + 2.73(E')^{0.75} + 3.51]^{-1} \qquad (0.1 < E' < 1.0)$$
$$w = \exp[-2.4191(E')^{0.25}] \qquad\qquad\quad (1.0 < E' < 8.0)$$

(11–16)

Differentiating equation 11–14 with respect to E gives, for the density of states,

$$N(E) = \frac{(E + aE_z)^{s-1}}{(s - 1)! \prod\limits_{i=1}^{s} h\nu_i}\left[1 - \beta\frac{dw(E')}{dE'}\right]$$

(11–17)

As shown in Table 11–1, equation 11–14 gives values for the sum of states that are in excellent agreement with the actual quantum mechanical values.

11.3 LINDEMANN-HINSHELWOOD THEORY OF THERMAL UNIMOLECULAR REACTIONS

Perrin[7] proposed in 1919 that reactant molecules acquire energy by absorbing radiation from the walls of the reaction vessel. Many textbooks on kinetics assert that this *radiation hypothesis* was refuted by Langmuir,[8] who showed that the amount of energy radiated to the gas would be insufficient to account for observed unimolecular reaction rates. Actually, a careful reading of Langmuir's paper reveals that the hypothesis is not disproved by his arguments. Langmuir calculated the radiation flux only for $h\nu = hc/\lambda > E_0$, i.e., for $\lambda < 400$ nm, corresponding to $E_0 \approx 300$ kJ/mol. Since the black-body radiation density is essentially zero for wavelengths below 400 nm, for a typical oven temperature of 1,000 K, there would of course be insufficient radiation to energize the molecules.

What Langmuir neglected in this analysis is that most molecules (with the exception of homonuclear diatomics, such as I_2) absorb strongly in the infrared region of the spectrum, and the black-body radiation curve actually peaks in the infrared, near $\lambda = 3$ μm at $T = 1,000$ K. To be fair, infrared absorption by molecules was essentially unknown in the early twentieth century, when this debate was being conducted. If Langmuir's calculation is redone for infrared absorption (see problem 11.4), it becomes clear that there is more than enough energy radiated into and absorbed by the gas to account for observed unimolecular decomposition rates.

The argument that *is* fatal to the radiation hypothesis, which apparently was overlooked by Langmuir, is a thermodynamic one. The hypothesis requires that one system (the reactor oven, represented by a black body at temperature T) do work on another system (the reaction mixture, also at temperature T) by inducing a chemical change in the latter. However, the second law of thermodynamics states that this is impossible unless there is a lower temperature reservoir to absorb excess heat. Physically, the gas would radiate as much energy back to the wall of the reactor as it absorbs. In order to produce net chemical work, we need a source at a much *higher* temperature than the gas. An infrared laser is, in fact, such a source, and irradiation with such lasers turns out to be a very effective way of initiating unimolecular reactions. This is discussed in chapter 14.

Despite the role that could be played by infrared radiation, the radiation hypothesis is in fact untenable for thermal unimolecular reactions. There are several reasons for this:

1. Langmuir's argument is applicable to reactions of homonuclear diatomic molecules such as iodine, which dissociates readily at 800–1,000 K but absorbs light only at wavelengths less than 600 nm. Since this case refutes the hypothesis, it cannot be generally applicable.
2. The radiation hypothesis predicts a rate dependence on the surface-to-volume ratio in the reactor, or on the presence of nonreactive absorbers, which is at variance with experimental results.
3. Most importantly, a radiation-driven reaction would have a rate truly independent of gas pressure. However, as discussed subsequently, there is a pronounced falloff of the rate at low gas pressures. It is primarily this observation which forces us to abandon the radiation hypothesis in favor of a collisional-activation mechanism developed by Lindemann and Hinshelwood.

In 1922, Lindemann[9] proposed a general theory for thermal unimolecular reactions which forms the basis for the current theory of thermal unimolecular rates. He proposed that molecules become energized by bimolecular collisions, as in equation 11–6, with a time lag between the moment of collisional energization and the time the molecule decomposes. Energized molecules could then undergo deactivating collisions before decomposition occurred. A major achievement of Lindemann's theory is its ability to explain the experimental finding that the reaction rate changes from first to second order in going from the high- to low-pressure limit.

According to these concepts, the mechanism for the unimolecular reaction may be written as

$$A + M \xrightarrow{k_1} A^* + M$$
$$A^* + M \xrightarrow{k_{-1}} A + M \qquad (11\text{--}18)$$
$$A^* \xrightarrow{k_2} \text{Products}$$

Here, A^* represents a molecule with sufficient energy to react and M is some collision partner. It is assumed that each $A^* + M$ collision is "strong" and thus leads to de-energizing of A; this is known as the *strong collision assumption* for de-energizing collisions. Therefore, k_{-1} is equated with the gas-kinetic collision number Z_1. If the steady-state hypothesis is applied to the concentration of A^*, the overall rate of reaction becomes

$$\text{rate} = k_{\text{uni}}[A] = k_2[A^*] = \frac{k_1 k_2 [A][M]}{k_{-1}[M] + k_2} \qquad (11\text{--}19)$$

Clearly, the reaction rate is first order at high pressures, where $[M] \to \infty$, and second order at low pressures, where $[M] \to 0$. The high- and low-pressure expressions for k_{uni} are

$$k_\infty = \frac{k_1 k_2}{k_{-1}} \qquad (11-20)$$

and

$$k_0 = k_1[M] \qquad (11-21)$$

The rate constant ratio k_1/k_{-1} is the equilibrium ratio $[A^*]/[A]$. Therefore, the high-pressure rate constant is simply the rate constant for unimolecular decomposition multiplied by the probability that the molecule is energized. At low pressure, the rate-determining step is bimolecular activation and the rate constant is linearly dependent on the pressure of $[M]$. The transition from the high-pressure rate constant $k_{uni} = k_\infty$ to the low-pressure linear decrease in k_{uni} is called the *fall-off region*. The pressure at which $k_{uni}/k_\infty = \frac{1}{2}$ is denoted by $p_{1/2}$. From equations 11–19 and 11–20, it follows that

$$p_{1/2} \propto [M]_{1/2} = k_\infty/k_1 \qquad (11-22)$$

A plot of k_{uni} vs. p is illustrated in Figure 11–7.

To calculate a theoretical Lindemann plot of k_{uni} vs. $[M]$, k_{uni} in equation 11–19 is written as

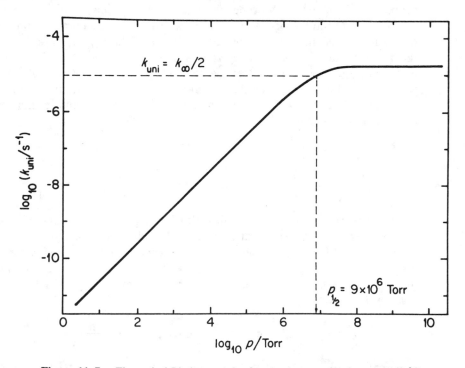

Figure 11–7. Theoretical Lindemann plot for *cis-* → *trans-* but-2-ene at 469 °C. [Adapted from P. J. Robinson and K. A. Holbrook, *Unimolecular Reactions* (New York: Wiley-Interscience, 1972). Reproduced with permission of John Wiley and Sons Limited.]

Unimolecular Reaction Dynamics   Chap. 11

$$k_{uni} = \frac{k_\infty}{1 + k_\infty/k_1[M]} \tag{11-23}$$

The value of k_∞, the first-order rate constant at high pressure, is found from experiment. A working assumption that can be made in applying the Lindemann theory is to calculate k_1 from the line-of-centers collision theory expression (see chapter 8, section 8.1.2)

$$k_1 = Z_1 \exp\left(-E_0/k_B T\right) \tag{11-24}$$

where E_0 is simply equated to the high-pressure activation energy E_∞. In all cases, such an interpretation of the Lindemann theory leads to a gross discrepancy with experiment. The theory predicts the fall-off in k_{uni} to occur at much higher pressures than what is observed experimentally. Since there can be no doubt about k_∞, which is an experimental quantity, the error must be in the estimation of k_1. The Hinshelwood modification to the Lindemann mechanism, described shortly, gives a much larger value for k_1, and theoretical fall-off curves that are in much better agreement with experiment are obtained.

The line-of-centers energization rate constant of equation 11–24 gives the frequency of collisions in which the relative kinetic energy along the line of centers is greater than or equal to E_0. No account is taken of the internal energy of the reactant molecule. In 1926 Hinshelwood[10] proposed that, in addition to the relative motion, the internal degrees of freedom can contribute to the threshold energy E_0. The probability that a molecule contains energy greater than or equal to E_0 clearly increases with the number of internal degrees of freedom. As a result, the energization rate constant k_1 is larger for a complex reactant molecule than for a simple one. For a reactant molecule with s classical degrees of freedom, Hinshelwood found that

$$k_1 = \frac{Z_1}{(s-1)!}\left(\frac{E_0}{k_B T}\right)^{s-1} \exp\left(-E_0/k_B T\right) \tag{11-25}$$

To derive the Hinshelwood expression for k_1, one starts with equation A2–31 in Appendix 2, which gives the classical probability that a molecule has internal energy in the range from E to $E + dE$. This probability also represents the fraction of molecules energized in this range. The differential energizing rate constant for formation of reactant molecules with energy lying between E and $E + dE$ is dk_1. At high concentrations (i.e., where $[M] \rightarrow \infty$), where dissociation of energized molecules is negligible, an equilibrium is maintained between the energizing and de-energizing steps in equation 11–18. Under these conditions dk_1/k_{-1} is the statistical fraction of molecules with internal energy between E and $E + dE$, so that

$$\frac{dk_1}{k_{-1}} = \frac{1}{(s-1)!}\left(\frac{E}{k_B T}\right)^{s-1} e^{-E/k_B T}\left(\frac{dE}{k_B T}\right) \tag{11-26}$$

Equation 11–26 is rigorously valid only in the high-pressure limit. To use it for all concentrations, the *strong collision assumption* must be made. The application of this assumption for de-energizing collisions has already been discussed. According to the strong collision assumption, large amounts of energy are transferred in the molecular collisions so that energizing and de-energizing may be viewed as single-step processes, in contrast to ladder-climbing processes. This assumption also requires the collision to be so violent that the distribution of energy resulting from energizing or de-energizing collisions is random. As a result, the distribution of energized molecules with energy in the range from E to $E + dE$ may be determined directly from statistical considerations; i.e., equation 11–26 (see also chapter 13, section 13.4.3).

To find the Hinshelwood expression for k_1/k_{-1}, dk_1/k_{-1} in equation 11–26 is integrated between the limits $E = E_0$ and ∞ to obtain

$$\frac{k_1}{k_{-1}} = \int_{E=E_0}^{\infty} \frac{dk_1}{k_{-1}} = \int_{E=E_0}^{\infty} \frac{1}{(s-1)!} \left(\frac{E}{k_B T}\right)^{s-1} e^{-E/k_B T} \left(\frac{dE}{k_B T}\right) \qquad (11\text{–}27)$$

For the normal situation where E_0 is much greater than $(s-1)k_B T$, the result of the integration is approximately [equation A2–33 in Appendix 2]

$$\frac{k_1}{k_{-1}} = \frac{1}{(s-1)!} \left(\frac{E_0}{k_B T}\right)^{s-1} e^{-E_0/k_B T} \qquad (11\text{–}28)$$

For $k_{-1} = Z_{-1} = Z_1$, the Hinshelwood expression for k_1, equation 11–25, results. Inserting equation 11–25 into equation 11–20 gives the high-pressure rate constant according to the Hinshelwood-Lindemann theory, i.e.,

$$k_\infty = \frac{k_2}{(s-1)!} \left(\frac{E_0}{k_B T}\right)^{s-1} e^{-E_0/k_B T} \qquad (11\text{–}29)$$

Equations 11–25 and 11–29 form the basis of what is usually called the Hinshelwood-Lindemann theory of thermal unimolecular dissociation. Unimolecular falloff curves calculated with this theory are far superior to those calculated with the Lindemann theory. However, significant differences are still found between theoretical and experimental k_{uni} curves versus [M], particularly at low pressures.

Since k_1 increases with s in the Hinshelwood-Lindemann theory, the dissociation rate constant k_2 decreases with s. This may be illustrated using the typical parameters $A_\infty = 10^{13}$ sec^{-1} and $E_\infty/RT = 40$. The value for k_2 is given by $k_\infty(k_{-1}/k_1)$; for $s = 1$, 5, and 10, $k_2 = 10^{13}$, $10^{9.5}$, and $10^{6.5}$ sec^{-1}, respectively. Thus, the lifetime $\tau = 1/k_2$ of the energized molecule increases when the molecule can store energy among a greater number of degrees of freedom. Intuitively, this is what is expected; however, one also expects k_2 to depend on the energy content of the energized molecule A*. According to the Hinshelwood-Lindemann theory, k_2 is a constant that is independent of energy.

In the next section we discuss how statistical theories may be used to calculate k_2 as a function of energy. Making k_2 energy-dependent, expressed as $k(E)$, requires a generalized Hinshelwood-Lindemann mechanism in order to consider energizing, de-energizing, and decomposition for an energy interval from E to $E + dE$:

$$A + M \xrightarrow{dk_1} A^*(E, E + dE) + M$$

$$A^*(E, E + dE) + M \xrightarrow{k_{-1}} A + M \qquad (11\text{–}30)$$

$$A^*(E, E + dE) \xrightarrow{k(E)} \text{products}$$

In this mechanism $k(E)$ has replaced the energy-independent rate constant k_2. Applying the steady-state approximation to the energized intermediate $A^*(E, E + dE)$ leads to the differential unimoleular rate constant

$$dk_{uni}(E, E + dE) = \frac{k(E)(dk_1/k_{-1})}{1 + k(E)/k_{-1}[M]} \qquad (11-31)$$

The total thermal unimolecular rate constant is obtained by integration thus:

$$k_{uni} = \int_{E_0}^{\infty} \frac{k(E)(dk_1/k_{-1})}{1 + k(E)/k_{-1}[M]} \qquad (11-32)$$

As discussed previously, it is assumed that for all pressures dk_1/k_{-1} represents the equilibrium probability that the reactant molecule A^* has energy in the range from E to $E + dE$. This probability may be denoted $P(E) \, dE$. Also, $k_{-1}[M]$ is the collision frequency ω between an A^* molecule and bath molecules. Making these two notational changes, equation 11–32 becomes the familiar expression for the thermal unimolecular rate constant,

$$k_{uni} = \omega \int_{E_0}^{\infty} \frac{k(E) \, P(E) \, dE}{k(E) + \omega} \qquad (11-33)$$

11.4 STATISTICAL ENERGY-DEPENDENT RATE CONSTANT k(E)

Two quite different approaches may be taken to determine an energy-dependent expression for k_2 denoted by $k(E)$. One is to consider the explicit nature of the intramolecular motion of highly energized molecules. This was the approach taken by Slater[11] in 1939, in which he pictured a molecule as an assembly of harmonic oscillators. Decomposition is assumed to occur when a critical coordinate (e.g., a bond length or bond angle) attains a critical displacement. A discussion and critique of the Slater theory, along with a general discussion of the intramolecular dynamics of highly energized molecules, is given in section 11-12.

In contrast to determining $k(E)$ in terms of intramolecular motion, an expression may be formulated totally on the basis of statistical assumptions. This is the approach used in the RRK (Rice-Ramsperger-Kassel) theory and its extension, which is referred to as the RRKM (Rice-Ramsperger-Kassel-Marcus) theory. These two theories picture a molecule as a collection of coupled harmonic oscillators which exchange energy freely under the assumptions that

1. All internal molecular states of A^* at energy E are accessible and will ultimately lead to decomposition products, and
2. Vibrational energy redistribution within the energized molecule is much faster than unimolecular reaction.

These two assumptions are embodied in the postulate of the RRK and RRKM theories that complete intramolecular vibrational energy redistribution (IVR) occurs on a time scale much shorter than $1/k(E)$.

Both the RRK and the RRKM theory assume that initially, when $\tau = 0$, there is a statistical population of reactant internal states at a fixed total energy. A collection of molecules forming such a distribution is called a *microcanonical ensemble*. The assumption requires that each state have equal probability of decomposing. Thus, a microcanonical ensemble will be maintained as the reactant molecules decompose, and, as a result, the unimolecular decomposition reaction will be described by only one time-independent rate constant $k(E)$, which is that for the microcanonical ensemble of reactant molecules.

The assumption of rapid and complete intramolecular vibrational energy redistribution implies that A* will have a random lifetime distribution given by[12]

$$P(\tau) = k(E) \exp \left[-k(E)\tau \right] \tag{11-34}$$

where $P(\tau)$ is the probability that A*, which contains energy E, has a lifetime τ. Equation 11–34 is similar to that for the probability of radioactive decay, which is what one expects if the decomposition of A* is purely statistical. With random lifetimes there is no particular tendency for all of the A* molecules to decompose soon after their formation, or even to exist for some particular length of time before decomposing.

Slater and Bunker[12,13] have shown that equation 11–33, which embodies the Lindemann-Hinshelwood theory with the strong collision assumption, requires the random lifetime distribution. To demonstrate this, let τ be the time required for a certain A* molecule with energy E to reach dissociation, by virtue of its natural motion in the absence of collisions. For an equilibrium distribution of molecular states at E, there will be a distribution $P(\tau)$ of these natural lifetimes. The probability that A* will *not* undergo a collision in time τ is $\omega \exp(-\omega\tau)$ if the average collision frequency is ω. In that case the rate of reaction at energy E is

$$k_{uni}(E) = \omega \int_0^\infty \exp(-\omega\tau) P(\tau) \, d\tau \tag{11-35}$$

and the complete thermal unimolecular rate constant is

$$k_{uni} = \int_0^\infty k_{uni}(E) P(E) \, dE \tag{11-36}$$

If we compare equation 11–36 with equation 11–33, we can see by trial that correspondence is complete if $P(\tau)$ is given by equation 11–34.

11.5 RRK THEORY

The theory which is now known as RRK theory was developed independently and nearly simultaneously by Rice and Ramsperger[14] and by Kassel[15] in 1927 and 1928. Both classical and quantum versions of RRK theory were developed, and in the limit of a large excitation energy E the two versions become identical.

The classical theory is based on the notion that the probability that a molecule of s classical oscillators with total energy E has energy greater than or equal to E_0 in one chosen oscillator, which is the critical mode leading to reaction. This probability is given by the number of ways to attain this particular distribution divided by the total number of ways to distribute E between the s oscillators. The latter number is given in turn by the classical density of states, i.e., $N(E)$ of equation 11–12. The former is the convolution of the density of states in which the critical oscillator contains energy $E' + E_0$ and the density of states for energy $E - E' - E_0$ in the remaining $s - 1$ oscillators. As a result, the probability for the critical oscillator having energy greater than or equal to E_0 is

$$\int_0^{E-E_0} \left[\left(\frac{(E - E' - E_0)^{s-2}}{(s-2)! \prod_{i=1}^{s-1} h\nu_i} \frac{dE'}{h\nu_s} \right) \right] \bigg/ \frac{E^{s-1}}{(s-1)! \prod_{i=1}^{s} h\nu_i} \tag{11–37}$$

which after integration becomes

$$\left(\frac{E - E_0}{E} \right)^{s-1} \tag{11–38}$$

The classical RRK rate constant is simply this probability multiplied by the vibrational frequency for the critical oscillator, i.e.,

$$k(E) = \nu \left(\frac{E - E_0}{E} \right)^{s-1} \tag{11–39}$$

The assumptions used to derive the quantum RRK rate constant are in essence very similar to those for the classical theory. In the quantum theory it is assumed there are s identical oscillators in the molecule, all having frequency ν. The energized molecule contains a total of j quanta, so that its energy is $E = jh\nu$. The critical oscillator must contain m quanta for dissociation to occur ($m = E_0/h\nu$). To find the quantum rate constant $k(E)$, two different statistical weights are required: the total number of ways to distribute the j quanta among the s oscillators, given by equation 11–9; and the total number of ways to attain the particular distribution which leads to reaction; i.e., a number of quanta in the critical oscillator which ranges from m to j. The latter statistical weight is the sum of states, equation 11–10, for a maximum of $j - m$ quanta in the remaining $s - 1$ oscillators, and is given by

$$\frac{(j - m + s - 1)!}{(j - m)!(s - 1)!} \tag{11–40}$$

The probability that a particular oscillator has m quanta and all s oscillators have j quanta is the ratio of these two statistical weights:

$$\frac{(j - m + s - 1)! j!}{(j - m)!(j + s - 1)!} \tag{11–41}$$

The quantum RRK rate constant is the product of this probability and the vibrational frequency ν and is given by

$$k(E = jh\nu) = \nu \frac{(j - m + s - 1)! \, j!}{(j - m)! \, (j + s - 1)!} \qquad (11\text{--}42)$$

In the classical limit, where $j \gg s$ and $j - m \gg s$, the quantum RRK rate constant becomes

$$k(E = jh\nu) = \nu \left(\frac{j - m}{j}\right)^{s-1} \qquad (11\text{--}43)$$

If both the numerator and denominator in this equation are multiplied by $(h\nu)^{s-1}$, the equation becomes the classical RRK rate constant in equation 11–39.

The RRK rate constant depends on both the energy in excess of the unimolecular threshold and the number of vibrational degrees of freedom. To illustrate, the classical RRK expression for $k(E)$ divided by ν is plotted for several values of s in Figure 11–8 for a model system for which $E_0 = 40.0$ kcal/mole. Clearly, the rate constant increases with energy approaching its limiting value of ν. For a particular value of E, the rate constant decreases as s is increased. The physical explanation of this is that for larger s there are more ways of distributing the energy and hence less chance for energy to be localized in the critical oscillator.

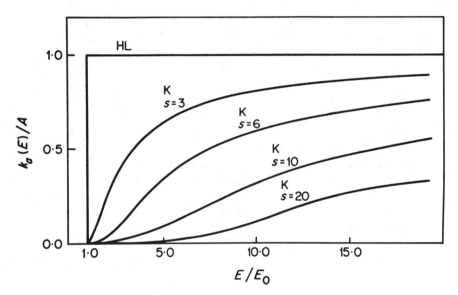

Figure 11–8. Comparison of $k_a(E)$ from Hinshelwood-Lindemann and classical Kassel theories. [Adapted from P. J. Robinson and K. A. Holbrook, *Unimolecular Reactions* (New York: Wiley-Interscience, 1972.) Reproduced with permission of John Wiley and Sons, Limited.]

There are both classical and quantum RRK expressions for the pressure-dependent thermal unimolecular rate constant k_{uni}, which result from using either the classical or quantum expression, respectively, for $dk_1/k_{-1} = P(E) \, dE$ and $k(E)$. The classical formulation of k_{uni} is the one most widely applied. It is found by inserting the classical mechanical expressions for $k(E)$ (equation 11–39) and $P(E)dE$ (equation 11–26) into equation 11–33. The resulting equation may be conveniently reduced by making the substitutions

$$x = \frac{E - E_0}{RT}$$

and

$$b = \frac{E_0}{RT}$$

which gives

$$k_{uni} = \frac{\nu e^{-E_0/RT}}{(s-1)!} \int_0^\infty \frac{x^{s-1} e^{-x} \, dx}{1 + \frac{\nu}{\omega}\left(\frac{x}{b+x}\right)^{s-1}} \qquad (11\text{--}44)$$

In the quantum mechanical formulation of k_{uni}, there is a summation rather than the integral of equation 11–44. As shown in Appendix 1, the quantum expression for $P(E)$ is

$$P(E) = \frac{W(E) \exp{(-E/k_B T)}}{Q_{vib}} \qquad (11\text{--}45)$$

where $W(E)$ is given by equation 11–9.

In the high-pressure limit ($\omega \to \infty$), both the classical and quantum RRK expressions for k_{uni} become the Arrhenius equation

$$k_\infty = \nu \exp{(-E_0/k_B T)} \qquad (11\text{--}46)$$

This is readily apparent for the classical case, since in the limit $\omega \to \infty$ the integral in equation 11–44 defines $(s-1)!$ Thus, according to the RRK model, the parameters E_0 and ν needed to evaluate $k(E)$ are simply given by the high-pressure Arrhenius equation. Note that, if equation 11–26 is used for dk_1/k_{-1}, equation 11–39 is the only expression that will give rise to equation 11–46.[16]

The parameters required to calculate a curve of k_{uni} vs. ω (i.e., pressure) by either the classical or quantum RRK theories are s, ν, and E_0. These theories are a major improvement over the Lindemann-Hinshelwood treatment. In fact, the RRK theories gave for the first time a treatment which could reproduce experimental results with reasonable accuracy. However, to obtain agreement with experiment with the classical theory, a value for s of about one-fourth to two-thirds the total number of normal modes in the molecule is needed, depending on the temperature. This feature arises from the serious errors that result from using classical statistical mechanics. The quantum RRK theory gives good results by equating s to the total number of normal modes.

The RRK theory assumes that the Arrhenius high-pressure thermal A-factor is given by the frequency for the critical oscillator, which is usually in the range of 10^{13} to 10^{14} sec^{-1}. However, as shown in Table 11–2 for many reactions, Arrhenius high-pressure thermal A-factors are in fact larger than 10^{14} sec^{-1}. The inability to ex-

TABLE 11–2 HIGH-PRESSURE ARRHENIUS PARAMETERS.

Reaction	Temperature, K	$\log A_\infty$	E_∞	Reference
$C_2H_6 \rightarrow 2CH_3$	840–913	16.72 ± 0.17	88.85 ± 0.68	a
Cyclopropane $\rightarrow$ propylene	690–1,038	15.50 ± 0.14	65.97 ± 0.50	b
$(CHO)_2 \rightarrow 2CO + H_2$	1,100–1,540	14.03	55.1	c
$CH_3OH \rightarrow CH_3 + OH$	—	15.97	89.9	d
Cyclobutane $\rightarrow 2C_2H_4$	1,308–1,498	15.6	62.5	e
Oxetan $\rightarrow C_2H_4 + H_2CO$	693–753	15.71 ± 0.31	63.02 ± 0.84	f
$C_3H_8 \rightarrow CH_3 + C_2H_5$	1,400–2,300	17.17	86.4	g
$(CH_3)_4C \rightarrow (CH_3)_3C + CH_3$	1,140–1,300	17.23	84.0	h

a. A.B. Trenwith, *Far. Trans. I 75*, 614 (1979).

b. H. Furue and P.D. Pacey, *Can. J. Chem. 60*, 916 (1982).

c. K. Saito, T. Kakumoto, and I. Murakami, *J. Phys. Chem. 88*, 1182 (1984).

d. H.G. Wagner, *Ber. Bunsenges. Phys. Chem. 86*, 2 (1982).

e. D.K. Lewis, A.S. Feinstein, and P.M. Jeffers, *J. Phys. Chem. 81*, 1887 (1977).

f. K.A. Holbrook and R.A. Scott, *Far. Trans. I 71*, 1849 (1975).

g. M.Z. Al-Alami and J.H. Kiefer, *J. Phys. Chem. 87*, 499 (1983).

h. D. Bernfeld and G.B. Skinner, *J. Phys. Chem. 87*, 3732 (1988).

plain these experimental results is one of the shortcomings of the RRK theory that is overcome by the RRKM theory.

11.6 RRKM THEORY

The RRKM theory was developed using the RRK model and extending it to consider explicitly vibrational and rotational energies and to include zero-point energies.[17] Several minor modifications of the theory have been made since its conception, primarily as a result of improved treatments of external degrees of freedom.[18–24] Detailed discussions of the RRKM theory are given in two excellent books, by Robinson and Holbrook[25] and Forst.[26]

RRKM theory is a microcanonical transition state theory,[27] and as such, it gives the connection between statistical unimolecular rate theory and the transition state theory of thermal chemical reaction rates discussed in chapter 10. Isomerization or dissociation of an energized molecule A* is assumed in RRKM theory to occur via the mechanism

$$A^* \xrightarrow{k(E)} A^\ddagger \longrightarrow \text{products} \tag{11–47}$$

where $A^\ddagger$ is the transition state. Bunker[19] has argued that *critical configuration* is a more appropriate name for $A^\ddagger$, since it represents a unique molecular configuration intermediate between reactant molecule and products. As discussed in chapter 10, trajectories passing from A* to products, or from products to A*, are assumed to pass through the transition state only once.

The energized molecule A* contains both vibrational and external rotational energies denoted by E_v and E_r, respectively. Their sum $E_v + E_r$ is E. Molecular motion from A* to A‡ involves a change in potential energy. As a result, the total energy of the critical configuration is

$$E^\ddagger + E_r^\ddagger = E - E_0 \qquad (11\text{--}48)$$

where E_0 is the potential energy difference between A* and A‡. Part of the energy $E^\ddagger$, denoted by $E_v^\ddagger$, is associated with vibration, while another part, denoted by $E_t^\ddagger$, consists of translational motion in the reaction coordinate. Their sum $E_v^\ddagger + E_t^\ddagger$ equals $E^\ddagger$. These energies are illustrated in Figure 11–9.

Figure 11–9. Diagram of the energies for the RRKM theory. [Adapted from W. L. Hase, in *Modern Theoretical Chemistry, Dynamics of Molecular Collisions, Part B*, edited by W. H. Miller (New York: Plenum, 1976), p. 121.]

The internal degrees of freedom are designated as either active or adiabatic. Active modes are assumed to exchange energy freely, while an adiabatic mode is one that remains in the same quantum state during the reaction. Conservation of angular momentum usually requires the external rotational modes to be treated as adiabatic. By contrast, the vibrational and internal rotational modes are considered to be active.

The RRKM rate constant $k(E)$ is the microcanonical transition state theory rate constant, equation 10–97, and is given by

$$k(E) = \frac{1}{h} \frac{G(E^\ddagger)}{N(E_v)} \qquad (11\text{--}49)$$

where $G(E^\ddagger)$ is the sum of states for the active degrees of freedom in the transition state and $N(E_v)$ is the density of states for the active degrees of freedom in the reactant. If the classical mechanical equations 11–11 and 11–12 are used for $G(E^\ddagger)$ and

$N(E_v)$, it is easily shown that the RRKM $k(E)$ becomes that of classical RRK theory, viz. equation 11–39. Note that the density in equation 11–49 is for all vibrational/internal rotational degrees of freedom in the reactant, while the sum does not include the reaction coordinate degree of freedom.

For many applications the RRKM rate constant is written as an explicit function of the rotational energy, i.e.,

$$k(E, E_r) = \frac{1}{h} \frac{G(E^{\ddagger})}{N(E - E_r)} \tag{11–50}$$

The external rotational modes are assumed to be adiabatic. One approach used to account for E_r is to use a classical mechanical model for rotation in which the moment of inertia about the reaction coordinate does not change during the unimolecular decomposition. This model is exemplified by the reaction $C_2F_6 \rightarrow 2CF_3$. The remaining two degrees of freedom have equal moments of inertia which depend on the reaction coordinate. Their rotational energy is $E_r = L^2/2I$, where L is the rotational angular momentum.

As shown in Figure 11–9, the energy for the sum of states in equation 11–50 may be written as

$$E^{\ddagger} = E_v - E_0 + E_r - E_r^{\ddagger} \tag{11–51}$$

For the classical rotational energy model,

$$E_r - E_r^{\ddagger} = \frac{L^2}{2I} - \frac{L^2}{2I^{\ddagger}} = E_r \left(1 - \frac{I}{I^{\ddagger}}\right) \tag{11–52}$$

since angular momentum is conserved during the unimolecular dissociation. Thus, the RRKM rate constant in equation 11–50 may be written as an explicit function of E_v and E_r:

$$k(E_v, E_r) = \frac{G(E_v + E_r[1 - I/I^{\ddagger}] - E_0)}{hN(E_v)} \tag{11–53}$$

To treat the external rotational energy of a nonlinear molecule quantum mechanically, the specific rotational energy levels must be considered. Since external rotation is treated as adiabatic, rotational quantum numbers do not change in proceeding from the reactant to the transition state. However, in general, the rotational energy spacings will change, since the rotational constants for the reactant and transition state usually differ. For symmetric top and "almost symmetric top" external rotors, the energy levels depend on the quantum numbers J and K. It has been suggested[21-24] that K should be treated as an active degree of freedom which interchanges energy with all the vibrational degrees of freedom.

To determine an RRKM rate constant requires evaluating the sum of states for the transition state and the density of states for the reactant molecule. The internal vibrational degrees of freedom are usually treated as quantum harmonic oscillators, and the sum and density are then evaluated by either state counting or the accurate Whitten-Rabinovitch approximation given by equations 11–14 through 11–17. Anharmonic corrections may be made to the harmonic sum and density.[21,28,29]

The information needed for calculating the sum and density of states in equation 11–53 includes the total reactant energy E and rotational energy E_r, the unimolecular threshold E_0, and harmonic vibrational frequencies and moments of inertia for both the reactant and the transition state. Usually E, E_r, and the reactants' moments of inertia and harmonic vibrational frequencies are known from experiment. However, obtaining the potential energy E_0 at the transition state and the transition state's structure (i.e., moments of inertia) and harmonic vibrational frequencies is usually less direct.

Determining the transition state's properties E_0, $I_i^\ddagger$, and $\nu_i^\ddagger$ depends on the nature of the transition state. There are two general situations: (1) a saddle point exists between the reactants and the products, at which the transition state is located; (2) either no saddle point exists, as in the reaction $CH_4 \rightarrow H + CH_3$, or the saddle-point region is very flat, so that the transition state's position on the potential energy surface is energy dependent and must be found by microcanonical variational transition state theory (see chapter 10). Finding transition state properties for the former situation is more straightforward and will be considered first.

The most widely used method of choosing properties for a transition state located at the saddle point is to adopt a model which reproduces the thermal high-pressure frequency factor A_∞ and activation energy E_∞. These experimental parameters are related to E_0 and the entropy difference $\Delta S^\ddagger$ via the transition state theory equations 10–49 and 10–54. Parameters for such model transition states are chosen using procedures such as BEBO concepts[30] or Benson's "thermochemical rules."[31] Although a unique transition state is not determined, it has been found that the RRKM rate constant is not particularly sensitive to the choice of vibrational frequencies and moments of inertia as long as the E_0 used is in agreement with E_∞ via equation 10–49 and $\Delta S^\ddagger$ agrees with A_∞ via equation 10–54.

To illustrate this approach, consider the decomposition of C_2H_5 to $H + C_2H_4$ and the four-centered elimination of HCl from C_2H_5Cl. For C_2H_5 decomposition, the transition state is depicted as

$$H$$
$$\backslash$$
$$\backslash$$
$$H_2 C ===== CH_2$$

The H—C stretch is taken as the reaction coordinate. The principal vibrational frequency changes in going from C_2H_5 to the transition state occur for the C—C stretch (an increase) and the two H—C_2H_4 bends (a decrease). If a transition state structure (i.e., a set of moments of inertia) is chosen, these three vibrational frequencies and E_0 may be adjusted to fit A_∞ and E_∞.

For HCl elimination from C_2H_5Cl, the proposed transition state structure is as follows:

$$H ----- Cl$$
$$| \qquad |$$
$$| \qquad |$$
$$H_2 C ===== C\, H_2$$

Using BEBO arguments, a transition state with ring bond orders of 1.9, 0.9, 0.1, and 0.1 for the C—C, C—Cl, C—H, and H—Cl bonds, respectively, is found to fit the experimental A_∞ frequency factor.[32] A ring breathing mode is assumed to be the reaction coordinate. (Problem 11.12 makes use of the transition state for this reaction.)

Ab initio calculations may also be used to determine properties of a fixed transition state at the saddle point. However, the calculated unimolecular threshold energy E_0 is usually several kJ/mol in error. For many cases a more accurate value for E_0 is found from the experimental activation energy using equation 10–21. The strength of the *ab initio* approach is that it gives a unique structure and set of vibrational frequencies for the transition state. Overall, thermal high-pressure frequency factors calculated from *ab initio* transition state structures and vibrational frequencies are in good agreement with experimental values. If *ab initio* values are also used for the reactant structure and vibrational frequencies, errors present in the *ab initio* results tend to cancel when the transition state's sum of states is divided by reactant's density of states in evaluating the RRKM $k(E)$.

To apply the variational version of RRKM theory (i.e., microcanonical variational transition state theory—see chapter 10, section 10-7) potential energy and vibrational frequencies are required along the reaction path in order to find the minimum in the sum of states (equation 10–10) which identifies the transition state.[33] These reaction path properties may be either derived from a postulated model or determined by an *ab initio* calculation. For reactions without saddle points, such as $C_2H_6 \rightarrow 2CH_3$, the transition state structure is strongly energy-dependent. This is illustrated in Figure 11–10. A major accomplishment of the variational form of RRKM theory is that it explains the "loose" transition states and large thermal

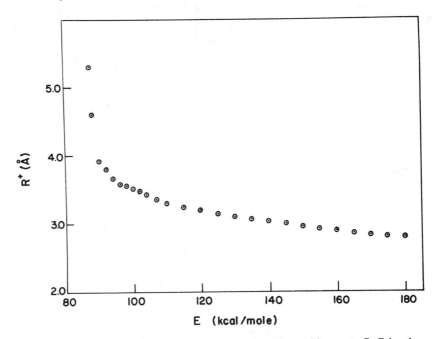

Figure 11–10. Relationship between the variational transition state C–C bond length $R^{\ddagger}$ and vibrational energy for a model of $C_2H_6 \rightarrow 2CH_3$ dissociation. [Adapted from W. L. Hase, in *Modern Theoretical Chemistry, Dynamics of Molecular Collisions, Part B,* edited by W. H. Miller (New York: Plenum, 1976).]

A-factors experimentally (see Table 11–2) for reactions without potential energy barriers.

To illustrate the details of an RRKM calculation with a fixed transition state structure, Table 11–3 makes use of the unimolecular isomerization reaction $CH_3NC \rightarrow CH_3CN$. The vibrational frequencies and moments of inertia for CH_3NC and the transition state are taken from the *ab initio* calculations of Saxe et al.[34] Calculated rate constants are listed versus E_v for $E_{rx} = E_{ry} = E_{rz} = 1.25$ kJ/mole. The rotational energy difference between the reactant and transition state is found from

$$E_r - E_r^{\ddagger} = \sum_{i=x,y,z} E_{ri}(1 - I_i/I_i^{\ddagger}) \qquad (11\text{–}54)$$

The transition state energy $E^{\ddagger}$ is given by equation 11–51; the sum and density of states are found from the Whitten-Rabinovitch approximation. The threshold energy E_0 is 157 kJ/mole.

TABLE 11–3 RRKM CALCULATION FOR $CH_3NC \rightarrow CH_3CN$[a]

Vibrational frequencies (cm^{-1})	
CH_3NC	Transition state
3014	3070
3014	3049
2966	2945
2166	1970
1467	1439
1429	1311
1129	975
1129	956
945	609
263	230
263	412i

Moments of inertia (amu $-$ Å^2)		
	CH_3NC	Transition state
I_x	3.20	11.80
I_y	50.12	32.87
I_z	50.12	41.42

E_v[b]	$E^{\ddagger}$[b]	$N(E_v)$[c]	$G(E^{\ddagger})$	$k(E_v)$[d]
167	9.6	1.502×10^3	5.547	3.321×10^8
209	51.5	7.515×10^3	8.798×10^2	1.053×10^{10}
251	93.3	3.031×10^4	1.927×10^4	5.717×10^{10}
335	177	3.135×10^5	1.302×10^6	4.257×10^{11}

[a]Frequencies and moments of inertia from reference 34.
[b]Energy in kJ/mol.
[c]Density of states in states/cm^{-1}.
[d]Rate constant in sec^{-1}.

11.7 APPLICATION OF RRKM THEORY TO THERMAL ACTIVATION

The RRKM rate constant for the overall thermal unimolecular reaction with external rotational energy treated classically is found by integrating equation 11–53 over Boltzmann distributions for E_v and E_r as defined by equation 11–33; i.e.,

$$k_{\text{uni}} = \int_0^\infty \int_0^\infty \frac{k(E_v, E_r)P(E_v)P(E_r)dE_vdE_r}{1 + k(E_v, E_r)/\omega} \tag{11–55}$$

The distributions $P(E_v)$ and $P(E_r)$ are given by

$$P(E_v) = \frac{N(E_v)\exp(-E_v/k_BT)}{Q_{\text{vib}}}$$

and

$$P(E_r) = \frac{N(E_r)\exp(-E_r/k_BT)}{Q_{\text{rot}}} \tag{11–56}$$

where $N(E_v)$ and $N(E_r)$ are the densities of vibrational and rotational quantum states (assumed to be continuous) and Q_v and Q_r are the molecular vibrational and external rotational partition functions. If the external rotational energy is treated quantum mechanically, a summation of the rotational energy levels is required instead of an integration over E_r.

Substituting equations 11–49 and 11–56 into equation 11–55 gives

$$k_{\text{uni}} = \frac{1}{hQ_vQ_r}\int_0^\infty \int_0^\infty \frac{G(E^\ddagger)N(E_r)\exp[-(E_v + E_r)/k_BT]dE_vdE_r}{1 + k(E_v, E_r)/\omega} \tag{11–57}$$

If the integration is performed over the energies of the critical configuration instead of those for the molecule, equation 11–57 becomes

$$k_{\text{uni}} = \frac{\exp(-E_0/k_BT)}{hQ_vQ_r}\int_0^\infty \int_0^\infty \frac{G(E^\ddagger)N(E_r^\ddagger)\exp[-(E^\ddagger + E_r^\ddagger)/k_BT]dE^\ddagger dE_r^\ddagger}{1 + k(E^\ddagger, E_r^\ddagger)/\omega} \tag{11–58}$$

At high pressures ($\omega \to \infty$),

$$k_\infty = \frac{\exp(-E_0/k_BT)}{h\,Q_rQ_v}\int_0^\infty N(E_r^\ddagger)\exp(-E_r^\ddagger/k_BT)dE_r^\ddagger$$

$$\times \int_0^\infty G(E^\ddagger)\exp(-E^\ddagger/k_BT)\,dE^\ddagger \tag{11–59}$$

The first integral in equation 11–59 is the external rotational partition function, and the second is the vibrational partition function times k_BT. As a result,

$$k_\infty = \frac{k_BT}{h}\frac{Q_r^\ddagger Q_v^\ddagger}{Q_rQ_v}\exp(-E_0/k_BT) \tag{11–60}$$

which is in agreement with transition state theory.

The quantity $k(E^\ddagger, E_r^\ddagger)$ in the denominator of equation 11–58 is

$$k(E^\ddagger, E_r^\ddagger) = \frac{1}{h} \frac{G(E^\ddagger)}{N(E^\ddagger + E_0 + E_r^\ddagger - E_r)} \qquad (11-61)$$

To extend equation 11–58 to the falloff region and low-pressure limit ($\omega \to 0$) and avoid the laborious double integration, Marcus[18] suggested that $E_r - E_r^\ddagger$ in equation 11–61 be replaced by the difference of the average rotational energies of reacting molecules, $\langle E_r \rangle$, and the critical configurations, $\langle E_r^\ddagger \rangle$. The quantity $\langle E_r^\ddagger \rangle$ may be found by averaging $E_r^\ddagger$ over $\exp(-E_r^\ddagger/k_B T)$; $\langle E_r^\ddagger \rangle = k_B T$. Because $E_r/E_r^\ddagger = I^\ddagger/I$, the result for two dimensions is

$$\Delta E_r = \langle E_r \rangle - \langle E_r^\ddagger \rangle = ((I^\ddagger/I) - 1)k_B T \qquad (11-62)$$

If we now define

$$k(E^\ddagger) = \frac{1}{h} \frac{G(E^\ddagger)}{N(E^\ddagger + E_0)} \qquad (11-63)$$

and introduce the factor

$$F = N(E^\ddagger + E_0 - \Delta E_r)/N(E^\ddagger + E_0) \qquad (11-64)$$

then, combining equations 11–61 through 11–64, inserting them into equation 11–58, and separating the terms containing $E_r^\ddagger$, one obtains

$$k_{\text{uni}} = \frac{Q_r^\ddagger \exp(-E_0/k_B T)}{hQ_r Q_v} \int_0^\infty \frac{G(E^\ddagger) \exp(-E^\ddagger/k_B T) \, dE^\ddagger}{1 + k(E^\ddagger)/F\omega} \qquad (11-65)$$

In the low-pressure limit ($\omega \to 0$),

$$k_{\text{uni}} = \frac{F\omega Q_r^\ddagger \exp(-E_0/k_B T)}{hQ_r Q_v} \int_0^\infty N(E^\ddagger + E_0) \exp(-E^\ddagger/k_B T) \, dE^\ddagger \qquad (11-66)$$

By performing exact numerical integrations of equation 11–58, Waage and Rabinovitch[35] achieved a very good approximation for F, viz,

$$F = \{1 + [(s + d/2 - 1)((I^\ddagger/I) - 1)k_B T]/(E_0 + aE_z)\}^{-1} \qquad (11-67)$$

where d is the number of internal rotational degrees of freedom. Other approximations[35] have been made for F that are similar to, but not as accurate as, that of Waage and Rabinovitch.

Different experimental techniques, including static pyrolysis,[21] carrier (flow) techniques,[36] shock tube methods,[37] and very-low-pressure-pyrolysis (VLPP),[38] have been used to measure the thermal unimolecular rate constant k_{uni} as a function of temperature and pressure. One of the most significant achievements of RRKM theory is its ability to match measurements of k_{uni} with pressure. The unimolecular falloff measurements made by Rabinovitch and co-workers of alkyl isocyanide isomerization[21,39–42] have provided some of the most thorough comparisons with the theory. The first reaction studied was that of equation 11–2. A comparison of the RRKM calculations and fall-off measurements for this reaction is shown in Figure 11–11. In addition to examining CH_3NC, Rabinovitch and co-workers also studied the isomer-

Figure 11–11. Pressure dependence of the unimolecular rate constant for CH_3NC isomerization. For clarity, the 260° and 200° curves are displaced one log unit to the left and right, respectively. [Adapted from F. W. Schneider and B. S. Rabinovitch, *J. Am. Chem. Soc. 84*, 4215 (1962).]

ization of CD_3NC, CH_2DNC, and C_2D_5NC and have found that the same model transition state that fits CH_3NC isomerization gives a good fit to the fall-off behavior of these molecules as well. Extensive comparisons between RRKM theory and experimental fall-off curves have also been made for cyclopropane isomerization, and again, a good fit has been found between theory and experiment.[36]

Agreement has also been found with experimental unimolecular rate constants when transition state structures derived from *ab initio* calculations have been used in RRKM calculations. For example, when used in RRKM theory the *ab initio* CH_3NC transition state structure in Table 11–3 gives a good fit to the thermal unimolecular rate constant.[34] Similarly, for the decomposition reaction $C_2H_5 \rightarrow H + C_2H_4$, the *ab initio* transition state fits both thermal and chemical activation (see next) rate constants.[43]

11.8 MEASUREMENT OF $k(E)$

One of the most sensitive tests of RRKM unimolecular rate theory is the direct determination of the microcanonical unimolecular rate constant $k(E)$ as a function of energy. Some of the most extensive studies of this type have involved the unimolecular dissociation of ions. An extremely powerful and generally applicable approach to the monoenergetic excitation of ions is that of photoelectron-photoion coincidence

(PEPICO).[44] Ions excited in this manner include those of the halobenzenes, aniline, benzonitrile, and phenol. In all cases the measured $k(E)$ values agree with those of RRKM theory. A comparison between the experimental and theoretical $k(E)$ for bromobenzene and deuterated bromobenzene ions[45] is given in Figure 11–12. No well-documented ionic reaction which occurs on the ground-state potential energy surface has been shown to have a rate constant in disagreement with that of RRKM theory. On the contrary, purported examples of non-RRKM behavior have been shown to involve direct dissociation from excited electronic, or repulsive, states.

Figure 11–12. Bromobenzene and deuterated bromobenzene ion-dissociation rates as a function of ion internal energy. The points are the experimental results, and the solid lines are the statistical-theory results [From T. Baer, *Adv. Chem. Phys.* **64**, 111 (1986). Copyright © 1986 by John Wiley & Sons, Inc. Reprinted by permission of John Wiley & Sons, Inc.]

Chemical activation and photoactivation techniques have been used to excite neutral molecules with much higher energies than is attained in thermal experiments and with narrow internal energy distributions. In chemical activation a chemical reaction is used to form molecules with sufficient energy to undergo unimolecular reactions.[46,47] Some of the more important chemical activation processes are H addition to olefins, CH_2 insertion and addition, and the radical recombination which is the reverse of equation 11–5. Photoactivation techniques, discussed in section 11.1, include internal conversion from an electronically excited state (equation 11–8) and direct overtone pumping. These techniques are similar to chemical activation, in that energized molecules are formed with a narrow energy distribution. The distribution is not completely sharp, however, because the thermal energy distribution in the initial reactants (typically a few kJ/mol in width) is just translated upward by the energy of the absorbed photon.

The molecule A* formed by chemical activation or photoactivation may either undergo a unimolecular reaction or be collisionally stabilized. (In the PEPICO experiments described, $k(E)$ is determined in the absence of collisions.) If the strong collision assumption is made and A* is assumed to be monoenergetically excited, we have

$$A^* \xrightarrow{k(E)} \text{decomposition products } (D)$$
$$A^* \xrightarrow{\omega} A_1 \qquad \text{stabilization } (S) \qquad (11\text{--}68)$$

where ω is the collision frequency. From this equation, it is plain that the rate constant $k(E)$ equals $\omega D/S$. When A* is formed with a distribution of energies $f(E)$ the average rate constant $\langle k(E) \rangle$ equals $\omega D/S$. The ratio D/S may be found by averaging over $f(E)$, and the following expression is obtained for $\langle k(E) \rangle$:

$$\langle k(E) \rangle = \omega \frac{\int \{ k(E)/[k(E) + \omega] \} f(E) \, dE}{\int \{ \omega/[k(E) + \omega] \} f(E) \, dE} \qquad (11\text{--}69)$$

At high pressures, i.e., where $\omega \gg k(E)$, we have

$$\langle k(E) \rangle_\infty = \frac{\int k(E) f(E) \, dE}{\int f(E) \, dE} \qquad (11\text{--}70)$$

Similarly, at low pressures, i.e., where $\omega \ll k(E)$, equation 11–69 becomes

$$\langle k(E) \rangle_0 = \frac{\int f(E) \, dE}{\int [f(E)/k(E)] \, dE} \qquad (11\text{--}71)$$

For some chemical activation processes it is possible to assume an equilibrium between the two reactants and A* in order to determine the form for the distribution function $f(E)$.

The most comprehensive applications of RRKM theory to chemical activation measurements have been for various homologous series. Agreement with the RRKM theory results when one transition state is suitable for the complete series. Rabinovitch and co-workers have made major contributions to studies involving H atom addition to olefins. Their most comprehensive studies have been of H + olefins to yield excited alkyl radicals that decompose by loss of CH_3, C_2H_5, C_3H_7, C_4H_9, and C_5H_{11} radicals. In all cases the results were found to be in satisfactory agreement with RRKM theory.[48] Agreement with RRKM theory has been found in many other chemical activation studies as well.[49-51]

In photoactivation the excited reactant is formed by

$$A + h\nu \xrightarrow{k_a} A^* \qquad (11\text{--}72)$$

The rates for the overall loss of reactant A and the formation of products are equal and, according to the mechanism of equations 11–68 and 11–72, are given by

$$\frac{-d[A]}{dt} = k(E)A^* \qquad (11\text{--}73)$$

Applying a steady-state treatment to A* leads to the rate law

Unimolecular Reaction Dynamics   Chap. 11

$$\frac{-d[A]}{dt} = \frac{k(E)k_a\rho_\nu[A]}{k(E) + \omega} \tag{11-74}$$

where ρ_ν is the photon number density. Hence, the apparent first-order rate constant for loss of A is

$$k = \frac{k(E)k_a}{k(E) + \omega} \tag{11-75}$$

Inversion of equation 11–75 gives

$$k^{-1} = k_a^{-1} + \frac{\omega}{k_a k(E)} \tag{11-76}$$

which is the standard *Stern-Volmer* equation. This equation is used in most treatments of collisional quenching of excited species. A plot of k^{-1} vs. pressure (i.e., ω)

Figure 11–13. Plot of $k(E)$ vs. E for cyclobutene to 1,3-butadiene isomerization. Excitation is by a CH overtone transition. E is defined as the photon energy plus the average thermal energy (246 cm^{-1}). The solid line is the RRKM curve. [Adapted from J. M. Jasinki, et al. *J. Chem. Phys.* **29**, 1312, (1983).]

should be linear with an intercept of k_a and a slope of $1/k_a k(E)$. The unimolecular rate constant $k(E)$ is then the intercept divided by the slope. Nonlinearity will arise in the plot if the energized molecules are not monoenergetic and/or if the strong-collision assumption for de-energization is invalid (see next section).

An example of a photoactivation experiment is the excitation of CH overtones in cyclobutene to induce isomerization to 1,3-butadiene.[52] The $k(E)$ values obtained in this experiment are compared with RRKM theory in Figure 11–13. Another sort of photoactivation is infrared multiple-photon excitation (IRMPE). In this case, however, a broad distribution of energies in the molecule is produced, rather than a nearly monochromatic excitation. The process is discussed in chapter 14, section 14.4.1.

11.9 INTERMOLECULAR ENERGY TRANSFER

The strong collision assumption can be tested experimentally for thermal unimolecular reactions by measuring k_{uni} in the second-order regime $(\omega \to 0)$[53]. The low-pressure unimolecular rate is simply the rate at which the molecules are excited to vibrational levels having energy greater than E_0, and the rate constant is one for bimolecular reaction, viz., k_{bi}. If k_{bi} is measured for a bath gas M, the pressure-for-pressure activation efficiency β_p of M relative to reactant A can be found from

$$\beta_p = \frac{(k_{bi})_M}{(k_{bi})_A} \qquad (11-77)$$

where $(k_{bi})_M$ is the slope of k_{uni} vs. the pressure of M with the reactant pressure kept constant, and $(k_{bi})_A$ is the slope of k_{uni} vs. the reactant pressure. This may be converted to a relative efficiency per collision by using the kinetic theory of gases to remove the M—A and A—A collision frequencies. The resulting expression for the relative activation efficiency of A is

$$\beta_c = \beta_p \left(\frac{\sigma_{AA}}{\sigma_{AM}}\right)^2 \left(\frac{\mu_{AM}}{\mu_{AA}}\right)^{1/2} \qquad (11-78)$$

where the μ's are reduced masses and the σ's are the collision diameters. The most extensive measurements of β_p and thus β_c are those of Rabinovitch and co-workers for methyl isocyanide isomerization.[54] These researchers measured the relative efficiencies of over 100 bath gases, and some of their results are shown in Table 11–4.

For the noble gases and smaller molecules, β_c relative to β_c of CH_3NC is much less than 1.00. As the molecular complexity increases, the collisional efficiencies approach a limiting value. It seems reasonable that the limiting efficiency is unity and that the strong collision assumption of the Lindemann mechanism is valid for large molecules at not too high temperatures. However, for smaller molecules with β_c less than 1, a multistep activation and deactivation Lindemann mechanism leads to the following master equation for each energy level:

$$\frac{dn_i}{dt} = \omega \sum_j P_{ij} n_j - \omega n_i - k_i n_i \qquad (11-79)$$

In this equation, P_{ij} is the probability of conversion to the state indicated by i from

TABLE 11-4 ENERGY TRANSFER EFFICIENCIES
FOR METHYL ISOCYANIDE ISOMERIZATION
IN THE SECOND-ORDER REGIME[a]

Activating gas	Relative efficiency (β_c)
CH_3NC	1.00
$n\text{-}C_4H_{10}$	1.01
$n\text{-}C_4F_{10}$	1.00
C_3H_8	0.79
C_2H_6	0.76
CH_4	0.61
NH_3	0.93
CO	0.46
Xe	0.232
Ar	0.279
He	0.243

[a]Data from Reference 54.

the state indicated by j and the quantities k_i and n_i are the unimolecular rate constant for state i and the number of molecules in state i, respectively.

Exact solutions to equation 11–79 are impossible because the individual P_{ij}'s and their time dependence are not known for complex molecules containing many quanta of vibrational energy. However, a formal approximate solution to the equation indicates that after an initial short induction period the energy levels reach a steady state $(dn_i/dt = 0)$.[55] If this steady-state approximation is made, a stochastic solution (see chapter 2, section 2.7.2) is possible if a form for the P_{ij}'s is assumed.[56,57] The stochastic analysis will yield a value of $\langle \Delta E \rangle$, the average amount of energy transferred per collision. The relative efficiency β_p at low pressures may be determined by calculating the total reaction rate from the steady-state distribution $\Sigma_i k_i n_{ij}$ and dividing it by the strong collision reaction rate given by equation 11–66.

Chemical activation experiments have also provided information about the dynamics of intermolecular energy transfer in unimolecular reactions.[58–60] The deactivation and unimolecular isomerization or decomposition of the chemically activated molecule A* may be regarded as the following cascade process proceeding through n energy levels which start at E and end at the first level below E_0:

Because A_n^* has energy less than E_0, it represents a stable molecule. The strong collision assumption means that the transition $A_i^* \rightarrow A_n^*$ occurs in one step. What is predicted by this cascade mechanism,[61] and what is found in many chemical activation experiments,[60] is decomposition or isomerization in excess of the strong collision prediction, because deactivation is a multistep process and there is a nonzero probability of unimolecular reaction after each of the deactivation steps until A^* has energy less than E_0.

Extensive tables of $\langle \Delta E \rangle$ for different energized reactants and deactivating molecules are available from analyses of thermal and chemical activation studies.[54,62,63] Overall, $\langle \Delta E \rangle$ values for these two different types of experiments are in agreement. However, these inferred values are considerably larger than values that are determined directly.[64,65] In the direct measurements an ensemble of vibrationally excited molecules is prepared by laser excitation at a particular energy. Then, the average energy of the ensemble versus time in the presence of a deactivating bath is monitored by ultraviolet "hot band" absorption or infrared fluorescence techniques. Values for $\langle \Delta E \rangle$ are found by comparing the average vibrational energy of the ensemble versus time with the collision frequency. The smaller $\langle \Delta E \rangle$ values obtained from the direct measurements are consistent with classical trajectory studies[66-69] and stochastic (random walk)[70] models of intermolecular energy transfer (see also chapter 13, section 13.4.3).

The theory of unimolecular reactions lacks a detailed microscopic picture of the collision dynamics of intermolecular energy transfer. What is needed is an understanding of how intermolecular energy transfer from highly energized molecules is influenced by the energizing level, the partitioning of the energy between vibration and rotation, and the relative translational energy between the colliding bath and energized molecules.

11.10 PRODUCT ENERGY PARTITIONING

Among the predictions of RRKM theory is that statistical distributions of relative translational, rotational, and vibrational energies will exist in the transition state. As an example, consider the relative translational energy distribution. If an ensemble of molecules is monoenergetically excited with a specific rotational energy, the rate constant for dissociation when the fragments have relative translational energy in the interval from $E_t^{\ddagger}$ to $E_t^{\ddagger} + dE_t^{\ddagger}$ is $k(E, E_t^{\ddagger})$ and, according to Eq. (10–94), is given by

$$k(E, E_t^{\ddagger})dE_t^{\ddagger} = \frac{N(E - E_t^{\ddagger})dE_t^{\ddagger}}{hN(E_v)} \tag{11–80}$$

where the N's are densities of states and the E's are as defined in Figure 11–9. The probability of a particular $E_t^{\ddagger}$ is then

$$P(E_t^{\ddagger})dE_t^{\ddagger} = \frac{k(E, E_t^{\ddagger})dE_t^{\ddagger}}{\int_0^{\infty} k(E, E_t^{\ddagger})dE_t^{\ddagger}} \tag{11–81}$$

After canceling terms and integrating, equation 11–81 becomes

$$P(E_t^\ddagger) = \frac{N(E^\ddagger - E_t^\ddagger)}{G(E^\ddagger)} \tag{11-82}$$

This equation applies to the situation where the energized molecules have one specified rotational energy. If instead the energized molecules have a distribution of rotational energies, equation 11–82 must be averaged over this distribution to obtain the final $P(E_t^\ddagger)$. Some of the most extensive applications of equation 11–82, with a correct treatment of rotational energy, have been for the decomposition of energized ions.[71-73]

If the unimolecular reaction does not have a barrier for the reverse recombination reaction (Figure 11–1), the energy distributions at the transition state may be used to predict those for the products. This can be done since there is a larger internuclear separation between the dissociation fragments at the transition state, so that the transition state resembles the products. If there is a barrier for recombination, as in reactions 11–3 and 11–4, the product energy distributions are usually nonstatistical and different from those at the transition state. This is because the potential energy released in going from the transition state to products is channeled nonstatistically to translation, vibration, and rotation. (Recall the discussion in chapter 9, section 9.4, concerning the relationship between energy partitioning and potential energy surfaces for direct bimolecular reactions.)

Two types of unimolecular dissociation potential energy curves and the resulting relative translational energy distributions are depicted in Figure 11–14. In (a) the product distribution is similar to that at the transition state; in (b) the barrier has broadened the distribution and shifted it to higher energies. These effects have been observed in numerous experimental and theoretical studies of unimolecular dissociation.

One of the predictions of RRKM theory is that in a ladder-climbing excitation process such as infrared multiphoton dissociation (IRMPD) it is difficult to produce unimolecular fragments with large vibrational energies. This is because the unimolecular dissociation rate rises rapidly with energy in excess of the threshold energy E_0. As a result, most molecules dissociate before they acquire any significant energy in excess of E_0. Large amounts of vibrational energy could appear in the products if there is a large potential energy release in going from the transition state to products that is preferentially directed to vibration. However, for most unimolecular reactions, the potential energy release is preferentially partitioned to translation. Products with large amounts of vibrational energy can be formed if the reactant is highly energized, as can be accomplished in chemical activation (section 11.7) and photoactivation (section 11.8).

To interpret energy partitioning correctly in unimolecular reactions, angular momentum constraints must be treated correctly. This is illustrated by the chemical activation process

$$F + C_2H_4 \longrightarrow C_2H_4F^* \longrightarrow H + C_2H_3F \tag{11-83}$$

which has been studied in crossed molecular beam experiments.[74] The distribution of rotational energy in the energized $C_2H_4F^*$ molecules results from the distribution of impact parameters (see chapter 8) leading to reaction. The initial orbital angular mo-

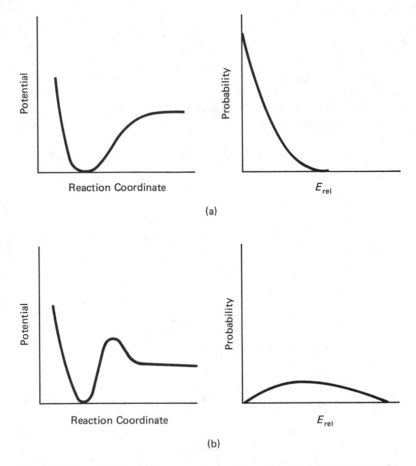

Figure 11–14. Relationship between the reaction path potential energy curve for a unimolecular dissociation reaction and the distribution of relative translational energy in the dissociation products.

mentum $l = \mu b v$ between the reactants is converted into rotational angular momentum J of $C_2H_4F^*$. However, since the dissociating H atom is very light, there is very little orbital angular momentum associated with the products H and C_2H_3F. Thus, conservation of angular momentum requires that l be almost entirely converted into rotational excitation j[75] of C_2H_3F, i.e., $l + j = J$. Extensive analyses[76] have been made of the role of angular momentum constraints in the partitioning of energy between unimolecular fragments. Such constraints should be included when using the RRKM theory to calculate energy distributions.

Other, non-RRKM-theory-based approaches to calculating product energy partitioning in unimolecular dissociation are phase space theory[77] and vibrationally/rotationally adiabatic theory.[78] Quack and Troe[79] have given a very useful semiempirical method for applying the latter theory.

11.11 APPARENT AND INTRINSIC NON-RRKM BEHAVIOR

In the last few sections, applications of RRKM theory to thermal, photoactivation, and chemical activation experiments have been illustrated. One of the major achievements of unimolecular rate theory is the extent of agreement realized between RRKM theory and such experiments. However, there are situations where the RRKM theory does not appear to be successful. Thus, at this point it is useful to review the basic postulates of RRKM theory and consider situations where they may be violated.[80]

The central assumption of RRKM theory is that isolated molecules behave as if all their accessible states were occupied with equal probabilities. This means that the probability of A* transforming to A‡ (equation 11–47) is due entirely to statistical considerations. Figure 11–15(a) illustrates the situation schematically and shows random transitions among states at some energy high enough for eventual reaction (toward the right). In reality, transitions between quantum states (though coupled) are *not* equally probable: some are more likely than others. Therefore, the molecular motion must be disorderly enough for the RRKM assumption to be mimicked, as crudely illustrated in Figure 11–15(b).

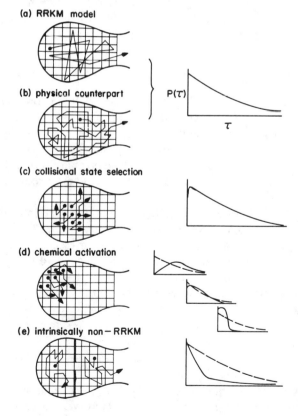

(a) RRKM model

(b) physical counterpart

$P(\tau)$

τ

(c) collisional state selection

(d) chemical activation

(e) intrinsically non-RRKM

Figure 11–15. Relation of state occupation (schematically shown at constant energy) to lifetime distribution for the RRKM theory and for various actual situations. Dashed lines in lifetime distributions for (d) and (e) indicate RRKM behavior. (a) RRKM model. (b) Physical counterpart of RRKM model. (c) Collisional state selection. (d) Chemical activation. (e) Intrinsically non-RRKM. [Adapted from D. L. Bunker and W. L. Hase, *J. Chem. Phys. 59*, 4621 (1973).]

The lifetime distribution will depend in part on the manner in which the energy needed for reaction is supplied. As discussed in section 11.3, the usual understanding of thermal unimolecular reactions assumes that the activating collisions randomly prepare the internal vibrational states of A*. However, it has been pointed out previously that thermal collisions surely at least approximately obey the Franck-Condon principle.[81] Thus, direct collisional preparation of a state with a large extension of some bond, in the direction of product formation, is suggested to be an implausible event. Consequently, one might expect some shortage of lifetimes less than one-fourth of a vibration period, as illustrated in Figure 11–15(c).

In many experiments, such as photoactivation and chemical activation, the molecular vibrational states are intentionally excited nonrandomly. Regardless of the pattern of initial energizing, the RRKM model in Figure 11–15(a) would require the distribution to become random in a negligibly short time. How well will the actual molecular motion approach this ideal? Three different possibilities are represented by Figure 11–15(d). The lifetime distribution may be similar to that of RRKM theory. In other cases, the probability of a short lifetime with respect to reaction may be enhanced or reduced, depending on the location of the initial excitation within the molecule. These are examples of *apparent* non-RRKM behavior arising from state selection. If there are very strong internal couplings, $P(\tau)$ will become that of RRKM theory after rapid intramolecular vibrational energy redistribution.

The effect of apparent non-RRKM behavior is to make the experimental unimolecular rate constant time-dependent according to the formula

$$k(E, t) = \frac{1}{N(t)} \frac{dN(t)}{dt} \tag{11–84}$$

Obviously, if the excitation is preferentially localized in the reaction coordinate, the initial $k(E,t)$ will be larger than the RRKM value. However, if there is strong intramolecular coupling, $k(E,t)$ will rapidly approach $k_{RRKM}(E)$. There is widespread agreement between RRKM rate constants and those derived from photoactivation and chemical activation experiments, which illustrates that for many molecules and experimental conditions the rate of intramolecular vibrational energy redistribution (k_{IVR}) is larger than $k_{RRKM}(E)$.

To detect the initial apparent non-RRKM decay in a photoactivation or chemical activation experiment, one has to monitor the reaction at short times. This can be performed by studying the unimolecular decomposition at high pressures, where collisional stabilization competes with the rate of IVR. The first successful detection of apparent non-RRKM behavior was accomplished by Rabinovitch and co-workers,[82] who used chemical activation to prepare vibrationally excited hexafluorobicyclopropyl-d_2:

$$^1CD_2 + CF_2{-}CF{-}CF = CF_2 \longrightarrow CF_2{-}CF{-}CF{-}CF_2^* \tag{11–85}$$

with bridging CH_2 on the left reactant and bridging CH_2 and CD_2 on the right product.

The molecule decomposes by elimination of CF_2, which should occur with equal probabilities from each ring when energy is randomized. However, at pressures in excess of 100 torr there is a measurable increase in the fraction of decomposition in the ring that was initially excited. Rabinovitch et al. also performed extensive studies on the series of chemically activated fluoroalkyl cyclopropanes:[83]

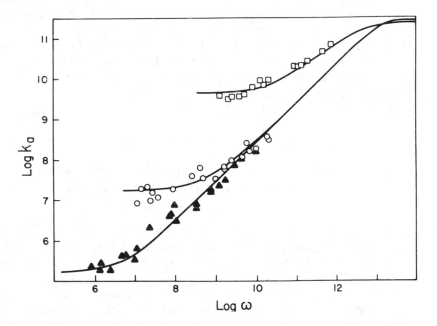

$$\text{(11–86)}$$

R, $\quad$ R $= CF_3, C_3F_7, C_5F_{11}$

The chemically activated molecules are formed by reaction of $^1CD_2$ with the appropriate fluorinated alkene. In all these cases apparent non-RRKM behavior was observed. As displayed in Figure 11–16, the measured unimolecular rate constants are strongly dependent on pressure. However, at low pressures each rate constant approaches the RRKM value.

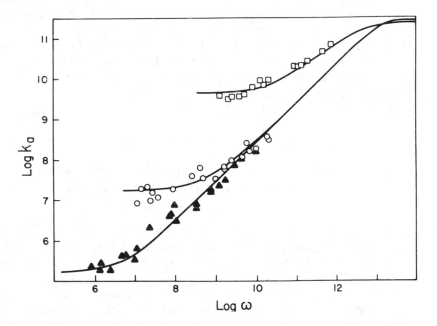

Figure 11–16. Chemical activation unimolecular rate constants vs. ω for fluoroalkyl cyclopropanes (see equation 11–86). [Adapted from J. F. Meagher, et al., *J. Phys. Chem. 78,* 2535 (1974). Copyright 1974 American Chemical Society.]

A situation that is both more interesting and completely distinct from apparent non-RRKM behavior is *intrinsic* non-RRKM behavior. By this, it is meant that A* has a nonrandom $P(\tau)$ even if the internal vibration states of A* are prepared randomly. This situation arises when transitions between individual molecular vibrational states are slower than transitions leading to products. As a result, the vibrational states do not have equal dissociation probabilities. Slow transitions between the states occur when there is at least one bottleneck in the molecular phase space other than the one defining the transition state. An example of such a bottleneck with a possible resultant $P(\tau)$ is shown in Figure 11–15(e).

One kind of an intrinsically non-RRKM model is Slater's theory of unimolecular reactions.[13] A general discussion of this theory and the nature of the classical mechanical motion of highly vibrationally excited molecules is presented in the next section: experimental examples of intrinsic non-RRKM behavior are given in the last section of the chapter.

11.12 CLASSICAL MECHANICAL DESCRIPTION OF INTRAMOLECULAR MOTION AND UNIMOLECULAR DECOMPOSITION

The first classical mechanical model of unimolecular decomposition, developed by Slater,[13] was based on the normal-mode harmonic oscillator Hamiltonian. Although the Hamiltonian is rigorously exact only for small displacements from the molecular equilibrium geometry, Slater extended it to the situation where molecules are highly vibrationally energized, undergo large amplitude motions, and decompose. Since there are no couplings in the normal-mode Hamiltonian, the energies in the individual normal modes do not vary with time. This is the essential difference from the RRK and RRKM theories, which treat a molecule as a collection of coupled normal modes which freely exchange energy.

The harmonic oscillator classical Hamiltonian is

$$H = \sum_{i=1}^{s} \frac{(p_i^2 + \lambda_i q_i^2)}{2} \tag{11-87}$$

where $\lambda_i = 4\pi^2 \nu_i^2$. From the classical equations of motion (equation 8–20), it is easily shown that each normal-mode coordinate varies with time according to

$$q_i = q_i^0 \cos (2\pi \nu_i t + \delta_i) \tag{11-88}$$

where q_i^0 is the amplitude and δ_i is the phase of the motion. Thus, if an energy $E_i = (p_i^2 + \lambda_i q_i^2)/2$ and phase δ_i are chosen for each normal mode, the complete intramolecular motion of the energized molecule may be determined for this particular initial condition.

For most unimolecular reactions the reaction coordinate is not a particular normal mode, but instead is an internal coordinate displacement such as a bond extension. In the normal-mode approximation the internal coordinate displacements **d** are related to the normal modes through the linear transformation

$$\mathbf{d} = \mathbf{L}_d \mathbf{q} \tag{11-89}$$

The transformation matrix $\mathbf{L}_d$ is obtained from a normal-mode analysis performed in internal coordinates. Thus, as the development with time of the normal-mode coordinates is evaluated from equation 11–88, the value for the reaction coordinate $q^\ddagger$, a particular internal coordinate displacement, is found from equation 11–89. The variation in $q^\ddagger$ with time results from a superposition of the normal modes. At a particular time the normal-mode vibrations will phase together so that $q^\ddagger$ exceeds a critical extension, at which point decomposition is assumed to occur.

The preceding discussion gives the essential details of the Slater theory. Since energy does not flow freely within the molecule, the theory predicts intrinsic non-

RRKM behavior. The attainment of the reaction coordinate critical extension is not a statistically random process as in RRKM theory, but depends on the energies and phases of the specific normal modes excited. To illustrate, consider the normal modes of the ethylene molecule, which consist of two separable groups, one for in-plane and one for out-of-plane vibrations. Thus, if in-plane normal vibrations are excited, only an in-plane dissociation path, such as C—H bond rupture, can occur. To promote an out-of-plane unimolecular path, such as rotation around the double bond (e.g., cis-trans isomerization in 1,2-dichloroethylene), the out-of-plane vibrations must be excited, but then displacements of the in-plane coordinates do not occur. Even if the molecule is excited randomly, in general its lifetime distribution will not be exponential and decay with the RRKM rate constant.

Overall, the Slater theory is unsuccessful in interpreting experiments. The theory is not compatible with photoactivation and chemical activation experiments where molecules are initially excited nonrandomly. Instead, the rate constants and unimolecular paths in these experiments are consistent with energy flowing randomly within the molecule. Nor does the theory adequately reproduce thermal unimolecular falloff curves. By an appropriate choice of parameters the high-pressure limit may be fit, but at lower pressures the unimolecular rate is severely underestimated. This is because certain patterns of normal-mode excitation prepared by collisional activation can never lead to reaction. The experimental results are consistent with ultimate dissociation if E is in excess of E_0, regardless of the excitation pattern.

If one considers the nature of classical Hamiltonians for actual molecules, at first glance it is not surprising that the Slater theory performs so poorly. The potential energy of actual molecules is anharmonic, and the modes are strongly coupled, not harmonic and separable as assumed in the normal-mode model. Also, for finite displacements of the atoms from their equilibrium positions, there will be couplings between the normal-mode momenta. Clearly, understanding the classical intramolecular motion of vibrationally excited molecules requires one to go beyond the normal-mode model.

The first classical trajectory study of unimolecular decomposition and intramolecular motion for realistic anharmonic molecular Hamiltonians was performed by Bunker.[12,19] Since this initial work, there have been numerous additional studies.[75,80,84-88] In some cases potential energy surfaces used in the trajectories have been derived from *ab initio* calculations. These studies have identified two distinct types of intramolecular motion: chaotic and quasiperiodic. Chaotic motion, depicted in Figure 11–17(a), results in RRKM behavior. Energy flows freely between the modes in the molecule, and thus the molecule does not exhibit regular vibrational motion as predicted by the normal-mode model. If the states in the molecular Hamiltonian are chosen randomly, exponential unimolecular decomposition occurs with a rate constant in agreement with that of RRKM theory. For nonrandom excitation initial apparent non-RRKM behavior is observed, but at longer times the probability of decomposition becomes that of RRKM theory.

The quasiperiodic motion depicted in Figure 11–17(b) is regular, like that of the normal-mode model. The vibrational motion can be represented by a superposition of individual vibrational modes, each containing a fixed amount of energy. For some cases these modes resemble normal modes, but they may be identifiably differ-

Figure 11–17. (a). Chaotic trajectory for a model HCC Hamiltonian. R_1 is the HC bond length, R_2 the CC bond length. Distance is in angstroms (Å).

ent. Thus, although actual molecular Hamiltonians contain potential and kinetic energy coupling terms, they may still exhibit regular vibrational motion with no energy flow between modes as predicted by the Slater theory.

At low levels of excitation, the motion associated with the molecular Hamiltonian is quasiperiodic. However, as the energy is increased, in most cases there is a gradual transition from quasiperiodic to chaotic motion, with both types present at intermediate energies. In many cases the motion becomes totally chaotic before the unimolecular threshold energy is reached. However, in other cases quasiperiodic (or nearly quasiperiodic) motion exists above the unimolecular threshold, and intrinsic non-RRKM lifetime distributions result. This type of behavior has been found for Hamiltonians with low unimolecular thresholds, widely separated frequencies, and/ or disparate masses.[12,86] Thus, classical trajectory simulations performed for realistic Hamiltonians predict that for some molecules the unimolecular rate constant may be strongly sensitive to the modes excited in the molecule, in agreement with the Slater theory. This property is called *mode specificity* and is discussed next.

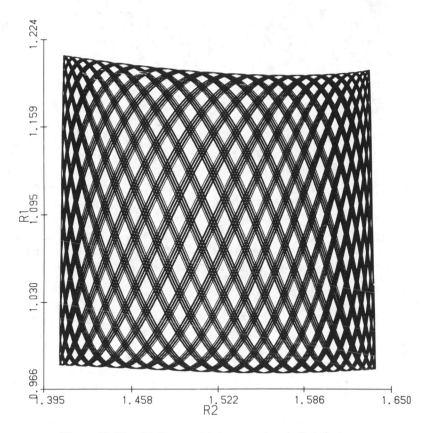

Figure 11–17. (b) Same as (a), but a quasi-periodic trajectory.

11.13 *MODE SPECIFICITY*

Mode specificity has been widely observed in the unimolecular decomposition of van der Waals molecules,[89,90] e.g.,

$$I_2 - He \longrightarrow I_2 + He$$

and

$$HF - HF \longrightarrow 2HF$$

A covalent bond (or particular normal mode) in the van der Waals molecule (e.g., the I_2 bond in $I_2 - He$) can be selectively excited, and what is usually observed experimentally is that the unimolecular dissociation rate constant is orders of magnitude smaller than the RRKM prediction. This is thought to result from weak coupling between the van der Waals bond and the other modes in the molecule.[91] In the $(HF)_2$ dimer the lifetimes associated with the two HF vibrational modes differ by a factor of 24. Other van der Waals molecules studied include $(NO)_2$,[92] $(C_2H_2)_2$,[93] and $(C_2H_4)_2$.[93]

There are fewer documented examples of mode specificity for the unimolecular decomposition of covalently bound molecules. One example is the decomposition of formaldehyde, viz.,

$$H_2CO \longrightarrow H_2 + CO$$

for which mode-specific effects have been measured experimentally.[94,95] In this reaction, individual vibrational/rotational states of highly excited formaldehyde are prepared by stimulated emission pumping (SEP) (see Figure 11–3). Although the excitation energy is less than that required to cross the saddle point separating the reactants from the products, the molecule may dissociate by tunneling through the potential energy barrier (see Figure 11–18). What is observed experimentally is that such states, with total energies that differ by only several cm^{-1}, may have rate constants that vary by more than an order of magnitude.[94,95] In fact, the rate constant may become significantly smaller if a state with a larger energy is excited. These results are contrary to those of RRKM theory, even with a tunneling correction included. RRKM theory predicts a *monotonic increase* in the rate constant with energy.

Figure 11–18. Dissociation of H$_2$CO vibrational/rotational states by tunneling through the potential energy barrier.

To explain these experimental results for H$_2$CO, first consider the situation where the barrier is of infinite width so that tunneling does not occur. The states prepared by SEP could not then dissociate and would simply be the highly excited vibrational/rotational states of bound formaldehyde. The extent of mode specificity in formaldehyde dissociation depends on the properties of these states. At low levels of excitation, it is known that wave functions ϕ_i for vibrational/rotational states are products of normal-mode vibration and rotational wave functions. If anharmonicity and vibrational-rotational coupling remained unimportant as the energy increased, such normal-mode rotational wave functions would also represent the high-energy states. Each state would then have a specific intramolecular vibrational property such as stretching, bending, rocking, twisting, etc., and would have different probabilities $\phi_i^2(q)$ in the coordinate space q.

The reaction coordinate for $H_2CO \rightarrow H_2 + CO$ dissociation is primarily a combination of an asymmetric HC stretch and HCH deformation. Each of the normal-mode rotational wave functions has a different probability amplitude along the reaction coordinate. For example, the probability would be large for a state with large amounts of quanta in the asymmetric HC stretch and HCH deformation. However, a nearly isoenergetic state with low excitation in these two modes and a large amount of excitation in the out-of-plane bend mode would have a small probability amplitude along the reaction coordinate. Thus, for a finite potential energy barrier separating H_2CO and $H_2 + CO$, these two nearly isoenergetic states would have different tunneling rates for unimolecular dissociation, since they have different probability amplitudes along the reaction coordinate.

In the absence of tunneling, the absorption spectrum for each of the highly excited states of formaldehyde would consist of a narrow line. As a result of the Heisenberg uncertainty principle, when tunneling occurs, the absorption line for each state is broadened. The relationship between the unimolecular lifetime of the excited vibrational/rotational state τ and the width at half-maximum Γ of its absorption spectrum is given by

$$\tau = \frac{\hbar}{\Gamma} \tag{11-90}$$

In the preceding discussion, we have considered the situation where highly excited vibrational/rotational states exhibit mode-specific effects akin to the prediction of the Slater theory. The situation should also be considered where mode-specific effects would not be observed. For many molecules, anharmonic and vibration/rotation coupling terms are important at high levels of excitation. As a result of these couplings, the vibrational-rotational states of the molecule cannot be defined by a single ϕ_i which is a product of normal-mode and rotational wave functions. Instead, the wave function ψ_j for each state is a linear combination of the ϕ_i, i.e.,

$$\psi_j = \sum_i c_{ji} \phi_i \tag{11-91}$$

If ψ_j is similar to a particular normal-mode rotational state, all except one of the coefficients c_{ji} are small. However, if the anharmonic and vibrational/rotational couplings are extensive, ψ_j will be a linear combination of many ϕ_i's. Thus, vibrational-rotational states which are nearly isoenergetic may have similar wave functions ψ_j. As a result, unimolecular rate constants may agree with RRKM theory, and mode specificity becomes unimportant.

Mode specificity is expected to be important for a bimolecular, as well as a unimolecular, reaction. Excited vibrational states in the reactants with individual in-

Figure 11–19. $C_2H_4 + F_2$: logarithm of the quantum yield for reaction as a function of the vibrational frequency and state. $\phi\,(\nu_7)$ indicates only an estimate of an upper limit, because no reaction could be observed. Error bars express uncertainties of relative quantum yields. [Adapted from H. Frei and G. C. Pimentel, *J. Chem. Phys. 78*, 3698 (1983).]

tramolecular motions may either suppress or enhance the reaction rate. One system that has been studied experimentally is the reaction between F_2 and C_2H_4 in a low-temperature matrix.[96] Although this reaction normally proceeds at an explosive rate, the low temperature and rigidity of the matrix slow it down to an extent whereby the rate can be measurably enhanced by vibrational excitation of one of the reactants, viz., the ethylene. Excitation is carried out using a series of infrared lasers emitting at frequencies corresponding to the C_2H_4 fundamental and combination bands. The monotonic rise in quantum yield between 1000 and 4000 cm^{-1} shown in Figure 11–19 is typical of RRKM or statistical behavior of $k(E)$; the small "blip" at the $\nu_2 + \nu_{12}$ band indicates mode-specific, and therefore intrinsically non-RRKM, behavior.

REFERENCES

1. D. W. Lupo and M. Quack, *Chem. Rev. 87*, 181 (1987).

2. K. V. Reddy and M. J. Berry, *Chem. Phys. Letters 52*, 111 (1977).

3. C. Kittrell, E. Abramson, J. L. Kinsey, S. A. McDonald, D. E. Reisner, R. W. Field, and D. H. Katayama, *J. Chem. Phys. 75*, 2056 (1981).

4. T. Beyer and E. F. Swinehart, *Commun. Assoc. Comput. Machin. 16*, 372 (1973); S. E. Stein and B. S. Rabinovitch, *J. Chem. Phys. 58*, 2438 (1973).

5. R. A. Marcus and O. K. Rice, *J. Phys. and Colloid Chem. 55*, 894 (1951).

6. G. Z. Whitten and B. S. Rabinovitch, *J. Chem. Phys. 41*, 1883 (1964); D. C. Tardy, B. S. Rabinovitch, and G. Z. Whitten, *J. Chem. Phys. 48*, 1427 (1968).

7. J. Perrin, *Ann. Phys. 11*, 1 (1919).

8. I. Langmuir, *J. Am. Chem. Soc. 42*, 2190 (1920).

9. F. A. Lindemann, *Trans. Faraday Soc. 17*, 598 (1922).

10. C. N. Hinshelwood, *Proc. Roy. Soc. (A) 113*, 230 (1927).

11. N. B. Slater, *Proc. Comb. Phil. Soc. 35*, 56 (1939).

12. Slater, *J. Chem. Phys. 40*, 1946 (1964).

13. N. B. Slater, *Theory of Unimolecular Reactions* (London: Methuen, 1959), chapter 9.

14. O. K. Rice and H. C. Ramsperger, *J. Am. Chem. Soc. 50*, 617 (1928).

15. L. S. Kassel, *J. Phys. Chem. 32*, 1065 (1928).

16. Slater, *op. cit.*, Chapter 2.

17. R. A. Marcus and O. K. Rice, *J. Phys. Colloid Chem. 55*, 894 (1951); R. A. Marcus, *J. Chem. Phys. 20*, 359 (1952); H. M. Rosenstock, M. B. Wallenstein, A. L. Wahrhaftig, and H. Eyring, *Proc. Natl. Acad. Sci. USA 38*, 667 (1952).

18. R. A. Marcus, *J. Chem. Phys. 43*, 2658 (1965); *52*, 1018 (1970).

19. D. L. Bunker and M. Pattengill, *J. Chem. Phys. 48*, 772 (1968).

20. D. L. Bunker, in *Proceedings of the International School of Physics*, Course XLIV, edited by C. Schlier (New York: Academic Press, 1970), p. 315.

21. F. W. Schneider and B. S. Rabinovitch, *J. Am. Chem. Soc. 84*, 4215 (1962).

22. J. H. Current and B. S. Rabinovitch, *J. Chem. Phys. 38*, 783 (1963).

23. M. Quack and J. Troe, *Ber. Bunsenges. Phys. Chem. 78*, 240 (1974).

24. W. H. Miller, *J. Am. Chem. Soc. 101,* 6810 (1979).

25. P. J. Robinson and K. A. Holbrook, *Unimolecular Reactions* (New York: Wiley-Interscience, 1972).

26. W. Forst, *Theory of Unimolecular Reactions* (New York: Academic Press, 1973).

27. *Ibid.,* Chapter 4.

28. J. Troe, *J. Phys. Chem. 83,* 114 (1979).

29. L. B. Bhuiyan and W. L. Hase, *J. Chem. Phys. 78,* 5052 (1983).

30. H. S. Johnston, *Gas Phase Reaction Rate Theory* (New York: Ronald, 1966), chapter 5.

31. H. E. O'Neal and S. W. Benson, *J. Phys. Chem. 71,* 2903 (1967).

32. K. Dees and D. W. Setser, *J. Chem. Phys. 49,* 1193 (1968).

33. W. L. Hase, *Acc. Chem. Res. 16,* 258 (1983).

34. P. Saxe, Y. Yamaguchi, P. Pulay, and H. F. Schaefer, III, *J. Am. Chem. Soc., 102,* 3718 (1980).

35. E. V. Waage and B. S. Rabinovitch, *Chem. Rev. 70, 377* (1970).

36. H. Furue and P. D. Pacey, *Can. J. Chem. 60,* 916 (1982).

37. W. Tsang, *J. Chem. Phys. 41,* 2487 (1964); W. Tsang, in *Shock Tubes in Chemistry,* edited by A. Lifshitz (New York: Dekker, 1981), p. 59.

38. G. N. Spokes, D. M. Golden, and S. W. Benson, *Agnes., Chem. Int. Ed. Engl. 12,* 534 (1973).

39. F. W. Schneider and B. S. Rabinovitch, *J. Am. Chem. Soc. 85,* 2365 (1963).

40. B. S. Rabinovitch, P. W. Gilderson, and F. W. Schneider, *J. Am. Chem. Soc. 87,* 158 (1965).

41. K. M. Maloney, S. P. Pavlou, and B. S. Rabinovitch, *J. Phys. Chem. 73,* 2756 (1969).

42. K. M. Maloney and B. S. Rabinovitch, *J. Phys. Chem. 73,* 1652 (1969).

43. W. L. Hase and H. B. Schlegel, *J. Phys. Chem. 86,* 3901 (1982).

44. T. Baer, *Adv. Chem. Phys. 64,* 111 (1986).

45. T. Baer and R. Kury, *Chem. Phys. Lett. 92,* 659 (1982).

46. B. S. Rabinovitch and M. C. Flowers, *Quart. Rev. 18,* 122 (1964).

47. B. S. Rabinovitch and D. W. Setser, *Advan. Photochem. 3,* 1 (1964).

48. C. W. Larson and B. S. Rabinovitch, *J. Chem. Phys. 52,* 5181 (1970).

49. W. L. Hase, R. L. Johnson, and J. W. Simons, *Int. J. Chem. Kinet. 4,* 1 (1972).

50. F. B. Growcock, W. L. Hase, and J. W. Simons, *Int. J. Chem. Kinet. 5,* 77 (1973).

51. K. Dees, D. W. Setser, and W. G. Clark, *J. Phys. Chem. 75,* 2231 (1971).

52. J. M. Jasinski, J. K. Frisoli, and C. B. Moore, *J. Chem. Phys. 79,* 1312 (1983).

53. H. S. Johnston, *J. Am. Chem. Soc. 75,* 1567 (1953); D. J. Wilson and H.S. Johnston, *J. Am. Chem. Soc. 75,* 5673 (1953).

54. S. C. Chan, B. S. Rabinovitch, J. T. Bryant, L. D. Spicer, T. Fujimoto, Y. N. Lin, and S. P. Pavlou, *J. Phys. Chem. 74,* 3160 (1970).

55. W. G. Valance and E. W. Schlag, *J. Chem. Phys. 45,* 4280 (1966).

56. J. Troe and H. G. Wagner, *Ber. Bunsenges. Phys. Chem. 71,* 937 (1967).

57. D. C. Tardy and B. S. Rabinovitch, *J. Chem. Phys. 48,* 1282 (1968).

58. R. E. Harrington, B. S. Rabinovitch, and M. R. Hoare, *J. Chem. Phys. 33.* 744 (1960).

59. G. H. Kohlmaier and B. S. Rabinovitch, *J. Chem. Phys. 48*, 1282 (1968).

60. D. W. Setser, in *MTP International Review of Sciences, Physical Chemistry*, Vol. 9, edited by J. Polanyi (Baltimore: University Park Press, 1972), p. 1.

61. M. Hoare, *J. Chem. Phys. 38*, 1630 (1963).

62. D. W. Setser and E. E. Siefert, *J. Chem. Phys. 57*, 3623 (1972).

63. Von S. H. Luu and J. Troe, *Ber. Bunsenges. Phys. Chem. 78*, 766 (1974).

64. J. R. Barker, *J. Phys. Chem. 88*, 11 (1984).

65. J. E. Dove, H. Hippler, and J. Troe, *J. Chem. Phys. 82*, 1907 (1985); H. Hippler, L. Lindemann, and J. Troe, *J. Chem. Phys. 83*, 3906 (1985).

66. A. J. Stace and J. N. Murrell, *J. Chem. Phys. 68*, 3028 (1978).

67. C. R. Gallucci and G. C. Schatz, *J. Phys. Chem. 86*, 2352 (1982).

68. N. J. Brown and J. A. Miller, *J. Chem. Phys. 80*, 5568 (1984).

69. W. L. Hase, N. Date, L. B. Bhuiyan, and D. G. Buckowski, *J. Phys. Chem. 89*, 2502 (1985).

70. K. F. Lim and R. G. Gilbert, *J. Chem. Phys. 84*, 6129 (1986).

71. C. E. Klots, *J. Chem. Phys. 58*, 5364 (1973).

72. C. E. Klots, *J. Chem. Phys. 69*, 4269 (1976).

73. W. J. Chesnavich and M. T. Bowers, *J. Chem. Phys. 66*, 2306 (1977).

74. J. M. Parson and Y. T. Lee, *J. Chem. Phys. 56*, 4658 (1972); J. M. Farrar and Y. T. Lee, *J. Chem. Phys. 65*, 1414 (1976).

75. W. J. Hase and K. C. Bhalla, *J. Chem. Phys. 75*, 2807 (1981).

76. W. B. Miller, S. A. Safron, and D. R. Herschbach, *Disc. Faraday Soc. 44*, 108 (1967).

77. J. C. Light, *Disc. Faraday Soc. 44*, 14 (1967).

78. R. A. Marcus, *J. Phys. Chem. 83*, 204 (1979).

79. M. Quack and J. Troe, *Ber. Bunsenges. Phys. Chem. 79*, 470 (1975).

80. D. L. Bunker and W. L. Hase, *J. Chem. Phys. 59*, 4621 (1973).

81. R. C. Baetzold and D. J. Wilson, *J. Phys. Chem. 68*, 3141 (1964).

82. J. D. Rynbrandt and B. S. Rabinovitch, *J. Phys. Chem. 75*, 2164 (1971).

83. J. F. Meagher, K. J. Chao, J. R. Barker, and B. S. Rabinovitch, *J. Phys. Chem. 78*, 2535 (1974).

84. C. S. Sloane and W. L. Hase, *J. Chem. Phys. 66*, 1523 (1977).

85. W. L. Hase and D. G. Buckowski, *J. Comput. Chem. 3*, 335 (1982); W. L. Hase, D. G. Buckowski, and K. N. Swamy, *J. Phys. Chem. 87*, 2754 (1983).

86. R. J. Wolf and W. L. Hase, *J. Chem. Phys. 72*, 316 (1980); R. J. Wolf and W. L. Hase. *J. Chem. Phys. 73*, 3779 (1980).

87. T. Uzer, J. T. Hynes, and W. P. Reinhardt, *J. Chem. Phys. 85*, 5791 (1986).

88. T. A. Holme and J. S. Hutchinson, *J. Chem. Phys. 83*, 2860 (1985); J. S. Hutchinson, *J. Phys. Chem. 91*, 4495 (1987).

89. K. E. Johnson, L. Wharton, and D. H. Levy, *J. Chem. Phys. 69*, 2719 (1978).

90. Z. S. Huang, K. W. Jucks, and R. E. Miller, *J. Chem. Phys. 85*, 3338 (1986).

91. G. E. Ewing, *J. Chem. Phys. 71*, 3143 (1979); G. E. Ewing, *J. Chem. Phys. 72*, 2096 (1980).

92. M. P. Casassa, A. M. Woodward, J. C. Stephenson, and D. S. King, *J. Chem. Phys.* *85,* 6235 (1986).
93. G. Fischer, R. E. Miller, P. F. Vohralik, and R. O. Watts, *J. Chem. Phys. 83,* 1471 (1985).
94. H. -L. Dai, R. W. Field, and J. L. Kinsey, *J. Chem. Phys. 82,* 1606 (1985).
95. D. R. Guyer, W. F. Polik, and C. B. Moore, *J. Chem. Phys. 84,* 6519 (1986).
96. H. Frei and G. C. Pimentel, *J. Chem. Phys. 78,* 3698 (1983).

BIBLIOGRAPHY

Theory of Unimolecular Reactions

BEYNON, J. H., and GILBERT, J. R. *Application of Transition State Theory to Unimolecular Reactions.* New York: Wiley, 1984.
BUNKER, D. L. *Theory of Elementary Gas Reaction Rates.* New York: Pergamon Press, 1966.
CHESNAVICH, W. J., and BOWERS, M. T. In *Gas Phase Ion Chemistry, Vol. 1,* edited by M. T. Bowers. New York: Academic Press, 1979, p. 119.
FORST, W. *Theory of Unimolecular Reactions.* New York: Academic Press, 1973.
HASE, W. L. In *Modern Theoretical Chemistry, Dynamics of Molecular Collisions, Part B,* edited by W. H. Miller. New York: Plenum, 1976, p. 121.
HOLBROOK, K. A. *Chem. Soc Rev. 12,* 163 (1983).
PRITCHARD, H. O. *The Quantum Theory of Unimolecular Reactions.* New York: Cambridge University Press, 1984.
QUACK, M., and TROE, J., In *Gas Kinetics and Energy Transfer, Vol. 2. Specialist Periodical Reports,* edited by P. G. Ashmore and R. J. Donovan. London: The Chemical Soc., Burlington House, 1977, p. 175.
QUACK, M., and TROE, J. *Int. Rev. Phys. Chem. 1,* 97 (1981).
ROBINSON, P. J., and HOLBROOK, K. A. *Unimolecular Reactions.* London: Wiley-Interscience, 1972.
SLATER, N. B. *Theory of Unimolecular Reactions.* Ithaca, NY: Cornell Press, 1959.
SMITH, I. W. M. *Kinetics and Dynamics of Elementary Gas Reactions.* London: Butterworths, 1980.
TROE, J. In *Physical Chemistry an Advanced Treatise, Vol. VI B,* edited by W. Jost. New York: Academic Press, 1975, p. 835.

Energy-Selected Dissociation

BAER, T. *Adv. Chem. Phys. 64,* 111 (1986).
CRIM, F. F. *Ann. Rev. Phys. Chem. 35,* 657 (1984).
LUPO, D. W., and QUACK, M. *Chem. Rev. 87,* 181 (1987).

Intermolecular Energy Transfer

TARDY, D. C., and RABINOVITCH, B. S. *Chem. Rev. 77,* 369 (1977).

Intramolecular Energy Transfer

BRUMER, P. *Adv. Chem. Phys. 47,* 201 (1981).

HASE, W. L. In *Potential Energy Surfaces and Dynamics Calculations,* edited by D. G. Truhlar. New York: Plenum, 1981, p. 1.

HASE, W. L., *J. Phys. Chem. 90,* 365 (1986).

HASE, W. L., and WOLF, R. J. In *Potential Energy Surfaces and Dynamics Calculations,* edited by D. G. Truhlar. New York: Plenum, 1981, p. 37.

NOID, D. W., KOSZYKOWSKI, M. L., and MARCUS, R. A.. *Ann. Rev. Phys. Chem. 32,* 267 (1981).

OREF, I., and RABINOVITCH, B. S. *Acc. Chem. Res. 12,* 166 (1979).

PARMENTER, C. S. *Faraday Discuss. Chem. Soc. 75,* 7 (1983).

QUACK, M. In *Energy Storage and Redistribution in Molecules,* edited by J. Hinzc. New York: Plenum, 1983, p. 493.

Mode Specificity

BISSELING, R. H., KOSLOFF, R., MANZ, J., MRUGALA, F., RÖMELT, J., and WEICHSELBAUMER, G. *J. Chem. Phys. 86,* 2626 (1987).

HEDGES, R. M., JR., SKODJE, R. T., BORONDO, F., and REINHARDT, W. P. In *Resonances in Electron Molecule Scattering, Van der Waals Complexes, and Reactive Chemical Dynamics.* Kansas City, ACS Symp. Ser. No. 263, 1984.

HOSE, G., and TAYLOR, H. S. *J. Chem. Phys. 76,* 5356 (1982).

HOSE, G., TAYLOR, H. S., and BAI, Y. Y. *J. Chem. Phys. 80,* 4363 (1984).

MIES, F. H., and KRAUSS, M. *J. Chem. Phys. 45,* 4455 (1966).

MIES, F. H., *J. Chem. Phys. 51,* 798 (1969).

MILLER, W. H., *Chem. Rev. 87,* 19 (1987).

SWAMY, K. N., HASE, W. L., GARRETT, B. C., MCCURDY, C. W., and MCNUTT, J. F. *J. Phys. Chem. 90,* 3517 (1986).

WAITE, B. A., GRAY, S. K., and MILLER, W. H. *J. Chem. Phys. 78,* 259 (1983).

WAITE, B. A., and MILLER, W. H. *J. Chem. Phys. 73,* 3713 (1980).

WAITE, B. A., and MILLER, W. H. *J. Chem. Phys. 74,* 3910 (1981).

PROBLEMS

11.1 Consider the three-oscillator harmonic system with frequencies of 800, 1000, and 1400 cm^{-1}. By direct count, determine the sum of states at E = 5000 cm^{-1}.

11.2 Use the Whitten-Rabinovitch approximation to determine the sum of states for cyclopropane at 40, 200, and 400 kJ/mol. Compare with the values in Table 11-1.

11.3 Compare $G(E)$ in Eqs. (11-11), (11-13) and (11-14) by applying these equations to CH_4 for E in the range of 0–400 kJ/mol. The harmonic vibrational frequencies for CH_4 are 3025, 1583(2), 3157(3), and 1367(3) cm^{-1}. Calculate the sum of states every 40 kJ/mol.

11.4 Using the Wien radiation law

$$E_\nu d\nu = \frac{2\pi h\nu^3}{c^2} e^{-h\nu/k_B T} d\nu$$

calculate the amount of infrared energy radiated in a 100 cm^{-1} bandwidth centered at 3000 cm^{-1} by a blackbody source at T = 1000 K. How does this compare with the amount of energy required to sustain a unimolecular reaction having a rate coefficient $k_{uni} = 10^{-2}$ sec^{-1} and an activation energy of 300 kJ/mole?

11.5 Using Eqs. (11–14) through (11–17) derive the following density of states expression for $(1.0 < E' < 8.0)$

$$N(E_V) = \frac{(E_V + aE_Z)^{s-1}}{(s-1)!\Pi h\nu_i}[1 + 0.60478\ \beta\ w\ (E')^{-0.75}]$$

11.6 Show that if classical mechanical expressions are used for the sum and density of states given by equations (11–11) and (11–12), RRKM theory gives the classical mechanical RRK expression, Eq. (11–39) for the unimolecular rate constant.

11.7 The high pressure A-factor for the dissociation of methane, i.e., $CH_4 \rightarrow CH_3 + H$, is 10^{16} sec^{-1}, and the activation energy is 400 kJ/mole. Using the RRK theory calculate the unimolecular rate constant for methane at an excitation energy of 500 kJ/mole.

11.8 Silane (SiH_4) unimolecularly decomposes to SiH_2. The frequencies for SiH_4 are four stretches of 2,000 cm^{-1} and five bends of 900 cm^{-1}.
(a) Propose a structure for the transition state.
(b) Estimate values for the transition state frequencies.
(c) Calculate the difference in harmonic zero-point energies between the transition state and reactant (83 cm^{-1} = 1 kJ/mol.).

11.9 What is the expression for the harmonic sum of states for a diatomic harmonic oscillator at energy E? Is the sum of states for the Morse oscillator at this energy larger or smaller? Why?

*11.10 Consider the thermal unimolecular rearrangement of hydrogen isocyanide, which involves only three atoms:

$$HNC \longrightarrow HCN$$

It is reasonable to assume that this reaction proceeds via a three-center transition state:

(a) Specify all the (non-translational) degrees of freedom of HNC and the transition state shown above. Note that HNC, being collinear, has two rotational degrees of freedom, while the transition state has three.
(b) Given the following values for the various degrees of freedom in HNC and the transition state, evaluate $Q^\ddagger$ and Q at T = 1000 K.
NOTE: HNC: $I = 11.3$ amu-$\mathring{A}^2$, $\nu = 3300$ cm^{-1} (stretch); 713 cm^{-1} (doubly degenerate bend), 2100 cm^{-1} (stretch); Transition state: $I^\ddagger = 1.02$, 10.5, and

12.0 amu-Å^2. $\nu = 1000$ cm^{-1} (bend), 2000 cm^{-1} (ring–breathing stretch), 3000 cm^{-1} (stretch). One of the reactant degenerate bends (corresponding to a bend of the transition state) we shall take to be the reaction coordinate.

*11.11 (a) Evaluate $k_B T/h$ at $T = 1000$ K, given $k_B = 1.38 \times 10^{-23}$ J K^{-1}, $h = 6.63 \times 10^{-34}$ J s.

(b) If we assume that $Q^{\ddagger}$ and Q are approximately the same for a particular reaction, give a numerical value for the frequency factor.

(c) Evaluate the frequency factor for the "tight" (three-center) transition state for the HNC + HCN reaction given above. Note: for a unimolecular reaction, the frequency factor is usually expressed notationally as A $= 10^{13.1}$ s^{-1} (or whatever the value is).

(d) Would the frequency factor for a "loose" transition state (where one of the vibrational degrees of freedom is of lower frequency in the transition state than in the reactant) be larger or smaller than $k_B T/h$?

11.12 Calculate the densities-of-states and RRKM dissociation rates for the unimolecular dissociation of chloroethane-d$_1$:

(a) Calculate $k(E_V)$ [sec^{-1}] for $E_V = 0$ to 50,000 cm^{-1} (0 to 594 kJ/mole), at 1000 cm^{-1} intervals, for the two reaction paths I (elimination of HCl) and II (elimination of DCl). The Whitten-Rabinovitch approximation should be used for calculating the sum, $G(E^{\ddagger})$, and density, $N(E_V)$, terms. The critical energies (E_0), rotational symmetry numbers, and vibrational frequencies are given in the following table:

CH$_2$DCH$_2$Cl		T.S. I (HC1 elimination)		T.S. II (DC1 elimination)	
g_i	ν_i, cm^{-1}	g_i	ν_i, cm^{-1}	g_i	ν_i, cm^{-1}
4	2940	3	3000	4	3000
1	2160	2	2200	1	2088
4	1340	2	1380	1	1380
2	1270	1	1115	1	1100
3	960	5	960	4	960
2	720	1	850	3	850
1	330	1	820	1	750
1	200	1	645	1	570
		1	403	1	501
s = 18		s‡ = 17		s‡ = 17	
critical energies (kJ/mole)			232.7		237.9
symmetry numbers	0.33		0.25		0.497

(Continued)

SUGGESTIONS: Make sure to count each vibration for a number of times equal to its degeneracy g_i (some of the frequencies, e.g., the C—H stretches, have been "grouped" to reduce computation). The symmetry numbers for the transition state are based on a vibrational transition state model (with hindered internal rotation treated as a torsion) due to Hassler and Setser; that for the reactant takes into account the entropy of mixing of *trans* and *gauche* forms. Assume that the moments of inertia of the transition states and of the parent molecule are approximately equal.

(b) Using transition state theory calculate k_∞, the net high-pressure limiting thermal unimolecular decomposition for C_2H_5Cl at 1000°K, and the predicted ratio $[HCl]_\infty/[DCl]_\infty$, assuming no secondary reactions.

11.13 The RRKM theory can be used to calculate the unimolecular rate constant as a function of E. The decomposition rate constant for $C_2H_6 \rightarrow 2CH_3$ is 5×10^9 sec^{-1} at $E = 481$ kJ/mole. E_0 for ethane dissociation is 355 kJ/mole and the frequencies in cm^{-1} for ethane are

2985		1190
2985		1190
2969		822 } CH$_3$ rocks
2969 } C—H stretches		822
2954		
2896		955 C—C stretch
		289 internal rotation
1469		
1469		
1468		
1468 } H—C—H bends		
1388		
1379		

Using computer program RRKM repeatedly, find a critical configuration that fits the above rate constant within a factor of two. Explain the frequency adjustments you made in your critical configuration.

11.14 Calculate rates for unimolecular dissociation of ethylsilane going through two possible transition states, *viz.*:

$$CH_3CH_2SiH_3$$

3-center elimination:
$$\left[CH_3—CH_2—SiH \begin{array}{c} \cdots \cdots H \\ \vdots \\ \cdots \cdots H \end{array} \right]^{\ddagger} \longrightarrow CH_3CH_2SiH + H_2$$

4-center elimination:
$$\left[\begin{array}{ccc} CH_2 & \cdots & CH_2 \\ \vdots & & \vdots \\ H & \cdots & SiH_3 \end{array} \right]^{\ddagger} \longrightarrow C_2H_4 + SiH_4$$

(a) In the first part of this problem, derive transition state parameters for both the 3-center and 4-center elimination mechanisms.

(i) The structure for ethylsilane and both transition states, as estimated from MNDO calculations, is given in Table 1. Using Benson's rules, the structural results, and Table 2 containing the generalized frequencies for polysilanes, derive frequencies for the transition state structures for 3-center and 4-center elimination pathways. *Hint:* Remember Pauling's rules which relate bond order to bond lengths and bond frequencies.

TABLE 1 USEFUL INFORMATION FOR RRKM CALCULATIONS

	ethylsilane	3-center	4-center
I_A (g cm^2 × 10^{-40})	35.5	29.5	50.7
I_B (g cm^2 × 10^{-40})	158.3	200.9	345.9
I_C (g cm^2 × 10^{-40})	183.4	219.1	383.6

The vibrational spectrum of ethylsilane is given in *Spectrochimica Acta 23A,* 2761 (1967).

Structural information (distances in Å)

ethylsilane	3-center	4-center
r(C-C) = 1.52	r(C-C) = 1.52	r(C $\cdots$ C) = 1.35
r(C-H) = 1.11	r(C-H) = 1.11	r(C $\cdots$ H) = 2.66
r(C-Si) = 1.81	r(C-Si) = 1.81	r(C—H) = 1.09
r(Si-H) = 1.38	r(Si-H) = 1.38	r(C $\cdots$ Si) = 3.20
	r(Si $\cdots$ H) = 1.39	r(Si $\cdots$ H) = 2.50
	r(H $\cdots$ H) = 0.95	r(Si—H) = 1.38

TABLE 2 OBSERVED AND GENERALIZED FREQUENCY ASSIGNMENTS IN METHYLSILANES*,a

Vibrations	CH$_3$SiH$_3$	(CH$_3$)$_2$SiH$_2$	(CH$_3$)$_3$SiH	HMDS[b]	Generalized
(C-H)$_{str}$	(?)				2960
(Si-H)$_{str}$	2(?),2169				2130
$\begin{array}{c}C\\(H \quad\diagup\diagdown\quad C)_{bend}\end{array}$	2(1412)	4(1380-1440)	6(1425-1467)		(1450)2
	1266	2(1260)	1257,1263		(1260)1
$\begin{array}{c}Si\\(H \quad\diagup\diagdown\quad C)\end{array}$	2(?)				690
	943				
(CH$_3$)$_{rock}$	2(870)	919,963	2(900)		850
		802,867	850,835		
(Si-C)$_{str}$	701	728,654	2(711),625		700
(SiH$_2$)$_{w,t}$		643,591			610
(SiH$_3$)$_{rock}$	2(539)				555
(SiH$_2$)$_{rock}$		467			460
$\begin{array}{c}Si\\(C \quad\diagup\diagdown\quad C)_{bend}\end{array}$		223	2(252,216)		225
Si-Si				184	185
$\begin{array}{c}Si\\(C \quad\diagup\diagdown\quad Si)\end{array}$				404	400
(Si-C)				638	2(685)
					1(640)

TABLE 2 (cont.) OBSERVED AND GENERALIZED FREQUENCIES IN POLYSILANES*

Vibration	Si_3H_8	n-Si_4H_{10}	i-Si_4H_{10}	neo-Si_5H_{12}	2,2-disilyl tetrasilane	2,3-disilyl	Generalized
(Si-H)$_5$	2147	2146	2145	2143	2144	2146	2130
Si⟋﹨ (H　　H)	920 873	930,917 900	932 895	926,909 845	925,913 891	927 893	2(920) 1(880)
(SiH$_2$)$_{w,t}$	698	682,652 744,714			679	679 743	700
Si⟋﹨ (H　　Si)			693				690
(SiH$_3$)$_r$	564	540	567	574	567	541	555
(SiH$_2$)$_r$	466	467			457	452	460
(Si-Si)	392	427 377	450 352	440 309	444,393 282,303	385 330	425 325
Si⟋﹨ (Si　　Si)	109	127 100	118 100	98	131 98	137 95	110

*Source: *J. Organometallic Chem.* **213**, 419 (1981).
[a] All frequencies in units of cm^{-1}.
[b] HMDS = Hexamethyldisilane.

(ii) The rates for 3-center elimination of H_2 for a series of alklysilanes show that a critical energy of dissociation (E_0) for the process is estimated for ethylsilane as 272.0 kJ mole^{-1} [Ring et al., *Organometallics* **2**, 1891 (1983)]. Estimate a critical energy for the 4-center elimination of SiH_4 for ethylsilane. Justify your estimates using results from analogous systems [Provide literature references.] You may use any other method of estimating E_0 so long as you justify your choice of value.

(b) Calculate $k(E_v)$ [sec^{-1}] for $E_v = 0$ to 60000 cm^{-1} at 1000 cm^{-1} intervals for the two reaction paths I (elimination of H_2) and II (elimination of SiH_4). Show results as a plot of k_I and k_{II} vs. E_v.

*11.15 Consider the simplest (i.e., 2-level) version of the Lindemann-Hinshelwood mechanism:

$$A + M \underset{k_{-1}}{\overset{k_1}{\rightleftharpoons}} A^* + M$$

$$A^* \xrightarrow{k_2} \text{products}$$

Let

$c_1(t) \equiv [A]$, $c_2(t) \equiv [A^*]$, and $M = [M]$ is the bath gas concentration.

The rate equations are

$$\dot{c}_1(t) = -k_1 M c_1(t) + k_{-1} M c_2(t)$$

$$\dot{c}_2(t) = k_1 M c_1(t) - (k_{-1}M + k_2)c_2(t)$$

Note that $k_1/k_{-1} = e^{-\beta\Delta\epsilon}$ where $\Delta\epsilon$ is the A → A* excitation energy.

(a) Solve the rate equations using the steady-state approximation (i.e., $\dot{c}_2 = 0$) to find the unimolecular rate constant

$$k_{uni} \equiv -\dot{c}_1/c_1.$$

(b) Solve the rate equations exactly by assuming a solution of the form

$$c_i(t) = A_i e^{-\lambda t};$$

substitute into the rate equations, find λ, and thus k_{uni}:

$$k_{uni} \equiv -\dot{c}_1/c_1 = \lambda$$

(c) Compare the unimolecular rate constants obtained in parts (a) and (b). In what limit are they equivalent, i.e., in what limit is the steady-state approximation valid?

11.16 The probability distribution of total thermal energy E is given by the Boltzmann distribution $P(E)$.

(a) Show that

$$P(E) = \frac{N(E)e^{-\beta E}}{Q(T)}$$

where Q is the canonical partition function and N the microcanonical density of states ($\beta = 1/k_B T$).

(b) For the case of s independent classical harmonic oscillators, show that

$$P(E) = \frac{1}{(s-1)!} \left(\frac{E}{k_B T} \right)^{s-1} \frac{e^{-\beta E}}{k_B T}$$

(c) For the $P(E)$ distribution in part (b), calculate the average energy,

$$\langle E \rangle \equiv \int dE\, P(E)E$$

and the width of the distribution about this average value,

$$\Delta E \equiv \sqrt{\langle E^2 \rangle - \langle E \rangle^2}$$

where

$$\langle E^2 \rangle \equiv \int dE\, P(E)\, E^2$$

11.17 Recall the following Lindemann expression for the thermal unimolecular rate constant, which assumes continuous energies

$$k_{uni}(\omega) = \int_0^\infty \frac{k(E)\omega\, P(E)\, dE}{k(E) + \omega} \tag{1}$$

where

$$P(E) = N(E)\exp(-\beta E)/Q$$

and

$$Q = \int_0^\infty N(E)\exp(-\beta E)\, dE$$

Since $k(E) \equiv 0$ for $E < E_0$ the lower limit of the integral in Eq. (1) is actually E_0, i.e.

$$k_{uni}(\omega) = \int_{E_0}^{E} \cdots \cdot dE \qquad (2)$$

(a) Show that Eq. (1) or (2) implies that a plot of $k_{uni}(\omega)$ versus ω (which is proportional to the pressure) has positive slope and negative curvature for all ω.

(b) For interpretational reasons it is useful to consider $k_{uni}(\omega)$ as the average of the quantity $k(E)\omega/[k(E) + \omega]$ over a normalized distribution function. The problem is that the function $N(E) \exp(-\beta E)/Q$ is *not* normalized over the interval (E_0, ∞). One thus introduces the constant C,

$$k_{uni}(\omega) = C \int_0^{\infty} dE \, \frac{N(E) \exp(-\beta E)}{CQ} \, \frac{k(E)\omega}{k(E) + \omega}, \qquad (3)$$

so that the distribution function $W(E)$

$$W(E) \equiv \frac{N(E) e^{-\beta E}}{CQ}, \qquad (4)$$

is normalized on (E_0, ∞); i.e.,

$$\int_{E_0}^{\infty} dE \, W(E) = 1;$$

show that this implies that

$$C = \int_{E_0}^{\infty} dE \, N(E) \, e^{-\beta E}/Q.$$

The expression for $k_{uni}(\omega)$ then reads

$$k_{uni}(\omega) = C \left\langle \frac{k(E)\omega}{k(E) + \omega} \right\rangle \qquad (5a)$$

where

$$\langle \cdots \rangle \equiv \int_{E_0}^{\infty} dE \, W(E)(\cdots) \qquad (5b)$$

Use Eq. (5) to show that in the high and low pressure limits, respectively, one has the limiting forms

$$k_{uni}(\omega) = C\langle k \rangle - C\langle k^2 \rangle (1/\omega) + \cdots$$
$$k_{uni}(\omega) = \omega C - \omega^2 C\langle 1/k \rangle + \cdots.$$

11.18 Some of the data of Schneider and Rabinovitch [*J. Am. Chem. Soc. 84*, 4215 (1962) on the thermal isomerization of methyl isocyanide, CH_3NC, are given below.

(a) Compare the pressure dependence with that which would be expected if the simple Lindemann-Hinshelwood theory held.

(b) From suitable plots and calculations, derive approximate values for $\langle k \rangle$ and for the standard derivation $\sqrt{\langle k^2 \rangle - \langle k^2 \rangle}$. [Hint: Convert the pressure P into the collision frequency ω by the relation

$$\omega = \bar{v}\pi d^2 C, \qquad (A)$$

where C is the concentration, given from the ideal gas law as $C = P/k_B T$. Viscosity measurements imply a value for d of 4.5 Å. Show that Eq. (A) gives the conversion ω (in sec^{-1}) = $9.03 \times 10^6 P$ (in mm Hg).

Pressure mm Hg	$k \times 10^6$ sec^{-1}	Pressure mm Hg	$k \times 10^6$ sec^{-1}	Pressure mm Hg	$k \times 10^6$ sec^{-1}
5980	75.0	570	69.8	24.8	30.0
5010	74.1	337	66.2	15.7	21.6
2888	74.5	208	60.2	12.7	19.8
2248	73.7	155.2	55.6	10.0	18.0
1780	73.5	120.1	55.8	9.07	16.6
1500	72.5	101.5	52.0	7.25	14.2
1143	73.0	72.3	45.0	5.21	13.2
948	67.9	35.7	34.0		

*These problems have been provided by Dr. R. G. Gilbert, Department of Theoretical Chemistry, University of Sydney, Australia.

CHAPTER 12

Dynamics Beyond the Gas Phase

Most reactions of interest to the chemical community occur either in solution or at the gas-surface interface. Much less is known about the reaction dynamics of these processes at the microscopic level than about the dynamics of gas phase reactions. For some situations, concepts derived from gas phase studies are helpful in interpreting the dynamics of reactions in solution and heterogeneous reactions. However, many dynamical characteristics of the latter processes are not prominent in the gas phase. An example is the manner in which solvent molecules can affect the molecular motion of a reactive chemical system in solution. Diffusion on a solid surface, a property important in many heterogeneous reactions, does not have a comparable analogue in gas phase reactions.

For reactions in solution, the solvent often has a dramatic effect on the potential energy surface of the reactive system. As an example, consider the generic S_N2 reaction

$$X^- + CH_3Y \longrightarrow XCH_3 + Y^-$$

Representative reaction path enthalpy profiles are shown in Figure 12–1 for reaction in the gas phase, and in dipolar aprotic and protic solvents. Due to the extensive delocalization of the electron in the transition state region of the potential, polar solvent molecules significantly stabilize the reactants and products in comparison to the transition state.[1] Indeed, rate constants for S_N2 reactions in aqueous solution can be up to 20 orders of magnitude smaller than in the gas phase!

Experimental and theoretical studies have provided much insight and detail into the dynamics of solution and gas-surface reactions. In this chapter, we shall consider models which have been developed for these two kinds of reaction.

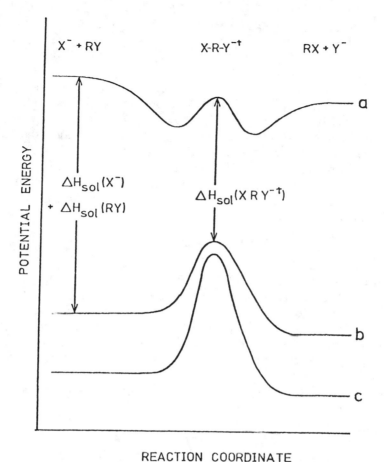

$$X^- + RY \qquad X\text{-}R\text{-}Y^{-\dagger} \qquad RX + Y^-$$

Figure 12–1. Representative reaction coordinate diagrams for a nucleophilic displacement reaction in (a) the gas phase and in (b) dipolar aprotic and (c) protic solvents. [Reproduced with permission from W. N. Olmstead and J. I. Brauman. *J. Am. Chem. Soc. 99*, 4219 (1977). copyright 1977, American Chemical Society.]

12.1 TRANSITION STATE THEORY OF SOLUTION REACTIONS

12.1.1 Static Solvent Effects: Thermodynamic Formulation

In the phenomenological treatment of solution reactions presented in chapter 4, a bimolecular solution reaction is modeled by equation 4–30, i.e.,

$$A + B \underset{k_{uni}}{\overset{k_D}{\rightleftharpoons}} AB \xrightarrow{k_2} \text{products}$$

Here, AB is the encounter complex of the reactants trapped inside a solvent cage. In contrast, transition state theory represents the reaction by the equation

$$A + B \longrightarrow AB^{\ddagger} \longrightarrow \text{products} \tag{12–1}$$

where $AB^{\ddagger}$ is the transition state.

There are two ways the solvent cage enters into the transition state theory of solution reactions. First, for fast diffusion-controlled bimolecular reactions, the process of diffusional approach of the reactants, i.e., cage-to-cage jumping, may be the rate-limiting elementary step. Second, the inner solvation shell interacts with the chemical reactive system as the AB encounter complex forms the reaction products.

In principle, transition state theory can be applied to either the physical diffusion process or the chemical reaction inside the solvent cage. For the former, the transition state would be a molecule in the act of making a jump from one position to another. The more common application of transition state theory is to the chemical transformation

$$AB \longrightarrow AB^{\ddagger} \longrightarrow \text{products} \tag{12–2}$$

It is this process which is treated here. If the chemical transformation is rate-determining, the transition state theory and experimental rate constants may be directly compared. The transition state theory of solution reactions can be developed so that the structure of the solvent and its interaction with the reactive system are explicitly considered. Another way to apply transition state theory is through its thermodynamic formulation. The latter approach is discussed next, while the former is considered in the next section.

From chapter 10, it is easily shown that the transition state theory rate for the reaction given by equation 12–2 is proportional to the concentration of transition states; i.e.,

$$\text{Rate} = \frac{k_B T}{h} [AB^{\ddagger}] \tag{12–3}$$

In solution, the equilibrium constant for formation of the transition state should be expressed in terms of activities instead of concentrations, so that

$$K^{\ddagger} = \frac{a^{\ddagger}}{a_A a_B} \tag{12–4}$$

where the a's are the activities. By introducing the concept of the activity coefficient γ (i.e., $a_A = \gamma_A[A]$) into equation 12–4, the concentration of transition states becomes

$$[AB^{\ddagger}] = K^{\ddagger} \frac{\gamma_A \gamma_B}{\gamma^{\ddagger}} [A][B] \tag{12–5}$$

If this equation is inserted into equation 12–3, one obtains the transition state theory expression for the rate constant for a reaction in solution in thermodynamic form:

$$k = \frac{k_B T}{h} K^{\ddagger} \frac{\gamma_A \gamma_B}{\gamma^{\ddagger}} \tag{12–6}$$

Note that both $K^{\ddagger}$ and the γ's must refer to the same standard state. Equation 12–6 is often written as

$$k = k_0 \frac{\gamma_A \gamma_B}{\gamma^\ddagger} \qquad (12-7)$$

where k_0 is the rate constant in the limit in which the activity coefficients become unity.

An important application of equation 12–7 is to the reaction between ions, i.e.,

$$A^{z_A} + B^{z_B} \longrightarrow [AB^\ddagger]^{z_A + z_B} \longrightarrow \text{products} \qquad (12-8)$$

From Debye-Hückel theory (see chapter 4, section 4.5) the activity coefficient for an ion in an aqueous solution at 25°C is given by

$$\log \gamma_i = -0.51 z_i^2 I^{1/2} \qquad (12-9)$$

where the ionic strength I is given by $I = \frac{1}{2} \Sigma\, m_j z_j$, in which the coefficient m_j is the molality of ionic species j in the solution. By inserting equation 12–9 into equation 12–7, one obtains

$$\log k = \log k_0 + \log \gamma_A + \log \gamma_B - \log \gamma^\ddagger$$
$$= \log k_0 - 0.51 I^{1/2}[z_A^2 + z_B^2 - (z_A + z_B)^2]$$

so that

$$\log k = \log k_0 + 1.02 z_A z_B I^{1/2} \qquad (12-10)$$

which is the same as equation 4–55 derived by the phenomenological approach.

Equation 12–7 is particularly useful for comparing reactions in the gas and liquid phases if a reference state is chosen so that k_0 equals the gas phase rate constant k_g. For a unimolecular reaction, the equation becomes

$$k_s = k_g \frac{\gamma_A}{\gamma^\ddagger}$$

Rate constants for many unimolecular reactions are the same in the gas and liquid phases. This is because the transition state, with a few bonds elongated and/or shortened, is often similar to the reactant. Consider, for example, the rearrangement of bicycloheptene to 1,3-cycloheptadiene:

$$(12-11)$$

The reactant itself is one of the best possible models for the transition state, and very similar interactions with the solvent are expected for both the reactant and the transition state. Thus, the activity coefficient ratio, $\gamma_A/\gamma^\ddagger$ should be unity and independent of the solvent. Accordingly, very similar rate constants and Arrhenius parameters are found for this reaction in the gas phase and in solution.[2] Differences arise between gas and liquid phase unimolecular rate constants when the reactant and transition state charge distributions (e.g., polarities) are different, so that a polar solvent preferentially stabilizes either the reactant or the transition state. Also, for a unimolecular dissociation reaction such as

$$C_2H_6 \longrightarrow 2CH_3 \tag{12–12}$$

the gas and liquid phase kinetics may differ if the products have difficulty escaping from the solvent cage and, thus, re-form the reactant. If diffusion effects are important, transition state theory as developed by equations (12–2) through (12–7) would not be applicable.

To consider differences between gas and liquid phase kinetics more explicitly, the activity coefficient can be expressed in terms of thermodynamic functions. At equilibrium the free energies of a species in solution and in the gas phase are equal, so that

$$G_s = G_g = G^0 + RT \ln C_g \tag{12–13}$$

where C_g is the concentration in the gas phase and G^0 is the free energy for one mole/liter of the gas. The gas is assumed to behave ideally, so that the activity in the gas phase is equal to the concentration, i.e., $a_g = C_g$. In solution, the activity is given by $a_s = \gamma C_s$. Since a_s and a_g are equal, equation 12–13 may be written as

$$G_s = G_g = G^0 + RT \ln \gamma C_s \tag{12–14}$$

which, for one mole/liter in solution, becomes

$$G_s = G_g = G^0 + RT \ln \gamma$$

Recognizing that $G^0 - G_s$ is the free energy of vaporization ΔG_{vap} for one mole of the species in solution, we see that the activity coefficient is given by

$$\gamma = \exp\left(-\Delta G_{vap}/RT\right) = \exp\left(-\Delta H_{vap}/RT\right) \exp\left(\Delta S_{vap}/R\right) \tag{12–15}$$

From the free-volume theory of liquids,[3] the term $\exp\left(\Delta S_{vap}/R\right)$ is related to the molar volume V_s^0 of the solution and free volume V_f per mole of the species by the expression

$$\exp\left(\Delta S_{vap}/R\right) = \frac{V_s^0}{V_f} \tag{12–16}$$

The free volume of the solute species is defined as the molar volume minus the actual volume of the molecules. Empirically, V_s^0/V_f turns out to be approximately 100.

Equations 12–15 and 12–16 are very useful for comparing bimolecular reactions in the gas and liquid phases. Inserting these equations into equation 12–7 and identifying k_0 as k_g gives

$$\frac{k_s}{k_g} = \exp\left(\Delta\Delta H_{vap}/RT\right) \frac{V_s^0}{V_{fA}} \frac{V_s^0}{V_{fB}} \frac{V_f^\ddagger}{V_s^0} \tag{12–17}$$

where $\Delta\Delta H_{vap}$ is the difference in the heats of vaporization of the transition state and the reactants. Unless there is a significant polarity change between the transition state and the reactants, $\Delta\Delta H_{vap}$ is approximately zero. Thus, from equation 12–17, the free-volume theory of liquids predicts the bimolecular rate constant to be approximately 100 times larger in solution than in the gas phase. Unfortunately, few bimolecular rate constants are known in both the gas and liquid phases.

It is of interest to consider the effect of pressure on a solution rate constant. With the thermodynamic formulation of transition state theory (see chapter 10, section 10.5), the rate constant may be written as

$$\ln k = \ln \frac{k_B T}{h} - \frac{\Delta G^{\ddagger}}{RT} \qquad (12\text{–}18)$$

where activity coefficients and other nonideal properties of the solution are absorbed in $\Delta G^{\ddagger}$. This can be done by choosing the appropriate reference state. The change in the rate constant with pressure at constant temperature is

$$\left(\frac{d \ln k}{dP}\right)_T = -\frac{1}{RT}\left(\frac{d\Delta G^{\ddagger}}{dP}\right)_T$$

$$= -\Delta V^{\ddagger}/RT \qquad (12\text{–}19)$$

since $(dG/dP)_T = V$.

The term $\Delta V^{\ddagger}$ is the standard change in volume in going from the reactants to the transition state, and is usually referred to as the *volume of activation*. A rate constant increases with increasing pressure if $\Delta V^{\ddagger}$ is negative. Conversely, the rate constant decreases with pressure if there is an increase in volume when the transition state is formed. For many cases $\Delta V^{\ddagger}$ is independent of pressure, so that integration of equation 12–19 gives

$$\ln k = \ln k_0 - \frac{\Delta V^{\ddagger}}{RT}P \qquad (12\text{–}20)$$

Some typical values of $\Delta V^{\ddagger}$ are listed in Table 12–1.

TABLE 12–1 VOLUMES OF ACTIVATION FOR SOME REACTIONS IN SOLUTION

Reaction	Solvent	$\Delta V^{\ddagger}(cm^3/mol)$
$OH^- + N(C_2H_5)_4^+ \longrightarrow CH_3OC_2H_5 + N(C_2H_5)_3$	methanol	+20.0
$(CH_3)_3COOC(CH_3)_3 \longrightarrow 2(CH_3)_3CO$	cyclohexane	+6.7
sucrose $+ H_2O \xrightarrow{H^+}$ glucose $+$ fructose	water	+2.5
$CH_3COOCH_3 + H_2O \xrightarrow{H^+} CH_3COOH + CH_3OH$	water	−8.7
$2C_5H_6(cyclopentadiene) \longrightarrow C_{10}H_{12}$	n-butyl chloride	−22.0

12.1.2 Static Solvent Effects: Formulation Based on Solvent Structure

In the foregoing application of transition state theory to solution reactions, the structures of the solvent and solvent-solute (i.e., reactant) interactions were not directly considered. More insight into the microscopic nature of the solution reaction is often revealed if solvent terms are explicit in the formulation of the transition state theory rate constant. Such an approach allows a direct calculation of the solution rate constant from the solvent-solvent, reactant-solvent, and reactant-reactant potential energies.

The relationship between the solution and gas phase transition state theory rate constants may be expressed as[4,5]

$$k_s = k_g e^{-(\Delta G_s^{\ddagger} - \Delta G_g^{\ddagger})/RT} \qquad (12\text{-}21)$$

where $\Delta G_s^{\ddagger}$ and $\Delta G_g^{\ddagger}$, calculated for the same standard-state convention, are the free energies of activation in solution and in the gas-phase, respectively. For the association process $A + B \rightarrow AB^{\ddagger}$, the free energy of activation in solution is given by

$$\Delta G_s^{\ddagger} = \Delta G_g^{\ddagger} + \Delta w^{\ddagger}(r^{\ddagger}) \equiv \Delta G_g^{\ddagger} + \Delta G_{\text{solv}}^{\ddagger} \qquad (12\text{-}22)$$

Here $\Delta w^{\ddagger}(r^{\ddagger}) = \Delta G_{\text{solv}}^{\ddagger}$ is the solvent contribution to the *potential of mean force;* i.e., the solvent contribution to the molar free energy change on bringing together A and B up to the transition state separation $r^{\ddagger}$ from infinite separation $r = \infty$ in the solvent.

From the equilibrium statistical mechanics of liquids[6,7] it is well-known that the above potential of mean force is related to the A-B pair distribution function $g_{A,B}(r)$ at any separation by

$$g_{A,B}(r) = e^{-U(r)/RT} e^{-\Delta w(r)/RT} \qquad (12\text{-}23)$$

where $U(r)$ is the gas-phase interaction potential at separation r. At the transition state equation (12–23) becomes

$$g_{A,B}(r^{\ddagger}) = e^{-E_0/RT} e^{-\Delta w^{\ddagger}(r^{\ddagger})/RT} \qquad (12\text{-}24)$$

To illustrate the evaluation of a radial distribution function, consider particles 1 and 2 separated by the distance r_{12} in a solution of N particles. The radial distribution function for particles 1 and 2 is defined by

$$g(r_{12}) = V^2 \int \cdots \int \frac{e^{-U(r_1, r_2, \cdots r_N)/k_B T} d^3 \mathbf{r}_3 \cdots d^3 \mathbf{r}_N}{Z} \qquad (12\text{-}25)$$

where V^2 is the square of the unit volumes corresponding to the coordinates for particles 1 and 2, $U(r_1, \ldots, r_N)$ is the potential energy for a particular configuration of the N particles, $d^3 r_N$ is the volume element $dx_N dy_N dz_N$ for particle N, and

$$Z = \int \cdots \int e^{-U(r_1, r_2, \cdots, r_N)/k_B T} d^3 \mathbf{r}_1 \cdots d^3 \mathbf{r}_N \qquad (12\text{-}26)$$

is the complete configuration integral over all $3N$ coordinates. The integral in equation 12–25 is a restricted configuration integral. That is, particles 1 and 2 are fixed at a separation r_{12} and the integration is over all possible configurations for the remaining $N - 2$ particles. Accurate evaluation of the multidimensional integral can be accomplished by Monte Carlo methods. By varying r_{12} in equation 12–25, the complete radial distribution function may be determined. Then, from equation 12–23, the potential of mean force is found as a function of the distance between particles 1 and 2.

Determining $g(r^{\ddagger})$ and $\Delta w^{\ddagger}(r^{\ddagger})$ for a reactive chemical system follows the description just given, except the reaction coordinate $r^{\ddagger}$ replaces r_{12}. The potential of mean force (i.e., free energy) is an equilibrium quantity. It is calculated by fixing the reaction coordinate and averaging over an equilibrium distribution of the solvent interacting with the reactive system. The potential of mean force is therefore analogous to Born-Oppenheimer electronic energies, which are determined by fixing the nuclear coordinates and averaging over distributions of the electrons.

If equations 12–21 and 12–22 are combined, the difference between the gas phase and liquid phase rate constants may be expressed as

$$k_s = k_g \exp\{-\Delta G_{\text{solv}}^{\ddagger}/RT\} \tag{12--27}$$

To illustrate the use of this equation, consider a bimolecular reaction in a high-density solvent. As shown in Figure 12–2, the solvent can favor the more compact transition state configuration as compared to the separated reactants. As a result, $\Delta G_s^{\ddagger}$ will be smaller than $\Delta G_g^{\ddagger}$. Calculations of potentials of mean force have shown that a nonpolar solvent can enhance bimolecular atom transfer reactions up to a factor of 35.[4] The diagram in Figure (12–2) is fairly typical for nonpolar reactants and a nonpolar transition state in a nonpolar solvent. In the thermodynamic formulation of transition state theory, the potential of mean force is absorbed in γ, ΔH_{vap}, and ΔS_{vap} (see equation 12–15).

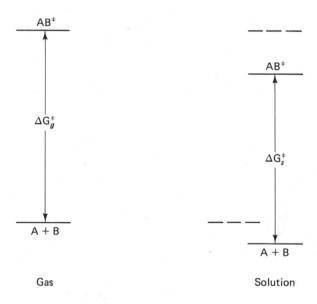

Figure 12–2. Static solvent effects for the A + B → AB‡ bimolecular reaction. The reactants are at a large separation. [Reprinted with permission from J. T. Hynes, in *Theory of Chemical Reaction Dynamics*, Vol. IV, edited by M. Baer (Boca Raton, Florida, CRC Press, 1985), p. 171. Copyright, CRC Press, Inc., Boca Raton, FL.]

Calculations of potentials of mean force have been applied to the S_N2 nucleophilic substitution reaction[8]

$$\text{Cl}^- + \text{CH}_3\text{Cl} \longrightarrow \text{ClCH}_3 + \text{Cl}^- \tag{12--28}$$

As illustrated earlier in Figure 12–1, a polar solvent is expected to have a pronounced effect on the kinetics of this reaction. Calculations of $\Delta G_s^{\ddagger}$ are shown in Figure 12–3 for reactions in aqueous and dimethylformamide (DMF) solutions and compared with the gas phase reaction coordinate potential energy. The Cl$^-$ anion is strongly stabilized in solution in comparison to the transition state. Also, in solution,

Figure 12–3. Calculated internal energies in the gas phase (short dashes) and the potential of mean force in DMF (long dashes) and in aqueous solution (solid curve) for the reaction of Cl^- with CH_3Cl as a function of the reaction coordinate r_c in angstroms. [Adapted from J. Chandrasekhar and W. L. Jorgensen, *J. Am. Chem. Soc. 107*, 2974 (1985)].

the pronounced ion-dipole complex is not observed and the overall potential energy barrier is significantly higher. The dipolar aprotic solvent DMF shows less anion solvating ability than does water.

12.2 KRAMERS' THEORY AND FRICTION

Most modern investigations of the validity of transition state theory in solutions are based on the classic paper by Kramers[9] on reactions in solution. Kramers viewed a reaction as a barrier crossing influenced by interactions with surrounding solvent molecules. Passage across the barrier then appears similar to Brownian motion, which is described via the Fokker-Planck (or Langevin)[10] equation. To account for the dynamical influence of the solvent, a damping force is introduced which is proportional to both the velocity along the reaction coordinate and a *friction constant* β. Although this friction constant is not clearly defined, it is expected to increase with solvent density, pressure, and viscosity.

The friction constant can have different influences on the solution rate constant, and Kramers identified several major kinetic regimes. (i) In one regime the requirement for validity of transition state theory is the same as for normal gas phase reactions. Thus, in the absence of intrinsic recrossing of the transition state, the equilibrium conditions assumed in the standard formulation of transition state theory are satisfied. Collisions with solvent molecules are sufficient to maintain an equilibrium distribution of energized reactants, but are not sufficiently frequent to perturb free passage over the potential energy barriers. (ii) As the friction is increased, the second regime is reached. Here, collisions with solvent molecules lead to recrossing of the potential barrier before a stable product molecule is formed. As a result, the rate is less than the transition state theory prediction, and the difference depends on the solvent friction. For broad, flat barriers and high friction, the rate becomes diffusion-controlled and is inversely proportional to the friction. Barrier recrossings are so frequent that the rate may become more than an order of magnitude less than the transition state theory value. (iii) A third regime, which occurs at low friction, represents a different breakdown of transition state theory. Because the reactive system is weakly coupled to the solvent, an equilibrium distribution of reactant states cannot be maintained. This phenomenon is similar to that responsible for the first-to-second-order transition in thermal unimolecular reactions (chapter 11). For extremely low friction, reactant activation becomes rate limiting and the rate is proportional to the solvent friction. In what follows, Kramers's treatment of the effects of solvent friction on solution reactions rates is outlined.

Kramers idealized the reactive system as a particle with effective mass m moving in the one-dimensional potential shown in Figure 12–4. The reaction is regarded as a Markov process, and the Fokker-Planck equation

$$\frac{\partial \rho}{\partial t} + v\frac{\partial \rho}{\partial t} + \frac{1}{m}\frac{\partial V}{\partial r}\frac{\partial \rho}{\partial v} = \beta v\frac{\partial \rho}{\partial v} + \beta\rho + \beta\frac{k_B T}{m}\frac{\partial^2 \rho}{\partial v^2} \qquad (12\text{–}29)$$

is solved for the distribution $\rho(r, v, t)$ of the particle's position and velocity vs. time. In this equation, $V = V(r)$ is the potential and β is the friction constant. To solve equation 12–29, $V(r)$ is assumed to be of the form

$$V(r) = \frac{1}{2}m\omega_0^2(r - r_0)^2 \qquad (12\text{–}30)$$

in the vicinity of the reactant, and of the form

$$V(r) = E_0 - \frac{1}{2}m\omega^{\ddagger 2}(r - r^{\ddagger})^2 \qquad (12\text{–}31)$$

near the transition state dividing the reactant from the product.

With the preceding potentials, Kramers solved the steady-state version of equation 12–29 for medium to high friction where energy equilibration is guaran-

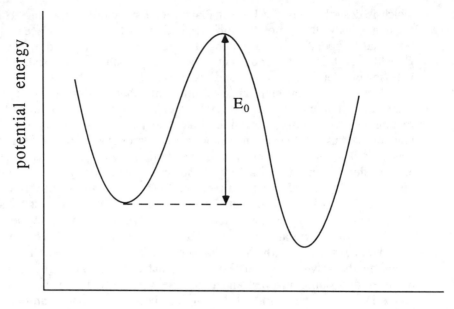

reaction coordinate

Figure 12–4. One-dimensional potential model for Kramers's theory.

teed. The resulting expression for the rate constant is

$$k = k^{TST} \kappa^{KR} \tag{12-32}$$

where k^{TST} is the transition state theory rate constant and $\kappa^{KR} \le 1$, the Kramers transmission coefficient, is given by

$$\kappa^{KR} = \left[1 + \left(\frac{\beta}{2\omega^{\ddagger}} \right)^2 \right]^{1/2} - \frac{\beta}{2\omega^{\ddagger}} \tag{12-33}$$

Deviation from transition state theory depends on the parameter $\beta/2\omega^{\ddagger}$. For intermediate friction, $\beta \ll 2\omega^{\ddagger}$, κ^{KR} is unity, and transition state theory results. In the high-friction limit, $\beta \gg 2\omega^{\ddagger}$ and equation 12–32 becomes

$$k = k^{TST} \left(\frac{\omega^{\ddagger}}{\beta} \right) \tag{12-34}$$

so that the rate constant decreases with increasing friction. Equation 12–34 can also be derived by a steady-state solution of the Smoluchowski diffusion equation.[5] Thus, k in that equation is a rate constant for diffusion-controlled passage across the potential energy barrier and may be expressed as

$$k_D = k^{TST} \left(\frac{\omega^{\ddagger} D}{k_B T} \right) \tag{12-35}$$

where $D = k_B T / \beta$ is the diffusion constant (see chapter 4) determined by the Einstein relation.[11]

The preceding equations show that substantial differences may exist between actual solution rate constants and those predicted by transition state theory. A plot of the Kramers transmission coefficient versus the frictional parameter $\beta/\omega^{\ddagger}$ is given in Figure 12–5. Transition state theory is approached for small values of friction, as predicted by equation 12–33.

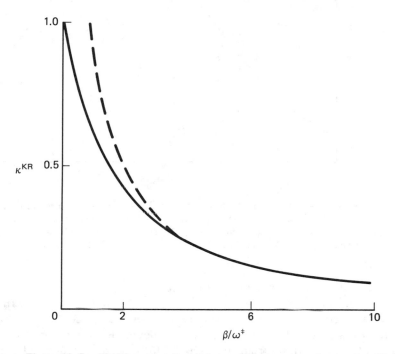

Figure 12–5. The Kramers transmission coefficient, as given by equation 12–33, vs. the frictional parameter $\beta/\omega^{\ddagger}$. The dashed curve is the high-friction diffusive limit expressed in equation 12–35. Transition state theory is approached for small values of friction. [Adapted from J. T. Hynes, in *Theory of Chemical Reation Dynamics*. Vol. IV. Edited by M. Baer (Boca Raton, FL: CRC Press, 1985).]

The foregoing approach is incomplete for a unimolecular reaction: at sufficiently low solvent friction, a thermal equilibrium no longer exists for the reactant energy states. The same situation occurs for unimolecular reactions in the falloff regime (see chapter 11, sections 11.2 and 11.3). It has been suggested[12] that for the unimolecular decomposition of a polyatomic molecule in solution the rate constant should be expressed as

$$k = \frac{k_1 k_2 [M]}{k_{-1}[M] + k_2} \kappa^{KR} \qquad (12\text{–}36)$$

The first term is the Lindemann-Hinshelwood expression (see chapter 11, section 11.3) for the unimolecular rate constant vs. pressure (i.e., [M]), while the second is the frictional transmission coefficient from equation 12–33. A high-pressure limit

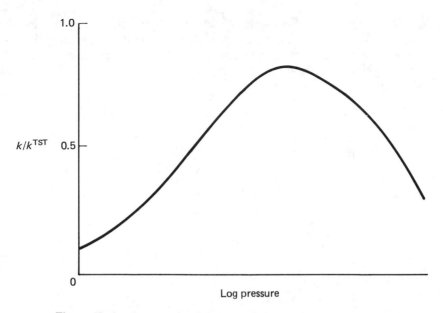

Figure 12–6. An example of the transmission coefficient vs. pressure for a unimolecular reaction. The plot shows that the transition state theory high-pressure limit is not attained. [Adapted from J. T. Hynes, in *Theory of Chemical Reaction Dynamics*. Vol. IV. Edited by M. Baer (Boca Raton, FL: CRC Press, 1985).]

plateau in the thermal unimolecular rate constant equal to $k_\infty = k_1 k_2 / k_{-1}$ is predicted by the Lindemann-Hinshelwood theory. For molecules with few internal degrees of freedom, there may not be a plateau in k vs. pressure. Since the frictional constant β increases with gas density and, thus, κ^{KR} decreases, a high-pressure limiting rate constant does not result from equation 12–36. Indeed, as illustrated in Figure 12–6, the rate constant decreases on approaching the high-pressure limit. However, with more internal degrees, there usually is a significant plateau, since the high-pressure limit for the first term in equation 12–36 is reached before solvent friction suppresses the rate.

If, as predicted by Kramers's theory, there is a breakdown in transition state theory, then less significance should be placed on thermodynamic kinetic parameters (section 12.1) determined by inappropriately fitting rate constants to transition state theory rate expression.[5] This is readily apparent for the parameter $\Delta V^\ddagger$, which could have a non-transition-state contribution if the transmission coefficient has a pressure (i.e., density) dependence. If the correct thermodynamic $\Delta V^\ddagger$ is negative, then, according to equation 12–20, the rate should increase with pressure. However, for the high-friction limit, where the rate decreases with pressure, the apparent $\Delta V^\ddagger$ would be positive.

Trajectory calculations as discussed in section 8.3 for gas-phase reactions have also been applied to reactions in solution.[13,14] Collinear models of the atom transfer reaction $A + BC \rightarrow AB + C$ and the S_N2 reaction $Cl^- + CH_3Cl \rightarrow ClCH_3 + Cl^-$ have been studied.

12.3 GAS-SURFACE REACTION DYNAMICS

Heterogeneous reactions, i.e., those that occur at the interface of two phases, are typical examples of catalysis. The field of heterogeneous catalysis has enormous practical applicability, and, therefore, heterogeneous chemical reaction kinetics has been extensively studied. The interest here is in a molecular-level (microscopic) description of gas-surface dynamics, to complement the more macroscopic approach in chapter 5.

Chemical reactions at a gas-surface interface involve the following sequence of steps:

1. Diffusion of reactants to the surface.
2. Adsorption of the reactants on the surface.
3. Reaction on the surface.
4. Desorption of products.
5. Diffusion of products from the surface.

Any one or a combination of these steps may be slow and, therefore, rate determining. Measurements of diffusion rates allow steps 1 and 5 to be characterized separately. Also, it may be possible to measure individual adsorption and desorption rates. Thus, it can often be established whether the actual reaction on the surface is rate limiting.

12.3.1 Adsorption and Langmuir Isotherms

Adsorption occurs on a solid surface and is due to the attractive forces between the surface and the adsorbed species (adsorbate). Two different types of adsorption can be distinguished. Heterogeneous catalysis is more intimately related to adsorption processes in which the adsorbed species forms a chemical bond with the surface (adsorbent). Such a process is called chemical adsorption, or *chemisorption,* as contrasted to the much weaker adsorption of gases due to van der Waals forces. This second kind of adsorption is more like a condensation process and is called physical adsorption, or *physisorption*. The dividing line between the two adsorption processes is not always sharp.

The enthalpy changes for chemisorption are usually substantially greater than those for physical adsorption. For chemisorption, once a monolayer of adsorbed gas molecules covers the solid's surface, no further chemical reaction between the gas and the solid can occur. For physical adsorption, once a monolayer has formed, intermolecular interactions between adsorbed molecules in the monolayer and gas-phase molecules can lead to the formation of a second layer of adsorbed gas. Thus, the enthalpy change for formation of the first layer of physically adsorbed molecules is determined by solid-molecule intermolecular forces, whereas the enthalpy change for formation of the second, third, etc., physically adsorbed layers is determined by molecule-molecule intermolecular forces and is about the same as ΔH in the condensation of a gas to a liquid. Although only one layer can be chemically adsorbed, physical adsorption of further layers on top of a chemisorbed monolayer sometimes occurs.

It should be recognized that there may be exceptions to the above general properties. In some cases the surface can influence the adsorption of the second layer. It is also possible for the second layer to be chemisorbed to the first layer.

The difference between physical and chemical adsorption is illustrated by the behavior of nitrogen on iron. At $-190°C$, nitrogen is liquid and nitrogen molecules are physisorbed on iron. The equilibrium amount of N_2 adsorbed in this manner decreases rapidly as the temperature rises until, at room temperature, nitrogen does not adsorb on iron at all. At high temperature, approximately $500°C$, nitrogen is chemisorbed on the iron surface. Physisorption is thus a very weak gas-surface interaction with no potential energy barrier and only occurs at low temperatures. In contrast, there may be activation energy for chemisorption, and a high temperature often is required. Since an appreciable activation energy may be associated with chemisorption, it is often called *activated adsorption*.

Adsorption and desorption may be represented by the chemical equation

$$A(g) + S \rightleftharpoons AS \qquad (12\text{--}37)$$

where $A(g)$ is the gaseous adsorbate, S is a vacant site on the surface, and AS represents an adsorbed molecule of A. A simple model can be developed for physisorption, since no chemical reaction is involved; in the case of chemisorption, a similar model is applicable for chemically stable adsorbates. The amount of adsorbed A is measured as a function of the temperature and pressure of $A(g)$; it is convenient to express this amount as the equivalent volume of adsorbed A, per gram of adsorbent, at the gas temperature and pressure. A plot of this equivalent volume V as a function of pressure is known as an *adsorption isotherm* and is illustrated in Figure 12–7. If only a monolayer is formed on the surface, as is the normal situation for chemisorption, V will approach a limiting value V_m as illustrated in the figure.

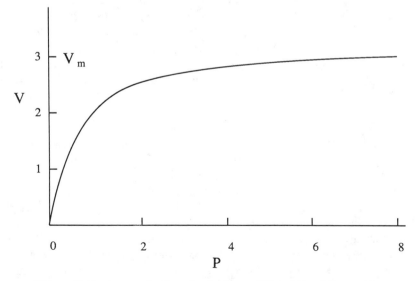

Figure 12–7. Langmuir adsorption isotherm. Units of V and P are arbitrary.

A simple model of adsorption was given by Langmuir[15] in 1916. In the model, all adsorption sites on the surface are the same, there are no interactions between one adsorbed molecule and another (i.e., sites are independent), and only a monolayer can be adsorbed. Now consider a surface with N absorption sites. If N_1 sites are occupied by adsorbate, then $N_0 = N - N_1$ are bare. At equilibrium, the rate of adsorption is equal to the rate of desorption and is proportional to the rate of collisions of gas-phase molecules with unoccupied adsorption sites, which in turn is proportional to the pressure P (see chapter 8). Thus,

$$\text{Rate of adsorption} = k_a P N_0 = k_a P (N - N_1)$$

The rate of desorption is proportional to the number N_1 of occupied sites. So

$$\text{Rate of desorption} = k_d N_1$$

Setting the two rates equal to each other and solving for N_1/N, we obtain

$$\frac{N_1}{N} = \theta = \frac{bP}{1 + bP} \qquad (12\text{–}38)$$

where θ is the fraction of surface covered and $b = k_a/k_d$. The limiting value V_m of Figure 12–7 corresponds to $\theta = 1$; hence, $V/V_m = \theta$, and we can write

$$V = \frac{V_m bP}{1 + bP} \qquad (12\text{–}39)$$

Equations 12–38 and 12–39 are known as the *Langmuir adsorption isotherm*.

The Langmuir isotherm may be interpreted in the context of statistical mechanics. The reaction in equation 12–37 has an equilibrium constant

$$K = \frac{[AS]}{[A][S]} \qquad (12\text{–}40)$$

For an ideal gas, $[A] = P/k_B T$. Also, for $[S]$ and $[AS]$, we have $(1 - \theta)N$ and θN, respectively. Thus, by substituting those quantities into equation 12–40 and rearranging terms, we obtain

$$\theta = \frac{(K/k_B T)P}{1 + (K/k_B T)P} \qquad (12\text{–}41)$$

which is equivalent to the Langmuir adsorption isotherm if

$$b = \frac{K}{k_B T} \qquad (12\text{–}42)$$

The statistical mechanical expression for the equilibrium constant (chapter 10) is

$$K = \frac{Q_{AS}}{Q_A Q_S} e^{-E_{ad}/RT} \qquad (12\text{–}43)$$

where the Q's are the partition functions and E_{ad} is the energy of adsorption, given by the difference in ground state energy between the molecules in the gaseous and adsorbed states. Thus, equations 12–42 and 12–43 provide one with a molecular interpretation of the constant b in terms of adsorption energy and the molecular quantities that determine the partition function, viz., vibrational and rotational energies, temperature, and masses.

The Langmuir model oversimplifies gas-surface interactions. In reality, the surfaces of most solids are not uniform, and the desorption rate depends on the location of the adsorbed molecule. Also, the force between adjacent adsorbed molecules is frequently substantial, as evidenced by changes in the heat of adsorption with increasing θ. To complicate matters further, there is substantial experimental and theoretical evidence that adsorbed molecules can move about on the surface and that this mobility is much greater for physically adsorbed molecules than for chemisorbed ones and increases as T increases. In addition, multilayer adsorption is quite common is physical adsorption. To account for all these effects, different adsorption isotherms may be used. The Freundlich isotherm[16] can be derived by modifying the Langmuir assumptions to allow for several kinds of adsorption sites on the solid, each having a different heat of adsorption.[17] The Brunauer, Emmett, and Teller isotherm[18] is a modification of Langmuir's model that allows for multilayer adsorption.

12.3.2 Adsorption/Desorption Kinetics and Transition State Theory

Two important kinetic parameters in gas-surface adsorption and desorption are the sticking probability s and the desorption rate constant k_d. The sticking probability is the fraction of the incident flux impinging on the surface which is completely accommodated (i.e., adsorbed). It is related to the adsorption rate R_a via

$$N = \frac{R_a}{N_g v} \qquad (12\text{--}44)$$

where N_g is the gas concentration and v is the component of the gas velocity that is perpendicular to the surface. For a thermal velocity distribution, s is a thermally averaged quantity. The sticking probability of NO on a Pt(111) surface is shown in Figure 12–8 as a function of surface coverage. Note that it decreases with coverage, since there is less probability that the NO molecule will find a chemisorption site before it desorbs.

Equation 12–44 represents the sticking probability per adsorption site. The adsorption rate per site is $R_a = k_a N_g$. According to transition state theory, k_a is given by equation 10–20, i.e.,

$$k_a = \frac{k_B T}{h} \frac{Q^{\ddagger}}{Q_A Q_S} e^{-E_0/k_B T}$$

In the limit of a loose transition state which resembles the reactants A and S (i.e, the separation between A and S approaches infinity in the transition state), it is straightforward to show that the sticking probability is unity. For a loose transition state, the only difference between the partition function $Q^{\ddagger}$ and the partition function product $Q_A Q_S$ is that $Q^{\ddagger}$ lacks the reaction coordinate degree of freedom which is a translation perpendicular to the surface. Thus,

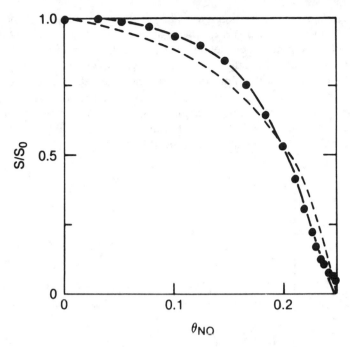

Figure 12–8. Sticking probability of NO on a Pt(111) surface vs. surface coverage. The surface temperature is 300 K, and S_0, the sticking probability for a clean surface, is 0.85. The solid curve is passed through the experimental points. The dashed curve is a model calculation. [Adapted from R. D. Levine and R. B. Bernstein, *Molecular Reaction Dynamics and Chemical Reactivity* (Oxford, New York, 1987).]

$$\frac{Q^{\ddagger}}{Q_A Q_S} = \frac{1}{\left(\dfrac{2\pi m k_B T}{h^2}\right)^{1/2}} \tag{12–45}$$

Since E_0 is equal to zero for a loose transition state, we have, upon substituting equation 12–45 into equation 10–20 reproduced above,

$$k_a = \left(\frac{k_B T}{2\pi m}\right)^{1/2} \tag{12–46}$$

This value for k_a is the same as the average thermal velocity of particles moving perpendicular to the surface. Thus, from equations 12–44 and 12–46, together with the relationship $R_a = k_a N_g$, the sticking coefficient is seen to equal unity.

As the transition state becomes tighter and acquires properties closer to those of the adsorbed state AS, the sticking probability becomes less than unity. This results from configurational (or entropic) effects which cause the gas to "turn back" before it crosses the transition state.

The desorption rate constant is given by

$$k_d = \frac{R_d}{N_a}$$

where R_d is the desorption rate and N_a is the concentration of the adsorbed species. If the adsorbed molecule is localized at one surface site, desorption is analogous to a gas phase unimolecular decomposition reaction. In Figure 12–9, two types of desorption potential energy curves are shown. Since the physisorption potential energy curve has no saddle point, variational transition state theory is needed to determine the temperature-dependent transition state location. For chemisorption the transition state may be tight, with the adsorbed species still strongly interacting with the surface. Thus, there may be a well-identified saddle point on the potential curve, at which the transition state will be positioned.

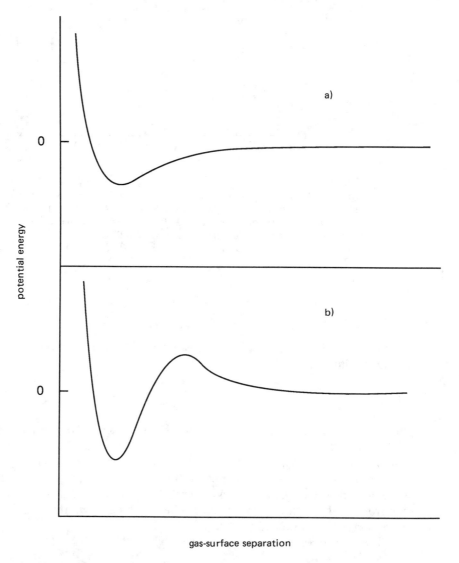

Figure 12–9. Two possible adsorption/desorption potential energy curves: (a) physisorption, (b) chemisorption.

Dynamics beyond the Gas Phase   Chap. 12

In Arrhenius form, the desorption rate constant is written as

$$k_d = Ae^{-E_a/RT}$$

For a physisorbed molecule or atom, the normal situation is a small activation energy and a large A-factor in the range of 10^{13} to 10^{15} sec^{-1}, similar to that for gas phase bond dissociations (e.g., $C_2H_6 \rightarrow 2CH_3$). The reason for a large A-factor is that in the transition state the desorbing molecule is free to rotate. The A-factor for the desorption of a chemisorbed molecule may be smaller if the desorbing molecule is not free to rotate in the transition state.

The thermodynamic transition state theory expression for the desorption rate constant is

$$k_d = \frac{k_B T}{h} e^{-\Delta G^{\ddagger}/RT}$$

where $\Delta G^{\ddagger}$ is the difference in free energy between the transition state and the adsorbed species. By combining this equation with equations 12–44 through 12–46, k_d can be written in terms of the sticking probability as

$$k_d = s\left(\frac{k_B T}{2\pi m}\right)^{1/2} e^{-\Delta G/RT} \qquad (12\text{–}47)$$

where ΔG is the difference in free energy between the gaseous and adsorbed states.

The inverse of the desorption rate constant is often referred to as the *average residence time* of the molecule on the surface.

12.3.3 Dissociative Adsorption and Precursor States

A process that is particularly important for catalysis is *dissociative adsorption,* in which a molecule does not adsorb intact but undergoes bond rupture. An example is the adsorption of H_2 on a Cu(100) surface, depicted by

$$H_2(g) + \cdot\cdot CuCu \cdot\cdot \longrightarrow \overset{\displaystyle H-H}{\overset{\displaystyle |\quad\;|}{\cdot\cdot Cu\,Cu \cdot\cdot}} \longrightarrow \overset{\displaystyle H\qquad H}{\overset{\displaystyle |\qquad\;|}{\cdot\cdot CuCu \cdot\cdot\cdot Cu \cdot\cdot}}$$

In the dissociative adsorption of a polyatomic molecule there can be total bond rupture so that only atoms are adsorbed, or there may be partial bond rupture as in the dissociative chemisorption of CH_4 on Ni(111) to an adsorbed methyl radical and a hydrogen atom.[19]

Mechanisms for dissociative chemisorption often include a *precursor state,* the notion of which is derived from the idea that before chemisorbing a molecule becomes trapped temporarily in a weakly bound molecular state which arises from long-range physisorption interactions. The existence of a precursor state means that there is an intermediate barrier (ΔE_b) to the chemisorbed state, as depicted in Figure 12–10. Two families of curves are illustrated that at large distances describe possible physisorbed states. The different curves represent different gas-surface systems, or different gas-surface adsorption sites for the same system. Only one curve is drawn for the short-range, strongly bound chemisorbed state. The barrier between

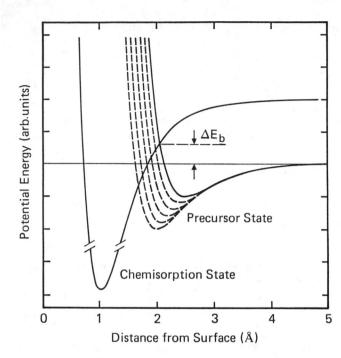

Figure 12–10. Schematic representation of potentials for mechanisms of dissociative chemisorption which include a precursor state. [Adapted from D. J. Auerbach and C. T. Rettner, in *Kinetics of Interface Reactions,* edited by M. Grunze and H. J. Kreuzer (Springer-Verlag, New York, 1987), p. 125.]

the precursor and chemisorbed states allows the molecule to become temporarily trapped on the surface in the precursor state before chemisorbing. The curve for the chemisorbed state depicts dissociative chemisorption. The asymptote ($r \rightarrow \infty$) of this curve gives the potential energy of the fragments, which are formed by chemisorption.

For some of the potential curves in the figure, there is an overall barrier ΔE_b for chemisorption. If the molecule is to become adsorbed to the surface, this barrier must be surmounted, or else the molecule will reside in the precursor state for a short time before desorbing. One way to detect this energy barrier, as shown in Figure 12–11, is to measure the sticking probability as a function of the component of the molecular translational energy normal to the surface, i.e., $E_n = E_1 \cos^2 \theta_i$, where θ_i is the angle of incidence. An increase in the sticking probability with E_n is indicative of a barrier for dissociative chemisorption.

As is found for gas phase reactions (see chapter 9), the effectiveness of translational and vibrational energy in surmounting the barrier for dissociative chemisorption may differ for different reactions. An example is the dissociative chemisorption of CO on Ni(111), for which a barrier of 30 kcal/mol is estimated. However, no dissociative adsorption is observed at a translational energy of 45 kcal/mol.[20] The conclusion is that CO must be vibrationally excited for chemisorption to occur.

12.3.4 Heterogeneous Reactions

Experimental studies of gas-surface heterogeneous reactions have provided considerable dynamical details of the elementary mechanisms of catalytic activity. An important aspect of these reactions is that a catalytic surface is *not* flat! Some examples

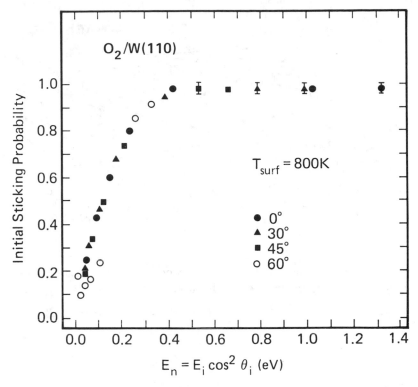

$O_2/W(110)$

$T_{surf} = 800K$

- ● 0°
- ▲ 30°
- ■ 45°
- ○ 60°

Initial Sticking Probability

$$E_n = E_i \cos^2 \theta_i \ (eV)$$

Figure 12–11. Initial dissociative chemisorption probability of O_2 on W(110) at a surface temperature of 800 K as a function of the kinetic energy of the incident molecules for angles of incidence between 0° and 60°. [Adapted from D. J. Auerbach and C. T. Rettner, in *Kinetics of Interface Reactions*, edited by M. Grunze and H. J. Kreuzer (Springer-Verlag, New York, 1987), p. 125.]

of surface imperfections are shown in Figure 12–12. Kinks and monatomic ledges (or steps) are particularly important in catalysis. Active adsorption sites which promote bond breaking are usually found at these irregular regions of the surface. Steps and kinks may have different specificities for bond rupture. For example, a step on a particular surface may preferentially rupture the C–H bond of a hydrocarbon, while C–C bond rupture may occur only at a kink.

Adatoms can modify surface reactivity by blocking binding at an adsorption site and/or altering the binding at sites not directly blocked by through-surface electronic effects. These electronic effects may alter the potential energy barriers for dissociative chemisorption and surface migration. If the adatom slows down the catalytic behavior of a surface, it is called a *poison;* if it speeds up catalytic behavior, it is called a *promoter*. A poison may physically block binding sites, increase potential barriers for dissociative chemisorption, or increase barriers for lateral motion (i.e., migration) on the surface. Lateral motion on the surface is important for the bimolecular recombination of adsorbed species. A promoter adatom might lower the potential barriers for chemisorption and/or migration.

Figure 12–12. Model of the heterogeneous surface depicting the different surface sites: terrace, kink, ledge, vacancy, ledge-adatom, and adatom. [Adapted from G. A. Somorjai, *Principles of Surface Chemistry*] (Englewood Cliffs, NJ: Prentice-Hall Inc., 1972).]

An important practical catalytic process is the oxidation of CO to CO_2. We may characterize the process by the mechanism[21]

$$
\begin{array}{rcl}
O_2 + {}^* & \underset{}{\rightleftharpoons} \; O_{2,ad} \; \xrightarrow{k_1} & 2O_{ad} \\[4pt]
CO + {}^* & \underset{k_{-2}}{\overset{k_2}{\rightleftharpoons}} \; CO_{ad} & \\[4pt]
CO_{ad} + O_{ad} & \xrightarrow{k_3} & CO_2 + 2^* \\[4pt]
CO + O_{ad} & \xrightarrow{k_4} & CO_2 + {}^*
\end{array}
\qquad (12\text{–}48)
$$

where * denotes a free adsorption site. The first step is the dissociative chemisorption of O_2, which proceeds through a molecular precursor state. In the second step carbon monoxide adsorbs molecularly. Next, there are two possible reaction steps for CO_2 formation. In the Langmuir-Hinshelwood[22] mechanism (the third step), chemisorbed CO and O recombine to form CO_2. In the fourth step, the Eley-Rideal mechanism,[23] CO_2 is formed by direct collision of CO from the gas phase with a chemisorbed O-atom. Experiments on Pd, Pt, Ir, and Rh surfaces have demonstrated the dominance of the Langmuir-Hinshelwood mechanism. A schematic potential energy diagram is given in Figure 12–13. All of the individual rate constants k_1, k_2, k_{-2}, and k_3 depend on temperature and surface coverage. The catalytic formation of CO_2 is only slightly influenced by the surface structure. Step sites apparently enhance O_2 dissociative chemisorption (step 1), but decrease the probability of CO_2 formation (step 3). Both effects nearly compensate each other.

Ammonia synthesis on an iron catalyst, $N_2 + 3H_2 \rightarrow 2NH_3$, (the Haber process),[24] illustrates the importance of promoters. The rate of ammonia formation is determined by the rate of N_2 dissociative chemisorption on the iron surface. Introducing potassium as an adatom promotes the reaction by decreasing the barrier for dissociative adsorption of N_2.

One of the most carefully studied heterogeneous reactions on a single crystal is the methanation reaction[25]

$$
3H_2 + CO \longrightarrow CH_4 + H_2O \qquad (12\text{–}49)
$$

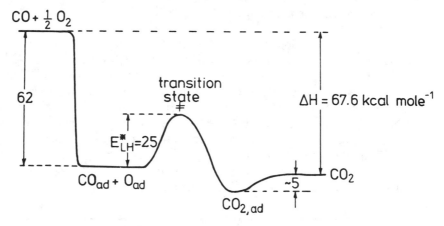

Figure 12–13. Schematic potential diagram illustrating the energetic changes associated with the individual reaction steps of equation (12–48) on Pd (111) at low coverages. [Adapted from T. Engel and G. Ertl, in *The Chemical Physics of Sold Surfaces and Heterogeneous Catalysis,* ed. D. A. King and D. P. Woodruff (New York: Elsevier, 1982).]

The mechanism for this reaction on a Ni(100) surface is represented by

$$H_2 + 2^* \longrightarrow 2H_{ad}$$

$$CO + ^* \longrightarrow CO_{ad} \longrightarrow C_{ad} + O_{ad} \qquad (12\text{–}50)$$

$$C_{ad} + 4H_{ad} \longrightarrow CH_4$$

where the last step represents a series of surface hydrogenation steps leading to the desorption of CH_4. Sulfur adatoms are a catalytic poison for this reaction as well as for others. The poisoning for this methanation reaction results from an inhibition in chemisorption of both hydrogen and CO. One sulfur atom deactivates approximately ten nickel atom adsorption sites.[25]

For many heterogeneous reactions, dissociative chemisorption is the rate-determining step. In these cases it is important to know how the dissociative chemisorption is affected by reactant translation, rotational, and vibrational energy. In some catalytic processes activated adsorption leads to unimolecular isomerization. The catalytic isomerization of cyclopropane to propylene has been studied on mica surfaces.[26] A model that explains the experimental results assumes a high barrier for adsorption that is surmounted most effectively by vibrational energy in the incident cyclopropane molecule.

12.3.5 Trajectory Studies

The classical trajectory theoretical approach (see chapter 8, section 8.3) has been particularly effective in simulating the dynamics of adsorption/desorption and of reactive gas-surface collisions. A potential energy function (chapter 7) which depends on the positions of all atoms is chosen to represent molecular and surface degrees of freedom, as well as molecule-surface and adsorbate-adsorbate interactions. Solving the classical equations of motion then gives a microscopic "picture" of the gas-surface event.

An important application of classical trajectories is the interpretation of angle-resolved secondary-ion-mass spectrometric (SIMS) measurements. As is evidenced in Figure 12–13, the surface structure can be a crucial factor in catalytic activity. A trajectory simulation[27] performed for a microcrystallite of approximately 240 atoms has been used to calculate the angular distributions of Ni^+ and Ni_2^+ ions ejected from a Ni(100) surface by 2-keV Ar^+ bombardment. Anisotropies in the experimental angular distributions are well reproduced by the dynamical calculations. A particularly significant discovery from the trajectories is that there is a dominant ejection mechanism for dimer formation, as shown in Figure 12–14. Atoms 1 and 3 are ejected through fourfold holes in the Ni surface and are moving parallel to each other. As

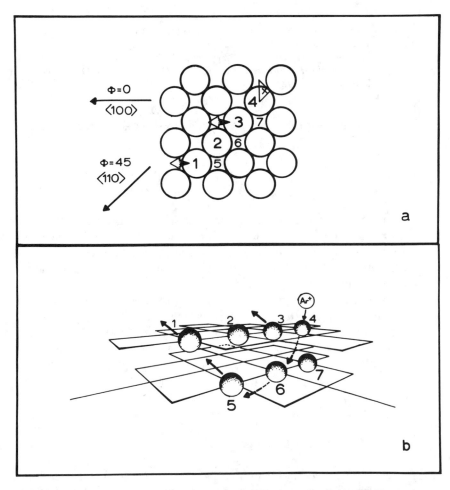

Figure 12–14. Elementary mechanism for dimer formation in the SIMS experiment of Ar^+ ion bombardment of a Ni (100) surface. The shaded spheres represent Ni atoms in the first and second layers. [Adapted from S. P. Holland, et al., *Phys. Rev. Lett. 44,* 756 (1980).]

such, they are in close proximity and susceptible to dimer formation. For this situation, the constituent atoms in the ejected dimer are known to form from surface atoms with a specific location relative to the primary impact point. Extrapolated to alloy surfaces such as CuNi, this finding would allow the determination of the relative placement of the alloy components on the surface.

In this trajectory simulation, a large number of atoms are treated explicitly to model the surface layers because a large number of atoms are directly involved in the gas-surface collision. For processes such as the scattering of gases from surfaces at thermal energies a method called the *stochastic classical trajectory approach*[28] is often used to reduce the number of surface degrees of freedom treated explicitly. Only atoms in the "primary" surface zone (usually the top layer of the surface) are treated explicitly and allowed to interact with gas molecules. Energy transfer between the "primary" surface zone and the rest of the lattice is included via the generalized Langevin technique, which introduces two additional terms into the classical equations of motion. The first, a generalized friction, describes the dissipation of energy from the primary zone to the rest of the lattice. The second, a fluctuating force, accounts for energy imparted to the primary atoms through thermal vibrations of the secondary atoms. The generalized Langevin technique has been used to simulate the effect of translational, rotational, and vibrational energies on the dissociative chemisorption of diatoms,[29] and to calculate adsorption/desorption probabilities.[30]

REFERENCES

1. W. N. Olmstead and J. I. Brauman, *J. Am. Chem. Soc. 99*, 4219 (1977).

2. G. R. Branton, H. M. Frey, D. C. Montague, and I. D. R. Stevens, *Trans. Faraday Soc. 62*, 659 (1966).

3. J. F. Kincaid, H. Eyring, and A. E. Stearn, *Chem. Rev. 28*, 301 (1941).

4. B. M. Ladanyi and J. T. Hynes, *J. Am. Chem. Soc. 108*, 585 (1986).

5. J. T. Hynes, in *Theory of Chemical Reaction Dynamics*. Vol. IV, edited by M. Baer (Boca Raton, FL.: CRC Press, 1985), p. 171.

6. T. L. Hill, *Statistical Mechanics* (New York: McGraw-Hill, 1956).

7. N. Davidson, *Statistical Mechanics* (New York: McGraw-Hill, 1962).

8. J. Chandrasekhar, S. F. Smith, and W. L. Jorgensen, *J. Am. Chem. Soc. 106*, 3049 (1984); J. Chandrasekhar and W. L. Jorgensen, *J. Am. Chem. Soc. 107*, 2974 (1985).

9. H. A. Kramers, *Physica* (The Hague) 7, 284 (1940).

10. D. A. McQuarrie, *Statistical Mechanics* (NY: Harper and Row, 1976); S. Chandrasekhar, *Rev. Mod. Phys. 15*, 1 (1943).

11. F. Daniels and R. A. Alberty, *Physical Chemistry* (New York: John Wiley, 1963).

12. J. Troe in *Physical Chemistry, An Advanced Treatise, Vol. VI B*, edited by W. Jost (New York: Academic Press, 1975), p. 835.

13. J. P. Bergsma, J. R. Reimers, K. R. Wilson, and J. T. Hynes, *J. Chem. Phys. 85*, 5625 (1986).

14. J. P. Bergsma, B. J. Gertner, K. R. Wilson, and J. T. Hynes, *J. Chem. Phys. 86*, 1356 (1987).

15. I. Langmuir, *J. Am. Chem. Soc. 38*, 2221 (1916); *40*, 1361 (1918).

16. H. M. F. Freundlich, *Kappilarchemie*, Leipzig, 1909.

17. R. Sips, *J. Chem. Phys. 18*, 1024 (1950); F. C. Tompkins, *Trans. Faraday Soc. 46*, 569 (1950).

18. S. Brunauer, P. H. Emmett, and E. Teller, *J. Am. Chem. Soc. 60*, 309 (1938).

19. M. B. Lee, Q. Y. Yang, and S. T. Ceyer, *J. Chem. Phys. 87*, 2724 (1987).

20. M. B. Lee, J. D. Beckerle, S. L. Tang, and S. T. Ceyer, *J. Chem. Phys. 87*, 723 (1987).

21. T. Engel and G. Ertl, in *The Chemical Physics of Solid Surfaces and Heterogeneous Catalysis, Vol 4*, edited by D. A. King and D. P. Woodruff (New York: Elsevier, 1982), p. 73.

22. C. N. Hinshelwood, *Kinetics of Chemical Change in Gaseous Systems* (Oxford: Clarendon Press, 1926), p. 145.

23. D. D. Eley and E. K. Rideal, *Proc. R. Soc. London A 178*, 429 (1941).

24. M. Grunze, in reference 21, p. 143.

25. R. D. Kelley and D. W. Goodman, in reference 21, p. 427.

26. L. Tsou, G. L. Haller, and J. B. Fenn, *J. Phys. Chem. 91*, 2654 (1987).

27. S. P. Holland, B. J. Garrison, and N. Winograd, *Phys. Rev. Letters 44*, 756 (1980).

28. S. Adelman and J. D. Doll, *J. Chem. Phys. 61*, 4242 (1974); *Acc. Chem. Res. 10*, 378 (1977); J. C. Tully, *J. Chem. Phys. 73*, 1975 (1980).

29. C. -Y. Lee and A. E. DePristo, *J. Chem. Phys. 87*, 1401 (1987).

30. J. C. Tully, *Ann. Rev. Phys. Chem. 31*, 319 (1980); J. C. Tully, *Acc. Chem. Res. 14*, 188 (1981); J. C. Tully and M. J. Cardillo, *Science 223*, 445 (1984); C. W. Muhlhausen, L. R. Williams, and J. C. Tully, *J. Chem. Phys. 83*, 2594 (1985).

BIBLIOGRAPHY

Thermodynamic Formulation of Transition State Theory

HAMMES, G. G. *Principles of Chemical Kinetics*. New York: Academic Press, 1978.

LAIDLER, K. J. *Chemical Kinetics*. New York: Harper and Row, 1987.

MOORE, J. W., and PEARSON, R. G. *Kinetics and Mechanism*. New York: Wiley, 1981.

WESTON, R. E., JR., and SCHWARZ, H. A. *Chemical Kinetics*. Englewood Cliffs, NJ: Prentice-Hall, 1972.

Kramers's Theory

HYNES, J. T. In *Theory of Chemical Reaction Dynamics, Vol. IV*, edited by M. Baer. Boca Raton, FL: CRC Press, 1985, p. 171.

KAPRAL, R. *Adv. Chem. Phys. 48*, 71 (1981).

TROE, J. In *Physical Chemistry, An Advanced Treatise, Vol. VI B*, edited by W. Jost. New York: Academic Press, 1975, p. 835.

TROE, J. *Ann. Rev. Phys. Chem. 38*, 163 (1987).

Gas-Surface Reaction Dynamics

GRUNZE, M., and KREUZER, H. J. *Kinetics of Interface Reactions*. New York: Springer-Verlag, 1987.

JORDAN, P. C. *Chemical Kinetics and Transport*. New York: Plenum, 1979.

KING, D. A., and WOODRUFF, D. P. *The Chemical Physics of Solid Surfaces and Heterogeneous Catalysis, Vol. 4, Fundamental Studies of Heterogeneous Catalysis*. New York: Elsevier, 1982.

LAIDLER, K. J. *Chemical Kinetics*. New York: Harper and Row, 1987.

PULLMAN, B., JORTNER, J., NITZAN, A., and GERBER, B. *Dynamics on Surfaces*. Boston: D. Reidel, 1984.

SOMORJAI, G. A. *Principles of Surface Chemistry*. Englewood Cliffs, NJ: Prentice-Hall, 1972.

PROBLEMS

12.1 A study of the rate of alkaline hydrolysis of propionamide as a function of hydrostatic pressure at 300 K gave the following results:

log k/k_0	0.10	0.21	0.31
P, lb per sq. in.	5,000	10,000	15,000

What is the volume of activation, $\Delta V^{\ddagger}$, for this reaction in units of liters/mole?

12.2 Outline the different assumptions made in deriving equation (12–10) as compared with equation (4–55).

12.3 A strong correlation has been found between volumes and entropies of activation for reactions in aqueous solution [K. J. Laidler and D. T. Y. Chen, *Can. J. Chem. 37,* 599 (1959)]. What is the nature of this correlation? What is its chemical significance?

12.4 The radial distribution function, $g_{AB}(r^{\ddagger})$, at the transition state for an association reaction in solution has a value of 5.69×10^{-3} at 300 K. The potential energy barrier for the association reaction in the gas-phase is 21 kJ/mol. What is the ratio between the gas-phase and solution association rate constants at 300 K? What is $\Delta G_{solv}^{\ddagger}$?

12.5 What is the mathematical relationship between the ω_0 and $\omega^{\ddagger}$ terms in equations (12–30) and (12–31) and the vibrational frequency?

12.6 The velocities of molecules desorbing from a surface are usually characterized by a Maxwell-Boltzmann distribution at the temperature of the surface. It might seem that the distribution of molecular velocities should be colder than the distribution of velocities of the molecules adsorbed on the surface by just the well depth of the interaction, since the molecules have to overcome a barrier to desorption. This problem will show that the former statement is correct.

Consider a one-dimensional square well of depth U. Let c represent the velocities inside the well (measured from the bottom of the well) and v represent the experimentally determinable velocities (the velocities of the desorbed particles).

(a) Draw a diagram showing the potential and the probability density of molecules adsorbed in the well.

(b) Write the mathematical expression for the probability density, and convert it to a quantity proportional to the probability *flux* of molecules in the well with velocities between c and $c + dc$.

(c) Using the diagram from part (a), transform the above expression to one in terms of the experimentally measurable velocities. Make sure to do the differentials (the Jacobian of the transformation) correctly.

(d) The answer to the original problem is clear from the expression found in part (c). The mean of the experimentally observed velocity distribution is

$$\left(\frac{\pi k_B T}{2m}\right)^{1/2} = \langle |v| \rangle$$

Derive this expression. Is the temperature of the desorbing molecules equal to or less than that of the surface? Why?

(e) The flux of molecules desorbing from a surface is characterized by a cosine angular distribution where the angle is measured from the surface normal. From the expression in part (b) or (c), identify what determines the rate at which the molecules escape from the surface. How does the rate relate to the angular distribution?

12.7 In order to study the properties of a metal-semiconductor interface one approach is to evaporate a monolayer of metal atoms onto the surface of the semiconductor in a vacuum chamber. Such an apparently simple procedure is, however, fraught with experimental difficulties. In particular, if the substrate is at room temperature, the metal adatoms will sometimes form "islands" rather than a continuous uniform layer. To circumvent this, the substrate is often cooled, the rationale being that on a sufficiently cold surface an atom will stick where it lands on the surface and a uniform coverage will result. We wish to examine this rationale quantitatively. Assume for simplicity that the substrate and adsorbate atoms are virtually the same size and that the adsorbed metal has the same crystalline arrangement as the semiconductor.

(a) If enough metal atoms are evaporated to give an average number of metal atoms/surface site $\overline{N}$ and there are b surface sites, what is the probability that a metal atom is adsorbed at a particular site b_0? What is the order of magnitude of b for a typical crystal surface area of 1 cm²?

(b) The probability that a site is covered by a layer N atoms deep can be calculated from the binomial distribution

$$P_N = \left(\frac{b(b-1)\cdots(b-N+1)}{N!}\right)\left(\frac{\overline{N}}{b}\right)^N\left(1 - \frac{\overline{N}}{b}\right)^{b-N} \tag{1}$$

However, this form is cumbersome as it includes b explicitly. Show that in the limit as $b \rightarrow \infty(b \gg N, b \gg \overline{N})$,

$$P_N \approx \frac{(\overline{N})^N e^{-\overline{N}}}{N!}$$

Hints: This derivation involves making approximations which are good in the limit $b \rightarrow \infty$. (Justify briefly the ones used.) Eliminate b from the first two factors of (1) alone. Then recognize a truncated Taylor series in the third factor.

(c) Suppose enough metal is evaporated to give an average surface coverage of 1 monolayer. What fraction of the substrate atoms are not covered by metal? By one metal atom exactly? By five metal atoms?

CHAPTER 13

Information-Theoretical Approach to State-to-State Dynamics

13.1 INTRODUCTION

The information-theoretical approach to the reaction kinetics of systems possessing a large number of internal levels was first introduced in the early 1970s by R. D. Levine, R. B. Bernstein, and J. L. Kinsey.[1] This approach represents an alternative to those taken up to this point in the book. We first considered a macroscopic, phenomenological approach in which we averaged over internal state distributions in the reactants and products. That approach was simply not detailed enough to describe experiments which employed state-selected reactants and/or state-resolved product detection, such as those described in chapter 9. Many instances are known of selective energy consumption in endothermic reactions and specific energy release in exothermic reactions which do not follow a simple, Boltzmann-like statistical distribution. A microscopic theory of reaction dynamics has been developed to deal with such experiments, but this turns out to be in many ways *too* detailed: the theory frequently requires the solution of many closely coupled equations and calculates far more pieces of information than could ever be observed. A middle ground between these extremes would be a statistical thermodynamic approach, analogous to that used for prediction of equilibrium thermodynamic properties. In such a theory, all of the dynamical variables are considered in principle, but suitable averages are taken over those variables that are not observed in any particular measurement. For reasons which will become apparent shortly, this approach has come to be called *information-theoretic*. Details of the early development of this theory, together with its practical implementations, may be found in several excellent reviews.[2,3]

13.2 THE MAXIMAL-ENTROPY POSTULATE

13.2.1 Basic Definitions

Consider an experiment having n possible distinguishable outcomes, e.g.,

$$F + H_2 \longrightarrow HF(v) + H$$

Since $\Delta H^0 + E_{act} + Q_{th} \approx 140$ kJ/mole, the vibrational states $v = 0, 1, 2,$ and 3 are energetically accessible. Thus, $n = 4$ for this system.* After N repetitions of this molecular process, we get a number of possible *sequences* representing the v level observed for each individual reaction. Such a sequence might be, for example,

$$\{v\} = 0\ 2\ 1\ 3\ 2\ 2\ 3\ 1\ 2\ 0\ 2\ 1\ 2\ 2\ 3\ \ldots$$

which is distinct from, say,

$$\{v\}' = 0\ 2\ 1\ 3\ 2\ 3\ 2\ 1\ 2\ 0\ 2\ 1\ 2\ 2\ 3\ \ldots$$

even though the first 15 members of the sequence include two reactions giving $v = 0$, three with $v = 1$, seven with $v = 2$, and three with $v = 3$. The number of possible sequences is clearly n^N, which can be expressed conveniently as $\exp(N \ln n)$. This is in general an enormous number; even in the present trivial case, $n^N = 4^{15} \approx 10^9$!

The complete specification of a sequence is entirely too detailed for practical use. What we are really interested in is the *set* of outcomes for a particular experiment, denoted by $\{N_i\}$, with $i = 1, 2, \ldots, n$, and normalized so that

$$\sum_{i=1}^{n} N_i = N \tag{13-1}$$

For example, for both of the preceding sequences, $\{N_i\} = (2, 3, 7, 3)$. The thermodynamic weight W associated with a set $\{N_i\}$ is the number of possible sequences which correspond to a distinct distribution or set $\{N_i\}$. This is just equivalent to the combinatorial problem of placing N balls in n boxes, which gives[4]

$$W(\{N_i\}) = \frac{N!}{\prod_{i=1}^{n} (N_i)!} \tag{13-2}$$

with the normalization

$$\sum_{\text{all sets}} W(\{N_i\}) = n^N \tag{13-3}$$

The unbiased or *a priori* probability of finding a particular set is then

$$P^0(\{N_i\}) = \frac{W\{N_i\}}{\sum W(\{N_i\})} = \frac{W(\{N_i\})}{n^N} \tag{13-4}$$

and clearly,

$$\sum_{\text{all distributions}} P^0\{(N_i)\} = 1 \tag{13-5}$$

Since typical experiments involve large N (on the order of 10^{23} or so), evaluating these expressions is facilitated considerably by use of the *Stirling approximation*[4,5]

$$\ln N! \equiv \log_e (N!) = N \ln N - N + \mathcal{O}\{\ln N\} \tag{13-6}$$

---

*For F + D_2, $v = 0, \ldots, 4$ can be populated, and $n = 5$.

For $N \approx 10^{23}$, the contribution of $\mathcal{O}(\ln N)$ relative to that of N is approximately

$$\frac{2.3 \times 23}{10^{23}} \approx \frac{50}{10^{23}} \approx 1 \text{ in } 10^{21}$$

so that $\mathcal{O}(\ln N)$ can be safely neglected. Using the Stirling approximation, the *a priori* probability can be evaluated from equation 13–4 as

$$\ln P^0(\{N_i\}) = N \ln N - N - \sum_i (N_i \ln N_i - N_i) - N \ln n$$

$$= -N \ln n - \sum_i (N_i \ln N_i - N_i \ln N)$$

$$= -N \left(\ln n + \sum_i X_i \ln X_i \right)$$

where

$$X_i = \frac{N_i}{N}, \text{ with } \sum_{i=1}^{n} X_i = 1 \qquad (13\text{–}7)$$

The *statistical entropy* associated with this distribution is given by

$$S(\{N_i\}) = \frac{1}{N} \ln W(\{N_i\})$$

$$= \frac{1}{N} \ln [n^N P^0(\{N_i\})]$$

$$= \frac{1}{N} \left(N \ln n - N \ln n - N \sum_i X_i \ln X_i \right)$$

so that

$$S(\{N_i\}) = -\sum_i X_i \ln X_i \qquad (13\text{–}8)$$

Equation 13–8 will look familiar, as the *entropy of mixing* in equilibrium thermodynamics, if X_i is the mole fraction and the entropy is expressed in e.u., or ergs mole^{-1} degree^{-1}, using the coefficient $S = (R/N_0) \ln W = k \ln W$.

We shall shortly prove that the entropy is a maximum for the *uniform distribution*, in which $X_i = 1/n$ for all i—that is, all possible outcomes are equally probable. From equation 13–8, this is

$$\max(S) = -\sum_{i=1}^{n} \frac{1}{n} \ln \left(\frac{1}{n} \right) = -\frac{n}{n} \ln \left(\frac{1}{n} \right) = \ln n \qquad (13\text{–}9)$$

For any other distribution, we may define an *entropy deficiency*

$$\Delta S = -\frac{1}{N} \ln (P\{N_i\}) = \ln n + \sum_i X_i \ln X_i \qquad (13\text{–}10)$$

The entropy S is just the difference between the maximum entropy of equation 13–9 and the entropy deficiency of equation 13–10, or

$$S = \max(S) - \Delta S$$

$$= \ln n - \left(\ln n + \sum X_i \ln X_i \right)$$

$$= -\sum_i X_i \ln X_i$$

so that

$$S = \frac{1}{N} \ln W(\{N_i\}) \qquad (13-11)$$

as before. For $X_i = 1/n$, i.e., $P(\{N_i\}) = P^0$, obviously $\Delta S = 0$. In communication theory,[6] this quantity is also called the *information content* of the distribution; the rationale is that a completely flat, unmodulated signal conveys no information, and only the departures from uniform amplitude (the entropy deficiency, according to equation 13–10) provide useful information to the receiver. A further important consequence of the expression given for the entropy in equations 13–8 and 13–11 can be seen from writing the statistical weight as $W(\{N_i\}) = \exp(NS)$. For large N (i.e., in bulk experiments), this means that a distribution possessing even a slightly smaller value of S than any other distribution will have an *overwhelmingly* smaller statistical weight; so, in effect, the experimentalist observes only a single outcome, the maximal-entropy outcome, every time the measurement is repeated. This is the basis of the *maximal-entropy postulate*, which is central to the development of the information-theoretic approach. It is also expressed, much more succinctly, by the Talmudic quotation with which Levine begins his 1978 review article.[7]

13.2.2 Constraints and the Variational Principle

While unique distributions are indeed observed in experiments, such as the vibrational distribution of the HF product in the $F + H_2$ reaction (chapter 9), these are hardly ever the *uniform* distribution with $X_i = 1/n$. This is because of the existence of constraints on the distribution.

What is meant by "constraints"? Constraints are conditions imposed on the distribution, e.g., *normalization*, given by

$$\sum_{i=1}^{n} X_i = 1 \qquad (13-12)$$

which is of course equivalent to both equation 13–1 and equation 13–5. Let us begin by finding the distribution which maximizes the entropy, subject only to the normalization constraint of equation 13–12. The technique for doing this is the *calculus of variations*, which is described in any standard calculus textbook.[8] We define a variational function

$$F = S - (\lambda_0 - 1) \sum_i X_i \qquad (13-13)$$

where the first term, S, is the function to be maximized (in this case, the entropy), and the second term is the statement of the constraint (normalization). The coefficient of the latter term is the *Lagrange multiplier*, which we write here as $\lambda_0 - 1$ for convenience, as we shall shortly see. We require that $\delta F / \delta X_i = 0$ for all variations δX_i, i.e., that F is an *extremum* (in this case, a maximum) in multidimensional $\{X_i\}$-space.

Substituting the definition of the statistical entropy, equation 13–8, into the variational function, equation 13–13, gives

$$F = \sum (-\ln X_i - \lambda_0 + 1)X_i$$

The variation δF is found by simple differentiation:

$$\delta F = \sum [-\ln X_i(\delta X_i) - X_i(\delta \ln X_i) - \lambda_0 \delta X_i + \delta X_i]$$

Recalling that $\delta \ln X_i = \delta X_i / X_i$, we have

$$\delta F = \sum_i (-\ln X_i - 1 - \lambda_0 + 1)\, \delta X_i$$

$$= \sum_i (-\ln X_i - \lambda_0)\, \delta X_i$$

Note the usefulness of writing the Lagrange multiplier as $\lambda_0 - 1$ here. In order for $\delta F / \delta X_i$ to be zero for all i, we require that the coefficient of δX_i be equal to zero, i.e.,

$$-\ln X_i - \lambda_0 = 0$$

or

$$X_i = e^{-\lambda_0}$$

But the normalization equation 13–12 requires that

$$\sum_{i=1}^{n} X_i = \sum_{i=1}^{n} e^{-\lambda_0} = ne^{-\lambda_0} = 1$$

Therefore, $n = e^{\lambda_0}$, $\lambda_0 = \ln n$, and $X_i = e^{-\lambda_0} = e^{-\ln n} = 1/n$, which is the uniform distribution defined on page 433. This means that normalization provides us with no "new information" about the distribution, which can be stated as a theorem:

When the only constraint is normalization of probabilities, the most probable distribution is the uniform one.

A lemma can also be stated:

Furthermore, the entropy of such a distribution is an absolute maximum.

Now suppose we have one more piece of information, such as the first moment of the distribution. Then

$$\sum_{i=1}^{n} iX_i = \langle i \rangle \qquad (13\text{--}14)$$

This furnishes an additional constraint on the distribution, and the variational function now becomes

$$F = S - (\lambda_0 - 1) \sum_{i=1}^{n} X_i - \lambda_1 \sum_{i=1}^{n} iX_i$$

or

$$F = -\sum_i X_i \ln X_i - (\lambda_0 - 1) \sum_i X_i - \lambda_1 \sum_i iX_i \qquad (13\text{--}15)$$

The variational quantity is thus

$$\delta F = \sum_i (-\ln X_i - 1 - \lambda_0 + 1 - \lambda_1 i) \, \delta X_i$$

and requiring that $\delta F / \delta X_i = 0$ gives

$$\ln X_i = -\lambda_0 - \lambda_1 i$$

or

$$X_i = e^{-\lambda_0 - \lambda_1 i} \qquad (13\text{--}16)$$

To find the Lagrange multipliers λ_0 and λ_1, we use a *sum rule* based on the constraint given by equation 13–14. Factoring out the common element e^{λ_0}, we can rewrite this equation as

$$\langle i \rangle = \frac{\sum\limits_i i \exp(-\lambda_1 i)}{\sum\limits_i \exp(-\lambda_1 i)} \qquad (13\text{--}17)$$

If $\langle i \rangle$ is known, equation 13–17 is sufficient to determine λ_1; however, because of the sums over exponentials, such determination in general must be done by numerical fitting. Once λ_1 is known, the normalization parameter λ_0 can be easily determined using equation 13–12 as follows:

$$\sum_i X_i = \sum \exp(-\lambda_0 - \lambda_1 i)$$

$$= \sum \exp(-\lambda_0) \exp(-\lambda_1 i) = 1$$

Therefore, $\exp(\lambda_0) = \sum_i \exp(-\lambda_1 i)$, and if the sum is taken over $n \to \infty$, this has the simple closed form

$$\exp(\lambda_0) = \frac{1}{1 - e^{-\lambda_1}} \qquad (13\text{--}18)$$

13.2.3 Example: Harmonic-Oscillator Levels at Thermodynamic Equilibrium

As an illustration of the variational principle discussed in the preceding section, let us consider $P(v)$, the equilibrium distribution among harmonic-oscillator levels. The variable i is the oscillator energy

$$E_v = v \hbar \omega_0 \qquad (13-19)$$

We take the zero of energy to be the zero-point level $\frac{1}{2} \hbar \omega_0$ and note that v can take any integer value between zero and infinity. (The sums thus have convenient closed forms.) The constraints we specify are normalization, viz.,

$$\sum_v P(v) = 1 \qquad (13-20a)$$

and fixed total energy, i.e.,

$$\sum_v E_v P(v) = \langle E_v \rangle = \text{constant} \qquad (13-20b)$$

The second condition is just that which specifies a canonical ensemble.[9] From the result previously obtained (equation 13–16),

$$P(v) = \exp\left[-\lambda_0 - \beta v\right] \qquad (13-21)$$

with the Lagrange multiplier $\lambda_1 = \beta \hbar \omega_0$. This quantity is determined by application of equation 13–17, the sum rule:

$$\langle E_v \rangle = \frac{\sum_v (v \hbar \omega_0) \exp(-\beta v \hbar \omega_0)}{\sum_v \exp(-\beta \hbar \omega_0)}$$

$$= \frac{\hbar \omega_0}{e^{\beta \hbar \omega_0} - 1}$$

The normalization constant is determined from

$$\exp(\lambda_0) = \frac{1}{1 - \exp(-\beta \hbar \omega_0)}$$

If we make the further connection with equilibrium thermodynamics, then β^{-1} must be proportional to a temperature T and $\beta = 1/k_B T$. Equation 13–21 is then the familiar Boltzmann distribution,

$$P(v) = [1 - e^{-\hbar \omega_0 / k_B T}] e^{-v \hbar \omega_0 / k_B T}$$

The purpose of this example was to illustrate the application of entropy maximization by means of the variational principle to a familiar situation, namely, the calculation of an equilibrium distribution. In the information-theoretical approach to chemical kinetics, the same procedures are applied to nonequilibrium systems, i.e., to calculation of rate coefficients, branching ratios, and reaction probabilities. While this approach cannot really be called a dynamical theory, since it does not actually predict the values of these quantities, numerous investigations have demonstrated its utility in describing reactive systems possessing a very large number of possible channels in terms of a much smaller number of Lagrange multipliers. In many instances, only a single such parameter is sufficient to represent all of the available information about such a system. It is by no means axiomatic, however, that a distribution possess only a single constraint. In general, there may be r independent constraints having the form

$$\sum_{i=1}^{n} g_m(i)X_i = \langle g_m \rangle \tag{13-22}$$

We have already considered constraints with

$$m = 0 \text{ (normalization):} \qquad g_0(i) = 1 \qquad \langle g_0 \rangle = 1$$

and

$$m = 1 \text{ (first moment):} \qquad g_1(i) = i \qquad \langle g_1 \rangle = \langle i \rangle$$

For $m = 2$, it is usually most useful to take $\langle g_2 \rangle = (\langle i^2 \rangle - \langle i \rangle^2)$, i.e., the *dispersion* of the distribution; but any form can be specified, as long as it is independent of all other constraints. The maximal-entropy distribution has the form

$$X_i = \exp\left[-\lambda_0 - \sum_{m=1}^{r} \lambda_m g_m(i)\right] \tag{13-23}$$

with

$$S = -\sum_i X_i \ln X_i$$

i.e.,

$$S = \lambda_0 + \sum_{m=1}^{r} \lambda_m \langle g_m \rangle \tag{13-24}$$

In the applications to be discussed in this chapter, we shall see numerous examples with $r \ll n$, which means that it is possible to describe a distribution over many possible outcomes with a very small number of moments of the distribution. An equivalent way of saying this is that there are only r independent pieces of information required to describe the experiment in full.

13.3 SURPRISAL ANALYSIS AND SYNTHESIS: PRODUCT STATE DISTRIBUTION IN EXOTHERMIC REACTIONS

We shall now proceed to apply these concepts to a nonequilibrium situation, specifically, the product state distribution in an exothermic reaction. The reaction we shall consider is one which has been discussed previously in the context of microscopic reaction dynamics and to which we shall return in the next chapter. This is the hydrogen + fluorine atom reaction,

$$F + H_2 \longrightarrow HF(v = n') + H$$

along with its isotopic variant,

$$F + D_2 \longrightarrow DF(v = n'') + D$$

We seek the *branching ratios* for this reaction, given by

$$P(n') = \frac{k(n')}{\sum\limits_{n'} k(n')} \qquad (13\text{--}25)$$

which take the place of the probability densities discussed in the context of equilibrium distributions. While in principle we may make a complete specification of product states, for the time being we shall consider just the total production rate into a specified vibrational state:

$$k(n') = \sum_{J'}\sum_{M_{J'}} k(n', J', M_{J}') \qquad (13\text{--}26)$$

The set $\{k(n')\}$ may incorporate one or more dynamic constraints, but we are generally not aware of that prior to carrying out the experiment. Therefore, we shall take the *prior rate* $k^0(n')$ to be simply proportional to the density of states available to the products. This is equivalent to the uniform distribution introduced earlier, and the result will be dominated by the translational density of states. (We shall take care of vibrational and rotational state degeneracies shortly.) The equation is shown in Appendix 2 to have the form

$$\rho_T(E_T') = \frac{(\mu')^{3/2}}{2^{1/2}\,\pi^2\hbar^3}(E_T')^{1/2} \qquad (13\text{--}26a)$$

or

$$\rho_T(E_T') = A_T(E_T')^{1/2} \qquad (13\text{--}26b)$$

We use the symbol $\rho(E)$ here, rather than $N(E)$ as in chapter 11, to facilitate reference to the extensive literature on information theory. Also, μ' is the reduced mass of the products, and E_T' is the translational energy available to the products, which is equal to the total reaction energy

$$E = -\Delta H^0(\text{reaction}) + E_{\text{act}} + E_{\text{thermal}}(\text{reactants})$$

less the energy in the product internal states, i.e.,

$$E_n = E_{\text{vib}}' + E_{\text{rot}}'$$

We assert that

$$k^0(n') = R\rho_T(E_T')\rho_n(E_n') = R\rho_T(E - E_n')\rho_n(E_n') \qquad (13\text{--}27)$$

with R a coefficient that provides the correct units for the rate coefficient. Note that $\rho_T(E_T') \propto (E_T')^{1/2}$, which is in turn proportional to the relative velocity of separation of the products. Thus, equation 13–27 can also be interpreted as stating that the prior rate coefficient is proportional to the net flux of products in a specified internal state out of the reaction zone.

By analogy with equation 13–25, we define

$$k^0 = \sum_{n'} k^0(n')$$

$$= R \sum_{n'} \rho_T(E - E'_n)\rho_n(E'_n)$$

so that

$$k^0 = R\rho(E) \qquad (13\text{–}28)$$

and the prior branching ratio

$$P^0(n') = \frac{k^0(n')}{k^0} = \frac{\rho_T(E - E'_n)\rho_n(E'_n)}{\rho(E)} \qquad (13\text{–}29)$$

The $P^0(n')$ are, of course, normalized to unity.

The observed distribution can be written in the maximal-entropy form (cf. equation 13–23)

$$P(n') = P^0(n') \exp\left[-\sum_{m=1}^{r} \lambda_m g_m(r')\right] \qquad (13\text{–}30)$$

We then define the "departure from the prior expectation,"

$$I(n') = -\ln \frac{P(n')}{P^0(n')} \qquad (13\text{–}31)$$

The quantity $I(n')$ has been called the *surprisal* of the product state distribution. Although the term "surprisal" has attracted a degree of opprobrium, its usage is quite reasonable: if you don't get what you expected beforehand, then by definition you are surprised. On the other hand, if $P(n') = P^0(n')$ for all n', then $I = 0$; in other words, the observed and prior distributions are identical, and that's no surprise at all! Indeed, the zero-surprisal situation implies *no* constraints beyond normalization, which is implicit in the definition of P (or P^0). In general, however, one finds nonzero surprisals which have the maximal-entropy form

$$I(n') = \sum_{m=1}^{r} \lambda_m g_m(n') \qquad (13\text{–}32)$$

Now let us apply this analysis to the special case of the HF product distribution. The most detailed specification includes both the vibrational state v' and the rotational state J', so that

$$P^0(v', J' | E) = \frac{(2J' + 1)\rho_T(E - E_{v'J'})}{\rho(E)} \qquad (13\text{–}33)$$

The factor $2J' + 1$ comes from having summed over $M_{J'}$ states in equation 13–26, while the quantity $P^0(v'J' | E)$ is a kind of *conditional* probability[4] which can be read as "the probability of finding a product molecule in state $v'J'$ given that a total energy E is available." We shall come across conditional probabilities often in the rest of this chapter.

The desired quantity, the net branching ratio into state v', is found by summing equation 13–33 over all accessible states J' up to the maximum $J^*(v')$ which is compatible with the energy $E_{v'}$ in vibration and the total energy E:

$$P^0(v'|E) = \sum_{J=0}^{J^*(v')} P^0(v'J'|E) \tag{13-34}$$

In order to evaluate equation 13–34, we need to choose a model for the vibration-rotation energy levels of the molecule. The most convenient model is the rigid rotor–harmonic oscillator (RRHO), given by

$$E_{v'J'} = E_R + E_v = BJ'(J'+1) + \hbar\omega_0 v' \tag{13-35}$$

We thus have, for the RRHO,

$$(2J'+1)\rho(v'J'E) = (2J'+1)\rho_T(E_T)$$

$$= (2J'+1)\rho_T(E - E_{v'J'})$$

$$= (2J'+1)\left(\frac{(\mu')^{3/2}}{2^{1/2}\pi^2\hbar^3}\right)(E - E_{v'J'})^{1/2}$$

and

$$\rho(E) = \sum_{v'=0}^{v^*(E)}\left[\sum_{J'=0}^{J^*(v')} \rho(v'J'E)\right]$$

$$\rho(E) = \frac{4}{15}\left(\frac{1}{\hbar\omega_0 B}\right)\left(\frac{(\mu')^{3/2}}{2^{1/2}\pi^2\hbar^3}\right)E^{5/2} \tag{13-36}$$

In equation 13–36, $v^*(E)$ is the *exothermic limit*—the vibrational state of the product in which all the energy E appears in vibration.

Combining equations 13–33 and 13–36, we have

$$P^0(v'J'|E) = (2J'+1)\hbar\omega_0 B\frac{15}{4}(E - E_v - E_R)^{1/2}E^{-5/2} \tag{13-37}$$

and

$$P^0(v'|E) = \sum_{J'=0}^{J^*(v')} P^0(v'J')|E)$$

so that

$$P^0(v'|E) = (\text{const})\frac{5}{2}(E - E_{v'})^{3/2}E^{-5/2} \tag{13-38}$$

It is convenient to rewrite these quantities in terms of the fraction of energy in product vibration.

$$f_v = \frac{E_{v'}}{E} \tag{13-39}$$

Thus,

$$P^0(f_v) \propto \frac{5}{2}E^{3/2}(1 - f_v)^{3/2}E^{-5/2}$$

or

$$P^0(f_v) \propto \frac{5}{2} \frac{(1 - f_v)^{3/2}}{E} \qquad (13\text{--}40)$$

In order to perform the calculation, we need to specify E. For F + H_2, $-\Delta H° = 130$ kJ mole^{-1}, $E_{act} = 4$ kJ mole^{-1}, and $\langle E_{thermal} \rangle \approx 6$ kJ mole^{-1}, for a total of $E = 140$ kJ mole^{-1}. Because of the large value of $-\Delta H°$, we can use a single value of E, i.e., we can treat the system as a microcanonical ensemble. In other situations, we have to carry out an average over energy distributions at some temperature (see section 13.4.1).

The analysis proceeds as follows for HF. We calculate the vibrational energy levels from the vibrational constants[10] and note the experimental values of $P(v')$, see chapter 9. Then the quantities f_v, $P^0(v')$, and $I(v')$ are readily calculated from equations 13–39, 13–40, and 13–21, respectively. The results are shown in Table 13–1(a). When $-I(v')$ is plotted against f_v, a straight line is obtained, as shown in Figure 13–1. If the calculation is now repeated for the reaction F + $D_2 \rightarrow$ DF(v') + D (Table 13–1(b)), the surprisal points are found to lie on the same straight line as previously found. Therefore, a *single dynamic constraint,* independent of isotopic substitution, is sufficient to determine energy disposal in this reaction. The F + (H_2, D_2) reaction system was among the first to be analyzed in this way,[11] and its success led to many further applications. It is worth pointing out that if we had possessed this key piece of information before actually doing the experiment, we could have justifiably incorporated it into the analysis together with our prior branching ratios $P^0(v)$; in that case we would have had $P(v) = P^0(v)$ for all v with $I(v) = -\ln [P^0(v)/P^0(v)] = 0$, and there would have been no surprise.

TABLE 13–1 SURPRISAL ANALYSIS FOR HF AND DF PRODUCT VIBRATIONAL DISTRIBUTIONS

v'	E_v-E_o kJ/mole	f_v	$(1$-$f_v)^{3/2}$	$P(v')$ experimental	$I(v')$
(a) F + $H_2 \rightarrow$ HF(v') + H					
0	0	0	1.00	(0.02)	(3.9)
1	47	0.355	0.54	0.15	1.28
2	92	0.657	0.20	0.57	−1.04
3	132	0.943	0.14	0.26	−2.9
(b) F + D_2 + DF(v') + D					
0	0	0	1.00	(0.01)	(4.6)
1	34.7	0.247	0.65	0.10	1.8
2	68.5	0.489	0.365	0.23	0.46
3	101.1	0.722	0.146	0.38	−0.95
4	132	0.944	0.013	0.28	−3.0

The obvious utility of the information-theoretical approach to this problem is that nine data points, giving the product branching ratios in the F + H_2 and F + D_2 reactions, can be reduced to a single parameter. In subsequent sections we shall see even more spectacular compaction of molecular collision data. In the meantime, let us consider how to obtain the single parameter from the data. Reducing equation

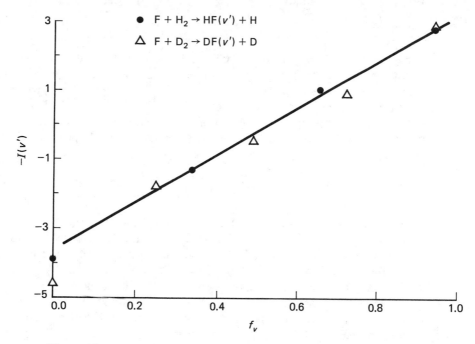

Figure 13–1. Surprisal plot for the reaction $F + H_2 \rightarrow HF\ (v') + H$ [●] and $F + D_2 \rightarrow DF\ (v') + D[\triangle]$.

13–30 to a single term and writing the normalization separately gives

$$k(v') = k_{\text{total}} P(v')$$

$$= \exp(-\lambda_0 - \lambda_v f_v)$$

so that

$$k(v') = k^0(v') \exp(-\lambda_v f_v) \tag{13–41}$$

and the surprisal is given by

$$I(v') = -\ln \frac{k(v')}{k^0(v')} = \lambda_v f_v \tag{13–42}$$

For this specific system, we obtain λ_v from the slope of the straight line in Figure 13–1:

$$\lambda_v = \frac{\Delta(I_v)}{\Delta(f_v)} = -7.2$$

A large value for $|\lambda_v|$ indicates a considerably nonstatistical distribution; moreover, when the value of λ_v is negative, as in the present case, it indicates preferential disposition of the energy in product vibration. Values of λ close to zero always indicate a near-statistical outcome.

Other useful measures of the reaction can be derived from the preceding. The average amount of the total available energy appearing as product vibration is

$$\langle f_v \rangle = \sum_v f_v P(v' | E) \qquad (13\text{--}43)$$

For HF, $\langle f_v \rangle = 0.66$. This quantity is the first moment of the distribution, but by itself it cannot specify $P(v' | E)$ until a linear-surprisal, maximal-entropy form is assumed for P. If such a form is assumed, then $\langle f_v \rangle$ can be used to find λ_v by means of a sum rule, which is discussed in section 13.5.

By analogy with the discussion of equilibrium distributions in section 13.2.3, we can define an effective *temperature* characterizing the HF or DF product distribution:

$$T_v = \frac{E}{R\lambda_v} = \frac{1.4 \times 10^5 \text{ J mole}^{-1}}{(8.3 \text{ J mole}^{-1} \text{ K}^{-1}) \times (-7.2)} \approx -2{,}400 \text{ K}$$

A large negative temperature, such as that just found, indicates a *population inversion* in the product distribution. An inverted population can be used to produce a laser,[12] and indeed, the F + (H_2, D_2) reaction has been used to power several models of high-energy chemical laser systems. Such a system can do thermodynamic work on any positive-temperature system to which it is coupled.

Many other applications of surprisal analysis to vibrational and rotational product distributions have appeared in the literature on kinetics.[2,3] Advances in experimental techniques have made it possible to carry out rotational surprisal analyses for the simple, prototypical H + H_2 reaction discussed in chapters 8 and 9. H atoms are produced by photolysis of HI, as in the early experiments of Kupperman and White (see chapter 9, problem 3); at the quadrupled Nd:YAG laser wavelength (266 nm) used in these experiments, a bimodal H atom velocity distribution is obtained, the result of two possible photolysis channels, viz.,

$$\text{HI} + h\nu\,(266 \text{ nm}) \longrightarrow \text{I}(^2P_{3/2}) + \text{H}$$
$$(\text{relative K.E.} = 1.30 \pm 0.04 \text{ eV} \approx 125 \text{ kJ/mole})$$

and

$$\text{HI} + h\nu\,(266 \text{ nm}) \longrightarrow \text{I*}(^2P_{1/2}) + \text{H}$$
$$(\text{relative K.E.} = 0.55 \pm 0.03 \text{ eV} \approx 53 \text{ kJ/mole})$$

The H atoms then react with D_2 according to the reaction

$$\text{H} + D_2 \longrightarrow \text{HD}(v, J) + \text{D}$$

and the resulting HD molecules are detected using a state-specific detection technique, such as resonant multiphoton ionization (REMPI)[13] or coherent anti-Stokes Raman spectroscopy (CARS)[14].

The maximal-entropy form for the rotational-state-specific rate coefficient is

$$k(v, J) = k^0(v, J) \exp(-\theta_0 - \theta_{\text{rot}} f_{\text{rot}})$$

with the prior rate coefficient given by

$$k^0(v, J) = (\text{const}) \cdot (2J + 1)(E_T)^{1/2}$$
$$= (\text{const}) \cdot (2J + 1)(E - E_v - E_{\text{rot}})^{1/2}$$

and $f_{rot} = E_{rot}/(E - E_v)$. The two techniques mentioned give somewhat different results for θ_R: using REMPI, Marinero et al.[15] obtained $\theta_{rot} = 5.3(v = 1)$, $\theta_{rot} = 5.0(v = 2)$, while Gerrity and Valentini,[16] using CARS, obtained $\theta_{rot} = 3.0(v = 0)$, $\theta_{rot} = 3.5(v = 1)$; their data for $v = 2$ did not fit a linear surprisal. Surprisal analysis has also been applied to theoretical calculations on the $H + H_2$ system.[17] In principle, all these methods should give the same results, so doubtless additional experiments and calculations will be carried out to clear up the discrepancies.

13.4 INFORMATION-THEORETICAL ANALYSIS OF ENERGY TRANSFER PROCESSES

As the next application of information theory to kinetics, let us consider *energy transfer processes*. These are state-to-state events in which no net chemical change occurs, but in which energy is exchanged among vibrational, rotational, and translational degrees of freedom. The most common types of energy transfer include the following:

	Example
Vibration-translation (V-T)	$HF(v = 1) + He \rightarrow HF(v = 0) + He$
	The amount of energy ΔE_T transferred into translation is equal to $E_{vib}(v = 1) - E_{vib}(v = 0) = 47$ kJ/mole.
Rotation-translation (R-T)	$HF(J = 9) + He \rightarrow HF(J = 8) + He$ $\Delta E_T = 2\ BJ \approx 5$ kJ/mole.
Vibration-(rotation & translation) (V-R,T)	$HF(v = 1, J = 0) + He \rightarrow HF(J = 8) + He$ In this case, $\Delta E_T = -\Delta E_{vib} + \Delta E_{rot} \approx (47 - 45) = 2.5$ kJ/mole.
Vibration-vibration (V-V)	$HF(v = 2) + HF(v = 0) \rightarrow HF(v = 1) + HF(v = 1)$ For harmonic oscillators, $\Delta E_T = 0$; but for real molecules, $\Delta E_T = 2\omega_e x_e(HF) \approx 150$ cm$^{-1} \approx 2$ kJ/mole.

13.4.1 Thermally Averaged Prior Rates

The new feature introduced in analyzing energy transfer processes is that we must now *average* a priori rates and probabilities over a distribution of initial energies. This becomes particularly important for *near-resonant* energy transfer processes, such as V-R, T and V-V, in which ΔE_T is on the order of or less than $k_B T$. In other words, upon specifying an initial temperature, we will have a Boltzmann distribution of collision energies at that temperature.*

The procedure for carrying out the desired averaging is quite straightforward.[18,19] Consider the general energy transfer process

$$AB(n) + M \rightarrow AB(n') + M$$

---

*This is not the case, however, in experiments with velocity-selected molecular beams.

where n and n' refer to the internal state (vibrational, rotational, electronic, etc.) of the molecule before and after the inelastic collision. We define the "state-to-state" rate coefficient $k(n \rightarrow n'; T)$. Then, to find the total rate of collisions removing molecules from state n, we sum over final states:

$$k(n; T) = \sum_{n'} k(n \rightarrow n'; T) \qquad (13\text{--}44)$$

The transition probability is a conditional probability (cf. equation 13–33):

$$P(n' \mid n; T) = \frac{k(n \rightarrow n'; T)}{k(n; T)} \qquad (13\text{--}45)$$

The mean energy transfer out of state n is simply the first moment of the P distribution:

$$\langle \Delta E_n \rangle = \sum_{n'} (E_{n'} - E_n) P(n' \mid n; T) \qquad (13\text{--}46)$$

To find the total inelastic collision rate, we must average the microscopic rates over all initial states of the molecule:

$$k(t) = \sum_n \rho(n, T) \sum_{n'} k(n \rightarrow n'; T)$$

or

$$k(t) = \sum_n \rho(n, T) k(n; T) \qquad (13\text{--}47)$$

We can also define a bulk-averaged energy transfer:

$$\langle\langle \Delta E \rangle\rangle = \sum_n \rho(n, T) \langle \Delta E_n \rangle \qquad (13\text{--}48)$$

This quantity is related to relaxation properties of the system, such as acoustic dispersion.

If the probability densities $\rho(n, T)$ in equations 13–47 and 13–48 are the Boltzmann equilibrium probabilities $\rho^{eq}(n, T)$, then the important *detailed balancing* condition applies:

$$\rho^{eq}(n, T) k(n \rightarrow n'; T) = \rho^{eq}(n'T) k(n' \rightarrow n; T) \qquad (13\text{--}49)$$

This is just a consequence of the definition of equilibrium, $\dot{\rho}^{eq}(n, T) = 0$ for all n, T. Equation 13–49 can be written

$$\frac{k(n \rightarrow n'; T)}{k(n' \rightarrow n; T)} = \frac{\rho^{eq}(n'T)}{\rho^{eq}(n, T)} = \frac{g(n')}{g(n)} e^{-(E_{n'} - E_n)/k_B T}$$

For a simple V-T process in a diatomic molecule, this comes down to

$$\frac{k(0 \rightarrow 1)}{k(1 \rightarrow 0)} = e^{-\hbar \omega_0 / k_B T}$$

As for the reactive case, we first find the prior rate for the energy transfer process at a specified total energy $E = E_n + E_T$, which is proportional to the total density of states at the final translational energy:

$$k^0(n \to n'; E) = R\rho(n', E)$$

$$= R\rho_T(E_{T'})$$

$$= R\left(\frac{\mu^{3/2}}{2^{1/2}\pi^2\hbar^3}\right)(E - E_{n'})^{1/2}$$

or

$$k^0(n \to n'; E) = RA_T(E - E_{n'})^{1/2} \tag{13-50}$$

The amount of energy transferred from internal degrees of freedom into translational energy is

$$\Delta E = E_n - E_{n'} = E_{T'} - E_T \tag{13-51}$$

We now have to carry out the thermal average over a distribution of collision energies. This can be expressed as

$$k^0(n \to n'; T) = \frac{\int k^0(n \to n'; E)e^{-E/k_BT}\rho(n, E)\, dE}{\int e^{-E/k_BT}\rho(n, E)\, dE}$$

or

$$k^0(n \to n'; T) = \frac{R\int \rho(n, E)\rho(n', E)e^{-E/k_BT}dE}{\int \rho(n, E)e^{-E/k_BT}dE} \tag{13-52}$$

To evaluate the foregoing integrals, we have to consider several specific cases. The most detailed possibility is full knowledge of the molecular vibrational and rotational states before and after the collision, as in the V-R,T case noted earlier:

$$AB(v, J) + M \to AB(v', J') + M$$

Making use of the RRHO approximation of equation 13–35 gives

$$E_{vJ} = \hbar\omega_0\left(v + \tfrac{1}{2}\right) + BJ(J + 1)$$

and we have

$$\rho(v, J; E) = A_T(2J + 1)(E - E_{vJ})^{1/2} \tag{13-53}$$

In many experiments, the rotational states are thermalized or not individually resolved. For such a V-T process.

$$AB(v) + M \to AB(v') + M$$

and we have, as previously discussed,

$$\rho(v; E) = A_T \sum_{J=0}^{J^*(v)} (E - E_{vJ})^{1/2}(2J + 1)$$

or

$$\rho(v; E) = \frac{A_T}{B\hbar c}\left(\frac{2}{3}\right)(E - E_v)^{3/2} \tag{13-54}$$

For the V-V case,

$$AB(v = n) + AB(v = m) \rightarrow AB(v = n') + AB(v = m')$$

and we have

$$\rho(nJ, mL; E) = A_T(2J + 1)(2L + 1)(E - E_{nJ} - E_{mL})^{1/2}$$

where J and L are the rotational angular momenta of the two molecules. Thus,

$$\rho(n, m; E) = \sum_{L=0}^{L^*(m)} \sum_{J=0}^{J^*(n)} \rho(nJ, mL; E)$$

$$= A_T\left(\frac{4}{15}\right)(E - E_n - E_m)^{5/2} \tag{13-55}$$

Calculation of $k^0(T)$ from equation 13–52 involves integrals over products of $\rho(E)$ having the form of equation 13–53, 13–54, or 13–55, depending on the case under consideration. All of these integrals can be expressed in terms of modified Bessel functions.[20] The general form is

$$\int_0^\infty x^{\nu-1}(2y + x)^{\nu-1}e^{-x}dx = \pi^{-1/2}(2y)^{\nu-1/2}e^y\Gamma(\nu)K_{\nu-1/2} \tag{13-56}$$

On examination, this can be seen to cover all of the possibilities noted if we let $x = E/kT$, $y = \Delta \equiv (E_n - E_{n'})/2k_BT$, and $\nu = 3/2$ for V-R, T, 5/2 for V-T, and 7/2 for V-V. $K_n(y)$ is called the *modified Bessel function of the second kind of order n;* the coefficient $\Gamma(\nu)$ is the *gamma function,* which is the analytic continuation of the factorial function, i.e., $\Gamma(n) = (n - 1)!$ for integer n.

The results for the three cases we have been considering are as follows:

$$V\text{-}R, T: \quad k^0(vJ \rightarrow v'J'; T) = RA_T(k_BT)^{1/2}\left(\frac{2}{\pi^{1/2}}\right)(2J' + 1)\Delta e^\Delta K_1(\Delta) \tag{13-57}$$

$$V\text{-}T: \quad k^0(v \rightarrow v'; T) = \frac{8RA_T(k_BT)^{3/2}}{3Bhc\pi^{1/2}}\Delta^2 e^\Delta K_2(\Delta) \tag{13-58}$$

$$\text{with } \Delta = \frac{E_v - E_{v'}}{2k_BT}$$

$$V\text{-}V: \quad k^0(nm \rightarrow n'm'; T) = \frac{32}{15}RA_T(k_BT)^{5/2}\pi^{-1/2}\Delta^3 e^\Delta K^3(\Delta) \tag{13-59}$$

$$\text{with } \Delta = \frac{(E_n + E_m) - (E_n - E_{m'})}{2k_BT}$$

Let us examine the properties of these prior rates, with reference to Figure 13–2, which shows the modified Bessel functions $K_\nu(x)$. First, we see that $K_\nu(x)$ is a

(a)

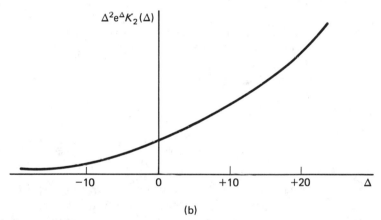

(b)

Figure 13–2. (a) Modified Bessel function $K_\nu(x)$ as a function of x, for $\nu = 1,2$, and 3. (b) Dependence of $k^0 (v \to v'; T)/A^0(T) = \Delta^2 e^{\Delta} K_2(\Delta)$ on $\Delta = (E_v - E_{v'})/2kT$.

monotonically decreasing function of x for $x > 0$ and any ν. Next, if we look at the behavior of the V-T prior rate, we see that, after dividing through by the Δ-independent factor, we have $k^0(v \to v'; T)/A^0(T) = \Delta \to 0(v' = v$, or $k_B T \gg \hbar\omega_0)$, and, using the asymptotic properties of Bessel functions,[20] we have

$$k^0(0; T) \to 2A^0(T) = T^{3/2}$$

which is independent of Δ. Similarly, for $\Delta \to \infty(k_B T \ll \hbar\omega_0)$, we have

$$k^0(\Delta; T) \to \left(\frac{\pi}{2}\right)^{1/2} A^0(T)\Delta^{3/2}$$

In other words, as is clear from Figure 13–2(b), the prior rate favors transferring as much energy as possible from internal motions into translation.

It is instructive to substitute $-\Delta$ for Δ in equation 13–58. Physically, this corresponds to vibrational activation ($T \to V$) as compared with deactivation ($V \to T$). Since K_2 is an even function,*

$$\Delta^2 K_2(\Delta) = (-\Delta)^2 K_2(-\Delta) = |\Delta|^2 K_2(|\Delta|)$$

We can rewrite equation 13–58 in terms of the absolute value of Δ as

$$k^0(v \to v'; T) = A^0(T)|\Delta|^2 K_2(|\Delta|) \exp [|\Delta| + (\Delta - |\Delta|)]$$

$$= k^0(|\Delta|; T) \exp (\Delta - |\Delta|)$$

The purpose of doing this becomes evident if we examine the last factor for $\Delta > 0$ ($v > v'$, i.e., deactivation $V \to T$) and for $\Delta < 0$ ($v < v'$, i.e., activation $T \to V$). For $\Delta > 0$, $\Delta = |\Delta|$ and $\exp (\Delta - |\Delta|) = \exp (0) = 1$. For $\Delta < 0$, $\Delta = -|\Delta|$ and $\exp (\Delta - |\Delta|) = \exp (-2|\Delta|)$. So for processes in which $\Delta v = 1$,

$$\frac{k^0_{\text{act}}}{k^0_{\text{deact}}} = \frac{k^0(v \to v + 1)}{k^0(v + 1 \to v)} = \exp \left(-2 \frac{E_{v+1} - E_v}{2k_B T} \right)$$

$$= e^{-\hbar\omega_0/k_B T}$$

and the correct detailed balancing relationship (equation 13–49) is obtained. The same holds true for all the other prior rates as well.

13.4.2 Surprisal Analysis

We next make the same maximal-entropy assertion as we did for the reactive case, namely,

$$k(n \to n') = k^0(n \to n') \exp \left[-\lambda_0 - \sum_{m=1}^{r} \lambda_m g_m(n, n') \right] \qquad (13\text{–}60)$$

where r may indeed be equal to unity (linear surprisal), but in principle may incorporate more than one independent piece of information. Now, what piece of information (i.e., what measure of the energy transfer) should we choose? The obvious one (for the V-T case) is $E_v - E_{v'} = 2k_B T\Delta$; since k^0 is symmetric in $|\Delta|$, the absolute value is especially convenient. Equation 13–60 then becomes

$$k(v \to v'; T) = k^0(v \to v'; T) \exp [-\lambda_0 - \lambda_v |E_v - E_{v'}|/k_B T - \ldots]$$

$$= k^0(\Delta; T) e^{-\lambda_0 - 2\lambda_v |\Delta|} - \ldots \qquad (13\text{–}61)$$

and the surprisal is

$$I(v, v'; T) = -\ln \frac{k(v \to v'; T)}{k^0(v \to v'; T)}$$

$$= I_0 + \frac{\lambda_v |E_v - E_{v'}|}{k_B T} + \ldots$$

$$= I_0 + \frac{\lambda_v |v - v'| \hbar\omega_0}{k_B T} + \ldots$$

*In general,[20] $K_\nu(-\Delta) = (-1)^\nu K_\nu(\Delta)$.

$$= I_0 + \Lambda_v \, |v - v'| + \dots \qquad (13\text{-}62)$$

Table 13–2 lists a number of systems for which surprisal analysis has been applied to V-T and V-V energy transfer data, with the plots for these systems given in Figures 13–3 through 13–7. It is clear from these and numerous additional examples that the "exponential gap law" (equation 13–62) provides a good representation of vibrational energy transfer properties. A large value of λ_v indicates a persistence of the original vibrational state, as is found in collisions with rare-gas atoms; small values of λ_v, on the other hand, are characteristic of a more nearly statistical distribution of final states, as may occur in reactive collisions (e.g., $Cl + Cl_2$, $Br + HBr$, $F + HF$). Note that the reduced surprisal parameter, $\Lambda_v = \hbar\omega_0\lambda_v/k_BT$, has values between 0.4 and 1.2 for all of the systems in Table 13–2; this suggests that a simple scaling behavior can be applied to V-T and V-V processes, which we shall exploit in the discussion of surprisal synthesis in section 13.5.

TABLE 13-2 SURPRISAL ANALYSES OF V-T AND V-V ENERGY TRANSFER

System	Figure	λ_v	$\Lambda_v = \dfrac{\hbar\omega_o}{k_BT}\lambda_v$	Reference
$I_2^*(v = 25) + M \rightarrow I_2^*(v') + M$ Trajectory calculations on a potential surface which match experimental data (reference 21): M = He,Ne,Ar,Kr,Xe	13–3	2.38	1.26	18,21
$Cl + Cl_2(v = 14) \rightarrow Cl_2(v') + Cl$ trajectory calculations; λ_v independent of T	13–4	0.51	0.38	19,23
$I + I_2(v) \rightarrow I_2(v') + I$ trajectory calculations		1.89	0.53	19,23
$Br + HBr (v = 5) \rightarrow Br + HBr (v')$ trajectory calculations, including both reactive and unreactive inelastic channels		0.088	1.12	19,24
$F + HF(v = 6) \rightarrow F + HF(v')$ trajectory calculations, including both reactive and unreactive inelastic channels	13–5(a)	0.11	1.33	19,25
$H + HF(v) \rightarrow HF(v - 1) + H$ trajectory calculations; Δ is linear in v because of anharmonicity*	13–5(b)			19,26
$Li_2^*(v = 4) + Ar \rightarrow Li_2^*(v') + Ar$ experimental data		1.32	0.57	19,27
$CO(v) + CO(0) \rightarrow CO(v - 1) + CO(1)$	13–6			19,28
$CO(v) + N_2(0) \rightarrow CO(v - 1) + N_2(1)$	13–7			19,29
$CO(v) + NO(0) \rightarrow CO(v - 1) + NO(1)$ experimental data; Δ is linear in $(v + 1)$ because of anharmonicity*	13–7			19,29

*See problem 13–1

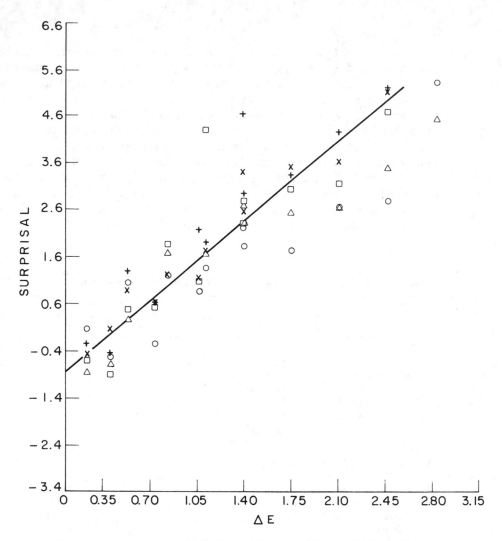

Figure 13–3. Surprisal plot for $I_2^*(v = 25) + M \rightarrow I_2^*(v') + M$ [M = He, Ne, Ar, Kr, Xe; $v' - v = \pm 1, \ldots, \pm 5$] from classical trajectory calculations from references 18 and 21. [From Rubinson and Steinfeld, *Chem. Phys. 4*, 467 (1974).]

13.4.3 Vibrational Deactivation in Polyatomic Molecules

Application of surprisal analysis techniques to systems more complex than diatomic molecules presents some additional difficulties since the more complex energy-level expressions lead to more complex prior rate coefficients. In the case of V-T deactivation of large polyatomic molecules, however, some simplifying approximations can be made which lead to closed-form expressions.[30] Since energy transfer processes of this type are important in many areas of kinetics, such as activation and deactivation

Figure 13–4. Surprisal plots for $Cl + Cl_2(v = 14) \rightarrow Cl_2(v') + Cl$ at $T = 700$ K, 1,100 K, and 1,500 K from classical trajectory calculations from references 19 and 23. [From Procaccia and Levine, *J. Chem. Phys. 63,* 4261 (1975).]

in unimolecular reactions (see chapter 11 and reference 31) and infrared multiple-photon dissociation (see chapter 14), we shall discuss these processes briefly here.

Consider a polyatomic molecule containing N atoms, initially excited to a high vibrational level by infrared laser excitation, chemical activation, or some other activation process. In collisions, some of this energy will be lost:

$$A^* + M \longrightarrow A^{(less^*)} + M$$

But the total energy

$$E = E_{vib} + E_{transl,rot} = E_{vib} + E_{T,R}$$

or

$$E = E'_{vib} + E'_{transl,rot} = E'_{vib} + E'_{T,R} \qquad (13–63)$$

is conserved, so that

$$\Delta E = E_{vib} - E'_{vib} = E'_{T,R} - E_{TR} \qquad (13–64)$$

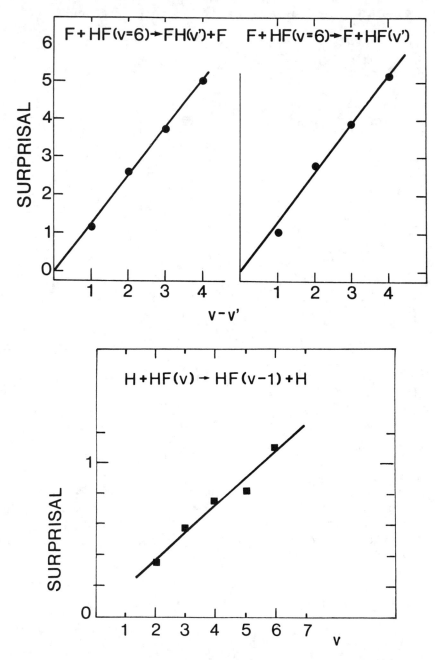

Figure 13–5. Surprisal plots for (a) F + HF(v = 6) → F + HF (v') and (b) H + HF(v) → HF(v − 1) + H from classical trajectory calculations from references 19, 25, and 26. [From Procaccia and Levine, *J. Chem. Phys. 63*, 4261 (1975).]

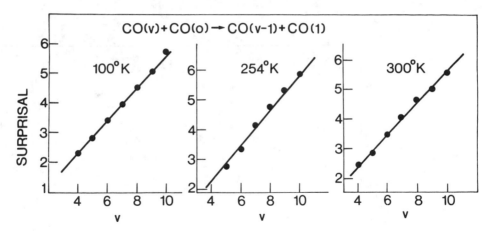

Figure 13–6. Surprisal plots for $CO(v) + CO(0) \rightarrow CO(v-1) + CO(1)$ at $T = 100$ K, 254 K, and 300 K; experimental results from reference 28. [From Procaccia and Levine, *J. Chem. Phys. 63*, 4261 (1975).]

For simplicity, suppose M is a structureless particle such as a rare-gas atom, so that the internal energy of M need not be considered. We seek to find the transfer probability at specified total energy $P(\Delta E \mid F_{\text{vib}}, E)$ or the thermally-averaged equivalent quantity, $P(E_{\text{vib}}, E'_{\text{vib}} \mid T)$.

As before, we assert that the prior probability is proportional to the final density of states, so that

$$P^0(\Delta E \mid E_{\text{vib}}, E) \propto \rho_{\text{transl,rot}}(E'_{T,R}) \cdot \rho_{\text{vib}}(E'_{\text{vib}})$$

For a diatomic molecule, $\rho_{\text{vib}} = 1$; but polyatomic molecules are characterized by a *vibrational density of states*, which can be approximated by a classical RRK expression (see Appendix 2),

$$\rho_{\text{vib}}(E) = \frac{1}{(s-1)!} \frac{E^{s-1}}{(h\nu)^s} \tag{13–65}$$

where $s = 3N - 6$ is the number of vibrational modes in the molecule. While it would be more accurate to use an exact or Whitten-Rabinovitch[32] density-of-states expression, use of equation 13–65 displays the essential features of the result and gives an approximate analytic expression for ρ_{vib}.

For an atom + polyatomic system,

$$\rho_{\text{transl,rot}}(E'_{T,R}) \propto (E'_{T,R})^2 \tag{13–66}$$

with one factor of $(E'_{T,R})^{1/2}$ coming from relative translation and three factors coming from three-dimensional rotation of the polyatomic molecule. Thus,

$$P^0(\Delta E \mid E_{\text{vib}}, E) \propto (E'_{T,R})^2 (E'_{\text{vib}})^{s-1}$$

or

$$P^0(\Delta E \mid E_{\text{vib}}, E) = (E'_{T,R})^2 (E - E'_{T,R})^{s-1} \tag{13–67}$$

Now consider the limiting case $E_{\text{vib}} \gg E_{T,R}$, which is appropriate for high vibrational

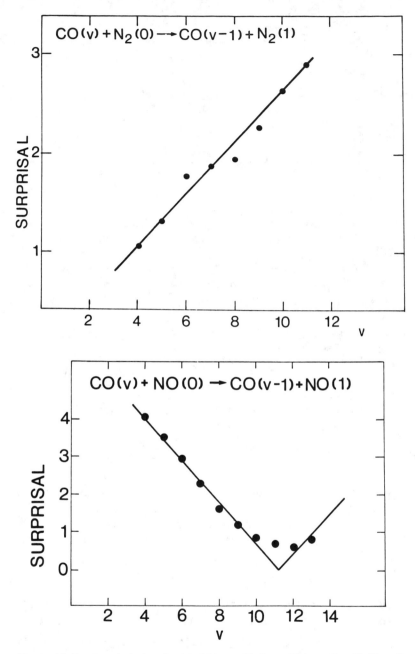

Figure 13–7. Surprisal plots for (a) $CO(v) + N_2(0) \rightarrow CO(v - 1) + N_2(1)$ and (b) $CO(v) + NO(0) \rightarrow CO(v - 1) + NO(1)$; experimental results from reference 29. [From Procaccia and Levine, *J. Chem. Phys. 63*, 4261 (1975).]

excitation at low to moderate kinetic temperatures. We then find by differentiating equation 13–67 that for the prior distribution,

$$E'_{T,R}(\text{most probable}) = \left(\frac{2}{s+1}\right) E_{\text{vib}} \qquad (13\text{–}68)$$

and

$$\langle E'_{T,R}\rangle = \int (E'_{T,R})^3 (E_{\text{vib}} - E'_{T,R})^{s-1} dE'_{T,R}$$

or

$$\langle E'_{T,R}\rangle = \left(\frac{3}{s+3}\right) E_{\text{vib}} \qquad (13\text{–}69)$$

Both of these moments of the distribution are much smaller than E_{vib} for large s. In effect, the large density of states associated with the vibrational degrees of freedom of a polyatomic molecule *biases* the prior probabilities toward energy remaining in vibration.

As before, we must take an average over a thermal distribution of collision energies, to give

$$P^0(\Delta E \mid E_{\text{vib}}, T) = \int_0^\infty p(E'_{T,R} \mid T) P^0(\Delta E \mid E_{\text{vib}}, E'_{T,R})\, dE'_{T,R}$$

With a Maxwellian distribution

$$p(E'_{T,R} \mid T) = N(T)(E'_{T,R})^2 e^{-E'_{T,R}/k_B T}$$

this yields

$P^0(\Delta E \mid E_{\text{vib}}, T)$

$$\propto \left[1 + \frac{|\Delta E|}{2k_B T} + \frac{1}{3}\left(\frac{|\Delta E|}{2k_B T}\right)^2\right]\left[1 - \frac{\Delta E}{E_{\text{vib}}}\right]^{s-1} \times \begin{cases} e^{\Delta E/2k_B T}, & \Delta E < 0 \\ 1, & \Delta E > 0 \end{cases} \qquad (13\text{–}70)$$

The last switching term in equation 13–70 simply gives the correct detailed-balancing ratio between activation ($\Delta E < 0$) and deactivation ($\Delta E > 0$). The dependence of $P^0(\Delta E \mid E_{\text{vib}}, T)$ on ΔE and E_{vib} is shown in Figure 13–8. Its functional form is very nearly that of a displaced and skewed Gaussian, and it is of interest to note that a functional dependence of this type has also been obtained from classical trajectory calculations of V-T exchange in polyatomics.[33] The mean $\langle \Delta E \rangle$ and the width of the distribution increase as the initial E_{vib} increases; note that, for any initial E_{vib}, there is a significant probability density at $\Delta < 0$, i.e., for additional vibrational excitation of the polyatomic molecule. This behavior is quite different from that of $P^0(\Delta) \propto \Delta^2 e^\Delta K_2(\Delta)$ for a diatomic molecule, also shown in Figure 13–8, which increases without limit for large Δ.

The prior distribution given in equation 13–70 predicts that $\langle \Delta E \rangle / E_{\text{vib}} \approx 0.1$. While early data cited in Jensen et al.[30] and Tardy and Rabinovitch[31] seemed to be consistent with this prediction, more recent, improved experimental techniques have

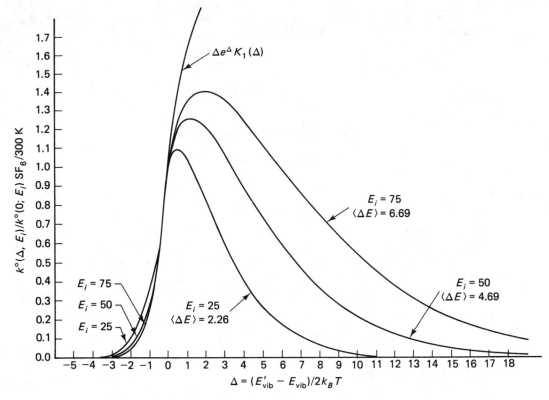

Figure 13–8. Prior vibrational energy transfer rates for SF_6 at $T = 300$ K; the rates are normalized by the rate for $\Delta E = 0$ at each E_i. Also shown is the unweighted prior, $\Delta e^{\Delta} K_1(\Delta)$, for comparison (from reference 30). [From Jensen, Steinfeld, and Levine, *J. Chem. Phys.* **69**, 1432 (1978).]

shown that energy transfer is clearly nonstatistical. A few of these results are summarized in the following table:

System	Initial E_{vib} kJ/mole	$< \Delta E >_{expt}$, kJ/mole	$\dfrac{\Delta E}{E_{vib}}$	λ_1	Reference
Azulene($s = 48$)	209	1.34	0.06	} $0.75 \sim 0.80$	34
	366	2.1	0.06		
Toluene($s = 39$)	619	1.7	0.03	≈ 1.0	35

The surprisal parameters in these experiments were derived from an exponential-gap representation,

$$P(\Delta E \mid E_{vib}, T) = P^0(\Delta E \mid E_{vib}, T) \exp\left(-\lambda_0 - \lambda_1 \frac{|E_{vib} - E'_{vib}|}{k_B T}\right) \qquad (13\text{--}71)$$

with the use of a sum rule,

$$\langle \Delta E_{vib} \rangle = \frac{\Sigma_{\text{final states}}\, P\,(\Delta E \mid E_{vib},\, T)(E_{vib} - E'_{vib})}{\Sigma_{\text{final states}}\, P\,(\Delta E \mid E_{vib},\, T)} \qquad (13\text{--}72)$$

13.4.4 Surprisal Analysis of Rotational Energy Transfer

State-to-state rotational energy transfer (R-T) has also been analyzed using information-theoretical methods.[36] By the usual prescription,

$$k^0(J \to J'; E) = R\rho\,(J', E)$$

$$= R(2J' + 1)\frac{\mu^{3/2}}{2^{1/2}\pi^2\hbar^3}(E_{T'})^{1/2}$$

or

$$k^0(J \to J'; E) = R(2J' + 1)A_T(E - E_{R'})^{1/2} \qquad (13\text{--}73)$$

The thermally averaged prior rate coefficient is then

$$k^0(J \to J'; T) = \int k^0(J \to J'; E)p\,(E; T)\,dE$$

or

$$k^0(J \to J'; T) = RA_T(k_BT)^{1/2}\left(\frac{2}{\pi^{1/2}}\right)(2J' + 1)\Delta e^{\Delta}K_1(\Delta) \qquad (13\text{--}74)$$

where $\Delta = (E_J - E_{J'})/2k_BT$. This is just a special case of the V-R,T prior rate of equation 13–57 in which v remains constant.

The maximum-entropy principle leads to an *exponential-gap law* (EGL),

$$k(J \to J'; E) = k^0(J \to J'; E)\exp\left[-\lambda_0 - \lambda_{rot}\left(\frac{\Delta E_{rot}}{E}\right)\right] \qquad (13\text{--}75a)$$

or

$$k(J \to J'; T) = k^0(J \to J'; T)\exp\left[-\lambda_0 - \lambda_{rot}\left(\frac{\Delta E_{rot}}{k_BT}\right)\right] \qquad (13\text{--}75b)$$

At first, the EGL appeared to give a satisfactory representation of R-T data for a number of systems.[18,21] However, as more precise data became available, significant deviations from the EGL were found. The first such case was in molecular beam studies of HF,[37] in which HF molecules in the molecular beam at specified translational energies (15–60 kJ/mole) were excited on a specific $(v' = 1, J') \leftarrow (v'' = 0, J'')$ transition by a HF chemical laser. The excited molecules were allowed to collide with low-pressure Ne, Ar, or Kr gas, and the resulting HF$(v' = 1, J'_F)$ distribution was monitored by resolved infrared fluorescence. The EGL predicts that the cross section for the $J' \to J'_F$ process is given by

Figure 13–9. Scaling representations of R-T cross-section data in HF + Ar collisions at $E = 37$ kJ/mole, from reference 37. (a) Exponential-gap law; (b) Statistical power-gap law. [From Barnes et al., *J. Chem. Phys.* **76**, 913 (1982).]

$$\sigma(J' \rightarrow J'_F; E) \propto (2J' + 1)\left(\frac{E_{T'}}{E_T}\right)^{1/2} e^{-\theta|\Delta E_{\text{rot}}|} \qquad (13\text{--}76)$$

where the factor $(E_{T'}/E_T)^{1/2}$ arises from the transformation of equation 13–75a from rate coefficients to cross sections. As can be seen from Figure 13–9(a), there are significant deviations from equation 13–76 at low $|\Delta E_{\text{rot}}|$. In fact, Polanyi[37] found that a much better representation of his data was given by a *statistical power-gap* (SPG) law,

$$\sigma(J' \rightarrow J'_f; E) \propto (2J' + 1)\left(\frac{E_{T'}}{E_T}\right)^{1/2} |\Delta E_{\text{rot}}|^{-\gamma} \qquad (13\text{--}77)$$

as shown in Figure 13–9(b), and similar conclusions were reached for a number of other systems, such as R-T exchange in electronically excited sodium dimer,[38] given by

$$Na_2^*(v', J') + M \longrightarrow Na_2^*(v', J'_f) + M$$

and studied by laser-excited fluorescence spectroscopy. It is interesting to note that the SPG (equation 13–77) can be readily obtained by applying entropy maximization to $\ln |\Delta E_{\text{rot}}|$ instead of to $|\Delta E_{\text{rot}}|$, but there is no *a priori* justification for doing so in the context of rotational energy transfer theory.

13.4.5 Angular-Momentum-Based Scaling Laws

The difficulty encountered in using the EGL for rotational energy transfer was alleviated by resorting to a scaling law based on angular momentum rather than energy constraints. Such a scaling law can be derived from the "infinite order sudden" (IOS) model,[39] which is based on "sudden approximation" in scattering theory. This approximation requires that the collision time τ_c, which can be estimated from

$$\tau_c \approx \frac{\text{collision distance } \ell_c}{\langle v \rangle_{\text{thermal}}}$$

be much less than the rotational period τ_{rot}. In this case, a suitable model is that of a nonrotating diatomic molecule colliding with an atomic collision partner, and the probability amplitudes $p(j_i = 0 \rightarrow j_f = \ell)$ can be calculated using strong-coupling theory.[40] The results are then phase-averaged over rotational angles to give

$$k^{\text{IOS}}(J_i \rightarrow J_f) =$$

$$(2J_f + 1)\exp\left[\frac{E_{J_i} - E_{J_f}}{k_B T}\right] \sum_\ell \left[\begin{pmatrix} J_i & \ell & J_f \\ 0 & 0 & 0 \end{pmatrix}^2 (2\ell + 1)k(\ell \rightarrow 0)\right] \qquad (13\text{--}78)$$

The exponential factor in equation 13–78 ensures detailed balancing, since for deactivation $(J_i > J_f)$, $J\rangle = J_i$ and $\exp(0) = 1$, while for rotational activation $(J_i < J_f)$, $J\rangle = J_f$ and the factor has the value $\exp(-\Delta E_{\text{rot}}/k_B T)$. The basis rates $k(\ell \rightarrow 0)$, which have the form $a[\ell(\ell + 1)]^{-\gamma}$, are multiplied by 3–$j$ or Clebsch-Gordan factors[41] to obtain rotational averaging. These in turn form the basis of the SPG law, since $E_{\text{rot}}/B = \ell(\ell + 1)$. There is also an "energy-corrected sudden" (ECS) approximation, which corrects for the finite duration of τ_c.

Figure 13–10. R-T data from reference 42 for $I_2^*(v_i = 13, j_i = 41, 81, 91,$ and $113) + Xe \rightarrow I_2^*(v_f = 13, j_f)$, compared with the EGL scaling relationship of equation 13–75b. [From J. Steinfeld and P. Ruttenberg, "Scaling Laws for Inelastic Collision Processes in Diatomic Halogens," JLIA Information Center Report No. 23 (Boulder, CO, 1983).]

The various scaling laws have been tested with very precise data on R-T energy transfer in electronically excited I_2.[42] As is clear from Figure 13–10, the EGL is really a forced fit to the actual data and deviates very badly at low $|J_f - J_i|$. The SPG and ECS laws, on the other hand, give a good representation of the data for all values of $J_f - J_i$ (see Figures 13–11, 13–12); in fact, the ECS law, which incorporates

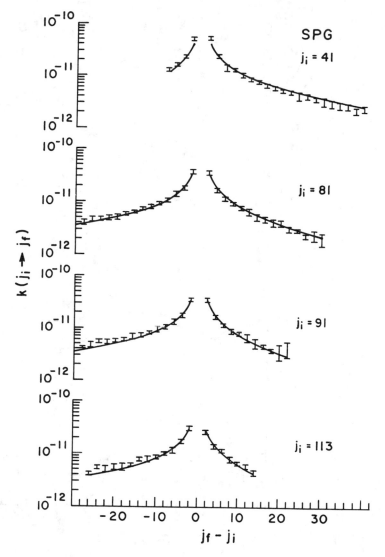

Figure 13–11. Data as in Figure 13–10, compared with the SPG scaling relationship. [From J. Steinfeld and P. Ruttenberg, "Scaling Laws for Inelastic Collision Processes in Diatomic Halogens," JLIA Information Center Report No. 23 (Boulder, CO, 1983).]

additional dynamical information about the range of the interaction potential, actually gives a slightly better statistical fit to the data. The angular-momentum-based scaling laws may be considered firmly established for R-T energy transfer. The fact that several hundred independently measured rate coefficients can be fit with two or three parameters is an impressive example of the power of the information-theoretical approach.

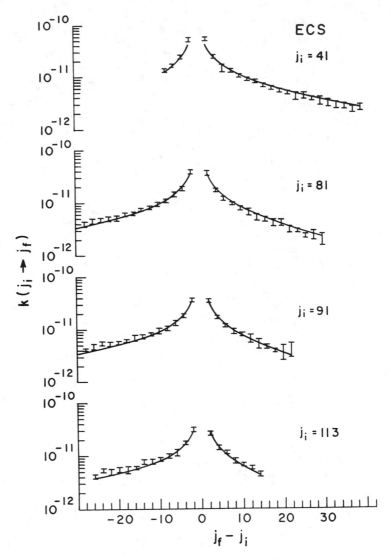

Figure 13–12. Data as in Figure 13–10, compared with the ECS scaling relationship [from reference 43]. [From J. Steinfeld and P. Ruttenberg, "Scaling Laws for Inelastic Collision Processes in Diatomic Halogens," JLIA Information Center Report No. 23 (Boulder, CO, 1983).]

13.5 SURPRISAL SYNTHESIS

At this point an issue may occur to the reader which has often been raised as a criticism of information theory, viz., that while a procedure such as surprisal analysis is clearly capable of systematizing and condensing a large body of data, it has essentially no predictive power. Rate coefficients or probabilities can indeed be expressed as products of prior probabilities and surprisal factors, as in equations 13–30 and

13–60, but the deviations from the prior probabilities contain an arbitrary number of constraints which can be known only after experimental data have been analyzed. The critical question is whether a predictive norm exists which will permit us to specify the constraints in advance of actually carrying out the measurements. The procedures for finding such a norm are known as *surprisal synthesis*, the last topic to be discussed in this chapter.

Consider, as a specific example, vibrationally inelastic atom-molecule collisions. From analyzing numerous data sets, we have found that the state-to-state probability can be represented as

$$P(v, v') = P^0(v, v') \exp\left[-\lambda_0 - \lambda_v \left| E_v - E_v' \right| / k_B T\right]$$

so that

$$I(v, v' \mid T) = -\ln \frac{P(v, v')}{P^0(v, v')} = \lambda_0(T) + \frac{\lambda_v \left| E_v - E_v' \right|}{k_B T}$$

The question we now address is whether the parameters λ_0 and λ_v can be obtained without reference to a set of measured $P(v, v')$.

The approach to such a synthesis involves the use of *sum rules*. For a given system $AB + M$, we have two such rules, viz.,

$$\sum_v \sum_{v'} \left| E_v - E_v' \right| P(v, v') = \text{some function of } T \qquad (13\text{--}79)$$

and normalization, i.e.,

$$\sum_v \sum_{v'} P(v, v') = 1 \qquad (13\text{--}80)$$

where $P(v, v') = k(v \to v'; T)/k(v; T)$. In the next chapter we shall derive a master equation for populations which has the form

$$\frac{dn_v(t)}{dt} = -\sum_{v'} k(v \to v'; T)n_v(t)\rho_M + \sum_{v'} k(v' \to v; T)n_{v'}(t)\rho_M$$

where ρ_M is the total density of colliders. If we then define a *bulk average vibrational energy*

$$\langle\!\langle V \rangle\!\rangle = \frac{\sum\limits_v E_v n_v(t)}{\sum\limits_v n_v(t)} = \frac{\sum\limits_v E_v n_v(t)}{N_{\text{total}}} \qquad (13\text{--}81)$$

then use of the master equation allows us to write

$$\frac{d}{dt} \langle\!\langle V \rangle\!\rangle = \frac{1}{\rho_M} \sum_v E_v \frac{dn_v(t)}{dt}$$

$$= -\sum_v \sum_{v'} E_v k(v \to v'; T)n_v(t) + \sum_v \sum_{v'} E_v k(v \to v'; T)n_{v'}(t)$$

$$= \sum_v \sum_{v'} (E_{v'} - E_v)k(v \to v'; T)n_v(t)^*$$

*The reduction from the preceding line is accomplished by exchanging v and v' in the second double summation.

Sec. 13.5   Surprisal Synthesis

or

$$\frac{d}{dt}\langle\langle V\rangle\rangle = \sum_v n_v(t)\,\frac{d}{dt}\,\langle V\rangle \qquad (13\text{–}82)$$

where $d\langle V\rangle/dt = \sum_v (E_{v'} - E_v)k(v \to v'; T)$.

We now appeal to experimental information on bulk systems and assert that $\langle V\rangle$ relaxes exponentially to some equilibrium value, i.e.,

$$\frac{d}{dt}\langle V\rangle = \alpha(\langle\langle V^{eq}\rangle\rangle - E_v) \qquad (13\text{–}83)$$

In that case,

$$\frac{d}{dt}\langle\langle V\rangle\rangle = \sum_v n_v(t)\cdot\alpha(\langle\langle V^{eq}\rangle\rangle - E_v) = -\alpha(\langle\langle V\rangle\rangle - \langle\langle V^{eq}\rangle\rangle) \qquad (13\text{–}84)$$

and α is seen to be equal to the reciprocal of τ_{vib}, the bulk vibrational relaxation time. We then write the sum rule, equation 13–79, as

$$\sum_{v'} (E_{v'} - E_v)k(v \to v'; T) = \beta - \alpha E_v$$

where $\beta/\alpha = \langle\langle V^{eq}\rangle\rangle$. There may be a number of different alternative sets of $k(v \to v'; T)$ that satisfy this sum rule, but the maximal-entropy postulate allows us to specify one of these sets uniquely. This is the set that maximizes the rate of entropy production in the systems undergoing relaxation.

The entropy deficiency, introduced on p. 433, can be associated with a set of rates or transition probabilities:

$$\Delta S[v'|v] = R\sum_{v'} P(v'|v)\ln P(v'|v) - R\sum_{v'} P(v'|v)\ln P^0(v'|v)$$

$$= R\sum_{v'} P(v'|v)\ln\frac{P(v'|v)}{P^0(v'|v)}$$

or

$$\Delta S[v'|v] = R\sum_{v'} P(v'|v)I(v'|v, T) \qquad (13\text{–}85)$$

We seek to *minimize* $\Delta S[v'|v]$, thereby maximizing the entropy of the final-state distribution, subject to the constraints

$$\sum_{v'} P(v'|v) = 1$$

and

$$\sum_{v'} |E_{v'} - E_v| P(v'|v) = \sum_{v'} |E_{v'} - E_v|\frac{k(v \to v'; T)}{k(v; T)} = F(v, T)$$

Using the variational principle on the Lagrange function

$$\mathcal{L} = \Delta S[v'|v] + R[(\lambda_0 - 1) \sum_{v'} P(v'|v) + \frac{\lambda_v}{k_B T} \sum_{v'} |E_{v'} - E_v| P(v'|v)]$$

gives

$$\delta\mathcal{L} = R \sum_{v'} \delta P(v'|v)\{\ln \frac{P(v'|v)}{P^0(v'|v)} + \lambda_0 + \frac{\lambda_v}{k_B T}|E_{v'} - E_v|\}$$

and we recover the linear-surprisal form,

$$-\ln \frac{P(v'|v)}{P^0(v'|v)} = I(v'|v, T) = \lambda_0 + \frac{\lambda_v |E_{v'} - E_v|}{k_B T}$$

or

$$P(v'|v) = P^0(v'|v) \exp[-\lambda_0 - \lambda_v |E_{v'} - E_v|/k_B T]$$

That is, the maximum-entropy set of rates which satisfies the sum rule given in equation 13–79 is just the linear-surprisal set which has appeared so often as a result of surprisal analysis. Since this sum rule also leads to exponential relaxation of bulk vibrational energy, we proceed to use that information to find the microscopic set of rates.

An example will best illustrate how this is done. Clearly, the quantity to be determined is the surprisal parameter λ_v. We can rewrite the sum rule expressed by equations (13–82) and (13–83) as a ratio of sums over v' for two different values of v:

$$R_{v, v''}(\lambda_v) = \frac{\sum\limits_{v'} (E_{v'} - E_v)k^0(v \to v'; T)e^{\lambda_v|E_{v'} - E_v|/k_B T}}{\sum\limits_{v'} (E_{v'} - E_{v''})k^0(v'' \to v'; T)e^{-\lambda_v|E_{v'} - E_{v''}|/k_B T}}$$

$$= \frac{\langle\langle V^{eq}\rangle\rangle - E_v}{\langle\langle V^{eq}\rangle\rangle - E_{v''}}$$

The energy levels E_v are known from the spectroscopy of the molecule, the prior rates k^0 may be calculated from equation 13–58, and the equilibrium $\langle\langle V^{eq}\rangle\rangle$ may be calculated from statistical mechanics. Since no simple analytic solution for λ_v can be extracted, a numerical or graphical solution is necessary.

Now consider the vibrational relaxation process

$$\text{Li}_2^*(v = 4) + \text{Ar} \to \text{Li}_2^*(v') + \text{Ar}(T = 900 \text{ K})$$

discussed in the preceding section. For the electronically excited $\text{Li}_2^* B\,^1\Pi_u$ state, the vibrational energies are given by the spectroscopic expression

$$E_v = \omega_e\left(v + \frac{1}{2}\right) - \omega_e x_e\left(v + \frac{1}{2}\right)^2$$

$$= 269.7\left(v + \frac{1}{2}\right) - 2.75\left(v + \frac{1}{2}\right)^2 \qquad (\text{in cm}^{-1})$$

The average energy $\langle\langle V^{eq}\rangle\rangle$ can be estimated well enough from the harmonic-oscillator expression:

$$\langle\langle V^{eq}\rangle\rangle = \frac{\theta_v}{2} + \frac{\theta_v}{\exp(\theta_v/T) - 1} \qquad \left(\theta_v = \frac{\hbar c\omega_e}{k_B}\right)$$

$$= 634\ cm^{-1}\ at\ T = 900\ K$$

Choose any two levels, say $v = 2$ ($E_v = 657.1\ cm^{-1}$) and $v'' = 4$ ($E_v = 1157.9\ cm^{-1}$). For these values,

$$R_{v,v''} = \frac{\langle\langle V^{eq}\rangle\rangle - E_v}{\langle\langle V^{eq}\rangle\rangle - E_{v''}} = \frac{634 - 657.1}{634 - 1157.9} = 0.044$$

The remaining step is to compute $R_{v,v''}(\lambda_v)$ as a function of λ_v for this system and find that value of λ_v which gives $R_{v,v''} = 0.044$. This is done in Figure 13–13, from which a value of $\Lambda_v = \hbar c\omega_e/k_B T = 0.545$ is obtained; the value obtained from surprisal analysis of experimental data is $\Lambda_v = 0.57$, which is quite close. The relative scaling of the V-T rates is determined in this way; to obtain an absolute magnitude, it is sufficient to appeal to the normalization condition,

$$\sum_{v'} k^0(v \rightarrow v'; T)e^{-\lambda_v|E_{v'}-E_v|/k_B T} = e^{\lambda_0}$$

along with the previously derived sum rule,

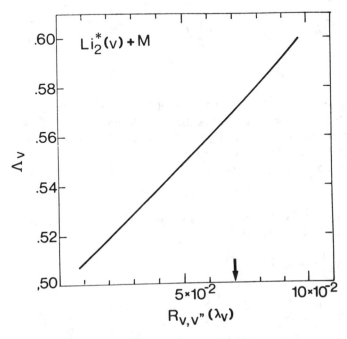

$$\sum_{v'} (E_{v'} - E_v)k(v \to v'; T) = \beta - \alpha E_v$$

with α set equal to the reciprocal of the bulk relaxation time τ_{vib}. If that quantity is not available, determination of a single V-T rate will suffice to set the value of the normalization factor, $\exp(\lambda_0)$.

It may be suggested that the accuracy of the surprisal synthesis for $Li_2^* + Ar$ is a fortuitous coincidence. A second example should alleviate such skepticism. Consider the system

$$Cl + Cl_2(v = 14) \longrightarrow Cl_2(v') + Cl \qquad (T = 1,100 \text{ K})$$

for which state-to-state rates are available from classical trajectory calculations[23] and τ_{vib} can be deduced from ultrasonic relaxation data. For this system,

$$E_v = 565\left(v + \frac{1}{2}\right) - 4\left(v + \frac{1}{2}\right)^2 \text{ cm}^{-1}$$

and

$$\langle\langle V^{eq}\rangle\rangle = 799 \text{ cm}^{-1}$$

and for $v = 5$, $v'' = 7$,

$$R_{v,v''} = \frac{799 - 2896}{799 - 4012} = +0.653$$

From the plot of $R_{v,v''}(\lambda_v)$ vs. λ_v in Figure 13–14, we obtain $\Lambda_v = 0.375$; surprisal analysis of the trajectory results yields $\Lambda_v = 0.38$.

We thus see that the relative scaling of V-T rates in atom–diatomic molecule

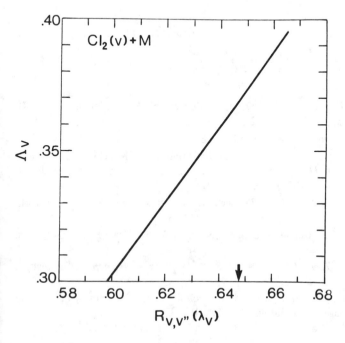

Figure 13–14. $R_{v, v''}$ vs. Λ_v plot for $Cl_2(v) + Cl$ collisions, for surprisal synthesis of V-T rates. [From Procaccia and Levine, *J. Chem. Phys.* **63**, 4261 (1975).]

collisions, which can be expressed by the single parameter λ_v, is quite well determined by the energy-level structure of the diatomic molecule. Knowledge of a single piece of kinetic data, such as a bulk relaxation time, then suffices to determine the magnitudes of all of the V-T rates.

13.6 CONCLUSION

We have seen how the judicious use of information theory, in the form of surprisal analysis and synthesis, can systematize a vast amount of molecular dynamics data and uncover correlations between seemingly unrelated phenomena. In many cases, though not all, selective energy consumption and specific energy release in state-to-state chemical reactions can be well represented by a single dynamical constraint superimposed on a statistical prior distribution. It is thus tempting to conclude that much of the vast amount of experimental and theoretical work in molecular dynamics described in earlier chapters has been unnecessary: since the results can be described by one- or two-parameter distributions, all that would have been necessary would be to measure one or two microscopic cross sections very accurately and then use surprisal synthesis to generate the rest of the data. It must be remembered, though, that information theory is not predictive, but is rather a methodology for systematizing and interpreting data. The experimental and theoretical results were essential for empirically establishing the validity of linear-surprisal representations, and instances where such a representation fails are certainly well known. On the basis of this prior work, we can now proceed to use information theory to organize and even predict dynamical results with a good degree of confidence, and innovative applications of the theory continue to be found. The bibliography lists numerous examples of such applications.

REFERENCES

1. J. L. Kinsey, *J. Chem. Phys. 54,* 1206 (1971); A. Ben-Shaul, R. D. Levine, and R. B. Bernstein, *J. Chem. Phys. 57,* 5427 (1972).

2. R. D. Levine, *Ann. Rev. Phys. Chem. 29,* 59 (1978).

3. R. D. Levine and J. L. Kinsey, "Information Theory Approach to Molecular Collisions," in *Atom-Molecule Collision Theory: A Guide for the Experimentalist,* edited by R. B. Bernstein (New York: Plenum, 1978), pp. 693–750.

4. W. Feller, *An Introduction to Probability Theory and its Applications,* 2d ed. (New York: Wiley, 1957).

5. P. Franklin, *Methods of Advanced Calculus* (New York: McGraw-Hill, 1944), pp. 263–266.

6. C. E. Shannon, *Bell Syst. Tech. J. 27,* 379 (1948); N. Wiener, *Cybernetics* (Cambridge, MA: MIT Press, 1948).

7. "Everything is determined even though freedom of choice exists." *Ethics of the Fathers* 3:19, Tractate *Nezikin,* Babylonian Talmud (ca. 200 B.C.E.).

8. F. B. Hildebrand, *Methods of Applied Mathematics* (Englewood Cliffs, N.J.: Prentice-Hall, 1952), pp. 120–226.

9. D. A. McQuarrie, *Statistical Thermodynamics* (Mill Valley, CA: University Science Books, 1985).

10. K. P. Huber and G. Herzberg, *Constants of Diatomic Molecules* (New York: Van Nostrand Reinhold, 1979), pp. 304–305.

11. A. Ben-Shaul, R. D. Levine, and R. B. Bernstein, *Chem. Phys. Letts. 15,* 160 (1964).

12. J. Steinfeld, *Molecules and Radiation,* 2d ed. (Cambridge, MA: MIT Press, 1984), pp. 308–313, 318–320.

13. Ibid., pp. 429–430.

14. Ibid., pp. 434–437.

15. E. E. Marinero, C. T. Rettner, and R. N. Zare, *J. Chem. Phys. 80,* 4142 (1984).

16. D. P. Gerrity and J. J. Valentini, *J. Chem. Phys. 81,* 1298 (1984).

17. S. H. SuckSalk, C. K. Lutrus, and D. A. Reago, Jr., *Phys. Rev. A35,* 1074 (1987).

18. M. Rubinson and J. I. Steinfeld, *Chem. Phys. 4,* 467 (1975).

19. I. Procaccia and R. D. Levine, *J. Chem. Phys. 63,* 4261 (1975).

20. M. Abramowitz and I. A. Stegun, *Handbook of Mathematical Functions* (New York: Dover Publications, 1965), pp. 374–379, 417–429.

21. M. Rubinson, B. Garetz, and J. I. Steinfeld, *J. Chem. Phys. 60,* 3082 (1974).

22. J. I .Steinfeld and W. Klemperer, *J. Chem. Phys. 42,* 3475 (1965).

23. D. L. Thompson, *J. Chem. Phys. 60* 455 (1974).

24. J. M. White and D. L. Thompson, *J. Chem. Phys. 61,* 719 (1974).

25. R. L. Wilkins, *J. Chem. Phys. 59,* 698 (1973).

26. R. L. Wilkins, *J. Chem. Phys. 58,* 3038 (1973).

27. G. Ennen and C. Ottinger, *J. Chem. Phys. 3,* 404 (1974).

28. I. W. M. Smith, "Vibrational Relaxation in Small Molecules," in *Molecular Energy Transfer,* edited by R. D. Levine and J. Jortner (London: Wiley, 1975), pp. 85–113.

29. G. Hancock and I. W. M. Smith, *Appl. Opt. 10,* 1827 (1971).

30. C. C. Jensen, J. I. Steinfeld, and R. D. Levine, *J. Chem. Phys. 69,* 1432 (1978).

31. D. C. Tardy and B. S. Rabinovitch, *Chem. Revs. 77,* 369 (1977).

32. G. Z. Whitten and B. S. Rabinovitch, *J. Chem. Phys. 38,* 2466 (1963).

33. R. G. Gilbert, *J. Chem. Phys. 80,* 5501 (1984).

34. M. J. Rossi, J. R. Pladziewicz, and J. R. Barker, *J. Chem. Phys. 78,* 6709 (1983).

35. H. Hippler, J. Troe, and H. J. Wendelken, *J. Chem. Phys. 78,* 6709 (1983.)

36. R. D. Levine, R. B. Bernstein, P. Kabana, I. Procaccia, and E. T. Upchurch, *J. Chem. Phys. 64,* 796 (1976).

37. J. A. Barnes, M. Keil, R. E. Kutina, and J. C. Polanyi, *J. Chem. Phys. 76,* 913 (1982).

38. T. A. Brunner, N. Smith, A. W. Karp, and D. E. Pritchard, *J. Chem. Phys. 74,* 3324 (1981).

39. A. DePristo, S. D. Augustin, R. Ramaswamy, and H. Rabitz, *J. Chem. Phys. 71,* 850 (1979).

40. D. Secrest, "Rotational Excitation I: The Quantal Treatment," in *Atom-Molecule Collision Theory, op. cit.,* pp. 265–299.

41. M. E. Rose, *Elementary Theory of Angular Momentum* (New York: Wiley, 1957); A. R.

Edmonds, *Angular Momentum in Quantum Mechanics* (Princeton, N.J.: Princeton University Press, 1957).

42. S. L. Dexheimer, M. Durand, T. A. Brunner, and D. E. Pritchard, *J. Chem. Phys. 78,* 4996 (1982).

43. J. I. Steinfeld and P. Ruttenberg, "Scaling Laws for Inelastic Collision Processes in Diatomic Halogens," *JILA Information Center Report No. 23* (Boulder, CO, 1983).

BIBLIOGRAPHY

Information Theory: Reviews, Fundamentals and Mathematical Techniques

ALHASSID, Y., and R. D. LEVINE, *J. Chem. Phys. 67,* 4321 (1977).

ALHASSID, Y., N. AGMON, and R. D. LEVINE, *Chem. Phys. Letts. 53,* 22 (1978).

AGMON, N., Y. ALHASSID, and R. D. LEVINE, *J. Comput. Phys. 30,* 250 (1979).

LEVINE, R. D., and A. BEN-SHAUL, "Thermodynamics of Molecular Disequilibrium," in *Chemical and Biochemical Applications of Lasers,* edited by C. B. Moore (New York: Academic Press, 1977), pp. 145–197.

Application to Reactive Processes

LEVINE, R. D., and R. B. BERNSTEIN, *Chem. Phys. Letts. 22,* 217 (1973).

LEVINE, R. D., and R. KOSLOFF, *Chem. Phys. Letts. 28,* 300 (1974).

LEVINE, R. D., and J. MANZ, *J. Chem. Phys. 63,* 4280 (1975).

POLLAK, E., and R. D. LEVINE, *Chem. Phys. Letts. 39,* 199 (1976).

LEVINE, R. D., "Energy Consumption and Energy Disposal in Elementary Chemical Reactions," in *The New World of Quantum Chemistry,* edited by B. Pullman and R. Parr (Dordrecht: Reidel, 1976), pp. 103–129.

POLANYI, J. C., and J. L. SCHREIBER, *Chem. Phys. 31,* 113 (1978).

Application to Electronically Nonadiabatic Processes

LEVINE, R. D., "The Formation of Electronically Excited Products in Chemical Reactions: An Information Theoretic Analysis," in *Electronic Transition Lasers,* edited by J. I. Steinfeld (Cambridge, Mass.: M.I.T. Press, 1976), pp. 251–267.

Application to Inelastic Collision Processes

ALEXANDER, M. H., *Chem. Phys. Letts. 40,* 267 (1976).

NANBU, K., *J. Chem. Phys. 66,* 136 (1977).

Application to Addition Reactions, Dissociation, and Ionization

ZAMIR, E., Y. HAAS, and R. D. LEVINE, *J. Chem. Phys. 73,* 2680 (1980).

ZAMIR, E., and R. D. LEVINE, *Chem. Phys. 52*, 253 (1980).

LAWRANCE, W. D., J. SILVERSTEIN, ZHANG FU-MIN, ZHU QING-SHI, J. S. FRANCISCO, and J. I. STEINFELD, *J. Phys. Chem. 85*, 1961 (1981).

SILBERSTEIN, J., and R. D. LEVINE, *Chem. Phys. Letts. 74*, 6 (1980).

Application to Nuclear Physics

LEVINE, R. D., S. G. STEADMAN, J. S. KARP, and Y. ALHASSID, *Phys. Rev. Letts. 41*, 1537 (1978).

Application to Spectral Intensities, Level Widths, and Transition Rates

ALHASSID, Y., and R. D. LEVINE, *Phys. Rev. Letts. 57*, 2879 (1986).

LEVINE, R. D., *Adv. Chem. Phys. 70*, 53 (1987).

ENGEL, Y. M., R. D. LEVINE, J. W. THOMAN, JR., J. I. STEINFELD, and R. I. MCKAY, *J. Chem. Phys. 86*, 6561 (1987).

PROBLEMS

13.1 In V-V processes such as

$$H + HF(v) \longrightarrow HF(v - 1) + H$$

and

$$CO(v) + AB(0) \longrightarrow CO(v - 1) + AB(1)$$

where AB = CO, N_2, or NO, linear surprisal plots are obtained, as shown in Figures 13–5(b), 13–6, and 13–7. Using the equation

$$E_v = \left(\frac{v + 1}{2}\right)\hbar\omega_e - \left(\frac{v + 1}{2}\right)^2 \hbar\omega_e x_e$$

for energy levels of an anharmonic oscillator, show that the exponential gap law (equation 13–61) predicts that a plot of the surprisal $I(v, v - 1)$ vs. v will be linear. Find the effective λ_v values for the given examples. Do you think that the use of the preceding energy-level equation is inconsistent with the use of the RRHO approximation to derive equation 13–53?

13.2 Derive equations 13–68 and 13–69, using equation 13–67 for the prior probabilities.

13.3 Show that the statistical power-gap scaling law given by equation 13–77 can be obtained by applying entropy maximization to $\ln |\Delta E_{rot}|$.

CHAPTER 14

Analysis of Multilevel Kinetic Systems

14.1 INTRODUCTION

Thus far, we have considered individual elementary kinetic processes and mechanisms consisting of a small number of such processes. In the latter case, the mechanism was often simplified to the point at which closed-form analytic solutions for the species concentrations could be obtained. The emphasis throughout was on the details of the individual steps, such as energy consumption and disposal.

For many kinetically controlled systems, however, the number of coupled reaction steps is so large that simple analytic solutions are just not possible. The complexity may be due to the necessity of keeping track of internal states of the reacting species, and the resulting set of equations is often referred to as a *master equation,* techniques for the solution of which have been developed. In very special cases, such as the harmonic oscillator, a closed-form solution can still be found, but in general, numerical solutions become essential. Practical examples of the use of master equations include infrared multiphoton dissociation and chemical laser kinetics.

Even more complex kinetic equations can arise when a large number of different chemical species are involved, as in atmospheric chemistry and combustion. In such instances, knowledge of both elementary kinetic processes and transport effects is essential to a complete understanding of the system. These large-scale systems will be discussed in chapter 15.

14.2 THE MASTER EQUATION

The master equation is actually a set of complex integro-differential equations that describe the evolution with time of a physical system. (For exact definitions and a discussion of the mathematical properties of the master equation, see Oppenheim et al.)[1,2] This set of equations can be written in a simple form for a system consisting of a set of N discrete levels; the canonical example is that of harmonic-oscillator vibrational levels. The state-to-state kinetic equations for such a system are

$$AB(v) + M \;\rightleftharpoons\; AB(v') + M$$

with a rate equal to

$$R(v \rightarrow v') = k(v \rightarrow v'; T)\rho_M$$

where ρ_M is the total density of collision partners inducing a change in oscillator state with rate coefficient $k(v \rightarrow v'; T)$.

The master equation for the level populations $n_v(t)$ $[v = 0, 1, \ldots, N]$ can be written as

$$\frac{dn_v(t)}{dt} = \rho_M \sum_{v' \neq v} [k(v' \rightarrow v; T)n_{v'}(t) - k(v \rightarrow v'; T)n_v(t)] \qquad (14\text{--}1)$$

The first set of terms in the summand of equation 14–1 represents the population of level $[v]$ from all other levels; the second term represents the depopulation of $[v]$ to all other levels. Let us define

$$-\sum_{v' \neq v} k(v \rightarrow v') \equiv k(v \rightarrow v) = k(v) \qquad (14\text{--}2)$$

in which we no longer write the temperature T explicitly. Then, with this definition, we have

$$\sum_{\text{all } v'} k(v' \rightarrow v) = 0$$

which ensures that $\sum_{\text{all } v} n_v(t) = $ constant and that the occupation probabilities are normalized and their sum is conserved, i.e., that

$$\sum_{\text{all } v} P_v(t) = \sum_v \frac{n_v(t)}{\sum_v n_v(t)} = \sum_v \frac{n_v(t)}{N} = 1$$

at all times. We can then write equation 14–1 formally as

$$\frac{dP_v(t)}{dt} = \rho_M \sum_{v' \neq 0}^{N} k(v' \rightarrow v)P_{v'}(t) \qquad (14\text{--}3)$$

The principle of detailed balancing is an important help in reducing the master equation to soluble form. This principle states that, as the system approaches equilibrium (time $\rightarrow \infty$), the equilibrium occupation probabilities no longer change in time; that is,

$$P_v(\infty) = P_v^{\text{eq}} = p(v|T)$$

and

$$\frac{dP_v(\infty)}{dt} = 0 \qquad (14\text{--}4)$$

Thus, from equation 14–1,

$$\sum_{v' \neq v} k(v' \rightarrow v; T)P_{v'}(\infty) = \sum_{v' \neq v} k(v \rightarrow v'; T)P_v(\infty)$$

Since this must be true for every v, the terms in the sum must be pairwise equal, and

$$\frac{k(v \to v')}{k(v' \to v)} = \frac{p(v'|T)}{p(v|T)} = \frac{g_{v'}}{g_v} e^{-(E_{v'} - E_v)/k_B T} \tag{14-5}$$

We use the Boltzmann distribution to calculate the equilibrium occupation probabilities $p(v|T)$.

A closed-form analytic solution can be found for the master equation 14–3 in only a few cases, the most important being the relaxation of harmonic oscillators with $E_v = (v + \frac{1}{2})\hbar\omega_0$ and $g_v = 1$. For the $k(v \to v')$, let us assume a Landau-Teller collision model[3-5], i.e., $\Delta v = v' - v = \pm 1$ (so that $\Delta E = \pm \hbar\omega_0$) and $k(v \to v - 1; T) = vk(1 \to 0; T)$. Then the detailed balancing equation 14–5 can be written as

$$\frac{k(v - 1 \to v; T)}{k(v \to v - 1; T)} = \frac{p(v|T)}{p(v - 1|T)} = e^{-\hbar\omega_0/k_B T} \equiv \alpha$$

and the equilibrium population distribution is

$$p(v|T) = \alpha^v(1 - \alpha)$$

The master equation can be then written as

$$\frac{dP_v(t)}{dt} = \frac{1}{\tau(1 - \alpha)}\{\alpha v P_{v-1}(t) - [v + \alpha(v + 1)]P_v(t) + (v + 1)P_{v+1}(t)\} \tag{14-6}$$

where the vibrational relaxation time

$$\tau = [(1 - \alpha)\rho_M k(1 \to 0; T)]^{-1}$$

$$= [\rho_M\{k_{10}(T) - k_{01}(T)\}]^{-1}$$

For $\hbar\omega_0/k_B T \gg 1$, $\alpha \to 0$ and $\tau \approx [\rho_M k_{10}(T)]^{-1}$.

The detailed solution of the set of N coupled linear first-order differential equations 14–6 depends on the initial conditions $P_v(t = 0) = P_v(0)$. If we take this to be the so-called canonical distribution

$$P_v(0) = \frac{\exp\left[-E_v/k_B T(0)\right]}{Q[T(0)]}$$

where

$$Q[T(0)] = \sum_v e^{-E_v/k_B T(0)}$$

then the solution is

$$P_v(t) = \frac{\exp\left[-E_v/k_B T(t)\right]}{Q[T(t)]}$$

That is, if the set of harmonic oscillators can be initially described by a temperature (in other words, if the set is in a Boltzmann distribution), then it can be described by a time-varying temperature $T(t)$ throughout its relaxation. In order to find the evolution with time of T, we calculate the average energy per oscillator, viz.,

Analysis of Multilevel Kinetic Systems   Chap. 14

$$\langle E_v(t) \rangle = \sum_v E_v P_v(t)$$

$$= \hbar\omega_0 \sum_v v P_v(t)$$

$$= \hbar\omega_0 \langle v(t) \rangle$$

so that

$$\langle E_v(\infty) \rangle = \sum_v E_v P(v \mid T)$$

$$= \hbar\omega_0 (1 - \alpha) \sum_v v \alpha^v$$

$$= \hbar\omega_0 \frac{\alpha}{1 - \alpha} = \hbar\omega_0 \langle v(\infty) \rangle *$$

Thus,

$$\frac{d}{dt}\langle E_v(t) \rangle = \frac{d}{dt} \sum_v E_v P_v(t)$$

$$= \sum_v E_v \frac{d}{dt} P_v(t)$$

$$= \sum_v \left\{ \frac{\hbar\omega_0 v}{\tau(1 - \alpha)} [\alpha v P_{v-1}(t) - \{v + \alpha(v + 1)\}P_v(t) + (v + 1)P_{v+1}(t)] \right\}$$

or

$$\frac{d}{dt}\langle E_v(t) \rangle = \frac{1}{\tau}[\langle E_v(\infty) \rangle - \langle E_v(t) \rangle] \qquad (14\text{--}7)$$

Integrating equation 14–7 gives

$$\langle E_v(t) \rangle = \langle E_v(\infty) \rangle + [\langle E_v(0) \rangle - \langle E_v(\infty) \rangle]e^{-t/\tau} \qquad (14\text{--}8)$$

That is, for harmonic oscillators, the energy content relaxes exponentially for any initial set of $P_v(0)$. If the initial distribution is canonical (i.e., if it is a Boltzmann distribution), then there is a temperature proportional to $\langle E_v \rangle$ that relaxes exponentially as well. That is,

$$T(t) = T(\infty) + [T(0) - T(\infty)]e^{-t/\tau} \qquad (14\text{--}8)$$

*Note that

$$\sum_v v\alpha^v = \alpha \frac{\partial}{\partial\alpha}\left(\sum_v \alpha^v\right) = \alpha\frac{\partial}{\partial\alpha}\left(\frac{1}{1 - \alpha}\right) = -\frac{\alpha}{(1 - \alpha)^2}\frac{\partial}{\partial\alpha}(1 - \alpha) = \frac{\alpha}{(1 - \alpha)^2}$$

when the sum over v goes from 0 to ∞.

Closed-form solutions can also be found for the case of relaxation of anharmonic oscillators (see Treanor et al.[6] and problem 1).

For system that cannot be simply described as involving the relaxation of harmonic (or anharmonic) oscillators, the situation is much more complicated. In general, the equation for such a system must be solved by numerical integration of the coupled differential equations, starting from some initial distribution $\{n_v(0)\}$. The numerical solution can be greatly facilitated by an eigenvalue technique, which we now proceed to develop.

Let us write equation 14–1 as

$$\frac{dn_v(t)}{dt} = \sum_{v'} A(v, v')n_{v'}(t) \tag{14–9}$$

where

$$A(v, v') = \rho_M[k(v' \to v; T) - \delta_{vv'}k(v \to v'; T)] \tag{14–10}$$

Similarly, equation 14–6 can be written in matrix form as

$$\dot{\mathbf{n}}(t) = \mathbf{A} \cdot \mathbf{n}(t) \tag{14–11}$$

where $\mathbf{n}(t)$ is the N-dimensional time-dependent population vector and $\mathbf{A}$ is the $N \times N$ array defined by equation 14–10. Equation 14–11 has the formal solution

$$\mathbf{n}(t) = \exp[\mathbf{A}t] \cdot \mathbf{n}(0) \tag{14–12}$$

in which the matrix exponential may be taken to be its Taylor-series representation, i.e.,

$$\exp[\mathbf{A}t] = 1 + \mathbf{A} + \frac{1}{2}(\mathbf{A} \cdot \mathbf{A})t^2 + \cdots$$

Equation 14–12 can be written out by components as

$$n_v(t) = n_v(\infty) + \sum_{j=1}^{N} C_j(v) \exp(\lambda_j t) \tag{14–13}$$

in which the λ_j are the N *eigenvalues* of the matrix $\mathbf{A}$, and the $C_j(v)$ are the corresponding eigenvectors, which depend on the set of initial conditions $\{n_v(0)\}$. Note that $\{n_v(\infty)\}$ is the unique eigenvector corresponding to $\lambda_0 = 0$; to see this, merely write

$$\frac{dn_v(\infty)}{dt} = \sum_{v'} A(v', v')n_{v'}(\infty) = \lambda_0 n_v(\infty) = 0$$

Since for $j > 0$ all the λ_j are negative, all the terms in equation 14–13 decay to zero and $n_v(\infty) = n(v|T) = Np(v|T)$. The series in this equation usually converges rapidly, so that the lowest one or two eigenvalues dominate the kinetics. Also, matrix diagonalization routines are readily available for most computers, so this procedure is a practical alternative to direct numerical solution of the coupled differential equations.

14.3 INFORMATION-THEORETICAL TREATMENT OF THE MASTER EQUATION

In chapter 13 we found that information theory could be used to reduce a large number of rate coefficients to a much smaller number of parameters which reflected constraints on the final-state distribution. This suggests a possible approach to simplification of the master equation: if the individual rate coefficients in, for example, equation 14–3 are not independent, then it should be possible to reduce the number of coupled differential equations which need to be solved. An even more drastic simplification is to bypass the rate coefficients entirely and use a maximum-entropy principle to compute the time-dependent populations $P_v(t)$ directly. This approach will be developed next.

To carry out the prescription, we make use of a sum rule, and consider some bulk property A of a system evolving with time. The average value of A is

$$\langle\langle A \rangle\rangle = \sum_v A_v P_v \qquad (14\text{--}14)$$

and both $\langle\langle A \rangle\rangle$ and P_v may be time-dependent. We have already considered the specific example of A being the energy and v the harmonic-oscillator level number, but we are by no means restricted to that example. The evolution with time of $\langle\langle A \rangle\rangle$ can be written in terms of the normalized, time-dependent probabilities as

$$\frac{d}{dt}\langle\langle A(t) \rangle\rangle = \sum_v A_v \frac{dP_v(t)}{dt} \qquad (14\text{--}15\text{a})$$

$$\frac{d}{dt}\langle\langle A(t) \rangle\rangle$$

$$= -\sum_v A_v \sum_{v'} P_v(t) \, k(v \rightarrow v') + \sum_v A_v \sum_{v'} P_{v'}(t) \, k(v' \rightarrow v) \qquad (14\text{--}15\text{b})$$

$$\frac{d}{dt}\langle\langle A(t) \rangle\rangle$$

$$= -\sum_v P_v(t) \sum_{v'} A_v k(v \rightarrow v') + \sum_v P_v(t) \sum_{v'} A_{v'} k(v \rightarrow v') \qquad (14\text{--}15\text{c})$$

$$\frac{d}{dt}\langle\langle A(t) \rangle\rangle = \sum_v P_v(t) \sum_{v'} [A_{v'} - A_v] k(v \rightarrow v') \qquad (14\text{--}15\text{d})$$

$$\frac{d}{dt}\langle\langle A(t) \rangle\rangle = \sum_v P_v(t) \frac{d\langle A_v \rangle}{dt} \qquad (14\text{--}15\text{e})$$

In going from equation 14–15a to equation 14–15b, we simply used equation 14–1 for $dP_v(t)/dt = N^{-1} dn_v(t) dt$; equation 14–15b is obtained by exchanging v and v' in the second set of terms. Equation 14–15e serves to define $d\langle A_v \rangle/dt$ as $\sum_{v'} [A_{v'} - A_v] k(v \rightarrow v')$.

We now introduce a time-dependent surprisal, in analogy with equations 13–31 and 13–32:

$$I(v, t) = -\ln \frac{P_v(t)}{P_v^0(t)} \qquad (14\text{–}16\text{a})$$

$$= \sum_{m=1}^{r} \lambda_m(t)\langle A_v^m\rangle \qquad (14\text{–}16\text{b})$$

Here, $P_v^0(t)$ is the Boltzmann equilibrium distribution with time-dependent temperature; i.e.,

$$P_v^0(t) = p(v \mid T(t))$$
$$= \frac{g_v e^{-E_v/k_B T(t)}}{\sum\limits_v g_v e^{-E_v/k_B T(t)}} \qquad (14\text{–}17)$$

The time-dependent *entropy deficiency* of the distribution (cf. equations 13–10 and 13–85) is

$$\Delta S[P(t)] = -R \sum_v P_v(t) I(v, t)$$

$$= R \sum_v P_v(t) \ln \frac{P_v(t)}{P_v^0(t)} \qquad (14\text{–}18)$$

As before, we seek to find a set $\{P_v(t)\}$ which is consistent with the available information (in this case, bulk relaxation of an observable $\langle\langle A\rangle\rangle$) and which minimizes the entropy deficiency $\Delta S(t)$, i.e., which maximizes the rate of entropy production in the relaxing system. Note that all reference to individual state-to-state rate coefficients or transition probabilities has now disappeared.

Proceeding as before, we first specify a set of constraints, which may now be explicitly time-dependent:

$$\langle\langle A^r(t)\rangle\rangle = \sum_v A_v^r P_v(t)$$

The first, as always, is normalization:

$$A^0 = 1, \langle\langle A^0\rangle\rangle = 1 \qquad \text{(not time-dependent)}$$

The first nontrivial constraint may just be the average energy:

$$A^1 = E_v, \qquad \langle\langle A^1\rangle\rangle = \langle\langle V(t)\rangle\rangle$$

Higher moments of the distribution may be introduced if needed. We use the now familiar calculus of variations, setting up a variational function

$$\mathcal{L} = \Delta S + (\lambda_0 - 1)(1) + \frac{\lambda_1}{k_B T}\langle\langle V(t)\rangle\rangle + \ldots, \qquad \delta\mathcal{L} = 0$$

The Lagrange multipliers $\lambda_0(t)$ and $\lambda_1(t)$ are now time-dependent. The solution is clearly

$$P_v(t) = P_v^0(t) \exp[-\lambda_0 t - \lambda_1(t) E_v/k_B T - \ldots] \qquad (14\text{–}19)$$

with the surprisal given by

$$I(v, t) = \lambda_0(t) + \lambda_1(t)\frac{E_v}{k_B T} + \dots \tag{14-20}$$

There is a possible ambiguity about which value of T to use in these equations; since the system of, say, harmonic oscillators is presumably in contact with a "bath gas" at a bulk temperature T^{eq}, $T = T^{eq}$ should be used.

To find the $\lambda(t)$'s, we employ the same procedure as for individual rate coefficients; that is, we make use of a sum rule for the experimentally known quantity, to obtain

$$\frac{d}{dt}\langle\langle V(t)\rangle\rangle = \sum_v P_v(t) \sum_{v'} (E_{v'} - E_v)k(v \to v')$$

$$= \sum_v P_v(t)\frac{d}{dt}\langle V\rangle$$

$$= \sum_v P_v(t)(\beta - \alpha E_v)$$

where $\beta/\alpha = \langle\langle V^{eq}\rangle\rangle = \langle\langle V(\infty)\rangle\rangle$ and α^{-1} is proportional to an experimental relaxation time. This can be clearly seen from

$$\frac{1}{\alpha}\frac{d}{dt}\langle\langle V(t)\rangle\rangle = \frac{\beta}{\alpha}\sum_v P_v(t) - \sum_v P_v(t)E_v$$

or

$$\frac{1}{\alpha}\frac{d}{dt}\langle\langle V(t)\rangle\rangle = \langle\langle V(\infty)\rangle\rangle - \langle\langle V(t)\rangle\rangle \tag{14-21}$$

Integrating the simple first-order equation 14–21 gives

$$\int_{t_0}^{t} \frac{d\langle\langle v(t)\rangle\rangle}{\langle\langle V(t)\rangle\rangle - \langle\langle V(\infty)\rangle\rangle} = -\alpha \int_{t_0}^{t} dt$$

$$\ln\left[\langle\langle V(t)\rangle\rangle - \langle\langle V(\infty)\rangle\rangle\right]_{t_0}^{t} = -\alpha(t - t_0)$$

or

$$\langle\langle V(t)\rangle\rangle = \langle\langle V(\infty)\rangle\rangle + \left[\langle\langle V(t_0)\rangle\rangle - \langle\langle V(\infty)\rangle\rangle\right]e^{-\alpha(t-t_0)} \tag{14-22}$$

But we also have the sum rule

$$\langle\langle V(t)\rangle\rangle = \frac{\sum\limits_v E_v P_v(t)}{\sum\limits_v P_v(t)} = \frac{\sum\limits_v E_v e^{-\lambda_1(t)E_v/k_B T}}{\sum\limits_v e^{-\lambda_1(t)E_v/k_B T}} \tag{14-23}$$

Equating the right-hand sides of equations 14–22 and 14–23 now furnishes an algorithm for finding $\lambda_1(t)$ and thus $P_v(t)$. Every piece of information is now specified: we have the initial $\langle\langle V(t_0)\rangle\rangle$, the bulk temperature determines $\langle\langle V(\infty)\rangle\rangle$, and the experimental relaxation time determines α. Consequently, by a plot similar to that of Figures 13–13 and 13–14, we can find $\lambda_1(t)$ at any time between $t = t_0$ and $t \to \infty$. The parameter $\lambda_0(t)$ is determined by normalization, i.e.,

$$\lambda_0(t) = \ln \left[\sum_v e^{-\lambda_1(t)E_v/k_B T} \right]$$

and $P_v(t)$ is calculated from equation 14–19. For a distribution of harmonic oscillators, this is identical to the exact analytic solution of the master equation; for other systems, this information-theoretical prescription for $P_v(t)$ is generally found to be indistinguishable from a numerical solution to the master equation.

A more difficult problem, which can also be handled quite easily by information theory, is that of V-V relaxation in an ensemble of anharmonic oscillators; viz.,

$$AB(v = v_1) + AB(v = v_2) \longrightarrow AB(v_1' = v_1 + \Delta v) + AB(v_2' = v_2 - \Delta v)$$

in which Δv may equal 1, 2, or more, but $v_1 + v_2 = v_1' + v_2'$, i.e., the total number of vibrational quanta is conserved. This is equivalent to neglecting V-T relaxation relative to the V-V process. For this case, the resulting distribution of oscillators is the *Treanor distribution* discussed in problem 1. Although the master equation is nonlinear, because of the terms which are quadratic in population density, the distribution may be readily derived using information theory by maximizing the entropy

$$S = -\sum_v P(v) \ln P(v)$$

subject now to the following three constraints:

(*i*) $$\sum_v P(v) = 1 \qquad \text{(normalization)}$$

(*ii*) $$\langle E \rangle_{R,T} + \sum_v P(v)E_v = \text{constant} \qquad \text{(conservation of total energy)}$$

and

(*iii*) $$\sum_v vP(v) = \langle v \rangle = \text{constant} \qquad \text{(conservation of total number of vibrational quanta)}$$

Note that since $E_v \neq v\hbar\omega_0$ for an anharmonic oscillator, conditions (*ii*) and (*iii*) are *independent* constraints. Maximization of S subject to these constraints, using the usual Lagrange multiplier procedure, yields the result

$$P(v) = \frac{n_v}{n_0} = \exp\left[-E_v/k_B T - v\gamma\right] \qquad (14\text{–}24)$$

A maximum-entropy procedure is also useful in dealing with the case of coupled V-V and V-R,T relaxation,[7] which can be written as

$$AB(m) + AB(n) \longrightarrow AB(r) + AB(s)$$

with no constraints on m, n, r, and s other than that they are integers. The resulting populations can be written as a time-dependent Treanor distribution

$$P(v, t) = \frac{\exp[-\lambda_1(t)E_v - \lambda_2(t)v]}{\sum_v \exp[-\lambda_1(t)E_v - \lambda_2(t)v]} \qquad (14\text{–}25)$$

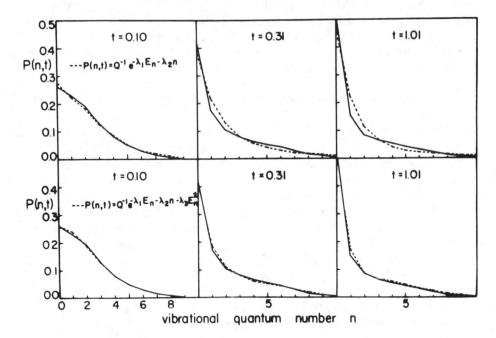

Figure 14–1. V-V and V-R,T relaxation in an ensemble of HF molecules. Comparison of exact numerical (full-line) distributions with (a–c) two-constraint (equation 14–25) and (d–f) three-constraint (equation 14–26) maximal entropy distributions at several times given in μsec. [From Tabor et al., *Mol. Phys. 37*, 141 (1979).]

and $\lambda_1(t)$ and $\lambda_2(t)$ can be evaluated by the use of sum rules on $\langle E_v(t)\rangle$ and $\langle v(t)\rangle$; note that as $t \to \infty$, $\lambda_1 \to (k_B T)^{-1}$ and $\lambda_2 \to 0$, so that a Boltzmann equilibrium distribution is obtained. The resulting $P(v, t)$, for a particular choice of initial conditions, are compared with exact numerical master-equation solutions in Figure 14–1(a)–(c). Much closer agreement can be obtained by adding one additional constraint to $P(v, t)$, viz.,

$$P(v, t) \approx \exp[-\lambda_1(t)E_v - \lambda_2(t)v - \lambda_3(t)E_v^2] \qquad (14\text{--}26)$$

There is, however, no *a priori* theoretical justification for the E_v^2 term. Nevertheless, the information-theoretical approach once again allows us to describe a many-level system in terms of a smaller number of time-dependent parameters.

14.4 SOME APPLICATIONS OF MASTER-EQUATION MODELING

The preceding discussion demonstrates that master-equation techniques are particularly useful for dealing with systems possessing a large number of kinetically coupled levels. In this section, we consider two examples of such systems: infrared multiple-photon excitation, which is a chemical reaction driven by a laser, and the hydrogen fluoride chemical laser, which is a laser driven by a chemical reaction. Both of these happen to be examples in which radiation (photons) is included as one of the reacting species, but this feature does not fundamentally change the treatment.

14.4.1 Infrared Multiple-Photon Excitation

In chapter 11 we noted that radiation could drive a molecular system to dissociation (the Perrin hypothesis) if the effective temperature of the radiation source were much higher than that of the molecular system. The molecule could then absorb multiple excitation photons from the radiation field and thus acquire sufficient energy to react. The first such reaction to be identified was actually driven by visible rather than infrared photons,[8] and so was actually a multiphoton photochemical process; the first observation of a reaction driven by actual multiple infrared photon absorption (IRMPA) was the decomposition of SiF_4.[9,10] Considerable attention was attracted by reports in the mid-1970s[11,12] of experiments in which the IRMPA-induced dissociation process was isotopically selective. These experiments used a high-powered CO_2 laser to dissociate SF_6; depending on the laser frequency employed, either $^{32}SF_6$ or the less abundant $^{34}SF_6$ could be dissociated. Such a reaction can be readily understood by noting that the excitation process occurs in the isolated molecule and that isotopic substitution results in frequency shifts of infrared absorption bands; if the pressure is kept low enough so that unimolecular dissociation proceeds faster than intermolecular collisions, the selectively excited isotopically labeled species will react preferentially. Following this early work, a large number of molecular species have been found to undergo IRMPD, and several comprehensive reviews of the subject are now available in the scientific literature.[13–16]

The details of the IRMPA process can be understood by reference to Figure 14–2. In the unexcited molecule (region I), the radiation with energy from the laser source must be resonant, or nearly so, with a vibration frequency of the molecule in order for absorption to take place. In principle, if the frequency interval remained constant for successive transitions, a sufficiently intense field at $h\nu_0$ could drive the system through a sequence of levels to arbitrarily high excitation. In actual molecules, however, anharmonicity causes the overtone levels to shift to a lower frequency, so that the resonance condition is lost. Multilevel excitation is possible through mechanisms such as rotational and/or anharmonic compensation, two- or three-photon absorption, etc.,* but eventually the excitation process becomes too off-resonant to proceed further.

At this point, the *density of states* in the molecule becomes significant (see Appendix 2). At an energy of several thousand cm^{-1}, or 40–50 kJ $mole^{-1}$, the vibrational state density $\rho(E)$ may be 10^3–10^5 or more (see Figure 11–6). This high density of states allows the laser-pumped vibrational levels to mix with a large number of other vibrational modes of the molecule, thus distributing the absorption strength over a range of transitions. In this dense manifold of states, called the *quasi continuum* or QC (region II in Figure 14–2), photons of energy $h\nu_0$ can be absorbed by a

---

*For a discussion of the spectroscopic aspects of IRMPA, see Francisco and Steinfeld[16] and Steinfeld.[17]

INFRARED MULTIPHOTON ABSORPTION:
POSTULATED MECHANISM

DISSOCIATIVE
CONTINUUM

REGION III

$k(E)$
⟿ PRODUCTS

INCOHERENT PUMPING
RATE EQUATIONS

$$R_{i,i\pm1} \sim \frac{\bar{g}X}{\Gamma} \frac{n}{2} a_{01}^2$$

QUASICONTINUUM
REGION II

$$R_{i,i\pm1} \sim \frac{1}{\Gamma} \frac{n}{2} a_{01}^2$$

$\Gamma(a\,1/T_1)$

$4\nu_p$

$$R_{i,i\pm1} \sim \frac{\Gamma}{\Delta^2} \frac{n}{2} a_{01}^2$$

$$(\Delta = \omega_p - \omega_L - 2n X_{pp})$$

PUMPED-MODE

$3\nu_p$

LADDER

COHERENT PUMPING

REGION I $2\nu_p$

a_{01} $\hbar\omega$

n-PHOTON RESONANCES
POWER BROADENING
ANHARMONIC BOTTLENECKS

$1\nu_p$

LADDER   BACKGROUND
                STATES

Figure 14–2.   Schematic model for infrared multiple-photon absorption (IRMPA). Excitation occurs by resonant transitions at frequency $h\nu_o = \hbar\omega_o$ in the discrete region (I), which connects to the quasi-continuum region (II). At sufficiently high energies, the true dissociative continuum (region III) is reached. [From J. I. Steinfeld, *Molecules and Radiation,* 2nd ed., (Cambridge, MA: MIT Press, 1985). Reproduced with permission.]

range of states, and this absorption can proceed until sufficient energy has been acquired to reach the threshhold of a dissociation or other chemical process. Above this threshhold (region III in Figure 14–2), the reaction rate can generally be estimated quite well by RRKM-type expressions (see chapter 11). Although non-RRKM behavior in such reactions has long been sought, there is little evidence to date for such behavior.

The rigorous mathematical treatment of IRMPD is known as the *optical Bloch equations*,[18] which represent a coupling of the Schrödinger equation for the molecule with the Maxwell equations for the radiation field. These equations, which treat the interaction between the complete set of molecular quantum states, the radiation field, and any dissipative processes such as collision or reaction, are generally far too detailed and difficult of solution for the desired objective, which is usually an ensemble-averaged quantity such as the total dissociation yield. The optical Bloch equations can be reduced to a more tractable master equation if certain assumptions are made, the most important of which is that the molecular levels are characterized only by their total energy content E. This is a valid description if *intramolecular vibrational relaxation* (IVR), or coupling between the various vibrational modes of the molecule, is an extremely rapid process compared with laser pumping, relaxation, or reaction. The "rapid IVR" assumption has been the subjct of some lively controversy (see chapter 11), but it seems to be appropriate for most cases. Using it, the equation of motion for the level populations $g(E,t)$ can be written as

$$\frac{dg(E,t)}{dt} = -[L_A(E) + L_S(E)]g(E) + L_A(E - h\nu_0)g(E - h\nu_0)$$

$$+ L_S(E + h\nu_0)g(E + h\nu_0) + \omega_{\text{coll}} \int [P(E,E')g(E') \quad (14\text{–}27)$$

$$- P(E',E)\, g(E)]\, dE' - k(E)g(E)$$

Let us consider each of the terms in this equation in turn. The first four terms represent the interaction of the molecular energy levels with the radiation field. The absorption coefficients $L_A(E)$ are given by

$$L_A(E) = \sigma_{\text{eff}}(E)\frac{I}{h\nu_0} \quad (14\text{–}28)$$

where the absorption coefficient σ_{eff} is assumed to be a slowly varying function of E in the quasi continuum and the radiation intensity I is typically a time-dependent quantity, as for a pulsed laser. The stimulated-emission coefficients are related to the absorption coefficient by the Einstein coefficient relationships[16] as follows:

$$L_S(E)\rho(E) = L_A(E - h\nu_0)\rho(E - h\nu_0) \quad (14\text{–}29)$$

Since $\rho(E)$ is a monotonically increasing function of E, as shown in Figure 11–6, the ratio $L_S(E)/L_A(E - h\nu_0)$ is less than unity, so that optical pumping tends to drive the system to higher energy, i.e., towards dissociation.

The collisional relaxation term $\omega P(E,E')$ is expressed as an integration over initial and final energies E', which makes analytical solution of equation 14–27 extremely difficult. The form of $P(E,E')$ has been discussed in chapter 11 and also from an information-theoretical viewpoint in chapter 13. Again, the density of states $\rho(E)$ tends to bias the final-energy distribution to net excitation, which also drives the system toward dissociation. The final term in equation 14–27, which represents molecular dissociation or isomerization with rate $k(E)$, can usually be estimated quite satisfactorily by an RRKM method (see chapter 11).

Before proceeding with a solution of equation 14–27, it is instructive to consider the magnitudes of the terms involved. To estimate the radiative pumping rate $L = \sigma I / h\nu_0$, we shall assume an infrared absorption cross section of 10^{-20} cm^2 and a photon energy $h\nu_0 = 2 \times 10^{-20}$ J/photon for $\nu_0/c = 1000$ cm^{-1} (CO_2 laser frequency). This gives $L = 0.5I$ sec^{-1}, where the intensity I has units of W/cm^2. In order to drive a molecule to dissociation, it is first necessary to *saturate* the absorbing transition, that is, to cause the molecule to absorb photons at a rate which exceeds spontaneous emission ($\tau_{rad}^{-1} \approx 10^3$ sec^{-1} for infrared transitions). This requires a saturation intensity I_{sat} greater than $10^3/0.5 \approx 2 \times 10^3$ W/cm^2. Saturation alone is not sufficient to induce multiple-photon absorption, however: in order to overcome the "anharmonic bottleneck," the intensity must be sufficiently high to cause power broadening of the resonance with a magnitude comparable to the spacing between vibrational levels, i.e., $\rho^{-1}(E)$. Since $\rho(E)$ may be on the order of several thousand states per cm^{-1} at energies corresponding to the anharmonic bottleneck, power broadening on the order of 10 MHZ is required, corresponding to $L \approx 10^7$ sec^{-1}. Also, the pumping rate must exceed collisional relaxation, which occurs at a rate of about 10^7 sec^{-1} at 1 torr. Thus, the intensity required for IRMPD may be estimated as $I > 10^7$ sec$^{-1}/0.5 \approx 2 \times 10^7$ W/cm^2.

For weakly bound species at very low pressures, where only one or a few infrared photons can induce dissociation and collisional relaxation is unimportant, dissociation has been found to occur at intensities of 10^2–10^3 W/cm^2, in accordance with the estimate from saturation. Some examples are van der Waals complexes in molecular beams[19] and ionic complexes in ion cyclotron resonance traps.[20] For most medium-sized polyatomic molecules, IRMPD threshhold intensities are on the order of 10^6–10^7 W/cm^2. Such intensities can be obtained from pulsed-discharge (TEA) CO_2 lasers, which typically generate 1-J pulses with 10^{-7} sec durations. The beams can be focused to 0.1 cm^2 or less to obtain the requisite intensity. Dissociation yields are often reported as a function of time-integrated pulse intensity, or fluence, given by

$$\phi = \int I(t)\, dt \qquad \text{J cm}^{-2} \qquad (14\text{–}30)$$

because yields tend to correlate best with fluence for a variety of pulse peak intensities and durations. It is now also clear why IRMPD products generally possess little excess internal excitation: $k_{RRKM}(E)$ rises rapidly with excess energy (see Figure 11–7) and exceeds the pumping rate L after only one or two infrared photons in excess of $E_0/h\nu_0$ are absorbed.

Analytic solutions of equation 14–27 can be approximated when collisional relaxation can be neglected ($\omega = 0$) and the extent of reaction is small. Alternatively, post-pulse reaction and relaxation can be solved by setting $I = 0$, a treatment equivalent to that of photoactivated unimolecular reactions presented in chapter 11. Solution of the complete equation must be carried out numerically, using matrix techniques.[21] One difficulty encountered in carrying out such a solution is that the energy E is a continuous variable. In the absence of collisions, this can be dealt with by recognizing that energy is added to or removed from the molecule in units of $h\nu_0$, the infrared photon energy; this results in an *energy-grained* master equation, with grain size $h\nu_0$. Collisional relaxation will tend to redistribute molecules over a wide range of energies, but it has been found that solutions are still practical in this case with a grain size equal to one-third or one-fourth $h\nu_0$. A computer program for solving the master equation for IRMPD is given by Dai.[22]

The object of solving the master equation is the prediction of an experimentally observed quantity, such as net dissociation yield. The yield can be found as the net fraction of molecules removed from populated energy states, given by

$$f = 1 - \lim_{t \to \infty} \frac{\int g(E,t)\, dE}{\int g(E,0)\, dE} \tag{14-31}$$

The IRMPD yield typically has a threshhold fluence on the order of 0.1 to 10 J cm^{-2}, below which dissociation does not occur; at high fluences, the yield saturates when all the molecules in the high-field region, which can interact with the laser, are dissociated. Some experimental examples are shown in Figure 14–3.

IRMPD is a useful technique for producing large densities of reactive free radicals in low-temperature environments, enabling the reaction rates of these radicals to be measured. It can also be used to generate reactive species in the proximity of a surface, at which etching or deposition reactions may then occur. One sought-after result has been the stimulation of "bond-selective" chemical reactions, in which the functional group of the molecule initially excited by the laser is the one that selectively reacts. That this has not been obtained is due to the rapid IVR process discussed earlier, which effectively redistributes the absorbed energy among all the vibrational modes of the molecule. Reaction then proceeds in an essentially statistical manner (see chapter 11).

Only limited information about the details of the IRMPA process can be obtained from highly averaged measurements, such as net dissociation yield. Additional information can be obtained by either studying systems which possess two closely spaced competitive unimolecular dissociation channels, studying the kinematics of the dissociated fragments using molecular-beam techniques,[26] or carrying time-resolved spectroscopic measurements on IRMP-excited molecules.[16,27]

14.4.2 The Hydrogen Fluoride Chemical Laser

The preceding section deals with a chemical reaction (unimolecular dissociation) driven by laser excitation. Here, we shall examine the converse situation, an operating laser system which derives its energy from a chemical reaction, viz., the hydrogen fluoride chemical laser.

We previously encountered the hydrogen-fluorine reaction in chapter 9, as a paradigmatic exothermic reaction in which the energy released is channeled preferentially into product (HF) vibrations. In the chemical laser system, the reaction is initiated by thermal decomposition of a fluorine-containing species such as F_2 or

NF$_3$, i.e.,

$$RF \longrightarrow R + F$$

The fluorine atom reacts with hydrogen to produce vibrationally excited HF:

$$F + H_2 \xrightarrow{k_f(v,J)} HF(v, J) + H \qquad (14\text{–}32)$$

The liberated H atom then reacts with an undissociated fluorine source (F$_2$, say) to produce more vibrationally excited HF:

$$H + F_2 \xrightarrow{k_f'(v,J)} HF(v, J) + F \qquad (14\text{–}33)$$

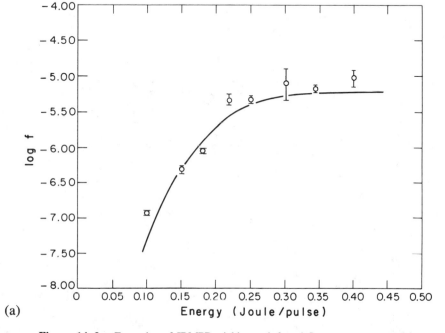

(a)

Figure 14–3. Examples of IRMPD yields vs. infrared fluence, as measured by several different experimental techniques. (a) Yield of total ethylene, measured by gas chromatograph techniques, in the dehydrohalogenation of chloroethane-d$_3$ by the 9P(14) CO$_2$ laser line. [From Francisco, Zhu, and Steinfeld, *J. Chem. Phys. 78,* 5339 (1983). Reproduced with permission.] (b) (see p. 490) Fluorine coverage on single-crystal silicon exposed to CF$_3$ radicals produced by IRMPD of C$_2$F$_6$ with the CO$_2$ 9R(30) laser line. The quantity measured is the integrated F(ls) X-ray photoelectron peak intensity [from Roop, et al., *J. Chem. Phys. 83,* 6012 (1985). Reproduced with permission]. (c) (see p. 490) In this experiment, bistrifluoromethyl peroxide (CF$_3$OOCF$_3$(BTMP)) is dissociated with the CO$_2$ 9R(30) laser line to produce CF$_3$O radicals which are further dissociated to CF$_2$O and F atoms by continued absorption of infrared photons. The F atoms react with added HI to produce vibrationally excited HF, which in turn fluoresces at 2.8 μm; producing an infrared emission that is detected. The threshhold fluence for this system appears to be very low or even zero, because the BTMP molecule is already in its vibrational quasi continuum under the conditions of the experiment. [From Zhang, Francisco, and Steinfeld, *J. Chem. Phys. 86,* 2402 (1982). Reproduced with permission.]

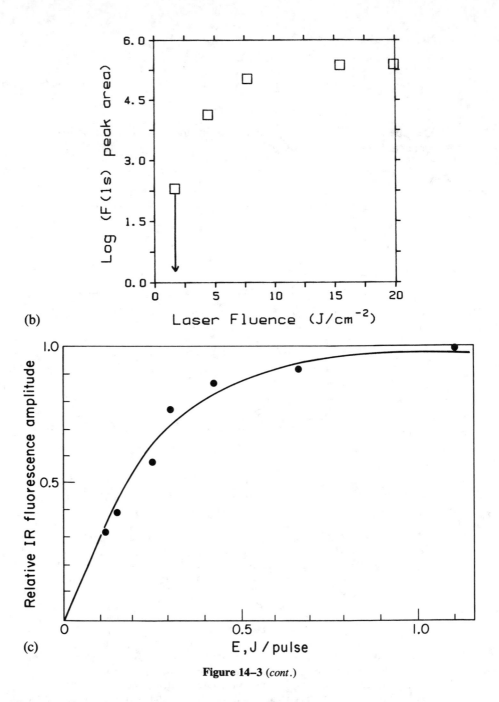

(b)

(c)

Figure 14–3 (*cont.*)

The net effect of the reaction equations 14–32 and 14–33 is a *chain reaction* producing "hot" HF. Some of this HF can react with H atoms to regenerate H_2, as noted in chapter 9:

$$HF\ (v > 3, J) + H \ \xrightarrow{k_R\ (v, J)} \ H_2 + F \qquad (14\text{–}34)$$

In addition to these reactive processes, there are a number of *vibrational* relaxation processes which may deactivate the newly formed HF molecules. These include the following:

(i) vibration-translation (V-T) relaxation

$$HF(v,J) + M \longrightarrow HF(v',J') + M \qquad (14\text{--}35)$$

(ii) vibration-vibration (V-V) relaxation between HF molecules

$$HF(v_a J_a) + HF(v_b J_b) \longrightarrow HF(v'_a J'_a) + HF(v'_b J'_b) \qquad (14\text{--}36)$$

There may also be energy exchange between vibration and rotation of the HF molecules (a V-R process). In equation 14–35, the collision partner M may be H and F atoms, H_2, F_2, and HF molecules, or a diluent gas (typically He).

The populations of HF in particular (v,J) states further interact by absorbing and emitting infrared photons in three possible ways:

$$HF(v,J) + h\nu \longrightarrow HF(v+1, J \pm 1) \qquad\qquad L^A(vJ\,|\,v'J')$$

$$HF(v,J) \qquad\longrightarrow HF(v-1, J \pm 1) + h\nu \qquad A(vJ\,|\,v'J')$$

$$HF(v,J) + h\nu \longrightarrow HF(v-1, J \pm 1) + h\nu \qquad L^S(vJ\,|\,v'J')$$

The quantities L^A and L^S represent the rates of photon absorption and stimulated emission, respectively, and are proportional to the absorption cross section (or, equivalently, the transition dipole strength) of the HF molecules and to the radiation density at frequency ν. A is the rate of spontaneous photon emission from the excited HF molecule.

When all of these processes are included, a master equation for HF vibrational populations $N(v,J)$ can be written as follows:

$$\frac{dN(v,J)}{dt} = k_f(v,J)[F][H_2] + k'_f(v,J)[H][F_2] - k_R(v,J)N_{v,J}[H]$$

$$+ \sum_M \sum_{v'J'} \rho_M k_{VT}(vJ\,|\,v'J')N_{v'J'}$$

$$+ \sum_{v'_a v'_b J'_a J'_b} \sum_{v_b J_b} k_{VV}(v_a J_a v_b J_b\,|\,v'_a J'_a v'_b J'_b)N_{v_a J_a}N_{v_b J_b}$$

$$- (L^A + L^S + A)N_{vJ} + L^A N_{v-1,J\pm1} + (L^S + A)N_{v+1,J\pm1} \qquad (14\text{--}37)$$

The radiation flux density at frequency ν is, in turn, given by

$$\frac{d\phi(\nu)}{dt} = L^S(N_{vJ} - N_{v-1,J\pm1}) - \frac{\phi(\nu)}{\tau_{\text{cavity}}} \qquad (14\text{--}38)$$

where τ_{cavity} includes both contributions from spontaneous radiation and cavity losses. The amount of power produced by the laser can be found by integrating the coupled sets of equations 14–37 and 14–38. Doing so requires accurate knowledge of the HF production and consumption rates $k_f(v,J)$, $k'_f(v,J)$, and $k_R(v,J)$; the product branching ratios into vibrational and rotational states; the energy transfer rates k_{VV} and k_{VT}; and the infrared transition strengths included in L^A, L^S, and A. A further complication is introduced by the fact that the laser is typically run not as a static system, but in a *supersonic flow,* as illustrated in Figure 14–4. Thus, hydrodynamic flow terms have to be added to equation 14–37 to gain a complete solution.

This daunting task has in fact been accomplished. Many (but by no means all) of the rate constants in equation 14–37 have been measured in laboratory experiments, and elaborate computer codes have been written to carry out the integrations. (A good overview of such efforts can be found in Gross and Bott.)[28] The impetus for doing so can be found in the high power at which the HF laser can be run. Since the exothermicity of the $F_2 + H_2$ reaction is 540 kJ mol^{-1}, and the molar weight of the reactants is 40 g mol^{-1}, each kilogram of reagents contains 13 MJ of energy. If the reagents can be made to flow through the laser cavity at rates on the order of kg sec^{-1} (which is typical of rocket engines), then, in principle, laser beams with average power levels in the megawatt range could be produced. This observation has led to proposals that "beam weapons," which could destroy missiles and satellites over vast distances in space, should be moved from the realm of science fiction[29] to that of military reality. The effort to accomplish this, known as the Strategic Defense Initiative (or "Star Wars") has engendered an enormous amount of controversy.[30,31] Among the many issues debated is the fact that chemical lasers simply do not operate at the theoretical efficiency implied in the preceding simple calculation. While

Figure 14–4. Schematic of supersonic-flow hydrogen fluoride chemical laser. The NF_3–C_2H_4 combination produces F atoms by thermal reaction. H_2 is introduced via a set of nozzles and trip holes set into the flow; vibrationally excited HF is produced almost immediately and flows through an optical cavity transversely to the flow direction. The N_2H_4 is added downstream to assist in exhausting the reaction products, thereby accelerating the flow rate. D_2 can be used in place of H_2 to produce longer wavelength radiation, which is transmitted more efficiently though the atmosphere.

devices based on the operating principle of Figure 14–4 (but far more elaborate in design) have run in excess of 1 MW average power, scaling such devices to the 20-MW average power apparently required for military applications appears to be difficult, if not impossible. There are several reasons for this, but kinetic limitations, including the deleterious effects of rapid vibrational relaxation, appear to be one of the fundamental constraints on scaling such lasers to higher powers. While chemical kinetics is, of course, only one of many aspects of the debate, it is nevertheless worth realizing that purely technical issues can, as they have in this instance, become critical elements of national policy controversies. Some other such examples are mentioned in the next chapter.

REFERENCES

1. I. Oppenheim, K. E. Shuler, and G. H. Weiss, *Adv. Mol. Relaxation Processes 1*, 13 (1967–68).

2. I. Oppenheim, K. E. Shuler, and G. H. Weiss, *Stochastic Processes in Chemical Physics: The Master Equation* (Cambridge, MA: M.I.T. Press, 1977).

3. G. G. Hammes, *Principles of Chemical Kinetics,* (New York: Academic Press, 1978), chapter 6.

4. J. T. Yardley, *Introduction to Molecular Energy Transfer,* (New York: Academic Press, 1980).

5. L. Landau and E. Teller, *Physik. Z. Sowjetunion 10,* 34 (1936).

6. C. E. Treanor, J. W. Rich, and R. G. Rehm, *J. Chem. Phys. 48,* 1798 (1968).

7. M. Tabor, R. D. Levine, A. Ben-Shaul, and J. I. Steinfeld, *Mol. Phys. 37,* 141 (1979).

8. G. Porter and J. I. Steinfeld, *J. Chem. Phys. 45,* 3456 (1966).

9. N. Isenor and M. C. Richardson, *Appl. Phys. Letts. 18,* 224 (1971).

10. N. Isenor, V. Merchant, R. Hallsworth, and M. Richardson, *Can. J. Phys. 51,* 1281 (1973).

11. R. V. Ambartzumian, Yu. A. Gorokhov, V. S. Letokhov, and G. N. Makarov, *J.E.T.P. Letts. 21,* 375 (1975).

12. J. L. Lyman, R. J. Jensen, J. Rink, C. P. Robinson, and S. D. Rockwood, *Appl. Phys. Letts. 27,* 87 (1975).

13. J. I. Steinfeld (ed.), *Laser-Induced Chemical Processes,* (New York: Plenum, 1981).

14. M. Quack, in *Adv. Chem. Phys. Vol. 50: Dynamics of the Excited State,* edited by K. P. Lawley (Chichester, England: Wiley, 1982), pp. 395–473.

15. D. W. Lupo and M. Quack, *Chem. Rev. 87,* 181 (1987).

16. J. S. Francisco and J. I. Steinfeld, in *Multiphoton Processes and Spectroscopy, Vol. 2,* edited by S. H. Lin (Singapore: World Publishing Co., 1986), pp. 79–173.

17. J. I. Steinfeld, *Molecules and Radiation,* 2d ed. (Cambridge, MA: MIT Press, 1985).

18. J. Ackerhalt and B. Shore, *Phys. Rev. A16,* 277 (1977).

19. M. P. Casassa, D. S. Bomse, J. L. Beauchamp, and R. C. Janda, *J. Chem. Phys. 72,* 6805 (1980): M. A. Hoffbauer, K. Liu, C. F. Giese, and W. R. Gentry, *J. Chem. Phys. 78,* 5567 (1983).

20. R. L. Woodin, D. S. Bomse, and J. L. Beauchamp, *J. Am. Chem. Soc. 100,* 3248 (1978); R. L. Woodin, D. S. Bomse, and J. L. Beauchamp, *Chem. Phys. Lett. 63,* 630

(1979); R. N. Rosenfeld, J. M. Jasinski, and J. L. Beauchamp, *Chem. Phys. Lett. 71,* 400 (1980).

21. W. D. Lawrance, A. E. W. Knight, R. G. Gilbert, and K. D. King, *Chem. Phys. 56,* 343 (1981).

22. H. L. Dai, Ph.D. dissertation, University of California, 1982.

23. J. S. Francisco, Zhu Qingshi, and J. I. Steinfeld, *J. Chem. Phys. 78,* 5339 (1983).

24. B. Roop, S. Joyce, J. C. Schultz, and J. I. Steinfeld, *J. Chem. Phys. 83,* 6012 (1985).

25. Zhang. Fumin, J. S. Francisco, and J. I. Steinfeld, *J. Phys. Chem. 86,* 2402 (1982).

26. P. A. Schulz, Aa. S. Sudbø, D. J. Krajnovich, H. S. Kwok, Y. R. Shen, and Y. T. Lee, *Ann. Rev. Phys. Chem. 30,* 379 (1979).

27. J. I. Steinfeld, in *Energy Storage and Redistribution in Molecules,* edited by J. Hinze (New York: Plenum Press, 1983), p. 1.

28. R. W. F. Gross and J. F. Bott, *Handbook of Chemical Lasers* (New York: Wiley-Interscience, 1976).

29. "Many think that in some way they [the Martians] are able to generate an intense heat in a chamber of practically absolute non-conductivity. This intense heat they project by means of a polished parabolic mirror of unknown composition, in a parallel beam against any object they choose . . . However it is done, it is certain that a beam of heat is the essence of the matter . . . Whatever is combustible flashes into flame at its touch, lead runs like water, it softens iron, cracks and melts glass, and when it falls on water incontinently that explodes into steam." [H. G. Wells, "The War of the Worlds," *Pearson's Magazine,* London (1897)]

30. American Physical Society Study Group (N. Bloembergen and C. K. Patel, co-chairmen), "Report to the American Physical Society of the Study Group on Science and Technology of Directed Energy Weapons," *Rev. Mod. Phys.* 59(3), Part II (July 1987).

31. P. M. Boffey, W. J. Broad, L. H. Gelb, C. Mohr, and H. B. Noble, *Claiming the Heavens* (New York: Time Books, 1988).

PROBLEMS

14.1 V-V relaxation in a gas of diatomic molecules involves the reaction

$$AB(v_1) + AB(v_2) \longrightarrow AB(v_3) + AB(v_4), \, k(v_1 v_2 \mid v_3 v_4)$$

in which the number of vibrational quanta is not changed, i.e., $v_1 + v_2 = v_3 + v_4$. The rates are related by detailed balancing as follows:

$$\frac{k(v_1 v_2 \mid v_3 v_4)}{k(v_3 v_4 \mid v_1 v_2)} = \exp[-(E_{v3} + E_{v4} - E_{v1} - E_{v2})/k_B T]$$

The quasi-steady distribution $n_v^{(0)}$ can be found by the Chapman-Enskog expansion procedure, which states that the exchange collision terms, when equated to zero, determine the quasi-steady distribution.

(a) Write the master equation for the evolution with time of the vibrational populations.

(b) For harmonic oscillators, $E_v^{\text{H.O.}} = v\hbar\omega$. Show that $n_v^{(0)} = n_0^{(0)} e^{-E_v/k_B T}$ is a solution to the master equation.

(c) For anharmonic oscillators, $E_v = v\hbar\omega(1 - x) - v^2 \hbar\omega x < E_v^{\text{H.O.}}$ $(x \ll 1)$. Show that $n_v^{(0)} = n_0^{(0)} e^{-v\gamma} e^{-E_v/k_B T}$ (equation 14–24) is the solution in this case.

(d) Find γ in terms of ω, x, and T.

14.2 The reaction scheme of the premixed HF chain-reaction laser involves the following steps:

$$\Delta H°(\text{kJ/mole})$$

$$F_2 + M \underset{k_2}{\overset{k_1}{\rightleftarrows}} 2F + M \qquad\qquad +159$$

$$F + H_2 \underset{k_4}{\overset{k_3}{\rightleftarrows}} HF + H \qquad\qquad -146.5$$

$$H + F_2 \overset{k_5}{\rightarrow} HF + F \qquad\qquad -443$$

(a) Calculate $\Delta H°$ for $H_2 + M \rightarrow 2H + M$ from the preceding reactions. Why is it neglected in the reaction scheme?

(b) Assume steady-state conditions for [F] and [H], and derive expressions for $d[HF]/dt$ and $d[H_2]/dt$.

(c) How can k_2 and k_4 be found if k_1 and k_3 are known?

*14.3 Suppose an enthusiastic colleague comes to you with the following proposal for a project. Correct the errors and suggest an explanation for the observed effect:

We have been obtaining thermal unimolecular rate coefficients for a large molecule A_h and its deuterated analogue A_d. While there is not much difference between the observed reaction rates for A_h and A_d, when we carry out a new experiment by shining an infrared laser on these molecules, the rate for A_d is greatly increased while that for A_h is apparently unchanged. This must be a tunneling effect. We want to carry out trajectory calculations to confirm this.

14.4 The IRMPD of chloroethane-d_1 has been the subject of recent investigation. The RRKM dissociation rates $k(E_V)$ for the two dissociation channels

were calculated in problem 12 of chapter 11 over the energy range 0 to 600 kJ/mole.

(a) Each CO_2 laser photon at 966 cm^{-1} carries 11.3 kJ/mole of energy. Make a plot of the expected lifetime of the multiphoton-activated CH_2DCH_2Cl as a function of the number of CO_2 laser photons absorbed. Also, calculate the predicted HCl/DCl branching ratio over the same range.

(b) At what level of excitation does the HCl/DCl branching ratio become equal to the thermal branching ratio calculated in part (b) of problem 12, chapter 11?

---

*These problems have been provided by Dr. R. G. Gilbert, Department of Theoretical Chemistry, University of Sydney, Australia.

Kinetics of Multicomponent Systems

In this chapter we shall analyze several chemical kinetic systems which are complicated by the simultaneous interactions of a large number of chemical species. The examples we shall consider are reactions that take place in the earth's atmosphere, particularly those that control the abundance of ozone, and the combustion of hydrogen and hydrocarbons. A complete analysis of such systems can be done only by means of a detailed numerical model. Because many of these problems are of great practical importance, extensive computer codes have been developed for analyzing such models. However, the essential features of these systems can often be understood by abstracting the most important reactions that take place and using them in a simple analytic model. We shall take this approach in what follows.

15.1 ATMOSPHERIC CHEMISTRY

15.1.1 Physical Structure of the Atmosphere

The earth's atmosphere is divided into regions primarily on the basis of temperature gradients, as shown in Figure 15–1. The *troposphere,* also referred to as the lower atmosphere, is characterized as the region where the air temperature falls with increasing altitude above the earth's sea level. The temperature reaches a minimum of around 210 K at the *tropopause.* The height of the troposphere varies from 8 km (25,000 ft) near the earth's poles to about 16 km (50,000 ft) in the tropics. Lying above the troposphere is the *stratosphere,* or upper atmosphere, which is located approximately 20 to 50 km above the earth's surface; here the temperature rises to a maximum value of about 280 K. Both of these regions are illustrated in Figure 15–1. The stratospheric region is virtually cloudless and is characterized by slow vertical circulation because the dense cooler air at the lower altitudes does not readily rise. Atmospheric pressure in this region decreases almost exponentially with increasing altitude; in fact, the pressure falls by about a factor of 100 from the interface of the stratosphere and troposphere to the top of the stratosphere. At the stratopause the temperature passes through a maximum of about 270 K, while the

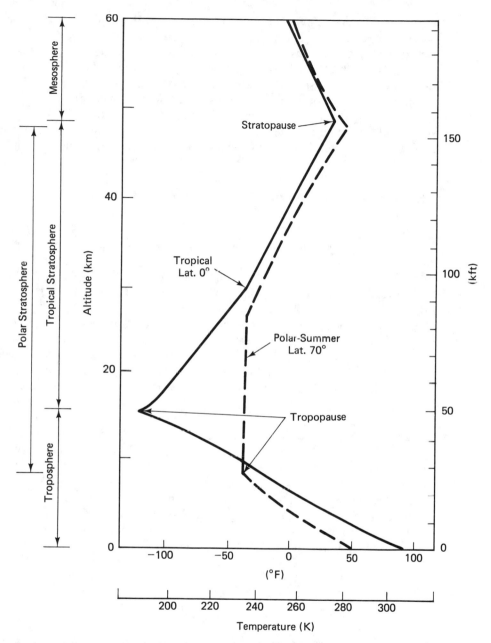

Figure 15–1. Physical structure of the atmosphere.

temperature in the stratosphere increases to this maximum. Gases within the stratosphere are transported and mixed mainly by eddy mixing rather than by molecular diffusion. Consequently, the rate of transport of gases in this region becomes independent of the nature of the molecule, and settling out of heavier molecules due to gravity is negligible.

Sec. 15.1   Atmospheric Chemistry   **497**

The chemistry involving most neutral species that enter the atmosphere takes place in basically two regions: the troposphere and the stratosphere. Two important factors arise from the physical structure of these two regions which have important consequences for chemical reactions occurring in the atmosphere: first, the exponentially decreasing pressure dependence generally results in a decrease in third-order recombination reaction rates; and second, the lower temperature in the upper troposphere and in the lower stratosphere reduces the rates of reactions that possess significant activation energies. Consequently, reactions with low activation energies are favored—in particular, recombination reactions with little or no activation energy.

15.1.2 Chemical Composition of the Atmosphere

Although the major constituents in the troposphere and stratosphere are molecular nitrogen and oxygen, the minor constituents play a major role in the overall atmospheric chemistry. For example, one minor constituent that plays a disproportionately important role is hydroxyl radicals. Their chemistry is largely controlled by free-radical reactions. These radicals dominate the chemistry of the troposphere, even though their concentrations are typically only in the range of 10^6–10^{10} molecules/cm^3, which is 10^{-13}–10^{-9} atm. The most important minor constituent from the point of view of the present discussion is ozone, which shields the earth from ultraviolet radiation that otherwise would reach the earth's surface. The concentration of ozone in the stratosphere does not exceed a few tenths of a part per million and is quite variable and diffuse there (see Figure 15–2). If the ozone in a column of the atmosphere were collected and compressed to standard temperature and pressure,* it would form a layer only 3 mm thick. Most of the ozone is contained in a layer between 20 and 50 km above the earth's surface. This ozone layer is maintained as a result of a dynamic balance between both chemical and photochemical formation and destruction processes for ozone. Ozone formation occurs predominantly at altitudes above 30 km (see Figure 15–3), where solar ultraviolet radiation with wavelengths less than 242 nm dissociates molecular oxygen into oxygen atoms. The oxygen atoms rapidly react with molecular oxygen to form ozone in the presence of another molecule represented by M (this could be O_2 or N_2). The presence of M acts to stabilize ozone by removing its excess energy. The essentials of the process are represented by the following set of reactions:

$$O_2 + h\nu \, (\lambda < 242 \text{ nm}) \longrightarrow O + O \tag{15–1}$$

$$2[O + O_2 + M \longrightarrow O_3 + M] \tag{15–2}$$

$$\text{net: } 3O_2 \longrightarrow 2O_3$$

This mechanism accounts for the formation of ozone in the stratosphere; in the mesosphere (above 80 km), where three-body recombination processes are slower, additional mechanisms may be significant.

Ultraviolet radiation can also dissociate ozone in the longer wavelength region of 240–320 nm according to the reaction

---

* 1 atm at 25°C (STP).

Figure 15–2. Concentration profiles of some major species in the atmosphere. The solid line represents concentration profiles during the day, the dashed lines profiles at night.

$$O_3 + h\nu \longrightarrow O_2 + O \tag{15–3}$$

It is this reaction that is responsible for shielding the earth from harmful ultraviolet radiation. The photolysis of ozone is not, however, considered a true destruction mechanism, since almost all the oxygen atoms produced by this process rapidly combine with molecular oxygen to re-form ozone (equation 15–2).

Balancing the formation of ozone (equations 15–1 and 15–2) are several ozone destruction processes. One such process is the reaction of ozone with atomic oxygen to produce molecular oxygen:

$$O + O_3 \longrightarrow O_2 + O_2 \tag{15–4}$$

The ozone and oxygen are referred to as "odd oxygen species" in this scheme, and the reactions of both equation 15–4 and equation 15–1 are slower than the processes represented by equations 15–2 and 15–3 which merely interconvert odd oxygen. This can be illustrated by considering a simplified model using equations 15–1 through 15–4. At altitudes between 30 and 50 km, the stratosphere can be treated in terms of a steady-state system of chemical reactions. Using a steady-state treatment leads to an estimation of the local concentrations of oxygen atoms and ozone. The rate of formation of [O] is given by

$$\frac{d[O]}{dt} = 2j_1[O_2] + j_3[O_3] - [O](k_2[O_2][M] + k_4[O_3]) \tag{15–5}$$

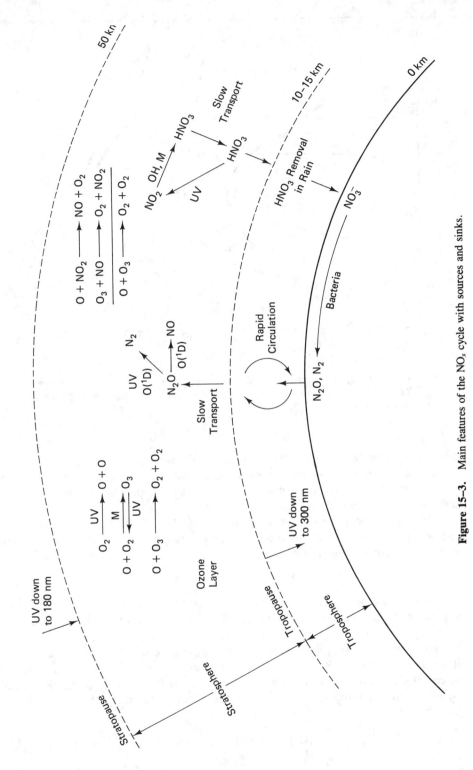

Figure 15–3. Main features of the NO_x cycle with sources and sinks.

where k_i is the chemical kinetic rate constant and j_i is the first-order rate constant for photolysis, given by

$$j_i \ (\text{sec}^{-1}) = \int \phi_i(\lambda) I_i(\lambda) \sigma_i(\lambda) d\lambda$$

where, in turn, $\phi_i(\lambda)$ is the quantum efficiency for photodissociation, $I_i(\lambda)$ is the intensity of the solar radiation, and $\sigma_i(\lambda)$ is the absorption cross section integrated over wavelength.

The rate of formation of ozone can be written as

$$\frac{d[O_3]}{dt} = k_2[O][O_2][M] - j_3[O_3] - k_4[O_3][O] \tag{15-6}$$

Under the steady-state approximation,

$$\frac{d[O]}{dt} = \frac{d[O_3]}{dt} = 0 \tag{15-7}$$

Solving equation 15–5 for [O] gives

$$[O] = \frac{2 j_l[O_2] + j_3[O_3]}{k_2[O_2][M] + k_4[O_3]} \tag{15-8}$$

and solving equation 15–6 for [O$_3$] gives

$$[O_3] = \frac{k_2[O][O_2][M]}{j_3 + k_4[O]} \tag{15-9}$$

Substituting equation 15–8 into equation 15–9 and solving the resulting quadratic equation for [O$_3$] gives

$$[O_3] = [O_2]\left(\frac{j_1}{2 j_3}\right)\left\{\left(1 + 4\frac{j_3}{j_1}\frac{k_2}{k_4}[M]\right)^{1/2} - 1\right\} \tag{15-10}$$

The magnitudes of [O$_3$] and [O] may then be calculated from data for rate constants and values for j_1 and j_3. Some typical values for these parameters at 30 km are $j_1 = 10^{-11.6}$ sec^{-1}, $j_3 = 10^{-3.5}$ sec^{-1}, $k_2 = 10^{-32.7}$ cm^6 molecule^{-2} s^{-1}, and $k_4 = 10^{-14.9}$ cm^3 molecule^{-1} s^{-1}. From the concentration profile shown in Figure 15–2, [O$_2$] is estimated to be $10^{+17.5}$ molecule cm^{-3}. With these estimates, we have

$$\frac{k_2[M]}{k_4} = \frac{10^{-15.1}}{10^{-14.9}} \approx 1$$

and

$$\frac{j_3}{j_1} \approx 10^{8.1}$$

which leads to the following simplification of equation 15–10:

$$[O_3] \approx [O_2] \cdot \left(\frac{j_1}{j_3}\frac{k_2}{k_4}[M]\right)^{1/2} \tag{15-11}$$

Evaluating equation 15–11 gives $[O_3] \approx 10^{12.8}$ molecules/cm^3, and substitution of this into equation 15–8 gives $[O] \approx 10^{7.4}$ atoms/cm^3. Rearranging equation 15–9 then gives the following expression for $[O_3]$:

$$[O_3] = \frac{k_2[O_2][M]/k_4}{((j_3/k_4[O]) + 1)} \tag{15–12}$$

and since

$$\frac{j_3}{k_4[O]} \approx \mathbb{O}(10^4) \gg 1$$

we have the result

$$\frac{[O_3]}{[O]} \approx \frac{k_2[O_2][M]}{j_3} \tag{15–13}$$

It can be seen from equation 15–13 that the reactions 15–2 and 15–3 serve merely to interconvert O and O_3. Furthermore, only reactions that destroy either O or O_3 without reproducing oxygen atoms or ozone will result in a direct reduction in $[O_3]$. In fact, the slowness of reaction 15–1 makes stratospheric ozone vulnerable to other removal processes. More precise estimates of the rate constants and more detailed calculations indicate that reaction 15–4 accounts for only 20% of the "odd oxygen" removal in the stratosphere, while other physical processes, such as the transport of ozone to the earth's surface, contribute only 0.5%. The general mechanism for ozone formation and destruction derived from only oxygen species was originally suggested by Chapman.[1] However, it has been learned over the last 30 years that there are other chemical processes that contribute to more significant removal of ozone.[2,3] Several of these are discussed next.

15.1.3 HO$_x$ Cycle

In the upper stratosphere, about 15% of the ozone removal rate occurs as a result of reactions involving hydrogen-containing species: H, HO, and HO$_2$. The importance of these species in reactions for ozone removal was first recognized by Bates and Nicolet.[4] The primary source of the species, which forms the basis of the HO$_x$ cycle, is from the reaction of excited atomic oxygen atoms (O(^{1}D)) with hydrogen-containing species such as naturally occurring water vapor (H$_2$O) and methane (CH$_4$). O(^{1}D) atoms are produced by the photolysis of ozone at wavelengths less than 310 nm according to the following reactions:

$$O_3 + h\nu\,(\lambda < 310 \text{ nm}) \longrightarrow O(^1D) + O_2 \tag{15–14}$$

$$O(^1D) + H_2O \longrightarrow OH + OH \tag{15–15}$$

$$O(^1D) + CH_4 \longrightarrow CH_3 + OH \tag{15–16}$$

The chief loss mechanism for O(^{1}D) atoms is not by either of reactions 15–15 or 15–16, because water and methane are only minor components of the stratosphere, but rather through quenching by N$_2$ and O$_2$ via the reaction

$$O(^1D) + M \longrightarrow O + M \qquad (15\text{–}17)$$

Methane transported from the troposphere is chiefly responsible for water vapor in the stratosphere, which results from the oxidation of methane by hydroxyl radicals $(OH\cdot)$ according to

$$OH + CH_4 \longrightarrow H_2O + CH_3 \qquad (15\text{–}18)$$

Other minor sources of OH radicals involve the photolysis of H_2O:

$$H_2O + h\nu\,(\lambda \le 190\ nm) \longrightarrow H + OH \qquad (15\text{–}19)$$

The removal of HO_x species from the stratosphere is by downward transport of H_2O to the troposphere, where they are involved in precipitation—usually cloud droplets and aerosol particles. The major catalytic cycles for ozone destruction involving HO_x species is attributed to the two-step reaction of hydroxy radicals with ozone and abstraction of oxygen atoms, the pair of which reactions regenerates oxygen as follows:

$$OH + O_3 \longrightarrow HO_2 + O_2 \qquad (15\text{–}20)$$

$$\underline{HO_2 + O \longrightarrow OH + O_2} \qquad (15\text{–}21)$$

$$\text{net:}\quad O + O_3 \longrightarrow 2O_2$$

In order for this overall reaction scheme to be effective, each reaction must be exothermic; rates for ozone destruction may then exceed those of the elementary reaction step 15–4, either because [OH] exceeds [O], or because the rate coefficient for reaction 15–20 exceeds that for reaction 15–4.

15.1.4 NO_x Cycle

The catalytic cycle involving nitric oxide (NO) and nitrogen dioxide (NO_2), collectively called NO_x, provides one of the most important destruction processes for ozone. Indeed, the process accounts for some 60–70 percent of the rate of destruction of ozone. Unlike the reactions involving HO_x species, which take place primarily in the upper stratosphere, the NO_x reaction occurs in the lower atmosphere. Crutzen[5] first highlighted the importance of naturally occurring nitrogen oxides in the ozone removal process due to the following reactions:

$$NO + O_3 \longrightarrow NO_2 + O_2 \qquad (15\text{–}22)$$

$$\underline{NO_2 + O \longrightarrow NO + O_2} \qquad (15\text{–}23)$$

$$\text{net:}\quad O + O_3 \longrightarrow 2O_2$$

Clearly, this process produces the same net effect as the reaction 15–21, but the overall rate for the NO_x cycle is faster.

The main source of stratospheric NO_x is the reaction of $O(^1D)$ atoms with nitrous oxide:

$$O(^1D) + N_2O \longrightarrow 2NO \qquad (15\text{–}24)$$

N_2O is produced at the earth's surface by bacteria in soil and water. Although most of the nitrous oxide is photodissociated into N_2 and O, about 5% reacts with excited oxygen atoms ($O(^1D)$) to yield nitric oxide, thereby activating the NO_x cycle. The role of NO_x species is illustrated in Figure 15–3. There are several other minor sources of NO_x. Cosmic rays which enter the earth's atmosphere contribute to NO production in the altitude range of 10–13 km. Ionization reactions involving N_2 also lead to production of NO. NO_x is removed by reacting with hydroxyl radicals via the reactions

$$OH + NO_2 + M \longrightarrow HNO_3 + M \qquad (15\text{–}25)$$

and

$$HNO_3 + h\nu \longrightarrow OH + NO_2 \qquad (15\text{–}26)$$

Reaction 15–26 binds about 40% of the NO_x into HNO_3, which is inactive in regard to ozone removal and is transported downwards into the troposphere where it is rapidly removed by rain, thereby returing the nitrate to the soil.

Let us consider the effect of HO_x and NO_x species on the Chapman mechanism. To assess the role of these reactions, we are primarily interested in those reactions which will change the total odd oxygen concentration, $[O] + [O_3]$. The significant reactions are as follows for the simple mechanism envisaged here:

$$O_2 + h\nu \longrightarrow 2O \qquad (1)$$
$$O + O_2 + M \longrightarrow O_3 + M \qquad (2)$$
$$O_3 + h\nu \longrightarrow O + O_2 \qquad (3)$$
$$O + O_3 \longrightarrow 2O_2 \qquad (4)$$
$$NO + O_3 \longrightarrow NO_2 + O_2 \qquad (5)$$
$$NO_2 + O \longrightarrow NO + O_2 \qquad (6)$$
$$NO_2 + h\nu \longrightarrow NO + O \qquad (7)$$
$$OH + O_3 \longrightarrow HO_2 + O_2 \qquad (8)$$
$$O + HO_2 \longrightarrow OH + O_2 \qquad (9)$$
$$O + OH \longrightarrow H + O_2 \qquad (10)$$

The rate expressions for [OH], [H] and [NO] are

$$\frac{d[OH]}{dt} = -k_8[OH][O_3] + k_9[O][HO_2] - k_{10}[O][OH]$$

$$\frac{d[H]}{dt} = k_{10}[O][OH]$$

$$\frac{d[NO]}{dt} = -k_5[NO][O_3] + k_6[NO_2][O] + j_7[NO]$$

and the total odd oxygen concentration is given by

$$\frac{d([\text{O}] + [\text{O}_3])}{dt} = 2j_1[\text{O}_2] - 2k_4[\text{O}][\text{O}_3]$$

$$-k_5[\text{NO}][\text{O}_3] - k_6[\text{NO}_2][\text{O}] + j_7[\text{NO}_2]$$

$$-k_8[\text{OH}][\text{O}_3] - k_9[\text{O}][\text{HO}_2] - k_{10}[\text{O}][\text{OH}]$$

If steady-state conditions for [OH], [H], and [NO] are assumed, the total oxygen concentration can be rewritten as

$$\frac{d([\text{O}] + [\text{O}_3])}{dt} = 2j_1[\text{O}_2] - 2[\text{O}](k_4[\text{O}_3] + k_6[\text{NO}_2] + k_9[\text{HO}_2])$$

The magnitude of $k_4[\text{O}_3]$ is calculated to be on the order of the sum $k_6[\text{NO}_2] + k_9[\text{HO}_2]$. The NO_x cycle is more important in limiting the dynamic balance of O_3 than the HO_x cycle, since the ratio $k_6[\text{NO}_2]/k_9[\text{HO}_2]$ which determines the relative importance of the nitrogen oxide and hydroxyl species is greater than unity.[2]

15.1.5 ClO_x Cycle

Many other substances can catalyze the destruction of ozone, but the HO_x and NO_x cycles constitute the main chemical and photochemical processes which determine the ozone concentration in the stratosphere. At present, a major public concern is the depletion of the ozone layer by photodissociation of halomethanes,[3,6] which are artificially produced substances such as CFCl_3 (Freon 11) and CF_2Cl_2 (Freon 12), as well as naturally occurring substances such as carbon tetrachloride (CCl_4), which also is artificially produced. These materials are inactive to reactions occurring in the troposphere and are transported into the stratosphere, where they are photolyzed according to the reaction

$$\text{CCl}_2\text{F}_2 + h\nu\,(180 \text{ nm} < \lambda < 220 \text{ nm}) \longrightarrow \text{CF}_2\text{Cl} + \text{Cl} \quad (15\text{--}27)$$

or they react with electronically excited oxygen atoms ($\text{O}(^1\text{D})$) to yield chlorine oxides (ClO_x):

$$\text{CCl}_2\text{F}_2 + \text{O}(^1\text{D}) \longrightarrow \text{CF}_2\text{Cl} + \text{ClO} \quad (15\text{--}28)$$

These products then participate in catalytic cycles analogous to the HO_x and NO_x cycles whose net effect is the removal of ozone and oxygen atoms:

$$\text{Cl} + \text{O}_3 \longrightarrow \text{ClO} + \text{O}_2 \quad (15\text{--}29)$$

$$\underline{\text{ClO} + \text{O} \longrightarrow \text{Cl} + \text{O}_2} \quad (15\text{--}30)$$

$$\text{net:}\quad \text{O} + \text{O}_3 \longrightarrow 2\text{O}_2$$

The major removal process for Cl atoms is reaction with methane to form HCl, which is only slowly attacked by hydroxyl radicals, allowing transport of HCl to the troposphere where it is removed by rain via the reactions

$$\text{Cl} + \text{CH}_4 \longrightarrow \text{HCl} + \text{CH}_3 \quad (15\text{--}31)$$

and

$$OH + HCl \longrightarrow H_2O + Cl \qquad (15\text{--}32)$$

At lower altitudes, the ClO_x cycle can link with the HO_x cycle in the following reactions:[3,6]

$$OH + O_3 \longrightarrow HO_2 + O_2 \qquad (15\text{--}33)$$

$$Cl + O_3 \longrightarrow ClO + O_2 \qquad (15\text{--}34)$$

$$HO_2 + ClO \longrightarrow HOCl + O_2 \qquad (15\text{--}35)$$

$$\underline{HOCl + h\nu \longrightarrow OH + Cl} \qquad (15\text{--}36)$$

net: $\qquad 2O_3 \longrightarrow 3O_2$

Rapid reaction of ClO with nitric oxide is an important reaction which links the NO_x to the ClO_x cycle:

$$ClO + NO \longrightarrow Cl + NO_2 \qquad (15\text{--}37)$$

Once initiated, the longer catalytic cycle that results in ozone destruction involving NO_x and ClO_x radicals occurs by the following chain of reactions:

$$Cl + O_3 \longrightarrow ClO + O_2 \qquad (15\text{--}38)$$

$$NO + O_3 \longrightarrow NO_2 + O_2 \qquad (15\text{--}39)$$

$$ClO + NO_2 \longrightarrow ClONO_2 \qquad (15\text{--}40)$$

$$ClONO_2 + h\nu \longrightarrow NO_3 + Cl \qquad (15\text{--}41)$$

$$\underline{NO_3 + h\nu \longrightarrow NO + O_2} \qquad (15\text{--}42)$$

net: $\qquad 2O_3 \longrightarrow 3O_2$

Taking account of all these reactions in model calculations based on the available rate data for such reactions indicates that continued release of halomethanes could eventually produce about a 6–8% depletion of ozone in the ozone layer.[7]

15.1.6 Modeling Studies of the Atmosphere

The task of atmospheric modeling studies is to understand to what extent gases produced at the earth's surface affect the stratospheric ozone. To accomplish this, several pieces of information are needed: (1) the identity and amounts of these gases which are released into the atmosphere; (2) their rates of transport to the stratosphere; (3) the absorption and photolytic properties of these compounds in the stratosphere; (4) the rates of reaction in ozone destruction catalytic cycles; and (5) the location and nature of sinks for removal of these species and their end products from the atmosphere.

Atmospheric models deal with over 160 chemical reactions and more than 40 reactive species. Each time a reaction rate is remeasured, or a new reaction or new species is discovered, the whole interlocking system is affected and must be remodeled. Models of the stratosphere are of three types: (1) a one-dimensional model which computes the composition of the atmosphere as a global average that varies only with altitude; (2) a two-dimensional model which calculates changes with both altitude and latitude, giving a longitudinal average; and (3) a three-dimensional model which includes altitude, latitude, and longitude as variables. Although a full three-dimensional model provides a more realistic picture, this sort of calculation is prohibitively expensive; and only a limited number of chemical reaction systems can be treated. Currently, most models of the upper atmosphere use one-dimensional models with altitude as the sole spatial variable.

The general problem of simulating the kinetics of the many species in the atmosphere even for a one-dimensional model is a mammoth task. It is clear from the required number of reactions (over 160) and the number of species (over 40) that, even for a modest treatment, the problem cannot be treated analytically with the methods discussed in chapter 2. Consequently, solutions must be obtained numerically. Attempts to produce such solutions using various numerical integration techniques such as Euler's method or Runge-Kutta methods (chapter 2) pose difficulties because of the wide range of rate constants which result in wide graduations in the time constants in the system of chemical reactions, thereby producing stiff equations. Use of a multivalue predictor-corrector method such as Gear's method (chapter 2) is found to be well suited for treating chemical systems numerically in atmospheric modeling. Edelson[8] has discussed this use of the Gear method for solving chemical kinetic rate equations.

15.1.7 Measurement Systems for the Atmosphere

Verification of existing atmospheric chemistry models is important for resolving problems concerning, for example, the likely causes of ozone depletion or the origin of acid rain. In assessing many of the models proposed for various reaction cycles such as HO_x, NO_x, ClO_x, etc., an important first step is the detection of species implicated in the mechanisms. Detection of the presence of various radicals in the atmosphere is the key to establishing the chemical structure of the atmosphere. Second, measurement of the elementary reaction rates for many species implicated in proposed cycles makes it possible to assess the reasonableness of complicated mechanisms describing certain regions of the atmosphere, which in turn allows concentration trends in the column levels of O_3 to be predicted. However, the most essential step in model verification is measurement of concentration levels of species implicated in atmospheric chemistry; information gained from these studies is necessary for evaluating the importance of various reactions in the atmosphere. Although ideally one would like to make *in situ* measurements, this is not always possible because of the lack of sensitive spectroscopic techniques for the detection of specific species. For this reason, most of the information we have about the chemistry of the atmosphere comes from laboratory observations—from kinetic studies—and not *in situ*. Studies aimed at species detection and rate constant measurement under laboratory conditions do, however, stimulate the development of experimental techniques for *in situ* measurements. Many of the techniques used in the laboratory and for *in situ* measurements have been discussed in chapter 3. In this section, we describe a few of the techniques used for *in situ* measurements.

Two of the most important measurement techniques currently used for investigating chemical processes in the upper atmosphere are LIDAR and reel-down methods. Basic LIDAR[9] systems employ the laser-induced fluorescence scheme (see chapter 3) for *in situ* measurement of various species in the atmosphere. The advantage of the LIDAR system is that spectral selection can be used to make detection

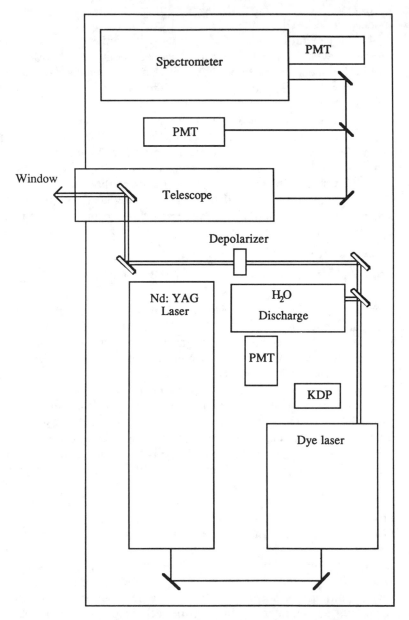

Figure 15–4. Schematic diagram of the LIDAR system. The abbreviations PMT and KDP denote *photomultiplier tube* and *potassium dihydrogen phosphate*, respectively. An Nd: YAG laser is *a neodymium: YAG laser*.

(a)

(b)

Figure 15–5. Diagram of the reel-down system, showing the winching system and suspended payload. (a) Geometry of the experiment in the docked position. (b) Schematic diagram of the suspended instrument system used to make *in situ* measurements. [Adapted from Anderson et al., *Science 228*, 1309 (1985).]

species-specific. A major disadvantage of the system is that fluorescence from aerosols and SO_2 excited by the laser and scattering from solar flux interfere with the measurements. Nevertheless, LIDAR systems have been used to make accurate measurements of OH and NO concentrations *in situ* with a sensitivity of a few parts per trillion. LIDAR system components are a laser source, transmitting optics, a receiver telescope, a detection system, and signal-processing electronics; a typical device for field measurements is 4.5 m long and 4 m high and weighs about 1,300 kg. A schematic diagram of a LIDAR system is shown in Figure 15–4.

The reel-down technique[10] uses resonance fluorescence or laser-induced fluorescence methods in balloon-parachute experiments. In this manner, concentrations of O, O_3, OH, HO_2, Cl, and ClO can be measured. As shown in Figure 15–5, the reel-down system consists of a payload which is repeatedly lowered and raised from a platform attached to a stationary balloon. Figure 15–6 shows the atmospheric oxygen atom concentration profile as a function of altitude as measured by the reel-down technique along with a typical calculated oxygen atom profile.

15.2 THE HYDROGEN-OXYGEN REACTION, AN EXPLOSIVE COMBUSTION PROCESS

As a further example of a multicomponent kinetic system, consider the reaction between hydrogen and oxygen. Although starting with only two reactants, the complete mechanism involves eight major species and at least 16 reactions for its description. It is also an important example of a *branched-chain reaction* leading to an explosion; we shall analyze this feature using a simplified reaction set which includes five reactions. Further discussion of the hydrogen-oxygen system may be found in the texts by Espenson[11] and Emanuel' and Knorre.[12]

The overall reaction

$$2H_2 + O_2 = 2H_2O \tag{15–43}$$

is exothermic by 228.7 kJ/(mol H_2O); but mixtures of gaseous hydrogen and oxygen are quite stable, because the rate of the four-center $H_2 + O_2$ reaction, or any other conceivable direct reaction between the two gases, is zero. (The reaction half-time has been estimated to be much larger than the age of the universe.) If, however, the reaction is initiated by some free-radical species, reaction 15–43 proceeds very rapidly and indeed violently. The radicals are typically H and O atoms arising from the dissociation of H_2 and O_2, respectively; since these species are consumed in the reaction, the radicals should be considered not catalysts (see chapter 5), but rather initiators of the reaction.

A simplified mechanism which represents the behavior of the $H_2 + O_2$ system at low pressures is the following:

Initiation: $H_2 \xrightarrow{w_0} 2H$

$O_2 \xrightarrow{w_0} 2O$

Chain branching: (2) $H + O_2 \xrightarrow{k_2} OH + O$

(3) $O + H_2 \xrightarrow{k_3} OH + H$ (15–44)

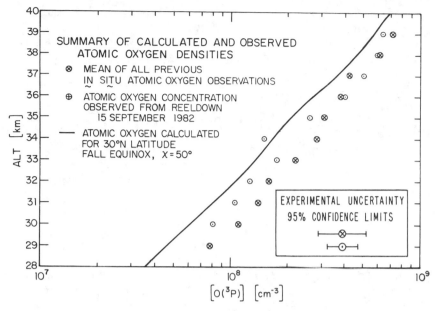

Figure 15–6. Comparison between the O(^{3}P) concentration profile from the reel-down experiment and earlier *in situ* measurements. [From Anderson et al., *Science* **228**, 1309 (1985).]

Chain propagation: (1) $OH + H_2 \xrightarrow{k_1} H_2O + H$

Chain termination: (4) $H + wall \xrightarrow{k_4} \frac{1}{2}H_2$

(5) $H + O_2 + M \xrightarrow{k_5} HO_2 + M$

In this mechanism, the initiation step is the dissociation of some amount of molecular species by a spark, flame, electric discharge, or other means; the step is characterized by a rate w_0 of free-radical production. To assess the contributions of steps 1 through 3, we must look at the enthalpy changes for these reactions. We can take the O=O bond energy to be approximately 5.1 eV, while for H-H it is 4.5 eV and for an O-H bond it is 4.4 eV. Thus, reaction 2, which involves breaking an O=O bond and making an O-H bond, is 0.7 eV endothermic (about 70 kJ/mol) and progresses slowly. Reactions 3 and 1, on the other hand, involve breaking an H-H bond and making an O-H bond, so the enthalpy change is +0.1 eV, or about 9 kJ/mol endothermic. This energy is on the order of a few $k_B T$, and so these reactions are relatively fast. The OH and O radicals are therefore rapidly consumed by the reactions, and the principal *chain carrier* is H atoms, the number of which we can set equal to the total free-radical density n. Reactions 2 and 3 are called *chain-branching* reactions, because two radical species are produced for every one consumed; reaction 1 is a propagation reaction which generates one radical for each one consumed. The principal termination steps are loss of H atoms at the wall (reaction 4 and three-body recombination of H and O_2 to form HO_2 (reaction 5). The latter is a termination step because of HO_2 species (which has been identified spectroscopically in hydrogen-oxygen flames) is unreactive with either H_2 or O_2, and therefore cannot propagate the chain reaction.

The analysis proceeds by finding the rate equation for the free-radical density n, which is taken to be the same as that for [H] atoms:

$$\frac{dn}{dt} = \frac{d[H]}{dt}$$

$$= w_0 - k_2[H][O_2] + k_3[O][H_2] + k_1[OH][H_2] - k_4[H] - k_5[H][O_2][M] \quad (15\text{–}45)$$

The other free-radical species are governed by the equations

$$\frac{d[OH]}{dt} = k_2[H][O_2] + k_3[O][H_2] - k_1[OH][H_2] \quad (15\text{–}46)$$

and

$$\frac{d[O]}{dt} = k_2[H][O_2] - k_3[O][H_2] \quad (15\text{–}47)$$

Since [O] and [OH] are both much lower than [H], we can make the steady-state assumption for these species; we must *not* do so for [H], however, if we are to obtain a solution. Accordingly, we set

$$\frac{d[OH]}{dt} = \frac{d[O]}{dt} = 0$$

giving

$$[O]_{ss} = \frac{k_2[H][O_2]}{k_3[H_2]}$$

and

$$[OH]_{ss} = \frac{k_2[H][O_2] + k_3[O][H_2]}{k_1[H_2]}$$

$$= \frac{k_2[H][O_2] + k_3[H_2]k_2[H][O_2]/k_3[H_2]}{k_1[H_2]}$$

$$= 2\frac{k_2[H][O_2]}{k_1[H_2]}$$

Combining all these relationships, we find, for the free-radical production rate,

$$\frac{dn}{dt} = w_0 - k_2[H][O_2] + k_3\frac{k_2[H][O_2]}{k_3[H_2]}[H_2]$$

$$+ k_1[H_2] \cdot \frac{2k_2[H][O_2]}{k_1[H_2]} - k_4[H] - k_5[H][O_2][M]$$

$$= w_0 + (2k_2[O_2] - k_4 - k_5[O_2][M])n$$

so that

$$\frac{dn}{dt} = w_0 + (f - g)n \quad (15\text{–}48)$$

In equation 15–48, the term $f = 2k_2[O_2]$ represents chain branching, which leads to an increase in the free-radical concentration, while the term $g = k_4 + k_5[O_2][M]$ represents chain termination, which removes free radicals.

Two classes of solutions of equation 15–48 for $n(t)$ are possible, depending on whether $g > f$ (termination exceeds branching) or, conversely, $g < f$. The former condition implies that $k_4 > (2k_2 - k_5[M])[O_2]$ so $g > f$ is guaranteed at sufficiently low O_2 pressures. The solution is then

$$n = \frac{w_0}{g - f}\{1 - e^{-(g-f)t}\} \qquad (15-49)$$

At short times, n increases linearly with slope $w_0 t$ and reaches a steady-state value $n_{ss} = w_0/(g - f)$, as shown in Figure 15–7(a). At higher O_2 pressures $g < f$, and the solution in this case, viz.,

$$n = \frac{w_0}{f - g}\{e^{+(f-g)t} - 1\} \qquad (15-50)$$

diverges, as shown in Figure 15–7(b). The free-radical concentration then increases exponentially, and, since the overall rate depends on the radical concentration, the reaction velocity also increases rapidly. This constitutes an explosion.

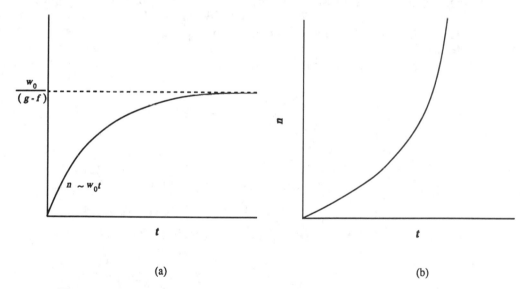

Figure 15–7. Time dependence of free-radical density n in the hydrogen-oxygen reaction. (a) below explosion limit, $g > f$. (b) above explosion limit, $g < f$.

The hydrogen-oxygen reaction behaves quite differently in different pressure regimes, as shown in Figure 15–8. At low pressures (region I), the reaction is wall-recombination limited, and a steady reaction occurs. At a critical pressure p_1^*, at which f begins to exceed g, the branched-chain reaction takes over and explosion ensues (region II). This pressure, known as the *first explosion limit*, depends on the temperature, initial $[H_2][O_2]$ ratio, presence of free-radical scavengers, and other

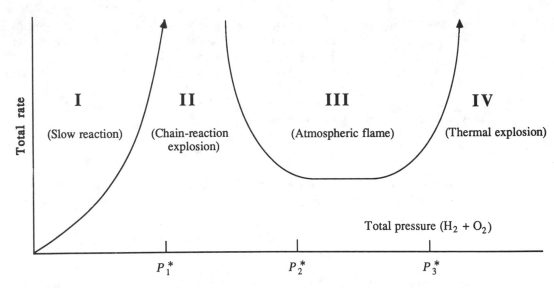

Figure 15–8. Explosion limits in the hydrogen-oxygen reaction.

parameters. At the pressure is further increased, the explosion is quenched and another region of steady reaction is encountered (region III, atmospheric oxy-hydrogen flame). In this region, the kinetics are dominated by the relatively unreactive hydroperoxyl free radical.[13] Other reactions that are important in this region, in addition to those in the mechanism given by equations 15–44, include the following:

$$2HO_2 \longrightarrow H_2O_2 + O_2$$

$$H_2 + HO_2 \longrightarrow H_2O_2 + H$$

$$H + HO_2 \longrightarrow 2OH$$

$$\longrightarrow H_2 + O_2$$

$$\longrightarrow H_2O + O$$

$$H_2O_2 + M \longrightarrow 2OH + M$$

$$H + H_2O_2 \longrightarrow H_2O + OH$$

$$\longrightarrow O_2 + H_2$$

$$OH + H_2O_2 \longrightarrow H_2O + HO_2$$

$$H + OH + M \longrightarrow H_2O + M$$

$$H + H + M \longrightarrow H_2 + M$$

At a still higher pressure p_3^*, the amount of heat liberated in the exothermic steps of the mechanism becomes larger than can be dissipated by conduction and other thermal transport processes, and the temperature of the reaction mixture rises. This, in turn, causes the rates of initiation and of the endothermic reactions to increase, liberating more heat, and so on, resulting in a *thermal explosion* (region IV).

We have noted that the hydrogen-oxygen reaction, once ignited, can proceed violently and explosively. This in fact has been the cause of several well-known major disasters. The hydrogen-filled dirigible *Hindenburg* was destroyed in 1937 by an atmospheric flame reaction (region III); newsreel photographs (see Figure 15–9(a)) show the ship burning but not actually exploding. The loss of the space shuttle *Chal-*

Figure 15–9. (Top) Explosion of the hydrogen-filled dirigible *Hindenburg* in Lakehurst, New Jersey, May 1937. [UPI/Bettman Newsphotos.] (Bottom) Explosion of the space shuttle *Challenger* at Cape Canaveral, Florida, January 1986. [AP/Wide World Photos.]

lenger in January, 1986 (Figure 15–9(b)), was most likely a thermal explosion (region IV), which occurred when the hydrogen- and oxygen-filled fuel tanks were ruptured and ignited by rocket exhaust penetrating a faulty O-ring seal.

15.3 THE METHANE COMBUSTION PROCESS

15.3.1 General Methane Oxidation Mechanism

The kinetics of the methane-oxygen system have been studied extensively over the years, because methane is the simplest hydrocarbon fuel and details of its combusion kinetics can be helpful in understanding the combustion of heavier hydrocarbon fuels. Further, methane is an important practical fuel which constitutes about 90% of the composition of natural gas. Like the hydrogen-oxygen system, the methane-oxygen system is complicated, involving at least 22 reaction steps and at least 12 species, depending on temperature and pressure. Also like the hydrogen-oxygen system, the methane-oxygen system is a branched chain reaction. However, before any branched chain reaction can take place, the reactions that generate radical intermediates must occur. This is not a facile process, since bonds must be broken. Methane oxidation is initiated by essentially two reactions, a hydrogen abstraction reaction from methane by oxygen, given by

$$CH_4 + O_2 \longrightarrow CH_3 + HO_2 \qquad (15\text{–}51)$$

and a collision-induced dissociation of methane, represented by

$$CH_4 + M \longrightarrow CH_3 + H + M \qquad (15\text{–}52)$$

The thermal C-H bond fission described by equation 15–52 is slower for CH_4 than for any other hydrocarbon because the C-H bond in CH_4 is stronger than that of most other hydrocarbons (e.g., 435 kJ/mol for methane as compared with 409 kJ/mol for n-propane or 385 kJ/mol for t-butane).

Following initiation, the reaction is propagated as a result of the following hydrogen abstraction reactions from CH_4 by radicals such as H, O, OH, and HO_2:

$$CH_4 + H \longrightarrow CH_3 + H_2$$
$$CH_4 + O \longrightarrow CH_3 + OH$$
$$CH_4 + OH \longrightarrow CH_3 + H_2O \qquad (15\text{–}53)$$
$$CH_4 + HO_2 \longrightarrow CH_3 + H_2O_2$$

The methyl radicals produced by the reactions in the mechanism given by equations 15–53 are oxidized to formaldehyde (CH_2O) and formyl radical (HCO) and eventually form carbon monoxide (CO). This is accomplished by the reaction of methyl radicals with O and O_2:

$$CH_3 + O_2 \longrightarrow CH_3O + O$$
$$CH_3 + O_2 \longrightarrow HCO + H_2O$$
$$CH_3 + O \longrightarrow CH_2O + H$$

These reactions form three new species, HCO, CH₂O, and CH₃O, which also partic-ipate in the methane-oxygen reaction by the following steps:

$$CH_3O + O \longrightarrow CH_3 + O_2$$

$$CH_3O + O_2 \longrightarrow CH_2O + HO_2$$

$$CH_3O + M \longrightarrow CH_2O + H + M$$

$$CH_2O + O \longrightarrow HCO + OH$$

$$CH_2O + H \longrightarrow HCO + H_2$$

$$CH_2O + M \longrightarrow HCO + H + M$$

$$HCO + O \longrightarrow CO + OH$$

$$HCO + HO_2 \longrightarrow CO + H_2O_2$$

Clearly, once methyl radicals are produced, they react to oxidize to the final product, carbon monoxide (CO). The oxidation process is terminated by the recom-bination reactions

$$H + OH + M \longrightarrow H_2O + M$$

$$HCO + OH \longrightarrow CO + H_2O$$

and

$$HCO + H \longrightarrow CO + H_2$$

The net reaction mechanism is

$$CH_4 + \tfrac{3}{2}O_2 \longrightarrow CO + 2H_2O$$

The exothermicity of this reaction is -543.8 kJ/mol methane.

15.3.2 Low-Temperature Oxidation of Methane

The mechanism for low-temperature combustion of methane differs from the general mechanism just given in that a number of reactions decrease in importance because reactions with high activation energies become less significant at lower temperatures. For example, of the initiation reactions, the collision-induced dissociation given by equation 15–52 is two orders of magnitude faster than equation 15–51 at a tempera-ture of 600 K. Consequently, the rate-limiting step at lower temperatures is the col-lision-induced dissociation of methane. After the initiation process, reaction is ac-celerated by radicals that enhance the rate of oxidation of methane to CO and H₂O. Among the ensuing radical-methane reactions, those which are most significant are

$$CH_4 + O \longrightarrow CH_3 + OH$$

and

$$CH_4 + OH \longrightarrow CH_3 + H_2O$$

since the activation energies for these two processes are about 8.4 and 31.9 kJ/mol, respectively. The latter reaction is the most important water-producing reaction in the methane combustion process. The other abstraction reactions produce little water at lower temperatures due to the fact that they have high activation energies. The principal reactions involving CH_3 are

$$CH_3 + O_2 \longrightarrow CH_3O + O$$

and

$$CH_3 + O \longrightarrow CH_2O + H$$

The former is more important in this mechanism than the reaction

$$CH_3 + O_2 \longrightarrow CH_2O + OH$$

Methoxyl radicals (CH_3O) react primarily by

$$CH_3O + M \longrightarrow CH_2O + H + M$$

with decomposition providing the major fraction of CH_2O. At lower temperatures the decomposition of CH_2O via the reaction

$$CH_2O + M \longrightarrow HCO + H + M$$

becomes insignificant because of the high activation energy about (339 kJ/mol). The most significant reaction of formaldehyde at lower temperatures is

$$CH_2O + O \longrightarrow HCO + OH$$

which has an activation energy of about 18 kJ/mol. This is followed by the production of CO by the reaction

$$HCO + O \longrightarrow CO + OH$$

which occurs at nearly gas-kinetic rates at low temperatures. The reaction responsible for the production of CO_2 is

$$CO + OH \longrightarrow CO_2 + H$$

This reaction has a low activation energy of 33 kJ/mol compared to 96 kJ/mol for the reaction of carbon monoxide with hydroperoxy radicals via

$$CO + HO_2 \longrightarrow CO_2 + OH$$

Consequently, the reaction of CO with OH is the most significant source of CO_2 in the combustion of methane at low temperatures.

Termination of the process follows by the radical-radical reactions

$$HCO + H \longrightarrow CO + H_2$$

and

$$HCO + OH \longrightarrow CO + H_2O$$

Thus, the number of reactions that take place is significantly reduced from that of the initial simplified mechanism when combustion occurs at lower temperatures. The reduced lower temperature mechanism can be summarized as follows:

$$CH_4 + O_2 \longrightarrow CH_3 + HO_2$$
$$CH_4 + O \longrightarrow CH_3 + OH$$
$$CH_4 + OH \longrightarrow CH_3 + H_2O$$
$$CH_3 + O_2 \longrightarrow CH_3O + O$$
$$CH_3O + M \longrightarrow CH_2O + H + M$$
$$CH_2O + O \longrightarrow HCO + OH$$
$$HCO + O \longrightarrow CO + HO$$
$$CO + OH \longrightarrow CO_2 + H$$
$$O_2 + M \longrightarrow 2O + M$$
$$H + O_2 \longrightarrow OH + O$$
$$H + H_2O \longrightarrow H_2 + OH$$
$$H + OH + M \longrightarrow H_2O + M$$
$$HO_2 + O \longrightarrow OH + O_2$$
$$HO_2 + OH \longrightarrow O_2 + H_2O$$
$$HCO + H \longrightarrow CO + H_2$$
$$HCO + OH \longrightarrow CO + H_2O$$

As mentioned earlier in this section, combustion of methane at low temperatures is slow. Usually there is an induction period lasting several minutes, or even hours, depending on temperature and pressure. Following the induction period, there is an increase in the reaction rate, but under conditions of low pressure and temperature the reaction rate may reach a maximum and then decrease. Typical pressure-vs.-time curves, illustrated in Figure 15–10, show that for methane there is an induction period followed by acceleration and then a slowdown in the oxidation rate. During this induction period, the reactions which are most significant are the following:

$$CH_4 + M \longrightarrow CH_3 + H + M \tag{1}$$
$$H + O_2 \longrightarrow OH + O \tag{2}$$
$$CH_4 + O \longrightarrow CH_3 + OH \tag{3}$$
$$OH + CH_4 \longrightarrow CH_3 + H_2O \tag{4}$$
$$CH_3 + O_2 \longrightarrow CH_3O + O \tag{5}$$
$$CH_3O + M \longrightarrow CH_2O + H + M \tag{6}$$
$$CH_3 + O \longrightarrow CH_2O + H \tag{7}$$

Since the concentrations of CH_4, O_2, and OH are approximately constant during the induction period, the rate equations describing the kinetics reduce to a set of

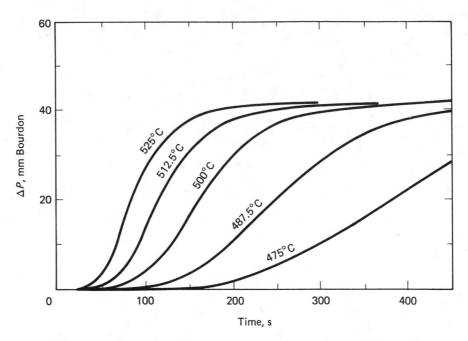

Figure 15–10. Typical pressure-vs.-time curves in the oxidation of CH_4, starting with 200 torr CH_4 and 100 torr O_3 in an HF-treated silica vessel (1.8 mm bourdon-1 torr). [From D. E. Hoare, "Low Temperature Oxidation," in *Combustion of Methane* (W. Jost, ed.) p. 125 (New York: Gordon and Breach, 1968). Reproduced with permission.]

solvable linear differential equations. Also, since the species CH_3O, H, O, and OH are produced in small concentrations, the steady-state approximation may be used. The rate equations for the mechanism given by equations 15–54 may be written as follows:

$$\frac{d[CH_3O]}{dt} = k_5[CH_3][O_2] - k_6[CH_3O][M] \tag{15–55a}$$

$$\frac{d[H]}{dt} = k_1[CH_4][M] - k_2[H][O_2] + k_6[CH_3O][M] + k_7[CH_3][O] \tag{15–55b}$$

$$\frac{d[O]}{dt} = k_2[H][O_2] - k_3[CH_4][O] + k_5[CH_3][O_2] - k_7[CH_3][O] \tag{15–55c}$$

$$\frac{d[OH]}{dt} = k_2[H][O_2] + k_3[CH_4][O] - k_4[OH][CH_4] \tag{15–55d}$$

$$\frac{d[CH_3]}{dt} = k_1[CH_4][M] + k_3[O][CH_4] + k_4[CH_4][OH] - k_5[CH_3][O_2]$$

$$- k_7[CH_3][O] \tag{15–55e}$$

Applying the steady-state approximation to equations 15–55(a)–(d), and simplifying the expression, we find that the rate equation for the methyl radical is

$$\frac{d[CH_3]}{dt} = 4k_1[CH_4][M] + 4k_5[CH_3][O_2] \qquad (15\text{--}56)$$

and if $[CH_4]$, $[O_2]$, and $[M]$ are approximately constant, this expression can be integrated to give

$$[CH_3]_{ss} = \frac{k_1[CH_4][M]}{k_5[O_2]}(e^{4k_5[O_2]t} - 1) + [CH_3]_0 e^{4k_5[O_2]t} \qquad (15\text{--}57)$$

The rate of consumption of methane can be written as

$$\frac{d[CH_4]}{dt} = -4k_1[CH_4][M] - 5k_5[CH_3][O_2] - k_7[CH_3][O] \qquad (15\text{--}58)$$

Assuming that $[CH_3][O]$ is negligible during the induction period, the last term in equation 15–58 can be ignored. Consequently, equation 15–58 simplifies to

$$\frac{d[CH_4]}{dt} = -4k_1[CH_4][M] - 5k_5[CH_3][O_2] \qquad (15\text{--}59)$$

Substituting the steady-state expression for $[CH_3]$ into this equation, assuming that $[CH_3]_0 = 0$, and simplifying we obtain

$$\frac{d[CH_4]}{dt} = -k_1[CH_4][M]\{5(e^{4k_5[O_2]t} - 1)\} \qquad (15\text{--}60)$$

Integration of this expression gives the fractional methane loss as

$$\frac{[CH_4]_0 - [CH_4]}{[CH_4]_0} = 1 - \exp\left\{\frac{5k_1[M]}{4k_5[O_2]}\left(1 - e^{4k_5[O_2]t} + \frac{4}{5}k_5[O_2]t\right)\right\} \qquad (15\text{--}61)$$

Defining the induction time t_i to be the time period during which 10% of the methane is consumed, we obtain the result

$$t_i = \frac{1}{4k_5[O_2]}\ln\left\{\frac{0.08k_5[O_2]}{k_1[M]} + 1 + \frac{5}{4}k_5[O_2]t_i\right\} \qquad (15\text{--}62)$$

This equation may be solved iteratively for t_i. The expression is valid until combustion has progressed to a point at which radical concentrations become significant, so that the steady-state approximation is no longer valid.

15.3.3 Higher Temperature Oxidation of Methane

At elevated temperatures ($T > 700$ K), the combustion mechanism for methane differs considerably from the low-temperature oxidation process described. The process becomes rapid, resulting in the following major changes in the reaction mechanism that lead to an accelerated rate of oxidation of methane:

1. The rates of the initiation processes increase.
2. Chain branching steps become more important. Because of the elevated temperatures, reactions with high activation energies, such as

$$CH_3 + O_2 \longrightarrow CH_3O + O$$

$$CH_3O + M \longrightarrow CH_2O + H + M$$

and

$$CH_3 + O \longrightarrow CH_2O + H$$

accelerate the oxidation process by increasing the radical pool of [H] and [O] species that propagate branched chain oxidation. Another important chain-branching step is the abstraction of hydrogen from methane by HO radicals, given by

$$HO + CH_4 \longrightarrow H_2O + CH_3$$

3. At the higher temperatures, CO plays an important role as an intermediate in the oxidation process. The reaction of interest is

$$CO + HO \rightleftharpoons CO_2 + H$$

The left-to-right reaction enhances chain branching by increasing the source of H atoms; conversely, the right-to-left reaction acts to reduce chain branching. The additional CO also promotes termination via the reaction

$$O + CO + M \longrightarrow CO_2 + M$$

At higher temperatures, the radical concentrations can be quite large. The rate constants for radical-radical reactions, which have virtually zero activation energies, do not themselves change significantly, but because of the high radical concentrations the rates of these reactions may become appreciable. In fact, it is these reactions which give rise to C_2H_6 by recombination, i.e.,

$$CH_3 + CH_3 + M \longrightarrow C_2H_6 + M$$

The ethane, in turn, is oxidized by H, O, and OH radicals. A summary of the high-temperature mechanism for the oxidation of methane is shown in Figure 15–11.

15.3.4 Atmospheric Oxidation of Methane

The major source of methane in the earth's atmosphere is not natural gas seepage; rather there are two major biological sources: bacterial fermentation processes that occur in swamps, lakes, and oceans; and bovine eructation, i.e., enteric fermentation in cattle. Once in the atmosphere, methane affects the chemistry of the troposphere and the stratosphere. In the troposphere, methane absorbs infrared radiation at 7.7 μm and traps this solar radiation as heat. Consequently, an increase in methane can lead to a rise in the temperature on the surface of the earth.[14] In the stratosphere, methane participates in catalytic cycles which deplete ozone that shields the earth's surface from harmful ultraviolet radiation. In order to assess the impact of increasing amounts of methane on the chemistry of the atmosphere, the underlying chemistry of methane must be well understood. How is methane oxidized in the atmosphere? Since temperatures in the lower atmosphere are below 300 K, kinetic processes which require little or no activation energy dominate the atmospheric

Figure 15–11. Mechanism of combustion of C_1/C_2 hydrocarbon species at high temperature. [After J. Warnatz, *Eighteenth Symposium (International) on Combustion* (Pittsburgh, PA: The Combustion Institute, 1981), pp. 369–384, and "Chemistry of Stationary and Non-stationary Combustion," in *Modeling of Chemical Reaction Systems,* edited by K. H. Ebert, P. Deuflhared, and W. Jager (Berlin: Springer-Verlag, 1981), pp. 162–188.]

oxidation of methane. Thus, the reactions

$$CH_4 + M \longrightarrow CH_3 + H + M$$

and

$$CH_4 + O_2 \longrightarrow CH_3 + HO_2$$

are insignificant in atmospheric oxidation. Methane oxidition in the troposphere occurs primarily by attack of hydroxyl radicals, i.e.,

$$CH_4 + OH \longrightarrow CH_3 + H_2O$$

a reaction which has a low activation energy (about 50 kJ/mole); by contrast, in the stratosphere attack by oxygen atoms is important:

$$CH_4 + O(^1D) \longrightarrow CH_3 + OH$$

The dominant propagation step is reaction with oxygen to form peroxyl radicals, i.e.,

$$CH_3 + O_2 \longrightarrow CH_3O_2$$

NO also plays a major role in the oxidation of methane via the reaction

$$CH_3O_2 + NO \longrightarrow CH_3O + NO_2$$

A further step is

$$CH_3O + O_2 \longrightarrow CH_2O + HO_2$$

The formaldehyde is finally oxidized to CO and CO_2. Figure 15–12 summarizes the major chemical reactions oxidizing methane in the atmosphere.

Figure 15–12.   Scheme for atmospheric oxidation of methane.

15.4 CONCLUSION

We have seen, in the preceding section, how a full description of an apparently simple reaction—the combustion of methane in oxygen—can lead to a very complicated mechanism, possessing numerous reactive intermediates and elementary reaction steps. Which of these intermediates and elementary processes are most important depends on the temperature and pressure at which the reaction takes place. One of the greatest strengths of modern chemical kinetics, especially when informed by an understanding of what is occuring at the molecular level, is the ability to provide detailed descriptions of such systems and to predict their behavior under a wide range of experimental conditions.

REFERENCES

1. S. Chapman, *Mem. Roy. Meteorol. Soc. 3*, 103 (1930).
2. H. S. Johnston, *Ann. Rev. Phys. Chem. 26*, 315 (1975).
3. M. J. Molina and F. S. Rowland, *Nature 249*, 810 (1974).
4. D. R. Bates and M. Nicolet, *J. Geophys. Res. 55*, 30 (1950).
5. J. P. Crutzen, *Quart. J. Roy. Meteorol. Soc. 96*, 320 (1970).
6. F. S. Rowland and M. J. Molina, *Rev. Geophys. Space Phys. 13*, 1 (1975).
7. R. T. Watson, ed., *World Meteorological Organization Report No. 16*, Global Ozone Research and Monitoring Project, Geneva (1985).
8. D. Edelson, *J. Chem. Educ. 52*, 642 (1975).
9. E. D. Hinkley, ed., *Laser Monitoring of the Atmosphere, Topics in Applied Physics*, Vol. 14 (Berlin: Springer Verlag, 1976).
10. J. G. Anderson, N. L. Hazen, B. E. McLaren, S. P. Lowe, C. M. Schiller, M. J. Schwab, L. Solsoman, E. E. Thompson, and E. M. Weinstack, *Science 228*, 1309 (1985).
11. J. H. Espenson, *Chemical Kinetics and Reaction Mechanisms* (New York: McGraw-Hill, 1981), section 7–7.
12. N. M. Emanuel' and D. G. Knorre, *Chemical Kinetics: Homogeneous Reactions*, transl. by R. Kondor (New York: J. Wiley and Sons/Halsted Press, 1973), section 8–4.
13. M. Kaufman and J. Sherwell, *Progr. Reaction Kinetics 12*, 1 (1983).
14. V. Ramanathan, R. J. Cicerone, H. B. Singh, and J. T. Kiehl, *J. Geophys. Res. 90*, 5547 (1985).

PROBLEMS

15.1 Some reactions interconvert O and O_3, while other reactions carry out a net change in the total "odd oxygen" concentration, expressed as the sum $[O] + [O_3]$. Express $d([O] + [O_3])/dt$ in terms of the relevant rates of reactions in the mechanism, and evaluate the relative importance of the reaction $O + O_3 \rightarrow 2O_2$ and the reactions of the NO_x, HO_x, and ClO_x catalytic cycles in the destruction of O and O_3.

15.2 In this problem we shall examine the importance of various reaction steps in the dynamic atmospheric ozone balance. Consider the Chapman mechanism given by

$$O + O_2 + M \longrightarrow O_3 + M \qquad k_1 = 4.41 \times 10^{-33}[M] cm^6 \ molecule^{-2} \ s^{-1}$$

$$O + O_3 \longrightarrow 2O_2 \qquad k_2 = 4.66 \times 10^{-16} \ cm^3 \ molecule^{-1} \ s^{-1}$$

$$O_2 \longrightarrow 2O \qquad k_3 = 5.00 \times 10^{-11} \ s^{-1}$$

$$O_3 \longrightarrow O + O_2 \qquad k_4 = 2.50 \times 10^{-4} \ s^{-1}$$

At time $t = 0$, the initial concentrations are as follows:

$$[O] = [O]_0 = 1.0 \times 10^6 \text{ atoms cm}^{-3}$$

$$[O_3] = [O_3]_0 = 1.0 \times 10^{12} \text{ molecule cm}^{-3}$$

$$[O_2] = [O_2]_0 = 3.7 \times 10^{16} \text{ molecule cm}^{-3}$$

We assume that $[O_2]$ remains constant.

(a) Plot the evolution with time of $[O]$ and $[O_3]$ over the time range 1.0×10^{-2} to 10^4 hours. Comment on the changes in $[O]$ and $[O_3]$ over this time range.

(b) Determine sensitivity coefficients for this system and plot the evolution with time of the coefficients. What is the sensitivity of $[O]$ and $[O_3]$ to uncertainties in the rate constants? What are the most important rate constants?

(c) Suppose that the experimental uncertainties for each of the four independent rate constants are one order of magnitude, so that

$$k_i e^{-2.303} \leq k_i \leq k_i e^{2.303}, \, i = 1, \ldots, 4$$

What is the sensitivity of $[O]$ and $[O_3]$ to these uncertainties in the rate coefficients? Suppose there is a two-order of magnitude uncertainty in k_2. What would your conclusions be then?

15.3 The relationship between the concentrations of NO, NO_2, and O_3 can be described by the following reaction sequence:

$$NO_2 + h\nu(\lambda < 400 \text{ nm}) \xrightarrow{j_1} NO + O(^3P) \quad (1)$$

$$O(^3P) + O_2 + M \xrightarrow{k_2} O_3 + M \quad (2)$$

$$NO + O_3 \xrightarrow{k_3} NO_2 + O_2 \quad (3)$$

NO_2 rapidly photolyzes in sunlight with $j_1 \approx 0.01 \text{ sec}^{-1}$ and $k_3 \approx 2.0 \times 10^{-12}$ $\exp(-4,000/RT) \text{ cm}^3 \text{ molecule}^{-1} \text{ sec}^{-1}$.

(a) Show that if reaction 3 is rapid, there is no net production of ozone.

(b) Under what conditions can reactions 1–3 lead to a net production of ozone?

15.4 The chemistry of HONO in the atmosphere is not well understood, but there are some aspects which are known. The principal sources of HONO are the reactions

$$HO + NO + M \xrightarrow{k_1} HONO + M$$

and

$$HO_2 + NO_2 \xrightarrow{k_2} HONO + O_2$$

The processes for the removal of HONO are

$$HONO + h\nu \xrightarrow{j_1} HO + NO$$

and

$$HO + HONO \xrightarrow{k_3} H_2O + NO_2$$

(a) Calculate the steady-state concentration of HONO from the preceding mechanism.

(b) Couple the reactions given with the ozone-NO_x cycle discussed in the text, and determine the total odd oxygen concentration. What effect could this cycle have on the net depletion of ozone in the stratosphere?

APPENDIX 1

Quantum Statistical Mechanics

According to the quantum mechanical Boltzmann distribution, the probability that a molecule has energy E_i is

$$P(E_i) = \frac{g_i \exp(-E_i/k_B T)}{\sum\limits_{i} g_i \exp(-E_i/k_B T)} \tag{A1-1}$$

where g_i is the degeneracy at energy E_i. The denominator in equation A1–1 is the partition function Q, so $P(E_i)$ becomes

$$P(E_i) = \frac{g_i \exp(-E_i/k_B T)}{Q} \tag{A1-2}$$

If the energy is assumed to be continuous, then $W(E)$, which represents the number of states in the energy interval $E \to E + dE$, is equivalent to g_i. $W(E)$ is related to the density of states $N(E)$ via the equation $W(E) = N(E)\, dE$. Using this relationship, the probability that a molecule has energy in the interval $E \to E + dE$ is given by

$$P(E)\, dE = \frac{N(E) \exp(-E/k_B T)\, dE}{Q} \tag{A1-3}$$

where

$$Q = \int_0^\infty N(E) \exp(-E/k_B T)\, dE \tag{A1-4}$$

Besides $W(E)$ and $N(E)$, another important term is $G(E)$, which represents the total number of states having energy in the range 0 to E. Calculation of $G(E)$, $N(E)$, and $W(E)$ for quantum harmonic oscillators is discussed in chapter 11.

APPENDIX 2

Classical Statistical Mechanics

A2.1 SUM AND DENSITY OF STATES

The classical sum of states, $G(E)$, is based on a theorem in classical statistical mechanics which says that the number of states for one degree of freedom with momentum p and coordinate q is

$$\frac{dpdq}{h} \tag{A2-1}$$

where $dpdq$ is the phase space volume associated with p and q, and h is Planck's constant. Thus, for one degree of freedom, the volume of phase space associated with a single state is h. The number of states, $G(E)$, is then the total phase space volume V divided by h, i.e.,

$$G(E) = \frac{V_1}{h} \tag{A2-2}$$

where the subscript 1 indicates that one degree of freedom is under consideration.

This example may be extended to any multidimensional classical system with coordinates q_i and conjugate momenta p_i. The total energy for the system is given by the classical Hamiltonian $H(p_1 \cdots p_s q_1 \cdots q_s)$. To find the total phase space volume for the Hamiltonian at energy E, the integral

$$V_s = \int_{H=0}^{H=E} \cdots \int dp_1 \cdots dp_s dq_1 \cdots dq_s \tag{A2-3}$$

is solved by considering possible combinations of p_i and q_i with the restriction that H lies between 0 and E. The classical sum of states is then

$$G(E) = \frac{V_s}{h^s} = \int_{H=0}^{H=E} \cdots \int \frac{dp_1 \cdots dp_s dq_1 \cdots dq_s}{h^s} \tag{A2-4}$$

The number of states for H in the range $E \rightarrow E + dE$, which is the analog to quantum mechanical degeneracy, is denoted by $W(E) = G(E + dE) - G(E)$ and is written as

$$W(E) = \int_{H=E}^{H+dH=E+dE} \cdots \int \frac{dp_1 \cdots dp_s dq_1 \cdots dq_s}{h^s} \qquad (A2-5)$$

This integral is a multidimensional volume of phase space enclosed by the two hypersurfaces defined by H and $H + dH$. Another way to express this multidimensional volume is to multiply the area of the hypersurface for $H = E$ by dH, so that $W(E)$ becomes

$$W(E) = \left[\int_{H=E} \cdots \int \frac{dp_1 \cdots dp_s dq_1 \cdots dq_s}{h^s} \right] dH \qquad (A2-6)$$

Ths density of states $N(E) = [G(E + dE) - G(E)]/dE$ is simply $W(E)/dH$ and is given by

$$N(E) = \int_{H=E} \cdots \int \frac{dp_1 \cdots dp_s dq_1 \cdots dq_s}{h^s} \qquad (A2-7)$$

Thus, the density of states is determined by a surface integral, while the sum of states (equation A2–4) is found from a volume integral.

A greater understanding of the expressions for the classical sum and density of states may be attained by considering translation, harmonic oscillators, and the rigid rotor.

A2.1.1 Three-Dimensional Translation

The Hamiltonian for the free translation of a particle inside a cube with sides of length ℓ is

$$H = \frac{(p_x^2 + p_y^2 + p_z^2)}{2m} \qquad (A2-8)$$

The potential energy V is zero inside the container and infinite at the walls. To obtain the translational sum of states, equation A2–4 must be solved for the Hamiltonian. The integration limits for each coordinate are 0 and ℓ. In contrast, the limits for the momenta are more complex, since the maximum value of a momentum term at a particular energy E is determined by the values for the remaining momenta. Thus, the sum of states is given by

$$G(E) = \int_{H=0}^{H=E} \cdots \int dp_x dp_y dp_z \frac{\int_0^\ell dx \int_0^\ell dy \int_0^\ell dz}{h^3} \qquad (A2-9)$$

The product of the integrals over the coordinates in equation A2–9 is simply ℓ^3, the volume of the container. Although the integral over the momentum terms may appear complex, it is in fact quite simple. As discussed, the integral represents the momentum phase space volume for H having values between 0 and E. However, since the translational Hamiltonian in equation A2–8 is the equation for a sphere, this volume is that for a sphere with a radius of $(2mE)^{1/2}$. Thus, the translational sum of states per unit volume ℓ^3 is

$$G(E) = \frac{4\pi \, (2mE)^{3/2}}{3h^3} \tag{A2-10}$$

To obtain the density of states $N(E)$, the sum $G(E)$ is differentiated with respect to E. The resulting $N(E)$ is usually written in terms of $\hbar$ instead of h, and is given by

$$N(E) = (2^{1/2} \pi^2 \hbar^3)^{-1} \, m^{3/2} \, E^{1/2} \tag{A2-11}$$

This result is used extensively in chapter 13.

A2.1.2 Harmonic Oscillators

As a further illustration of the classical sum and density of states, consider a single harmonic oscillator for which the classical Hamiltonian is

$$H(p,q) = \frac{p^2}{2} + \frac{\lambda q^2}{2} \tag{A2-12}$$

where $\lambda = 4\pi^2 \nu^2$. The phase space volume is that enclosed by energy E and is formally given by equation A2-3, which for this example is

$$V_1 = \int_{H=0}^{H=E} \int dp \, dq \tag{A2-13}$$

This is simply the area of an ellipse, logical enough since the Hamiltonian for the harmonic oscillator is the equation for an ellipse. The area of an ellipse is πab, where a and b are the semiaxes. For the harmonic oscillator Hamiltonian, the semiaxes equal $(2E)^{1/2}$. Thus, the phase space volume for the harmonic oscillator is

$$V_1 = \frac{E}{\nu} \tag{A2-14}$$

and the classical number of states is

$$G(E) = \frac{E}{h\nu} \tag{A2-15}$$

The harmonic oscillator energy levels and their associated phase space volumes are illustrated in Figure A2-1.

For s harmonic oscillators, the classical number of states is given by equation (A2-4). The Hamiltonian for s harmonic oscillators is

$$H(p_i q_i) = \sum_{i=1}^{s} \left(\frac{p_i^2}{2} + \frac{\lambda_i q_i^2}{2} \right) \tag{A2-16}$$

Evaluation of the phase space volume for s harmonic oscillators is straightforward if one recognizes that equation A2-16 is the equation for a $2s$-dimensional ellipsoid with semiaxes along p_i and q_i. The volume for this ellipsoid with $H(p_i, q_i) = E$ is given by

$$V_s = \frac{E^s}{s! \displaystyle\prod_{i=1}^{s} \nu_i} \tag{A2-17}$$

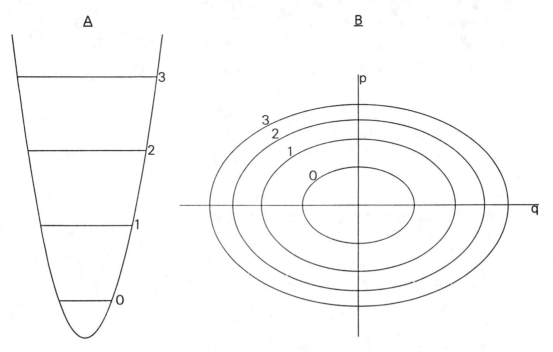

Figure A2-1. (a) The $n = 0$, 1, 2, and 3 energy levels for a harmonic oscillator. (b) Phase space volumes for the $n = 0$, 1, 2, and 3 harmonic oscillator energy levels.

The sum of states is then

$$G(E) = \frac{E^s}{s! \prod_{i=1}^{s} h\nu_i} \qquad (A2-18)$$

and the similarity with equation A2-15 should be noted. The density of states $dG(E)/dE$ is

$$N(E) = \frac{E^{s-1}}{(s-1)! \prod_{i=1}^{s} h\nu_i} \qquad (A2-19)$$

A2.1.3 Rigid Rotor

The last example is a rigid rotor with two degrees of freedom, for which the Hamiltonian is

$$H(p_\theta, p_\phi, \theta, \phi) = \frac{p_\theta^2}{2I} + \frac{p_\phi^2}{2I \sin^2 \theta} \qquad (A2-20)$$

where θ varies from 0 to π and ϕ varies from 0 to 2π. This Hamiltonian is also the equation for an ellipse with semiaxes $a = (2IE)^{1/2}$ and $b = (2IE \sin^2 \theta)^{1/2}$. Therefore, for fixed values of θ and ϕ, the phase space volume is $V_2(\theta, \phi) = 2IE \sin \theta$.

Integration over θ and ϕ is necessary to find the total phase space volume

$$V_2 = 8\pi^2 EI \qquad \text{(A2-21)}$$

Thus, the expression for the classical number of states for a rigid rotor is

$$G(E) = \frac{8\pi^2 EI}{h^2} \qquad \text{(A2-22)}$$

and the classical density of states is

$$N(E) = \frac{8\pi^2 I}{h^2} \qquad \text{(A2-23)}$$

A2.2 PARTITION FUNCTION AND BOLTZMANN DISTRIBUTION

As shown in equation A2–4 of Appendix 1, for the situation where the energy is continuous, as in classical mechanics, the partition function is

$$Q = \int_0^\infty N(E)e^{-E/k_B T}\, dE$$

If the classical expression for $N(E)$ in equation A2–7 is used, the partition function becomes

$$Q = \int_0^\infty \left[\int \cdots \int_{H=E} \frac{dp_1 \cdots dp_s dq_1 \cdots dq_s}{h^s} \right] e^{-H/k_B T}\, dH \qquad \text{(A2-24)}$$

This multiple integral is over all possible combinations of coordinates and momenta, with each combination weighted by $e^{-H/k_B T}$. Equation A2–24 is usually written as

$$Q = \frac{1}{h^s} \int dp_1 \cdots dp_s dq_1 \cdots dq_s e^{-H/k_B T} \qquad \text{(A2-25)}$$

where the single integral denotes integration over all possible values of coordinates and momenta. The integral in equation A2–25 is called the *phase integral*.

The classical expression for the Boltzmann distribution is

$$P(H)dH = \frac{\left[\int \cdots \int_{H=E} dp_1 \cdots dp_s dq_1 \cdots dq_s/h^s \right] e^{-H/k_B T}\, dH}{\int dp_1 \cdots dp_s dq_1 \cdots dq_s e^{-H/k_B T}/h^s} \qquad \text{(A2-26)}$$

where the denominator is the partition function and the integral in the numerator is the density of states. As an illustration of equation A2–26, consider the rigid rotor equations A2–20 through A–23. The partition function for the rigid rotor is

$$Q = \int_0^\infty \frac{8\pi^2 I}{h^2} e^{-H/k_B T}\, dH$$

$$= \frac{8\pi^2 IkT}{h^2} \qquad \text{(A2-27)}$$

so that the Boltzmann distribution becomes

$$P(H)dH = \frac{1}{k_B T} e^{-H/k_B T} dH \qquad (A2\text{--}28)$$

This distribution may be used to calculate the average energy for the rigid rotor:

$$\langle E \rangle = \int_0^\infty H P(H) dH \qquad (A2\text{--}29)$$

Using equation A2–28 for $P(H)dH$, we find that

$$\langle E \rangle = \frac{1}{k_B T} \int_0^\infty H e^{-H/k_B T} dH = k_B T \qquad (A2\text{--}30)$$

A particularly important classical Boltzmann distribution is that for s harmonic oscillators. By inserting equation A2–19 for the classical density of states into equations A2–24 through A2–26, the classical probability that s harmonic oscillators contain energy E is found to be

$$P(E)dE = \frac{1}{(s-1)!} \left(\frac{E}{k_B T} \right)^{s-1} e^{-E/k_B T} \left(\frac{dE}{k_B T} \right) \qquad (A2\text{--}31)$$

A molecule must possess some minimum energy before a reaction can be expected to occur. The classical mechanical value for the fraction of molecules which have energy greater than E_0 is found by integrating equation A2–31 between the limits E_0 and ∞; i.e.,

$$F(E \geq E_0) = \int_{E_0}^\infty \frac{1}{(s-1)!} \left(\frac{E}{k_B T} \right)^{s-1} e^{-E/k_B T} \left(\frac{dE}{k_B T} \right) \qquad (A2\text{--}32)$$

For the special case where $E_0/k_B T \gg 1$, this fraction becomes

$$F(E \geq E_0) = \frac{1}{(s-1)!} e^{-E_0/k_B T} \left(\frac{E_0}{k_B T} \right)^{s-1} \qquad (A2\text{--}33)$$

BIBLIOGRAPHY

BLINDER, S. M. *Advanced Physical Chemistry*. London: Macmillan, 1969.

COURANT, R., and JOHN, F. *Introduction to Calculus and Analysis*. New York: Wiley, 1974.

DAVIDSON, N. *Statistical Mechanics*. New York: McGraw-Hill, 1962.

FORST, W. *Theory of Unimolecular Reactions*. New York: Academic Press, 1973.

HASE, W. L. *J. Chem. Ed. 60*, 379 (1983).

KAUFMAN, E. D. *Advanced Concepts in Physical Chemistry*. New York: McGraw-Hill, 1966.

McQUARRIE, D. A. *Statistical Thermodynamics*. New York: Harper and Row, 1973.

PROBLEMS

A2–1 The probability distribution $P_T(E)$ of the total energy E in a Boltzmann distribution is the Boltzmann of $\delta(E - H)$.

    (a) Show that

$$P_T(E) = \frac{N(E)e^{-E/k_B T}}{Q(T)}$$

    where Q is the canonical partition function and N is the microcanonical density of states.

    (b) For the case of F independent harmonic oscillators, show that

$$P_T(E) = \frac{1}{(F-1)!}\left(\frac{E}{k_B T}\right)^{F-1}\frac{e^{-E/k_B T}}{k_B T}$$

    Sketch a graph of this function.

    (c) For the system of part (b), calculate the average energy

$$\langle E \rangle \equiv \int dE\, P_T(E)E$$

and the width

$$\Delta E \equiv \sqrt{\langle E^2 \rangle - \langle E \rangle^2}$$

of the distribution about this average value, where

$$\langle E^2 \rangle \equiv \int dE\, P_T(E)E^2$$

APPENDIX 3

Data Bases in Chemical Kinetics

In order to carry out chemical kinetics modeling and calculations, it is necessary to have available accurate values of rate constants for a large number of elementary reactions over a wide range of temperatures. Until recently, the data base of rate constants has been rather inadequate. This is now being remedied, with the appearance of critically evaluated compilations of kinetic data. Much more work remains to be done in the systemization of such data bases and their transfer to electronic networks. In this Appendix, we list some of the more useful compilations that are now available in the literature.

1. Literature through 1966 (somewhat out-of-date, but comprehensive on earlier literature and includes charged species and electron-neutral collision data):

HOCHSTIM, A. R. (ed.), "Bibliography of Chemical Kinetics and Collision Processes," IFI/Plenum, New York (1969).

Also covering earlier literature:

TROTMAN-DICKINSON, A. F., and G. S. MILNE, "Tables of Bimolecular Gas Reactions," NSRDS-NBS-9, National Bureau of Standards, Washington, DC (1967).

More recent compilations:

KERR, J. A., and S. J. MOSS, "CRC Handbook of Bimolecular and Termolecular Gas Reactions," CRC Press, Boca Raton, FL. Volume I (1981), Volume II, (1981), Volume III (1987).

2. Data surveys for modelling atmospheric chemistry:

GARVIN, D., and R. F. HAMPSON (eds.), "Chemical Kinetics Data Survey. VII. Tables of Rate and Photochemical Data for Modelling of the Stratosphere," NBSIR-74-430, National Bureau of Standards, Washington, DC (1975).

WATSON, R. T., "Chemical Kinetics Data Survey. VIII. Rate Constants of ClO$_x$ of Atmospheric Interest," NBSIR-74-516, National Bureau of Standards, Washington, DC (1974).

HAMPSON, R. F., JR., and D. GARVIN (eds.), "Chemical Kinetic and Photochemical Data for Modelling Atmospheric Chemistry," NBS Technical Note 866, National Bureau of Standards, Washington, DC (1977).

HAMPSON, R. F., JR., and D. GARVIN (eds.), "Reaction Rate and Photochemical Data for Atmospheric Chemistry—1977," NBS Special Publication 513, National Bureau of Standards, Washington, DC (1977).

HERRON, J. T., R. E. HUIE, and J. A. HODGESON (eds.), "Chemical Kinetic Data Needs for Modelling the Lower Troposphere," NBS Special Publication 557, National Bureau of Standards, Washington, DC (1979).

Most recent update:

DEMORE, W. B., et al. (eds.), "Chemical Kinetics and Photochemical Data for Use in Stratospheric Modelling, Evaluation Number 7," JPL Publication 85–37, Jet Propulsion Laboratory, Pasadena, CA (1985).

3. Standardized data sheets for a number of reactions have been published:

COHEN N., and K. R. WESTBERG, "Chemical Kinetic Data Sheets for High-Temperature Chemical Reactions," *J. Phys. Chem. Ref. Data 12*, 531–590 (1983).

Reactions included are:

$$F + D_2 \longrightarrow D + DF$$

$$F + H_2 \longrightarrow H + HF$$

$$F + HD \longrightarrow D + HF, H + DF$$

$$H + F_2 \longrightarrow F + HF$$

$$D + F_2 \longrightarrow F + DF$$

$$O + OH \longrightarrow H + O_2$$

$$H + OH \longrightarrow O + H_2$$

$$OH + H_2 \longrightarrow H + H_2O$$

$$OH + OH \longrightarrow O + H_2O$$

$$H + H + M \longrightarrow H_2 + M \ (M = H_2, H, Ar, N_2, H_2O)$$

$$OH + CH_4 \longrightarrow CH_3 + H_2O$$

$$OH + C_2H_6 \longrightarrow C_2H_5 + H_2O$$

$$OH + C_3H_8 \longrightarrow C_3H_7 + H_2O$$

$$OH + n\text{-}C_4H_{10} \longrightarrow C_4H_9 + H_2O$$

$$OH + i\text{-}C_4H_{10} \longrightarrow C_4H_9 + H_2O$$

$$\mathrm{OH} + \mathrm{c\text{-}C_4H_8} \longrightarrow \mathrm{C_4H_7} + \mathrm{H_2O}$$

$$\mathrm{OH} + \mathrm{c\text{-}C_5H_{10}} \longrightarrow \mathrm{C_5H_9} + \mathrm{H_2O}$$

$$\mathrm{OH} + \mathrm{c\text{-}C_6H_{12}} \longrightarrow \mathrm{C_6H_{11}} + \mathrm{H_2O}$$

$$\mathrm{OH} + \mathrm{(CH_3)_4C} \longrightarrow \mathrm{CH_2C(CH_3)_3} + \mathrm{H_2O}$$

$$\mathrm{OH} + \mathrm{(CH_3)_2CHCH(CH_3)_2} \longrightarrow \mathrm{C_6H_{13}} + \mathrm{H_2O}$$

$$\mathrm{OH} + \mathrm{(CH_3)_3CCH(CH_3)_2} \longrightarrow \mathrm{C_7H_{15}} + \mathrm{H_2O}$$

$$\mathrm{OH} + \mathrm{(CH_3)_3CC(CH_3)_3} \longrightarrow \mathrm{CH_2C(CH_3)_2C(CH_3)_3} + \mathrm{H_2O}$$

$$\mathrm{Al} + \mathrm{O_2} \longrightarrow \mathrm{O} + \mathrm{AlO}$$

$$\mathrm{B} + \mathrm{O_2} \longrightarrow \mathrm{O} + \mathrm{BO}$$

4. Selected classes of reactions have been reviewed in *Journal of Physical and Chemical Reference Data, International Journal of Chemical Kinetics,* and various review journals. Examples include:

COHEN, N., "A Reevaluation of Low Temperature Experimental Rate Data for the Reactions of O Atoms with Methane, Ethane, and Neopentane," *Intl. J. Chem. Kinetics 18,* 59–82 (1986).

COHEN, N., "The O + NH$_3$ Reaction: A Review," *Intl. J. Chem. Kinetics 19,* 319–362 (1987).

STEINFELD, J. I., S. M. ADLER-GOLDEN, and J. W. GALLAGHER, "Critical Survey of Data on the Spectroscopy and Kinetics of Ozone in the Mesosphere and Thermosphere," *J. Phys. Chem. Ref. Data 16,* 911–951 (1987).

WESTBROOK, C. K., and F. L. DRYER, "Chemical Kinetic Modeling of Hydrocarbon Combustion," *Progr. Energy Combustion Sci. 10,* 1–57 (1984). (Includes rate data on 335 reactions, also 494 references.)

5. A computer-searchable data base of chemical reaction rate coefficients is now available from the National Institute of Standards and Technology (formly National Bureau of Standards), Gaithersburg, Md. 20899, U.S.A. Rate parameters for over 2000 gas phase reactions are stored in the form $k(T) = A(T/298)^n \exp(-E_a/R)/T$, and can be displayed either in tabular form, or as Arrhenius plots if a standard graphics card is available. Literature citations for each measurement include pressure and temperature range, experimental method, and experimental uncertainties. The reference is NIST Standard Reference Database No. 17, "NIST Chemical Kinetics Database" (1989).

Index

Eley-Rideal mechanism, 196, 424
Encounters, in solution, 156–157
Encounter complex, 163, 403
Endothermic reaction, 15, 43, 431
Energy of activation. *See* Activation energy
Energy conservation, 252, 254, 482
Energy-gained master equation, 488
Energy partitioning, 376–378
Energy transfer. *See* Intramolecular energy transfer; Relaxation processes
Ensembles, 331
  canonical, 331
  microcanonical, 331
Enthalpy
  of activation, 179, 322–323
Enthalpy diagram, 15–16, 179, 403
Entropy, 179
  of activation, 322–323
    relation to steric factor p, 250, 336
  of mixing, 433
  statistical, 433
Entropy deficiency, 433, 466, 480
Entropy maximization, 434
Entropy production, 466, 480
Enzyme-catalyzed reactions, 190–193, 195, 200
Equilibrium
  and catalysis, 178–180
  definition, 1
  and detailed balance, 23
Errors
  in kinetic measurements, 133–136, 142
    random, 134–135
    systematic, 133–134
  in numerical integration, 59–62, 66
Errors, propagation of, 135–136, 141
Ethyl acetate. *See* $CH_3COOC_2H_5$
Euler methods, 58–59
  in atmospheric modeling, 507
  compared with Runge-Kutta method, 62
  computer program for, 92–97
Exchange integral, 232
Exothermic reactions, 15, 43, 285, 431, 441
Explosion, 513–515
Explosion limits, 513–514
Exponential-gap law, 451, 459

$F + H_2$ reaction, 293–297
  chemical laser, 301, 444, 488–492
  classical trajectory calculations, 268
  possible outcomes, 431–432
  potential energy surface, 225–229, 263, 294
  product distribution, 294
  selective energy consumption (H + HF), 300
  specific energy release, 295

$F + H_2$ reaction (*Contd.*)
  surprisal analysis, 438–444
  transition-state theory calculation, 328–330
Fall-off region, 354, 369–370
Fick's law of diffusion, 158
First-order reactions
  analytic solutions for, 6–8, 114, 144
  half-life for, 14
  pseudo-first-order, 41, 51, 115
Fischer-Tropsch synthesis, 194, 196–198
Flash photolysis, 117–119
Flooding, 41
Flow kinetic systems, 113–115, 369
  error analysis, 135–136
Fluctuations, 67, 71–73
Fokker-Planck equation, 411
Force constants, 216, 218
Formaldehyde. *See* CH_2O
Formyl radical. *See* HCO
Four-center reaction, 509
  for $H_2 + I_2$, 45
Fourier Transform Doppler spectroscopy, 293
Franck-Condon principle, 328, 380
Free energy
  of activation, 322
  solvent effects on, 169–171, 405–406
  of vaporization, 406
Free-radical reaction, 42, 498ff.
Free-volume theory of liquids, 406
Freons. *See* Chlorofluorocarbons
Frequency factor, 15, 314
Freundlich isotherm, 418
Friction constant, 411ff.

Gamma function, 448
Gas-kinetic collision rate, 247
  Lennard-Jones correction, 260–261
Gas-surface interfaces, 194–196, 415
Gear routine, 67
  in atmospheric models, 507
Generalized-valence-bond(GVB) method, 225
Gibbs free energy. *See* Free energy
Gibbs-Helmholtz equation, 322
Gillespie algorithm, 71, 97
Global error
  in classical trajectory calculations, 263
  in numerical integration, 59
Green's function method, 146, 147
Greenhouse effect, 522

HCO, 148, 516–517
H(D) + H_2 reaction, 268 *n.*, 290, 291, 292, 340
  classical trajectory calculations, 266–268
  potential energy surface, 225, 229

Index

Pressure jump, 174
Primary salt effect, 167, 405
Prior rate, 439, 445–450
Probability of reaction, 247, 253, 290
Progress variable, 9, 11, 26, 44
Promoter, for catalyzed reactions, 423
Propagation, in free-radical reactions, 43, 511
Propagation of errors, 135–136, 141
Pseudo-first-order reaction, 41, 51
Pyrolysis, 369

Quantum mechanics, 235
   effects in transition-state theory, 318–321
   of reactive scattering, 268–272
      coupled-channel theory, 271
Quasicontinuum (QC), 484–485
Quasi-equilibrium hypothesis, 311, 317
Quasi-periodic trajectory, 383–384

Radiation hypothesis. *See* Perrin hypothesis
Radiative recombination, 328
Rainbow scattering, 259
Raman spectroscopy, 292, 444
Random lifetime distribution, 358
Rate coefficient, 6, 97, 109
   relation to cross section, 204–205
   state-to-state, 205, 207, 300, 446, 480
Rate constant, 4
   for desorption, 419
   diffusion-limited, 161–163
   effect of ionic strength, 164–168
   temperature dependence, 14–16
   units, 4
Rate-determining step, 40, 178
   in solution reactions, 163, 404
Rate equation, 2
Rate laws, integrated, for elementary reactions, 6–12
Rate-limiting step. *See* Rate-determining step
Rate of reaction, 2
Reaction coordinate, 221, 312, 343–344
Reaction mechanism. *See* Mechanism
Reaction order. *See* Order
Reaction path, 221, 236–237, 309–310
Reaction probability. *See* Probability of reaction
Reaction rate. *See* Rate of reaction
Reaction types
   consecutive, 27–33, 127
   elementary, 2, 5–12
   parallel, 33–38, 52
   reversible, 22, 26, 50–52
Reactions on surfaces. *See* Adsorption; Heterogeneous reactions

Reactive hard sphere model, 248–250
Recombination reactions, 10, 15, 118
Reduced mass μ, 252,
Redundant coordinates, 217
Reel-down technique, 509, 510
Relaxation methods, 171–174
Relaxation processes
   intramolecular, 382–385
   master equation, 476–483
   rotation-translation, 445, 459–464
   surprisal analysis, 445–449
   vibration-translation, 374–376, 445, 451–458, 465–469, 491
   vibration-vibration, 445, 448, 491
REMPI. *See* Multiphoton ionization
Residence time, 421
Resonance, quantum-mechanical, 272
Reversible reactions, 22–26
Reynolds number, 115
Rigid rotor, 531
   partition function, 325–326
Rise time, 28, 123
rms deviation, 138
Rotating sector method, 127
Rotational energy transfer. *See* Relaxation
RRHO approximation, 441, 447
RRK theory, 357–362
   Arrhenius A-factor, 361
RRKM theory, 362–370
   ab initio calculations, 366
   active modes, 363
   adiabatic modes, 363
   example calculation, 367
   measurements of k(E), 370–374
   non-RRKM behavior, apparent and intrinsic, 379–381
   product energy partitioning, 376–378
   rotation energy, 364, 369
   thermal activation, 368–370
   transition state properties, 365
   variational version, 366
Runge-Kutta method, 59–63
   for atmospheric models, 507
   compared with Euler method, 62
   computer program for, 92–97
Rydberg-Klein-Rees method, 235

"S" curve, 183
S_N2 reaction, 402, 409, 414
Saddle point, 221, 236, 309–310
Salt effect,
   primary, 167, 405
   secondary, 182
Sampling, of initial conditions, 265